D0594202

A Bibliography of Astronomy 1970-1979

A Bibliography of Astronomy 1970-1979

ROBERT A. SEAL
University of Virginia Library

SARAH S. MARTIN
Librarian
National Radio Astronomy Observatory

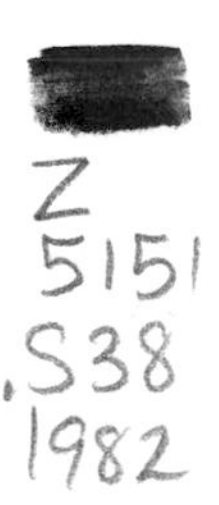

1982
LIBRARIES UNLIMITED, INC.
Littleton, Colorado

LIBRARIES UNLIMITED, INC.
P.O. Box 263
Littleton, Colorado 80160

Library of Congress Cataloging in Publication Data

Seal, Robert A.
A bibliography of astronomy, 1970-1979.

Includes index.
1. Astronomy--Bibliography. I. Martin, Sarah S., 1945- . II. Title.
Z5151.S38 [QB43.2] 016.52 81-20877
ISBN 0-87287-280-7 AACR2

Libraries Unlimited books are bound with Type II nonwoven material that meets and exceeds National Association of State Textbook Administrators' Type II nonwoven material specifications Class A through E.

TABLE OF CONTENTS

ACKNOWLEDGMENTS

The authors would like to express their great appreciation to their institutions for moral and other support during this long and difficult project. In particular, Mr. Seal would like to thank his former employer, the University of Virginia Library, which provided him with extensive leave with pay as well as travel expenses to visit libraries for research purposes. Without this support, the project could not have been completed within a reasonable amount of time. Ms. Martin would like to thank Morton S. Roberts, Director of the National Radio Astronomy Observatory, and NRAO in general for their support during the project.

Both authors also wish to thank colleagues for their suggestions, advice, and assistance. There have been many who have assisted us, but in particular two librarians should be acknowledged for their expert help: Fran Waranius of the Lunar and Planetary Institute, and Brenda Corbin of the U.S. Naval Observatory Library.

Finally, to our families and friends, your support, endurance, and patience are appreciated most of all.

RAS
Director of Library Public Services
University of Oklahoma Libraries
Norman, Oklahoma

SSM
Librarian
National Radio Astronomy Observatory
Charlottesville, Virginia

October 1981

INTRODUCTION

In 1970, a landmark in scientific and astronomical bibliography appeared in the form of D. A. Kemp's *Astronomy and Astrophysics: A Bibliographical Guide.*[1] This volume, a compendium of more than 3,700 entries covering approximately 1950 to 1968, provided the first comprehensive sourcebook in the field for librarians, astronomers, and researchers. Its entries, arranged chronologically in 75 separate subject sections, provided broad coverage in almost all areas. The author received an award of merit from the Physics-Astronomy-Mathematics Division of the Special Libraries Association in 1977.

Since 1970, no work of a similar nature has been undertaken until now. Robert Seal's *A Guide to the Literature of Astronomy*[2] was published in 1977 as an introduction to the field, with a selected list of 578 entries. Although this annotated bibliography was well received, it was not intended for the same use or audience as Kemp's work, nor did it provide comprehensive, in-depth coverage in all areas. It was intended as an introduction, overview, and selection aid for all types of libraries and users, not as a comprehensive reference tool.

Astronomy and Astrophysics: A Bibliographical Guide, though now out of date by a dozen years, is still often used by astronomy and other science librarians. But as the amount of astronomical information continued to increase rapidly during the 1970s, Kemp's book began to lose its usefulness, like other works of its type. In 1977, the present authors (Seal and Martin) presented the idea of a new bibliography of astronomy to colleagues, both informally at conferences and formally by questionnaire. The response was overwhelmingly positive: a new work was needed to fill the gap in the literature between Kemp's book and the present. Thus, work on the present volume began.

More than four years in the making, it has consumed hundreds of hours of research, writing, typing, editing, and indexing. Each item (with a very few exceptions) was examined first-hand by the authors. It is believed that the effort has been worthwhile, and that the reader will find this work useful, whether for quick reference, methodical study, or review. To be sure, there are very likely some omissions, but hopefully, these small gaps will not detract from the work's intended goal: to help the user find as many key papers and other works as possible on a given subject.

General Overview

This volume is primarily intended to present a fairly comprehensive guide to bibliographical sources in astronomy and astrophysics—sources which review, summarize, and lead the reader to further information in the field. By consulting a particular subject section, or by using the topical index, the reader should be able to find most, if not all, major review papers, monographs, and proceedings on a particular topic for the decade of the 1970s.

Primary emphasis has been explicitly placed on papers and works containing a large number of references, with additional emphasis placed on identifying conference proceedings in the field. For state-of-the-art reviews, one should consult especially the IAU symposia and colloquia, as well as the *IAU Transactions. Reports on Astronomy.*

Complete lists of the IAU symposia and colloquia which took place in the late 1960s and all of the 1970s are included in the appendices, and all are cited individually in the text. In addition, the more inclusive review articles and summaries from these volumes have been selected for this compilation, but the reader should keep in mind that many individual papers from proceedings volumes of all types are not included here because they are not review-type articles. Nevertheless, such articles would be very useful to someone conducting in-depth research or review in a specific field, and they should be examined at some length for the sake of completeness.

Since it is fairly comprehensive in its coverage of review and bibliographic sources, the book should also prove useful for ready-reference type work: to verify a source, to check a citation, or to locate specific information on a conference. To some extent, the volume might also be useful for book selection. While the notes are more descriptive than evaluative and are often very brief, the book's completeness will allow the librarian to check selected subject sections in the book against library holdings to determine what gaps might exist in a given area in the collection. The authors recommend book reviews and advice from library users and colleagues as the best sources of information on selection, especially for the more expensive or questionable items. The two appendices of IAU publications, mentioned earlier, may also be used as a collection checking device or as a bibliographic source.

The volume is aimed at a variety of potential users: astronomers, graduate students and advanced undergraduates, science librarians (observatory librarians in particular), physicists, science historians, and the educated layperson. Each group will have a different reason for consulting this work.

Finally, the book is a selected record of key works published in the field during the decade of the 1970s. As is true of similar bibliographical works, the volume is dated almost before it is off the press. Nonetheless, it is felt that it fills a broad gap in the reference literature and will therefore prove useful for years to come.

Scope

Years. Although the title indicates coverage of 1970 through 1979, the book also includes a large number of entries for 1969, and selected entries for 1967, 1968, and 1980. The 1967 and 1968 references are items not in Kemp's work, while the 1980 entries are primarily limited to conference proceedings volumes for meetings which took place in the 1970s, and selected important reference works. There are numerous 1969 entries because Kemp's book, though its imprint is 1970, included only a few items from that year.

Subject coverage. Generally, the topical coverage of this volume is restricted to those subjects found in *Astronomy and Astrophysics Abstracts*, whose subject arrangement and headings have been used for the majority of this work. In short, all areas of the field have been covered, including physical and mathematical papers directly related to astronomy and astrophysics. It is perhaps more appropriate to speak of exclusions, which are primarily those items only indirectly related to astronomy, such as general optics, plasma physics, time, astronautics, etc. Items in these categories have been cited only when they included material directly relevant to the field. For example, a basic optics text would be excluded, but a book on optics for x-ray astronomy would not. Likewise, items discussing astrophysical plasmas would be cited, while books on plasma physics in general would not. Space astronomy is covered, but astronautics is not. History of astronomy is covered only in terms of bibliographies of that material. Other examples also occur.

Types of material. As noted earlier, this volume concentrates on bibliographic and review sources. Specifically, there are included collections of and individual review articles, bibliographies, non-elementary texts, conference proceedings and papers, data compilations, major atlases and catalogs,[3] advanced monographs, and reference works such as standard handbooks, dictionaries, and directories. Excluded are non-review journal articles (with a few exceptions: e.g., articles about astronomical data, literature, and nomenclature), individual book chapters (except in the case of review volumes), elementary texts (college and secondary school), children's books, popular works, and theses and dissertations.

In addition to these standard types of sources of bibliographic and numerical information, there are also included some additional sources which have emerged in the 1970s as important resources of a different nature. These include machine-readable data files, microform compilations, data centers, and computer databases. Individual data files that are available from data centers such as the ones at Strasbourg or Greenbelt are not included, but instead, several citations which describe these files and their availability are listed. In addition, several computer bibliographic databases that include some of the astronomical literature available from the commercial database vendors are cited.

Arrangement

The book consists of four parts: reference works, subject sections, appendices, and indices, with the majority of information presented in the subject sections. As noted previously, *Astronomy and Astrophysics Abstracts*' subject headings and numerical designations have been used. The Astronomisches-Rechen Institut in Heidelberg, publisher of *Astronomy and Astrophysics Abstracts*, was kind enough to permit use of this arrangement, an idea suggested by a reviewer of Kemp's work, speculating on future efforts of that nature. It is felt that this arrangement will facilitate the use of the book, since many of its potential users will be familiar with *Astronomy and Astrophysics Abstracts* (hereafter referred to as *AAA*), the single most important reference source in the field.

Even without a knowledge of the *AAA* subject scheme, the volume should be fairly easy to use. The table of contents and subject index should provide easy access to any entry in the book. Entries in each section, reference and subject, are arranged by author, and are numbered in the same fashion as *AAA*, except for reference works which use a similar number format but are arranged by the authors' own scheme. The latter was done because the *AAA* arrangement for reference materials is not subdivided into specific groups.

Briefly, all *AAA* headings and subject divisions from 021 through 162 are employed. Certain of these categories contain a great many entries, while others contain very few or, in rare cases, none at all. A paucity of entries does not indicate a lack of research in those particular areas; it merely means that there is a lack of review material or data compilations, the type of sources on which this book concentrates. To find materials on these sparse areas, and all other areas, the reader is directed to *Astronomy and Astrophysics Abstracts*. In *AAA*, monographic works and conference proceedings are placed under two general headings: Books (003) and Proceedings of Colloquia, Congresses, Meetings, and Symposia (012), with cross-references from the subject sections. However, the authors felt it more useful to cite these works in their appropriate subject sections. This was not always an easy task, since the *AAA* subject sections are not always geared to citing monographic works, especially those covering several topics in one volume.

Consequently, there was not always a satisfactory subject division in which to place a book or conference volume. Therefore, if the reader is puzzled by the placement of a particular item in a particular section, it is probably because there was no easy choice for its placement. However, items with multiple subjects are cited in up to three different subject sections. There will be one complete entry (numbered), with one or two additional references appearing in other sections with a reference to the primary citation number. This method is used by *AAA* quite successfully.

Two slight modifications have been made to the nomenclature used by *AAA* in their subject division headings. To section 081 has been added "Geophysics," and to section 080 has been added "Solar Physics," because there seemed to be no logical place for these frequently seen, general, important topics. Other changes could have been made, but it was felt that they should be kept to a minimum.

The numbering scheme also follows *AAA*. A typical entry number consists of the subject category number followed by a decimal, followed by the number of the item within that section. For example, the fourteenth book in section 162 would be numbered 162.014. The only exception to this is that entry numbers in the reference section are preceded by a capital R (e.g., R002.005). Entries that are conference proceedings will have a capital C appearing in the left-hand margin, while entries for atlases and catalogs will have a capital A to the left. These designations are not part of the number itself; they are meant only as quick methods of identifying these special materials.

Description of Bibliographic Information

Each entry contains standard bibliographic information and a descriptive note or paragraph. The following describes each type of information in detail.

Author(s): Up to three authors, editors, or compilers are given. If there are four or more persons responsible for a given work, the first author, followed by "et al.," is used. For proceedings volumes and other monographic works, the full title-page name is used; for journal articles and conference papers, last name and initials only are cited. The author index uses last name and initials only for all entries to avoid double listings for the same author. The full name is given for books because the reader using a card catalog to locate an item would need the full name for searching purposes.

Title: If the title is in a language other than English, the English translation will follow in brackets. If the work is an English translation of a foreign-language work, the original title is given in the descriptive note if it was easily discernible from the work itself or from standard bibliographic sources. Russian titles have been transliterated according to the Library of Congress transliteration scheme.[4] Titles are given as they appeared in print, with one exception: certain conferences that occur on a regular basis and which have multiple entries in this work, e.g., *6th Lunar Science Conference* is cited as "*Lunar Science Conference*. 6th." Other monographs such as reference works whose titles begin with an ordinal number are not inverted because usually only one (the latest) edition is cited. An example is *Fourth Consolidated Guide to International Data Exchange through the World Data Centres*.

Place: Place of publication includes city, state or province, and country, unless the city is very well known. Examples are Bellingham, WA; Strasbourg, France; Chicago; London.

Publisher: The publisher is abbreviated in cases of familiar or well-known companies. For example, D. Reidel Publishing Company becomes simply D. Reidel. If there are both a publisher and a distributor, or two publishers, both are given when known.

Dates: If a monograph has more than one date (e.g., a copyright and a printing date), both are given when known. If the date is unknown, then "[?]" is used. Journal article citations use year only: months or days are not given except in the case of journals lacking continuous pagination through a numerical volume.

Pagination: For monographic works, the number of pages is simply stated: e.g., 561p. Individual papers in a monograph have pagination following the imprint date (e.g., Washington, DC: NASA; 1980: 54-67). Pages are never abbreviated (e.g., 142-145 not 142-5).

Journal article citations: In citing volume, pages, and date for journal articles, this book uses the format employed in *American National Standard Z39.29-1977 for Bibliographic References*. An example is 17: 29-103; 1978 (volume: pagination; year). Issue numbers of parts are used only when the pagination is not continuous throughout or when no page information is available. Examples are 17(2): 15-34; 1972, or v. 6, no. 3, 1969.

Series: The series statement, if any, appears in parentheses following pagination. Series statements (and journal titles) are abbreviated according to the scheme used by *AAA*, which is primarily based on the UNISIST/ICSU-AB Working Group on Bibliographic Descriptions (1970) guidelines. Series abbreviations and their full forms are included in section R007, along with the list of journal abbreviations. Volume number for each series is included in the series statement if appropriate.

Report numbers: Technical report numbers, if any, appear in parentheses following the series statement, or in place of it if no series exists. If there is more than one report number associated with a given item, they are separated by semicolons.

References: The number of references have been included for all papers, journal articles, and book chapters. Most, but not all, monographic works citations include the number of references if easily discerned. Very large numbers of references (i.e., more than 1,000) are given as estimates—e.g., ~ 2,000 where ~ is the symbol for "approximately," or >1,000 where > is the symbol for "greater than."

Special terminology has been employed for references in the case of the *IAU Transactions. Reports on Astronomy*. In many of the *IAU Reports*, the references are scattered throughout the text and in many cases are merely *Astronomy and Astrophysics Abstracts'* numbers. Some have references both in the text and at the end, which makes counting difficult, if not impossible. To simplify the reporting of the numbers of references in these special cases, "+ refs" or "- refs" are used instead of an actual number. The former indicates a substantial number (i.e., an important bibliographic source), while the latter indicates a very small number (usually less than 10)—not a major source, but potentially useful nonetheless.

Dates of references: In most cases, a date range for the references is given so that the reader will know roughly what period is treated in the item. These should be taken with a grain of salt, however. For example, a review article summarizing recent results may begin with two or three references to classic works in the field, which may be quite old, giving a misleading date range. The date range, for instance, may be 1922-79, when in fact the article concentrates on 1968-79. The annotations should be helpful in most cases in overcoming this minor difficulty.

Papers/Chapters: "Papers" can mean several things: actual conference papers, lectures, brief communications, summaries, and occasionally abstracts. The latter are usually cited separately following the number of papers. For example, "40 papers, 23 abstracts" means that there are 63 items in the volume, 40 of which are full papers (with or without abstracts) and 20 of which are abstracts only. Book chapters are given when appropriate to give an indication of the scope of the work.

Conference data: It was mentioned earlier that a capital C would appear in the left-hand margin if a given item were a volume of conference proceedings. Such items include conferences, workshops, seminars, colloquia, lecture series, so-called summer school sessions, etc. The conference title, if different from the book title, will appear as the first element in the annotation.

The site of a meeting is stated in terms of city and country (and state or province for the United States and Canada)—e.g., Washington, DC, USA; Toronto, ON, Canada; Paris, France; Pacific Palisades, CA, USA; etc. Individual place (e.g., Harvard University) is occasionally, but not normally, given. For simplicity, England, Scotland, and Wales are designated by "UK" for United Kingdom.

The dates of a conference are given in inverted form: e.g., 18 Jul 1978; 23-30 Nov 1967; 30 Jun—3 Jul 1975; Aug 1971; etc. Months are abbreviated to the standard three-letter form.

Annotation: In almost all cases, these notes are descriptive rather than evaluative. Their purpose is to convey the emphasis, audience, format, subject matter, and special features of the item. They also occasionally include internal cross-references to related entries in the book.

Cross-references: At the end of most sections are found cross-references (the "see also" type), which are intended to lead the user to related materials. These references consist of title and reference number only. They are used in cases of monographs and other works which have more than one major subject emphasis.

Notes

[1]Published jointly by Macdonald Technical & Scientific and the Shoe String Press.

[2]Published by Libraries Unlimited, Inc.

[3]This book emphasizes atlases and catalogs published after Mike Collins' comprehensive *Astronomical Catalogues 1951-1975* (INSPEC, 1977). Only major works published before Collins' book are included here.

[4]Published in *Library Resources & Technical Services*, Spring 1980, p. 172.

LIBRARIES USED

No bibliography can be compiled without a library. In this case, 15 libraries were used to gather the information presented in this work. They are listed below in alphabetical order. Of those listed, those with asterisks were the major sources for this book, especially the authors' home libraries. We also wish to acknowledge the grateful assistance of the interlibrary loan departments of the University of Virginia Library and the University of Oklahoma Libraries. They were very instrumental in locating a number of hard-to-find items that were essential to this compilation.

David Dunlap Observatory Library (University of Toronto)

* Harvard College Observatory Library (Cambridge, Massachusetts)

* Kitt Peak National Observatory Library (Tucson, Arizona)

* Library of Congress (Washington, DC)

Lunar and Planetary Laboratory Library (University of Arizona)

NASA Headquarters Library (Washington, DC)

* National Radio Astronomy Observatory Library (Charlottesville, Virginia)

* Smithsonian Astrophysical Observatory Library (Cambridge, Massachusetts)

Steward Observatory Library (University of Arizona)

* U.S. Naval Observatory Library (Washington, DC)

University of California, Berkeley, Library

University of Oklahoma Libraries (Norman, Oklahoma)

University of Toronto Library

* University of Virginia Library (Charlottesville, Virginia)

* University of Waterloo Library (Waterloo, Ontario)

REFERENCE SOURCES
Bibliographies, Bibliographic Publications, Library Catalogs

R001 Bibliographies (General), Library Catalogs

R001.001 **A Catalog of NASA Special Publications.** Washington, DC: NASA; 1981. 104p. (NASA SP-449). Available from NTIS.

A complete listing from 1961 to 1980 of NASA technical reports in the SP, RP, and CP series. Entries contain title, author(s), brief note, report number, NASA accession number, number of pages, year, and availability. The following types of reports are covered: general; handbooks and data compilations; histories and chronologies; technology utilization; management evaluation and analysis standards; bibliographies; space vehicle design criteria; reference publications; conference publications.

R001.002 **Catalog of the Naval Observatory Library, Washington, D.C.** Boston: G. K. Hall; 1976. 6v.

A substantial source of data in the field, this set lists the holdings of one of the most important and extensive observatory library collections anywhere. Approximately 75,000 volumes are presented in this book catalog, which is a photographic reproduction of the library's dictionary card catalog. The only publication to list an entire astronomy library's materials, this work reflects the library's strong holdings in the history of astronomy, observatory publications, popular and technical periodicals, astrometry, almanacs, ephemerides, and navigation. Prepared under the direction of Brenda G. Corbin, librarian.

R001.003 Collier, Beth A.; Aveni, Anthony F. **A Selected Bibliography on Native American Astronomy.** Hamilton, NY: Department of Physics and Astronomy, Colgate University; 1978. 148p. 1,480 refs, 1850-1977.

Of interest to historians of astronomy and science in general, this compilation covers a broad range of subjects in an attempt "to examine evidence of the practice of astronomy among the Native American Civilizations." As a result, the book cites sources from archaeology, ethnology, geography, and astronomy in producing the most complete work of its kind. Arranged alphabetically by author, this list covers both North and South American Indian cultures. There is a subject index but no abstracts. Both books and journals are cited.

R001.004 Collins, M., comp. **Astronomical Catalogues 1951-1975.** London: INSPEC; 1977. 325p. (INSPEC Bibliogr. Ser., 2).

A bibliographical work containing nearly 2,500 items, each of which contains full reference data and usually a brief abstract. The catalogs are listed under seven major headings: solar system; stars; nebulae; galaxies; electromagnetic wave sources; miscellaneous objects and sources; and other topics. Each section is subdivided into narrower categories, and entries are listed chronologically under each. There are three indexes: designation, author, and corporate author.

R001.005 Dickson, Katherine Murphy, comp. **History of Aeronautics and Astronautics: A Preliminary Bibliography.** Washington, DC: NASA; 1968. 413p. (NASA TM-X-61708; HHR-29; N69-28385). 921 refs, ca. 1908-1967.

Aimed at the science historian, but also useful for the student, researcher, and librarian, this substantial reference work covers five major areas: 1) the meaning of space exploration; 2) the evolution of space technology; 3) the rise of space science; 4) impact and applications of space exploration; 5) history of related institutions. A wealth of infôrmation, including 64 references under astronomy, this volume is not strictly historical since it lists general sources such as indexes. Entries are annotated, and there are author, title, and subject indexes.

R001.006 Dupuy, D. L. A bibliography of recent astronomy education artícles. **J. R. Astron. Soc. Canada** 72: 221-222; 1978. 24 refs, 1976-78.

A brief list emphasizing teaching aids and methods for elementary undergraduate astronomy courses. Most entries include a descriptive note.

R001.007 Grassi, Giovanna, comp. **Union Catalogue of Printed Books of the XV and XVI Centuries in Astronomical European Observatories.** Rome: Rome Astronomical Observatory Library; 1977. 105p.

A work aimed at providing quick and easy access to rare astronomical materials and their bibliographic data. Containing books held in 37 observatory libraries, the catalog also includes a chronological list of titles and an index to publishers and printers. Entries are formatted in a concise way, following the short-title method of the British Museum, and are arranged by author, then title. Additional features: holding library information, bibliographic sources used in verification, and cross-references from variant forms of names.

R001.008 Kemp, D. A. **Astronomy and Astrophysics: A Bibliographical Guide.** London: Macdonald Technical & Scientific; Hamden, CT: Archon Books, Shoe String Press; 1970. 584p. (Macdonald Bibliogr. Guides Ser.).

A classic in the field of science bibliography, this extensive volume is aimed at the scientist and researcher who need to find articles and books in 75 different areas. Each of the more than 3,700 entries contains complete bibliographic data, numbers of references and their dates, and a descriptive note. Primarily covering 1950 to 1968, with some earlier citations to important works, the book also presents useful supplementary information such as lists of periodicals and abbreviations. Entries are arranged chronologically in each subject section, and there are subject and author indexes.

R001.009 Lavrova, N. B. **Bibliografii͡a russkoĭ astronomicheskoĭ literatury 1800-1900 gg.** Moscow: Izdatel'stvo Moskovskogo Universiteta; 1968. 386p. (Tr. Gos. Astron. Inst. P. K. Sternberga, 37). In Russian.

A bibliography of Russian astronomical literature of the nineteenth century, including 4,250 citations to books, pamphlets, journal articles, and observatory publications. Arranged chronologically within subject categories, the entries include both Russian and Western sources. There is an author index.

R001.010 Marienfeld, Horst. **GRS-Dokumentation. Bibliographien 1960-1968.** Darmstadt-Arheilgen, Germany: Gesellschaft für Regelungstechnik und Simulationstechnik GmbH; 1969. 5v. German/English.

Aimed at the engineer and scientist working in the field of aerospace and aeronautical science and technology, this work contains 2,782 entries covering bibliographical works, literature surveys, and review works. Citations include basic bibliographic data, number of references, and usually an abstract (some very lengthy) in English. Drawing upon the abstract journals such as *Scientific and Technical Aerospace Reports* (*STAR*) and *International Aerospace Abstracts* (*IAA*), as well as individual institutions and firms, for its source material, the guide contains 131 citations in the field of astronomy and 91 in the geosciences. The introduction and headings are in both German and English.

R001.011 Müller, Edith A., ed. **A List of Astronomy Educational Material.** Sauverny, Switzerland: Observatoire de Genève; 1970. unpaged.

Sponsored by the IAU Commission 46 (Teaching of Astronomy), the main purpose of this work is to guide persons or institutions who wish to build an astronomy library. A secondary purpose is to provide a list of materials for those interested in learning about astronomy. Includes references in the following categories: books; atlases and catalogs, dictionaries, encyclopedias, handbooks; magazines, journals, yearbooks; lecture notes, exercises, examination problems; films; filmstrips, photographs, slides. Under each section, materials are arranged by language, and each item has an indication of intended readership, very brief comment, bibliographic information, and price. A thorough compilation with a wide variety of material from around the world. There have been three addenda: 1973, 1976, and 1979.

R001.012 Seal, Robert A. **A Guide to the Literature of Astronomy.** Littleton, CO: Libraries Unlimited; 1977. 306p.

Aimed at the student, librarian, and layperson, this survey of the various sources in astronomy and astrophysics can also be used by the non-astronomer scientist wishing an introduction and overview. Each of the 578 entries includes complete bibliographic data and a lengthy annotation. The book is divided into four sections: reference sources; general materials; descriptive astronomy; and special topics. Also includes a bibliography of basic reference materials for all types of libraries.

R001.013 Smyth, M. F. I.; Smyth, M. J. **Supplement to the Catalogue of the Crawford Library of the Royal Observatory Edinburgh.** Edinburgh: Royal Observatory; 1977. 112p.

"A computer-produced index of the earlier works of the Crawford Collection published before the year 1800, and of manuscripts, broadsheets and works on comets published up to the early 1900s." There are author, date, comet, and pamphlet indexes.

R001.014 University of Oklahoma Libraries. **The Catalogue of the History of Science Collections of the University of Oklahoma Libraries.** London: Mansell; 1976. 2v. v.1: 584p.; v.2: 608p.

A printed dictionary catalog (authors, titles, subjects, etc.) of the most important collection of its kind in the world. A large number of astronomical books, including works by or about Galileo, Kepler, Copernicus, Brahe, and others, are found in this significant bibliographical source. Compiled and edited by Duane H. D. Roller (curator) and Marcia M. Goodman (librarian), the catalog includes about 40,000 printed and 10,000 microform volumes.

R002 Abstracting and Indexing Journals

R002.001 **Astronomischer Jahresbericht.** Leipzig, Germany: W. de Gruyter. v.1-68, 1899-1968. Annual. Superseded by *Astronomy and Astrophysics Abstracts.*

The first regularly published index to the literature of astronomy, this yearly publication provided complete coverage of books, book reviews, journal articles, observations, and conference papers. An indispensable source for retrospective searching, it includes about 7,000 entries per year and is primarily in German, except for the conference papers. Some entries have abstracts.

R002.002 **Astronomy and Astrophysics Abstracts.** Berlin; New York: Springer-Verlag. v.1- , 1969- . Semi-annual. Continues *Astronomischer Jahresbericht.*

The primary source of bibliographic information in the field, this comprehensive source covers books, conference proceedings, atlases and catalogs, data compilations, journal articles, and much more. Arranged under 112 subject categories, the entries contain full bibliographic data and a short abstract. About 90% of the entries are in

English; most of the rest are German and French, with some Russian. Cross-references are included at the end of each subject division to lead the user to related entries. This title's subject arrangement was used in the present volume by Seal and Martin.

R002.003 **Astronomy and Astrophysics Monthly Index.** Sierra Madre, CA: Olivetree Associates. v.1- , 1976- . Monthly.

This publication provides fairly fast coverage of approximately 67 journals and an equal number of observatory publications. Entries contain author(s), title, journal or publication, volume, pagination, and year. Access to entries is two-fold: an author index and a permuted title index. Its timeliness and wide coverage are its most important aspects.

R002.004 **Bulletin Signalétique 120: Astronomie, Physique Spatiale, Géophysique.** Paris: Centre de Documentation du CNRS. v.32- , 1971- . Monthly. Supersedes *Bulletin Signalétique 120: Astronomie et astrophysique: physique du globe.*

The third major world index to astronomical literature, this publication is entirely in French and therefore is not used much in the United States. Part of a series of abstracts and indexes covering a wide variety of topics, this section comprehensively addresses astronomy, astrophysics, space physics, and geophysics. It covers all the major Western journals and many others worldwide. Entries are arranged according to a classified subject arrangement, and include a bibliographic citation in the original language and a French abstract.

R002.005 **Current Physics Index.** New York: American Institute of Physics. v.1- , 1975- . Quarterly.

A secondary source for astronomical journal articles, this publication's emphasis is on physics journals, in particular the more than 40 journals published by AIP. Of these titles, three are astronomy-oriented: *Astronomical Journal*, *Soviet Astronomy*, and *Soviet Astronomy Letters.* Entries are arranged according to the AIP Physics and Astronomy Classification Scheme, and contain bibliographic data and an author abstract. Publication is among the fastest in the field, so it is a very up-to-date work; some entries are only a month old.

R002.006 **International Aerospace Abstracts.** New York: American Institute of Aeronautics and Astronautics. v.1- , 1961- . Semi-monthly.

The non-technical report counterpart to *Scientific and Technical Aerospace Reports* (*STAR*), *IAA* provides worldwide coverage in the astronomy-related fields of aeronautics, space science, and technology. Entries, which include bibliographic data and an abstract, cover periodical articles, books, conference proceedings, and journal article translations. Format and subject headings are identical to *STAR*, and astronomical entries can be found under "Space Sciences."

R002.007 **Meteorological and Geoastrophysical Abstracts.** Boston: American Meteorological Society. v.1- , 1950- . Monthly.

Not a primary source of information on astronomy and astrophysics, this abstracting journal concentrates on weather. Coverage of astronomy is limited to the solar system, primarily the effects of solar emissions on the atmospheres (and weather) of the Earth and other planets. Entries include bibliographic data, a lengthy abstract, subject headings, and very selected locations, such as the Library of Congress. Both foreign and domestic literature are covered.

R002.008 **Physics Abstracts.** London: Institution of Electrical Engineers. v.69- , 1966- . Bi-weekly. Continues *Science Abstracts, Series A* and its volume numbering.

Although this major abstracting and indexing journal is primarily devoted to the many aspects of physics, it does contain, in each issue, a fairly substantial section devoted to astronomy and astrophysics. Consequently, it is a major source of information, second only to *Astronomy and Astrophysics Abstracts* in the United States and the United

Kingdom, and much more timely. Coverage is worldwide and includes books, conference proceedings, review papers, journal articles, dissertations, patents, etc. Entries contain full bibliographic information and a brief abstract, as well as the number of references associated with the item.

R002.009 **Referativnyi Zhurnal: Astronomiia.** Moscow: VINITI. 1963- . Monthly.

One of the primary sources of astronomical information worldwide, this publication is but one section of a series of abstracting and indexing journals. Coverage is quite complete, with a majority of the entries from Western journals. Topics: general; astronomical instruments; observatories; theoretical astronomy; astrometry; astrophysics; solar system; the Sun; stars; nebulae and interstellar matter; galactic and extragalactic astronomy. Each entry has the title in both Russian and the original language, the citation in the original language, and a Russian abstract. Each issue has about 800 to 1,000 entries.

R002.010 **Science Citation Index.** Philadelphia, PA: Institute for Scientific Information. 1961- . Bi-monthly.

One of the most comprehensive and powerful indexes ever published, *SCI* covers over 2,500 journals worldwide, including 40 astronomy and astrophysics titles. Strictly an indexing service (no abstracts), the serial is divided into three parts: Source Index, Citation Index, and Permuterm Subject Index. The first part is an index of current journal articles with author, title, references, and volume/paging/date information. The Citation Index is an unusual but highly useful feature: it is an index of articles cited by other authors. An expensive but very useful reference tool, *SCI* provides excellent coverage of astronomy.

R002.011 **Scientific and Technical Aerospace Reports.** Washington, DC: NASA. v.1- , 1965- . Semi-monthly.

More commonly known as *STAR*, this publication emphasizes aeronautics and astronautics, but does provide coverage of NASA and other technical reports in astronomy and its various subfields. Emphasis within the sections on astronomy, not surprisingly, is on lunar and planetary exploration and space astronomy. Entries are found under "Space Sciences" and include bibliographic data, an abstract, report numbers, etc. A variety of indexes help access the information: author (personal and corporate), subject, contract number, report/accession number.

Ten years of "Astronomy and Astrophysics Abstracts." *R011.029.*

R003 Bibliographic Databases

During the 1970s, the use of online bibliographic databases became widespread in academic, special, and larger public libraries. In most cases, these computer files of bibliographic information were the machine-readable equivalents of hard-copy abstracting and indexing journals commonly used by reference librarians and library patrons to locate information in author or subject searches. In a few cases, however, the databases had no hard-copy equivalent at all—they were totally new sources of information, and in some cases were non-bibliographic files.

To state that the computerized form was the same as the paper copy was not quite true, though. While both versions usually contained basically the same information (although online files often had more data), the computer-accessible version provided much faster and more powerful searching capabilities. Both forms were searchable by author and subject, but the computer database could also be searched in many other ways: by report number, title words, abstract words, multiple subjects, etc. Long, complicated searches that would be impossible or extremely difficult manually could be accomplished in minutes using a computer terminal.

Three database vendors provided access to dozens of bibliographic and non-bibliographic files in the mid-1970s and into the early 1980s: Dialog Information Retrieval Service (Palo Alto, California); System Development Corporation (SDC) (Santa Monica, California); and Bibliographic Retrieval Service (BRS) (Scotia, New York). Dialog provided the greatest number (142) and variety of data files at the time of this writing. Although there were no astronomy-only databases in 1981, there were nine files which contained a varying amount of bibliographic data in the field. This does not include non-public databases like NASA RECON and others. Because of the importance of computer-accessible bibliographic data in the 1970s and 1980s, this volume includes a section describing the appropriate available files for astronomy and their contents. Each librarian or searcher will have to choose the file or files which best meet the needs of the literature search.

It should be noted that all of the nine databases described below were available (in 1981) from at least one of the three vendors cited above, and some are available from all three. Since database availability varies from year to year, it was thought best not to cite which vendor offered which file. Also, the number of records can vary from supplier to supplier, and will, of course, have changed dramatically by the time this book is printed. The numbers of records in each file are as of mid-1981. For detailed information on the specific databases, the reader is directed to the vendors and their extensive literature (available for the asking). For further reading on bibliographic searching, the reader is directed to the library literature, which abounds in articles on technique, theory, and analysis.

R003.001 **Comprehensive Dissertation Index.** Ann Arbor, MI: University Microfilms. Hardcopy equivalent: *Comprehensive Dissertation Abstracts.* Coverage: 1861-present. Citations: 730,000+. Updates: monthly.

An author, title, subject index to nearly all American dissertations from accredited institutions since 1861, as well as selected Canadian and other foreign entries. Provides access to astronomy graduate work, i.e., Ph.D. dissertations.

R003.002 **Conference Papers Index.** Washington, DC: Cambridge Scientific Abstracts. Coverage: 1973-present. Citations: 850,000+. Updates: monthly.

Provides access to over 100,000 scientific and technical papers annually from major regional, national, and international meetings. Information includes paper titles, authors, and authors' addresses if available. Provides coverage of astronomical papers as well as those of many other scientific disciplines.

R003.003 **GPO Monthly Catalog.** Washington, DC: U.S. Government Printing Office. Hardcopy equivalent: *Monthly Catalog* (GPO). Coverage: July 1976-present. Citations: 100,000+. Updates: monthly.

Includes comprehensive coverage of U.S. publications from all branches of government. Coverage of astronomy is very limited, however, usually including items in spectroscopy (National Bureau of Standards), popular items on the space program (NASA publications), ephemerides and astrometry (U.S. Naval Observatory), and some others.

R003.004 **INSPEC.** London: Institution of Electrical Engineers. Hardcopy equivalents: *Physics Abstracts; Electrical and Electronic Abstracts; Computer and Control Abstracts.* Coverage: 1969-present. Citations: 1,660,000+. Updates: monthly.

Although the main emphasis of this database is physics, it is a major source of astronomical information among all databases. Both English- and foreign-language items are included in the citations to journal articles, conference proceedings, books, technical reports, theses, etc. Coverage is not complete by any means, but it is one of the best sources available at present; it covers all the major astronomical journals.

R003.005 **Meteorological and Geoastrophysical Abstracts.** Washington, DC: American Meteorological Society and NOAA. Hardcopy equivalent: same title. Coverage: 1972-present. Citations: 78,000+. Updates: irregular.

A secondary source for astronomy, this database contains information on solar-terrestrial relations (e.g., solar wind, aurora, etc.). Coverage includes technical journals, monographs, proceedings, reviews, etc.

R003.006 **NTIS.** Springfield, VA: National Technical Information Service, U.S. Department of Commerce. Hardcopy equivalent: *Government Reports Announcements and Index.* Coverage: 1964-present. Citations: 830,000+. Updates: biweekly.

A key to U.S. government research and development reports, this database includes selected astronomy items (usually from NASA) which emphasize space, planetary, and lunar science. Other topics covered include the environment, energy, urban and regional planning, statistics, social sciences, etc.

R003.007. **SciSearch.** Philadelphia, PA: Institute for Scientific Information. Hardcopy equivalent: *Science Citation Index.* Coverage: 1974-present. Citations: 3,100,000+. Updates: monthly.

A massive, multidisciplinary index to 2,600 scientific and technical journals including 40 in astronomy, it covers 90% of the world's significant scientific literature. For more details, see *Science Citation Index* (*R002.010*).

R003.008 **SPIN (Searchable Physics Information Notices).** New York: American Institute of Physics. Coverage: 1975-present. Citations: 141,733+. Updates: monthly.

Covers the world's most significant physics journals, including all the AIP publications. A small number of astronomy journals are included.

R003.009 **SSIE Current Research.** Washington, DC: Smithsonian Science Information Exchange. No hardcopy equivalent. Coverage: previous two years. Citations: approx. 160,000. Updates: monthly.

Contains reports of government and privately funded scientific research projects currently in progress or initiated and completed during the past two years. Includes project descriptions from more than 1,300 organizations, with coverage of both physical (including astronomy and astrophysics) and social sciences.

REFERENCE SOURCES
Serials and Periodicals

R004 Review Journals and Series

R004.001 **Advances in Astronomy and Astrophysics.** New York: Academic. v.1-9, 1962-72.

The aim of the series was to provide a "semi-periodic" forum for publication of review articles on different aspects of astronomy and astrophysics which would be too long for ordinary scientific periodicals but too short for separate monographs, and to help bridge the gap between astronomy and neighboring fields of science. Emphasis on the Sun and Moon and the solar system.

R004.002 **Annual Review of Astronomy and Astrophysics.** Palo Alto, CA: Annual Reviews. v.1- , 1963- . Annual.

Articles are basically state-of-the-art reviews on varying aspects of astronomy and astrophysics, including observations, theory, and instrumentation. Authors are usually recognized experts in their fields who write so that articles are comprehensible to non-specialists in the specific topic under consideration, although a fairly high level of scientific expertise is assumed. Good bibliographies plus well-written articles make this the logical starting place for study of a specific topic.

R004.003 **Annual Review of Earth and Planetary Sciences.** Palo Alto, CA: Annual Reviews. v.1- , 1973- . Annual.

Like others in the excellent Annual Reviews series, this title includes reviews of two types: those that serve a tutorial function for persons in related fields, and those that expand the knowledge of the specialist. Each volume usually includes one or two papers of interest to the astronomical community.

R004.004 **Reviews of Geophysics and Space Physics.** Washington: American Geophysical Union. v.1- , 1963- . Quarterly. Through v.7, 1969, entitled *Reviews of Geophysics.*

Attempts to distill and place in perspective previous scientific work in currently active subject areas of geophysics and space physics. Aimed at specialists, graduate students, and scientists in related fields, it includes both full-length and topical reviews.

R004.005 **Space Science Reviews.** Dordrecht, Holland: D. Reidel. v.1- , 1962- . Monthly.

Contains invited review papers that provide a contemporary synthesis of the situation in various branches of space science. Primarily concerned with purely scientific aspects, but includes some few papers on instrumental and technical aspects. Papers are principally in English, but may be in French, German, or Russian with English abstracts.

R004.006 **Vistas in Astronomy.** Oxford: Pergamon. v.1-18, 1955-75. Annual.

Established initially to present original articles and reviews reflecting fully and critically the present situation and recent advances in astronomy and related fields. Aimed at both the expert and the less specialized reader wishing to gain a general idea of the latest developments. Toward the end of its run, articles included more history and therefore less current material, culminating in a pair of volumes on Kepler and Copernicus. Continued by a review journal of the same title.

R004.007 **Vistas in Astronomy: An International Review Journal.** Oxford: Pergamon. v.19- , 1975- . Quarterly.

Continues monographic series of the same title. Purports to provide a complete survey of contemporary astronomy and its allied sciences, containing articles by internationally famous authors. However, articles tend to be uneven in scope and presentation, and the publication schedule has been erratic. Began using camera-ready copy in 1977.

R005 Cumulative and Selected Indexes

R005.001 Abt, H. A.; Biggs, E. S., comps. **The Astrophysical Journal. General Index to Volumes 146-165 (1966 October to 1971 May) and to the Supplement Series Volumes 14-22 (1966 November to 1971 April).** Chicago: University of Chicago Press; 1972, c1971. 344p.

See also *R005.010.*

A cumulative author and subject index. Author entries include all authors (each is indexed), title, volume, and beginning page. Subject entries are arranged chronologically under each of the subject terms and phrases listed in a brief thesaurus at the beginning of the subject index. Entries are listed more than once if more than one subject term is appropriate. "See" and "see also" references are included.

R005.002 Abt, Helmut A.; Biggs, Eleanor S., comps. **The Astronomical Journal. General Index by Subjects and by Authors. Volumes 51-80, 1944-1975.** New York: American Institute of Physics; 1979. 444p.

Published for the American Astronomical Society by AIP, this index is the follow-up to the first *AJ* cumulative index published in 1948. Entries include author(s), title, volume, beginning page, and year. The subject index is arranged according to major topics and their subdivisions, for example: Planets—Pluto or Sun—flares.

R005.003 American Meteorological Society. **Cumulated Bibliography and Index to Meteorological and Geoastrophysical Abstracts, 1950-69: Classified Subject and Author Arrangements.** Boston: G. K. Hall; 1972. 9v. 7,056p.

Some 150,000 bibliographical references are presented in this set, consisting of four volumes arranged by Universal Decimal Classification (UDC) number and five arranged by author. The entries are photographic reproductions of the first five lines of each original entry from *Meteorological and Geoastrophysical Abstracts*, and they include the following information: UDC number, location within the abstracts, author and title, date, publisher, affiliation if known, and a brief portion of the abstract. Most entries also include location symbols to "major libraries where an item might be obtained."

R005.004 **Astronomicheskiĭ T͡Sirkuliar: avtorskiĭ ukazatel' k N°N° 400-599 (1967-1971 gg.).** Moscow: Bi͡uro Astronomicheskikh Soobshcheniĭ Akademii Nauk SSSR; 1974. 46p. In Russian.

A cumulative author index. Each entry includes author, title, circular number, and page.

R005.005 **Astronomy and Astrophysics Abstracts, Volume 15/16: Author and Subject Indexes to Volumes 1-10, Literature 1969-1973.** Berlin: Springer-Verlag; 1976. 655p.

Cumulative, computer-produced lists. Author indexes have names followed by appropriate *AAA* entry numbers. Subjects are indexed on two levels, general and more specific. For example, "Planetary Nebulae" and "Planetary Nebulae—Infrared Spectra."

R005.006 **Astronomy and Astrophysics Abstracts, Volume 23/24: Author and Subject Indexes to Volumes 11-14 and 17-22, Literature 1974-1978.** Berlin: Springer-Verlag; 1979. 1,127p.

See previous entry.

R005.007 **Astrophysics and Space Science: An International Journal of Cosmic Physics. Index to Volumes 1-45 (1968-1976).** Dordrecht, Holland; Boston: D. Reidel; 1977. 44p.

An author index whose entries include title, volume, and first page. Articles, letters, and research notes are indexed.

R005.008 **Astrophysics and Space Science: An International Journal of Cosmic Physics. Index to Volumes 46-60 (January 1977-February 1979).** Dordrecht, Holland; Boston: D. Reidel; 1979. 16p.

See previous entry.

R005.009 Author index: 1921-1969. **Bull. Astron. Inst. Netherlands** 20: 337-361; 1969.

A cumulative index covering the life of the journal. Entries include authors, title, volume, and first page. Each author is indexed separately.

R005.010 Fox, R. A., comp. **The Astrophysical Journal. General Index to Volumes 166-202 (15 May 1971 to 15 December 1975) and to the Supplement Series Volumes 23-29 (April 1971 to December 1975).** Chicago: University of Chicago Press; 1978. 466p.

See *R005.001.*

R005.011 Fraknoi, Andrew. A subject index to astronomy articles in *Scientific American* magazine (1960-1976). **Mercury** v.6, Jan-Feb 1977. pp. 20-21.

A listing of 185 articles arranged into 33 subject categories. Each entry consists of author, title, and date.

R005.012 Harper, W. E., comp. **General Index to the Publications of the Astronomical and Physical Society of Toronto 1890-1899, Toronto Astronomical Society 1900-1901, Royal Astronomical Society of Canada 1902-1931.** Toronto: University of Toronto Press; 1931. 122p.

An author/title/subject index to the *Journal of the Royal Astronomical Society of Canada* and its predecessors. Citations include author, title, volume, and first page. See also *R005.017.*

R005.013 **Index to Special Reports** (251-325). Cambridge, MA: Smithsonian Astrophysical Observatory; 1972. 43p.

A listing in report number order of the SAO Special Reports issued between October 13, 1967 and October 15, 1970. Entries include lengthy abstracts and the bibliographic data; there are also author, title, and subject indexes.

R005.014 **Index to Special Reports** (326-360). Cambridge, MA: Smithsonian Astrophysical Observatory; 1974. 25p.

See previous entry. This edition includes reports issued between October 16, 1970 and April 15, 1974. There is no subject index.

R005.015 Lavrova, N. B.; Petrova, N. D., comps. **Astronomicheskiĭ T͡Sirkuliar. Ukazatel' soderzhaniia k N° N° 151-400 (1954-1967 gg.).** [Astronomical Circular. Table of Contents to Nos. 151-400 (1954-1967).] Moscow: Biuro Astronomicheskikh Soobshchenii Akademii Nauk SSSR; 1972. 135p. In Russian.

A cumulative index to authors, subjects, and individual celestial objects, such as comets. Author entries contain author, title, circular number, and page.

R005.016 **The Moon and the Planets. Index to Volumes 1-20.** Dordrecht, Holland; Boston: D. Reidel; 1979. 14p.

An author index, primarily, to *The Moon: An International Journal of Lunar Studies* (v.1-17), and *The Moon and Planets: An International Journal of Comparative Planetology* (v.18-20). The volumes cover 1970-1979. Entries include all authors, article title, volume, and first page. Also indexes letters to editor, research notes, bibliographies, book reviews, conferences, etc.

R005.017 Northcott, Ruth J., comp. **General Index to the Journal of the Royal Astronomical Society of Canada 1932-66.** Toronto: Royal Astronomical Society of Canada; 1971. 171p.

Covering volumes 26-60, this is an author/title/subject index following the same format as the earlier edition (*R005.012*). Available from the RASC, 252 College St., Toronto, ON, Canada.

R005.018 Petri, W., comp. **Zeitschrift für Astrophysik: Generalregister für die Bande 51-69 (1960-1968).** Berlin: Springer-Verlag; 1969. 61p.

Cumulative index to the journal, with author, subject, and review paper access.

R005.019 **Scientific American: Cumulative Index 1948-1978: Index to the 362 Issues from May, 1948 through June, 1978.** New York: Scientific American; 1979. 565p.

An eight-part guide to the best-known general science periodical, this work offers access to nearly 3,000 regular articles and some 3,500 "Science and the Citizen" items. Since articles on astronomy and astrophysics are common in the journal, this volume is a very useful source for students, researchers, and librarians in astronomy and space sciences. Included are an Index to Topics (a rotated key-word index); Listing of Tables of Contents (chronologically); Index to Authors; Index to Titles (by first and other key words); Index to Book Reviews; Index to Mathematical Games; Index to "The Amateur Scientist"; and Index to Proper Names (of those persons mentioned throughout the journal).

R005.020 **Solar Physics: A Journal for Solar Research and the Study of Solar Terrestrial Physics. Index to Volumes 11-25 (1970-1972).** Dordrecht, Holland: Boston: D. Reidel; 1972[?]. 29p.

An index to authors, book reviews, and reports from solar institutes. No subject access is included.

R005.021 **Solar Physics: A Journal for Solar Research and the Study of Solar Terrestrial Physics. Index to Volumes 26-40 (1972-1975).** Dordrecht, Holland; Boston: D. Reidel; 1976[?]. 29p.

See previous entry.

R005.022 van Houten, C. J., comp. Subject index: 1921-1969. **Bull. Astron. Inst. Netherlands** 20: 362-394; 1969.

See *R005.009*.

Index of comets: 1921-1969. *102.021*.

Index of minor planets. *098.016*.

Index of variable stars. *123.007*.

R006 Periodical Lists, Observatory Publications

R006.001 Jackson, George; Bradley, Olive, comps. **Periodicals on Astronomy Held by the SRL.** London: British Library, Science Reference Library; 1976. 26p.

A guide to the astronomical serial resources and related services (including interlibrary loan) of the British Library's Science Reference Library. The list is arranged by title, and includes 386 periodicals and observatory publications and 18 abstracting services and bibliographies. A good checklist for reference work and interlibrary loan, its entries include holdings volumes, type of item, language, and subject classification. All areas of astronomy and astrophysics and UFOs are covered.

R006.002 Smulders, P. A. H., ed. **Bibliography of Non-commercial Publications of Observatories and Astronomical Societies.** 4th ed. Utrecht, The Netherlands: Sonnenborgh Observatory; 1981. unpaged (approx. 240).

Lists publications of some 210 institutions, providing current address, discontinued series, current series, and availability status of each, as well as frequent helpful notes about name changes of various observatories. Arrangement is by city location, with a cross-reference list from observatory name to city. The listing for each is only as complete as the information provided to the editor, and not all observatories worldwide are included; nonetheless, this is an invaluable source of information.

R007 Periodicals, Serials, and Series Cited

What follows are lists of serial publications and publishers' series used in this volume. Part 1, the list of journals and serials, includes ISSN (International Standard Serial Number) when known. Part 2, the list of publishers' series, contains the country of publication and the publisher.

With regard to the list of journals, this compilation cannot in any sense be construed as a complete list of astronomy and astronomy-related publications, for, as was pointed out in the introduction, the authors were very selective in choosing only those entries that included substantial bibliographical material. However, it does represent almost all the titles available for 1969-80 that published review-type or bibliographical material in astronomy and astrophysics.

All abbreviations are based on the format used in *Astronomy and Astrophysics Abstracts.*

PART 1: LIST OF JOURNALS AND OTHER SERIALS

ABH. HAMBURGER STERNW.	Hamburger Sternwarte. Abhandlungen. 0374-1583
ACTA ASTRON.	Acta Astronomica. An international quarterly journal. 0001-5237
ACTA ASTRON. SINICA	Acta Astronomica Sinica. 0001-5245
ACTA ASTRONAUT.	Acta Astronautica. Journal of the International Academy of Astronautics. [v.1, 1974-] 0094-5765
ACTA COSMOLOGICA	Acta Cosmologica. 0137-2386
ADV. ASTRON. ASTROPHYS.	Advances in Astronomy and Astrophysics. 0065-2180
ADV. ASTRONAUT. SCI.	Advances in the Astronautical Sciences. 0065-3438

ADV. ELECTRON. ELECTRON PHYS.	Advances in Electronics and Electron Physics. 0065-2539
ADV. PHYS.	Advances in Physics. 0001-8732
ADV. PLASMA PHYS.	Advances in Plasma Physics. 0065-3187
ADV. SPACE SCI. TECHNOL.	Advances in Space Science and Technology. 0065-3365
AMERICAN J. PHYS.	American Journal of Physics. 0002-9505
AMERICAN MINERAL.	American Mineralogist. 0003-004X
AMERICAN PHILOS. SOC. MEM.	American Philosophical Society. Memoirs. 0065-9738
AMERICAN SCI.	American Scientist. 0003-0996
ANN. GÉOPHYS.	Annales de Géophysique. 0003-4029
ANN. PHYSIQUE	Annales de Physique. 0003-4169
ANN. TOKYO ASTRON. OBS.	Annals of the Tokyo Astronomical Observatory. 0082-4704
ANNU. REV. ASTRON. ASTROPHYS.	Annual Review of Astronomy and Astrophysics. 0066-4146
ANNU. REV. EARTH PLANET. SCI.	Annual Review of Earth and Planetary Sciences. 0084-6597
ANNU. REV. FLUID MECH.	Annual Review of Fluid Mechanics. 0066-4189
ANNU. REV. NUCL. SCI.	Annual Review of Nuclear Science. 0066-4243
ANNU. REV. PHYS. CHEM.	Annual Review of Physical Chemistry. 0066-426X
APPL. OPT.	Applied Optics. A monthly publication of the Optical Society of America. 0003-6935
APPL. PHYS.	Applied Physics. 0340-3793
APPL. SPECTROSC. REV.	Applied Spectroscopy Reviews. 0570-4928
ASTROFIZ. ISSLED. IZV. SPETS. ASTROFIZ. OBS.	Akademiia nauk SSSR. Astrofizicheskie Issledovaniia. Izvestiia Spetsial'naia Astrofizicheskaia Observatoriia.
ASTROFIZIKA	Astrofizika. 0571-7132
ASTROMETR. ASTROFIZ.	Astrometriia i Astrofizika. 0582-8201
ASTRON. ASTROPHYS.	Astronomy and Astrophysics. A European Journal. 0004-6361
ASTRON. ASTROPHYS. SUPPL. SER.	Astronomy and Astrophysics Supplement Series. 0365-0138
ASTRON. J.	Astronomical Journal. 0004-6256
ASTRON. NACHR.	Astronomische Nachrichten. 0004-6337
ASTRON. PAP.	Astronomical Papers prepared for the use of the American Ephemeris and Nautical Almanac. 0083-243X

ASTRON. VESTN.	Astronomicheskii Vestnik. 0320-930X
ASTRONAUT. ACTA	Astronautica Acta. [v.1-18, 1955-73]. 0004-6205
ASTROPHYS. J.	Astrophysical Journal. 0004-637X
ASTROPHYS. J. LETT.	Astrophysical Journal Letters to the Editor.
ASTROPHYS. J. SUPPL. SER.	Astrophysical Journal Supplement Series. 0067-0049
ASTROPHYS. LETT.	Astrophysical Letters. 0004-6388
ASTROPHYS. SPACE SCI.	Astrophysics and Space Science. 0004-640X
ASTROPHYSICS.	Astrophysics. 0004-6396
AUSTRALIAN J. PHYS.	Australian Journal of Physics. 0004-9506
AUSTRALIAN J. PHYS., ASTROPHYS. SUPPL.	Australian Journal of Physics, Astrophysical Supplement. 0572-1164
BOL. INST. TONANTZINTLA	Boletin del Instituto de Tonantzintla. 0303-7584
BULL. ACAD. SCI. USSR, PHYS. SER.	Bulletin of the Academy of Sciences of the USSR, Physical Series. 0001-432X
BULL. AMERICAN ASTRON. SOC.	Bulletin of the American Astronomical Society. 0002-7537
BULL. ASTRON. INST. NETHERLANDS	Bulletin of the Astronomical Institutes of the Netherlands.
BULL. ASTRON. SOC. INDIA	Bulletin of the Astronomical Society of India.
BULL. BUR. INT. INF. EPHEMERIDES ASTRON.	Bulletin of the International Information Bureau on Astronomical Ephemerides (IAU-COSPAR).
BULL. INF. CENT. DONNÉES STELLAIRES	Bulletin d'Information du Centre de Données Stellaires.
BYULL. ABASTUMAN. ASTROFIZ. OBS.	Abastumanskaia Astrofizicheskaia Observatoriia, Gora Kanobili. Biulleten'. 0375-6644
BYULL. INST. TEOR. ASTRON. AKAD. NAUK SSSR.	Akademiia nauk SSSR. Institut teoreticheskoi astronomii. Biulleten'.
CALIFORNIA INST. TECHNOL. DIV. GEOL. PLANET. SCI. CONTR.	California Institute of Technology, Pasadena. Division of Geological and Planetary Sciences. Contribution.
CARTER OBS. ASTRON. BULL.	Carter Observatory Astronomical Bulletin.
CASE STUD. AT. COLLISION PHYS.	Case Studies in Atomic Collision Physics. 0300-4503
CELESTIAL MECH.	Celestial Mechanics. 0008-8714
CHINESE ASTRON.	Chinese Astronomy. 0146-6364
CIEL TERRE	Ciel et Terre. 0009-6709
COMMENTS ASTROPHYS.	Comments on Astrophysics. Comments on Modern Physics, Part C. 0146-2970
COMMUN. LUNAR PLANET. LAB.	Communications of the Lunar and Planetary Laboratory of the University of Arizona. 0066-7595

CONTEMP. PHYS.	Contemporary Physics. 0010-7514
CONTRIB. ASTRON. OBS. SKALNATÉ PLESO.	Contributions of the Astronomical Observatory Skalnaté Pleso. 0583-466X
CONTRIB. KITT PEAK NATL. OBS.	Contributions from the Kitt Peak National Observatory. 0075-6253
CZECHOSLOVAK J. PHYS. B	Czechoslovak Journal of Physics, Section B. 0011-4626
DUDLEY OBS. REP.	Dudley Observatory Report. 0070-7449
EARTH EXTRATERR. SCI.	Earth and Extraterrestrial Sciences. 0070-7902
EOS TRANS. AMERICAN GEOPHYS. UNION	EOS Transactions of the American Geophysical Union. 0096-3941
ESA SCI. TECH. REV.	ESA Scientific and Technical Review. 0378-2204
ESO BULL.	European Southern Observatory Bulletin. 0071-3104
ESO SCI. PREPR.	European Southern Observatory Scientific Preprint.
FIVE COLLEGE OBS. CONTR.	Five College Observatory Contribution.
FIZ. ELEM. CHASTITS AT. YADRA	Fizika Elementarnykh Chastits i Atomnogo Yadra.
FORSCHR. PHYS.	Forschritte der Physik. 0015-8208
FUNDAM. COSMIC PHYS.	Fundamentals of Cosmic Physics. 0094-5846
GEN. RELATIV. GRAVITATION	General Relativity and Gravitation. 0001-7701
GEOCHIM. COSMOCHIM. ACTA SUPPL.	Geochimica et Cosmochimica Acta Supplement.
GEOMAGN. AERON.	Geomagnetizm i Aeronomii͡a. 0016-7940 [Translation in: Geomagnetism and Aeronomy. 0016-7932]
GEOPHYS. SPACE DATA BULL.	Geophysics and Space Data Bulletin.
HIGHLIGHTS ASTRON.	Highlights of Astronomy as presented at the General Assembly of the International Astronomical Union.
IAU COLLOQ.	International Astronomical Union Colloquium.
IAU SYMP.	International Astronomical Union Symposium. 0074-1809
IAU TRANS. REP. ASTRON.	International Astronomical Union Transactions. Reports on Astronomy. 0080-1372
ICARUS	Icarus. International Journal of Solar System Studies. 0019-1035
IEEE TRANS. AEROSP. ELECTRON. SYST.	IEEE Transactions on Aerospace and Electronic Systems. 0018-9251

IEEE TRANS. GEOSCI. ELECTRON.	IEEE Transactions on Geoscience Electronics. 0018-9413
IEEE TRANS. NUCL. SCI.	IEEE Transactions on Nuclear Science. 0018-9499
INDIAN J. RADIO SPACE PHYS.	Indian Journal of Radio and Space Physics. 0367-8393
INFRARED PHYS.	Infrared Physics. An International Research Journal. 0020-0891
INST. THEOR. ASTROPHYS., BLINDERN-OSLO, REP.	Institute of Theoretical Astrophysics, Blindern-Oslo, Report. 0078-6780
INTERDISCIPLINARY SCI. REV.	Interdisciplinary Science Reviews. 0308-0188
INT. COMET Q.	International Comet Quarterly.
IRISH ASTRON. J.	Irish Astronomical Journal. 0021-1052
ITOGI NAUKI TEKH. SER. ASTRON.	Itogi Nauki i Tekhniki. Seriia Astronomiia. 0579-1421
ITOGI NAUKI TEKH. SER. ISSLED. KOSM. PROSTRANSTVA.	Itogi Nauki i Tekhniki. Seriia Issledovanie Kosmicheskogo Prostranstva. 0202-0734
IZV. AKAD. NAUK SSSR, SER. FIZ.	Izvestiia Akademii Nauk SSSR. Seriia Fizicheskaia. 0367-6755
IZV. ASTRON. EHNGEL'GARDT. OBS.	Izvestiia Astronomicheskoi Ehngel' gardtovskoi Observatorii.
IZV. VUZ RADIOFIZ.	Izvestiia Vysshikh Uchebnykh Zavedenii, Radiofizika. 0021-3462
J. ATMOS. SCI.	Journal of the Atmospheric Sciences. 0022-4928
J. ATMOS. TERR. PHYS.	Journal of Atmospheric and Terrestrial Physics. 0021-9169
J. BRIT. INTERPLANET. SOC.	Journal of the British Interplanetary Society. 0007-084X
J. GEOPHYS. RES.	Journal of Geophysical Research. 0148-0227
J. INST. MATH. APPL.	Journal of the Institute of Mathematics and Its Applications. 0020-2932
J. PHOTOGR. SCI.	Journal of Photographic Science. 0022-3638
J. PHYS. CHEM. REF. DATA	Journal of Physical and Chemical Reference Data. 0047-2689
J. PHYSIQUE	Journal de Physique. 0302-0738
J. QUANT. SPECTROSC. RADIAT. TRANSFER	Journal of Quantitative Spectroscopy and Radiative Transfer. 0022-4073
J. R. ASTRON. SOC. CANADA	Journal of the Royal Astronomical Society of Canada. 0035-872X

KLEINE VERÖFF. REMEIS-STERNWARTE BAMBERG	Kleine Veröffentlichungen der Remeis-Sternwarte Bamberg. 0404-7125
KVANTOVAYA ELEKTRON., MOSKVA	Kvantovaia Elektronika. 0368-7147
KYOTO UNIV. INST. ASTROPHYS. OBS. CONTR.	Contributions from Kyoto University Institute of Astrophysics and Kwasan Observatory. 0451-1514
LICK OBS. BULL.	Lick Observatory Bulletin. 0075-9317
LICK OBS. CONTR.	Lick Observatory Contribution. 0457-7833
LOUISIANA STATE UNIV. OBS. CONTR.	Louisiana State University Observatory Contribution. 0459-8776
LOWELL OBS. BULL.	Lowell Observatory Bulletin.
LUNAR PLANET. SCI. INST. CONTR.	Lunar and Planetary Science Institute Contribution.
LUNAR SCIENCE INST. CONTR.	Lunar Science Institute Contribution.
MEM. FAC. SCI. KYOTO UNIV.	Memoirs of the Faculty of Science, Kyoto University. Series of Physics, Astrophysics, Geophysics, and Chemistry. 0368-9689
MEM. R. ASTRON. SOC.	Memoirs of the Royal Astronomical Society. 0369-1829
MEM. SOC. ASTRON. ITALIANA	Memorie della Società Astronomica Italiana. 0037-8720
MEM. SOC. R. SCI. LIÈGE	Mémoires de la Société Royale des Sciences de Liège.
MERCURY	Mercury. The Journal of the Astronomical Society of the Pacific. 0047-6773
METEORITICS	Meteoritics. The Journal of the Meteoritical Society. 0026-1114
MIT PLANET. ASTRON. LAB. CONTR.	Massachusetts Institute of Technology Planetary Astronomy Laboratory Contribution.
MITT. ASTRON. GES.	Mitteilungen der Astronomischen Gesellschaft, Hamburg. 0374-1958
MITT. KARL-SCHWARZSCHILD-OBS. TAUTENBURG	Mitteilungen des Karl-Schwarzschild-Observatoriums Tautenburg der Deutschen Akademie der Wissenschaften der DDR. 0496-9618
MON. NOT. R. ASTRON. SOC.	Monthly Notices of the Royal Astronomical Society. 0035-8711
MOON	The Moon. An International Journal of Lunar Studies. 0027-0903
MOON PLANETS	The Moon and the Planets. An International Journal of Comparative Planetology. 0165-0807
NATL. RES. COUNCIL CANADA. ASTROPHYS. BRANCH PUBL.	National Research Council of Canada. Astrophysics Branch Publication.

NATURE	Nature. 0028-0836
NATURE PHYS. SCI.	Nature Physical Science. 0300-8746
NATURWISSENSCHAFTEN	Die Naturwissenschaften. 0028-1042
NAVAL RES. LAB. REP.	United States Naval Research Laboratory Report.
NRAO REPR.	National Radio Astronomy Observatory Reprint.
NRC	National Research Council of Canada [Reprint]
NUCL. ENERGY INF. CENT. REV. REP.	Nuclear Energy Information Center Review Report. [Warsaw]
NUOVO CIMENTO, RIV.	Rivista del Nuovo Cimento. 0035-5917
NUOVO CIMENTO, SUPPL.	Supplemento al Nuovo Cimento. 0550-3868
OPT. ENG.	Optical Engineering. 0091-3286
OPT.-MEKH. PROM.-ST.	Optiko-Mekhanicheskaia Promishlennost. 0474-3105
OPT. SCI. CENT. UNIV. ARIZONA TECH. REP.	Optical Science Center, University of Arizona, Technical Report.
ORIGINS LIFE	Origins of Life; An International Journal Devoted to the Scientific Study of the Origin of Life. 0302-1688
OSS. MEM. OSS. ASTROFIS. ARCETRI	Osservazioni e Memorie dell'Osservatorio Astrofisico di Arcetri.
PEREM. ZVEZDY	Peremennye Zvezdy.
PHILOS. TRANS. R. SOC. LONDON, SER. A	Philosophical Transactions of the Royal Society of London. Series A, Mathematical and Physical Sciences. 0080-4614
PHYS. CHEM. EARTH	Physics and Chemistry of the Earth. 0079-1946
PHYS. EARTH PLANET. INTER.	Physics of the Earth and Planetary Interiors. 0031-9201
PHYS. REP.	Physics Reports. [Formerly: Phys. Rep. Phys. Lett. C]. 0370-1573
PHYS. REP. PHYS. LETT. C.	Physics Reports. Physics Letters Section C.
PHYS. SCR.	Physica Scripta. 0031-8949
PHYSICA	Physica. 0378-4371
PLANET. SPACE SCI.	Planetary and Space Science. 0032-0633
POSTEPY ASTRON.	Postepy Astronomii. 0032-5414
PROBL. KOS. FIZ.	Problemy Kosmichskoǐ Fiziki. 0555-2796
PROC. AMERICAN PHILOS. SOC.	Proceedings of the American Philosophical Society. 0065-972X
PROC. ASTRON. SOC. AUSTRALIA	Proceedings of the Astronomical Society of Australia. 0066-9997

PROC. IEEE	Proceedings of the Institute of Electrical and Electronics Engineers. 0018-9219
PROC. R. SOC. LONDON, SER. A	Proceedings of the Royal Society of London. Series A: Mathematical and Physical Sciences. 0080-4630
PROC. (TR.) P. N. LEBEDEV PHYS. INST.	Proceedings (Trudy) of the P. N. Lebedev Physics Institute.
PROG. MATER. SCI.	Progress in Materials Science. 0079-6425
PUBL. ASTRON. BÜRO	Publications of the Astronomisches Büro (Vienna).
PUBL. ASTRON. SOC. PACIFIC	Publications of the Astronomical Society of the Pacific. 0004-6280
PUBL. CZECH. ACAD. SCI. ASTRON. INST.	Publications of the Czechoslovak Academy of Sciences, Astronomical Institute. 0528-6336
PUBL. DAVID DUNLAP OBS.	Publications of the David Dunlap Observatory.
PUBL. DEP. ASTRON., UNIV. BEOGRAD	Publications of the Department of Astronomy, University of Beograd, Faculty of Sciences.
PUBL. DOMINION ASTROPHYS. OBS.	Publications of the Dominion Astrophysical Observatory, Victoria, B.C.
PUBL. GOETHE LINK OBS.	Goethe Link Observatory. Publication.
PUBL. LEANDER McCORMICK OBS.	Publications of the Leander McCormick Observatory, Charlottesville, VA.
PUBL. OBS. ASTRON. BEOGRAD	Publications de l'Observatoire Astronomique de Beograd.
PUBL. OBS. GENÈVE, SER. B.	Publications de l'Observatoire de Genève, Série B. 0435-2939
PUBL. R. OBS. EDINBURGH	Publications of the Royal Observatory, Edinburgh.
PUBL. SPEC. CENT. DONNÉES STELLAIRES	Publication Speciale du Centre de Données Stellaires.
PUBL. UNITED STATES NAVAL OBS.	Publications of the United States Naval Observatory. 0083-2448
PUBL. WARNER SWASEY OBS.	Publications of the Warner and Swasey Observatory.
Q. J. R. ASTRON. SOC.	Quarterly Journal of the Royal Astronomical Society. 0035-8738
Q. J. R. METEOR. SOC.	Quarterly Journal of the Royal Meteorological Society. 0035-9009
R. GREENWICH OBS. BULL.	Royal Greenwich Observatory Bulletin. 0080-438X [LC]; 0308-5074 [item]
R. OBS. ANN.	Royal Observatory Annals. 0080-4371
R. OBS. BULL.	Royal Observatory Bulletin (= Royal Greenwich Observatory Bulletin)

RADIO ELECTRON. ENG.	Radio and Electronic Engineer. 0033-7722
RADIO SCI.	Radio Science. 0048-6604
RADIOPHYS. QUANTUM ELECTRON.	Radiophysics and Quantum Electronics. 0033-8443
REP. PROG. PHYS.	Reports on Progress in Physics. 0034-4885
REV. GEOPHYS.	Reviews of Geophysics. 0034-6853
REV. GEOPHYS. SPACE PHYS.	Reviews of Geophysics and Space Physics. 0034-6853
REV. MEXICANA ASTRON. ASTROFIS.	Revista Mexicana de Astronomia y Astrofisica. 0185-1101
REV. MOD. PHYS.	Reviews of Modern Physics. 0034-6861
REV. PHYS. APPL.	Revue de Physique Appliquée. 0035-1687
RIC. ASTRON.	Ricerche Astronomiche. 0081-3591
SCI. PROG.	Science Progress. 0036-8504
SCIENCE	Science. 0036-8075
SMITHSONIAN ASTROPHYS. OBS. SPEC. REP.	Smithsonian Astrophysical Observatory Special Report. 0081-0320
SOL. PHYS.	Solar Physics. A Journal for Solar Research and the Study of Solar Terrestrial Physics. 0038-0938
SOOBSHCH. SPETS. ASTROFIZ. OBS.	Akademiia nauk SSSR. Spetsial'naia Astrofizicheskaia Observatoriia. Soobshcheniia.
SOVIET J. OPT. TECHNOL.	Soviet Journal of Optical Technology. 0038-5514
SOVIET J. PART. NUCL.	Soviet Journal of Particles and Nuclei. 0090-4759
SOVIET J. QUANTUM ELECTRON.	Soviet Journal of Quantum Electronics. 0049-1748
SOVIET PHYS. USP.	Soviet Physics-Uspekhi. 0038-5670
SPACE LIFE SCI.	Space Life Sciences. 0038-6286
SPACE SCI. INSTRUM.	Space Science Instrumentation. An International Journal of Scientific Instruments for Aircraft, Balloons, Sounding Rockets, and Spacecraft. 0377-7936
SPACE SCI. REV.	Space Science Reviews. 0038-6308
SPECTROCHIM. ACTA B	Spectrochimica Acta B, Atomic Spectroscopy. 0584-8547
SPECTROSC. LETT.	Spectroscopy Letters. 0038-7010
SPIE PROC.	Proceedings of the Society of Photo-Optical Instrumentation Engineers. 0583-9572
STERNE	Die Sterne. Zeitschrift für alle Gebiete der Himmelskunde. 0039-1255
STUD. APPL. MATH.	Studies in Applied Mathematics. 0022-2526

TR. GLAV. ASTRON. OBS. PULKOVO	Trudy Glavnoi Astronomicheskoi Observatorii v Pulkove.
TR. GOS. ASTRON. INST. P. K. SHTERNBERGA	Trudy Gosudarstvennogo Astronomiches-Kogo Instituta imeni P. K. Shternberga. 0041-3453 [LC]; 0371-6791 [item]
U.S. NAVAL OBS. CIRC.	United States Naval Observatory Circular. 0097-0336
UNIV. NEW MEXICO PUBL. METEORITICS	University of New Mexico Publications in Meteoritics. 0085-3968
UNIV. TEXAS PUBL. ASTRON.	University of Texas Publication in Astronomy. [ser. I] 0563-2579; [ser. II] 0563-2587
USP. FIZ. NAUK	Uspekhi Fizicheskii Nauk. 0042-1294
VERENIGING VOOR STERRENKUNDE MEM.	Vereniging voor Sterrenkunde Memoirs (Brussels).
VERÖFF. ASTRON. RECH-INST.	Veröffentlichungen des Astronomischen Rechen-Instituts Heidelberg.
VESTN. KHAR'KOV. UNIV.	Vestnik Khar'kovskogo Universiteta. Seriia Astronomicheskaia. 0453-8056
VESTN. KIEV. UNIV.	Vestnik Kievskogo Universiteta. Seriia Astronomii. 0453-8587
VILLANOVA UNIV. OBS. CONTR.	Villanova University Observatory Contribution.
VISTAS ASTRON.	Vistas in Astronomy. An International Review Journal. 0083-6656

PART 2: LIST OF PUBLISHERS' SERIES

AAAS SEL. SYMP.	American Association for the Advancement of Science. Selected Symposium. USA: AAAS.
AAS SCI. TECHNOL. SER.	American Astronautical Society Science and Technology Series. USA: AAS.
AIP CONF. PROC.	American Institute of Physics Conference Proceedings. USA: AIP.
ADDISON-WESLEY SER. PHYS.	Addison-Wesley Series in Physics. USA: Addison-Wesley.
ADV. ASTRONAUT. SCI.	Advances in Astronautical Sciences. USA: AAS.
ADV. SPACE EXPLORATION	Advances in Space Exploration. UK; USA: Pergamon.
ANN. INT. YEARS QUIET SUN	Annals of the International Years of the Quiet Sun. USA: MIT Press.

ANN. NY ACAD. SCI.	Annals of the New York Academy of Sciences. USA: New York Academy of Sciences.
ASTRON. ASTROPHYS. SER.	Astronomy and Astrophysics Series. USA: Pachart.
ASTROPHYS. SPACE SCI. LIBR.	Astrophysics and Space Science Library. Holland; USA: D. Reidel.
BENCHMARK PAP. GEOL.	Benchmark Papers in Geology. USA: Dowden, Hutchinson & Ross.
CAMBRIDGE MONOGR. PHYS.	Cambridge Monographs in Physics. UK: Cambridge University Press.
COLLOQ. INT. CENT. NATL. RECH. SCI.	Colloque International du Centre National de la Recherche Scientifique. France: CNRS.
COMMONWEALTH INT. LIBR. SEL. READINGS PHYS.	The Commonwealth and International Library. Selected Readings in Physics. UK; USA: Pergamon.
DEV. PETROL.	Developments in Petrology. Holland; USA: Elsevier.
DEV. SOL. SYST. SPACE SCI.	Developments in Solar System and Space Science. Holland; USA: Elsevier.
ELECTRICAL SCI. SER. MONOGR. TEXTS	Electrical Science; a series of monographs and texts. USA; UK: Academic.
ENCYCLOPEDIA EARTH SCI. SER.	Encyclopedia of Earth Sciences Series. USA: Reinhold.
ESA SP.	European Space Agency Special Publication. Holland: ESA.
ESA SR.	European Space Agency Special Report. Holland: ESA.
ESRO SP.	European Space Research Organization Special Publication. Holland: ESRO.
GEOPHYS. ASTROPHYS. MONOGR.	Geophysics and Astrophysics Monographs. Holland; USA: D. Reidel.
GEOPHYS. MONOGR.	Geophysical Monographs. USA: American Geophysical Union.
GRUNDLEHREN MATH. WISS. EINZELDARSTELLUNGEN	Die Grundlehren der mathematischen Wissenschaften in Einzeldarstellungen. Germany; USA: Springer-Verlag.
HARVARD BOOKS ASTRON.	Harvard Books on Astronomy. USA: Harvard.
HEIDELBERG SCI. LIBR.	Heidelberg Science Library. Germany; USA: Springer-Verlag.
IEEE PRESS SEL. REPR. SER.	IEEE Press Selected Reprint Series. USA: Institute of Electrical and Electronics Engineers [distributed by Wiley].

IPST ASTROPHYS. LIBR.	Israel Program for Scientific Translations Astrophysics Library. Israel: ISPT.
INSPEC BIBLIOGR. SER.	INSPEC Bibliography Series. UK: Institution of Electrical Engineers, INSPEC.
INT. ASTROPHYS. SER.	International Astrophysics Series. UK: Chapman and Hall.
INT. GEOPHYS. SER.	International Geophysics Series. USA: Academic.
INT. SER. EARTH SCI.	International Series in Earth Sciences. UK; USA: Pergamon.
INT. SER. MONOGR. ANAL. CHEM.	International Series of Monographs in Analytical Chemistry. UK; USA: Pergamon.
INT. SER. MONOGR. NAT. PHILOS.	International Series of Monographs in Natural Philosophy. UK; USA: Pergamon.
INT. SER. MONOGR. PHYS.	International Series of Monographs on Physics. UK: Clarendon.
INT. SER. NAT. PHILOS.	International Series in Natural Philosophy. UK; USA: Pergamon.
INT. SER. PURE APPL. PHYS.	International Series in Pure and Applied Physics. USA: McGraw-Hill.
INTERMEDIATE SHORT TEXTS ASTROPHYS.	Intermediate Short Texts in Astrophysics. USA: Pachart.
INTERSCI. MONOGR. TEXTS PHYS. ASTRON.	Interscience Monographs and Texts in Physics and Astronomy. USA: Wiley.
INTERSCI. TRACTS PHYS. ASTRON.	Interscience Tracts on Physics and Astronomy. USA: Wiley.
LECTURE NOTES PHYS.	Lecture Notes in Physics. Germany: Springer-Verlag.
MACDONALD BIBLIOGR. GUIDES SER.	Macdonald Bibliographical Guides Series. UK: Macdonald Technical and Scientific.
MATH. LECTURE NOTE SER.	Mathematics Lecture Note Series. USA: W. A. Benjamin.
METH. EXP. PHYS.	Methods of Experimental Physics. USA; UK: Academic.
MINER. ROCKS.	Minerals and Rocks. Germany; USA: Springer-Verlag.
MINN. STUD. PHILOS. SCI.	Minnesota Studies in the Philosophy of Science. USA: University of Minnesota Press.
MONOGR. ASTRON. SUBJ.	Monographs on Astronomical Subjects. UK: Adam Hilger and Oxford University Press.

MONOGRAFIEEN ASTRON. ASTROFYSICA	Monografieen over Astronomie en Astrofysica. Belgium: Volkssterrenwacht Urania.
NAS PUBL.	National Academy of Sciences Publications. USA: NAS.
NASA CP.	National Aeronautics and Space Administration Conference Publications. USA: NASA.
NASA CR.	National Aeronautics and Space Administration Contractor Reports. USA: NASA.
NASA RP.	National Aeronautics and Space Administration Reference Publications. USA: NASA.
NASA SP.	National Aeronautics and Space Administration Special Publications. USA: NASA.
NASA TM.	National Aeronautics and Space Administration Technical Memoranda. USA: NASA.
NASA TT F.	National Aeronautics and Space Administration Technical Translations. USA: NASA.
NATO ADV. STUD. INST. SER., SER. B, PHYS.	NATO Advanced Study Institutes Series, Series B, Physics. USA: Plenum.
NATO ADV. STUD. INST. SER., SER. C, MATH. PHYS. SCI.	NATO Advanced Study Institutes Series, Series C, Mathematical and Physical Sciences. Holland; USA: D. Reidel.
NBS MONOGR.	National Bureau of Standards Monographs. USA: NBS.
NBS SP.	National Bureau of Standards Special Publications. USA: NBS.
NSRDS-NBS.	National Standard Reference Data Series, National Bureau of Standards. USA: NBS.
1:5,000,000 MAP SER.	The 1:5,000,000 Map Series. USA: NASA.
OXFORD PHYS. SER.	Oxford Physics Series. UK: Clarendon; Oxford University.
OXFORD SCI. PUBL.	Oxford Science Publications. UK: Oxford University.
OXFORD STUD. PHYS.	Oxford Studies in Physics. UK: Oxford University.
PART. FIELDS SUBSER.	Particles and Fields Subseries (AIP Conference Publications). USA: American Institute of Physics.
PERGAMON INT. LIBR.	Pergamon International Library. UK; USA: Pergamon.

PERGAMON INT. LIBR. SCI. TECHNOL. ENG. SOC. STUD.	Pergamon International Library of Science, Technology, Engineering, and Social Studies. UK; USA: Pergamon.
PERGAMON INT. POPULAR SCI. SER.	Pergamon International Popular Science Series. UK; USA: Pergamon.
PHYS. CHEM. SPACE	Physics and Chemistry in Space. Germany; USA: Springer-Verlag.
PONTIFICIAE ACAD. SCI. SCR. VARIA	Pontificiae Academiae Scientiarum scripta varia. Holland: North Holland; New York: American Elsevier.
PRINCETON SER. ASTROPHYS.	Princeton Series in Astrophysics. USA: Princeton University Press.
PRINCETON SER. PHYS.	Princeton Series in Physics. USA: Princeton University Press.
PROGRESS ASTRONAUT. AERONAUT.	Progress in Astronautics and Aeronautics. USA: MIT Press.
PURE APPL. MATH. SER. MONOGR. TEXTBOOKS	Pure and Applied Mathematics: A Series of Monographs and Textbooks. USA; UK: Academic.
PURE APPL. PHYS.	Pure and Applied Physics. USA; UK: Academic.
REF. WORKS ASTRON.	Reference Works in Astronomy. USA: Pachart.
REND. SEMIN. FAC. SCI. CAGLIARI SUPPL.	Rendiconti del Seminario della Facoltà di Scienze dell'Università di Cagliari. Supplement. Italy: Graficoop Soc. Tipografica Editoriale.
SCANDINAVIAN UNIV. BOOKS.	Scandinavian University Books. Norway: Universitetforlaget.
SER. BOOKS ASTRON. ASTROPHYS.	A Series of Books in Astronomy and Astrophysics. USA: W. H. Freeman.
SER. EXTRATERR. CHEM.	Series in Extraterrestrial Chemistry. USA: Gordon and Breach.
SOURCE BOOKS HIST. SCI.	Source Books in the History of the Sciences. USA: Harvard.
SPACE SCI. TEXT SER.	Space Science Text Series. USA: Wiley.
SPEC. LIT. SURV.	Special Literature Survey. USA: TRW Systems Group.
SPRINGER TRACTS MOD. PHYS.	Springer Tracts in Modern Physics. Germany; USA: Springer-Verlag.
STARS STELLAR SYST.	Stars and Stellar Systems. USA: University of Chicago Press.
STUD. GEOPHYS.	Studies in Geophysics. USA: National Academy of Sciences.

TEXTS MONOGR. PHYS.	Texts and Monographs in Physics. Germany; USA: Springer-Verlag.
TOP. APPL. PHYS.	Topics in Applied Physics. Germany; USA: Springer-Verlag.
TOP. ASTROPHYS. SPACE PHYS.	Topics in Astrophysics and Space Physics. USA: Gordon and Breach.
WYKEHAM SCI. SER.	The Wykeham Science Series. UK: Wykeham.

REFERENCE SOURCES
Monographs and Monographic Works

R008 Directories and Biographical Information

R008.001 **American Men and Women of Science: Physical and Biological Sciences.** 14th ed. New York: R. R. Bowker; 1979. 8v.

This standard reference work presents brief biographical information on 130,500 scientists: place and date of birth, subject field, degrees, positions held, present research interests, mailing address, etc. Astronomers are listed under the two general headings astronomy and astrophysics, and under the narrower subdivisions of astrogeology, cosmology, planetary atmospheres, radio astronomy, celestial mechanics, cosmochemistry, and theoretical astrophysics. Besides the main alphabetical listings, there is also a geographic breakdown by city within states and provinces.

R008.002 **American Men and Women of Science: Physics, Astronomy, Mathematics, Statistics, and Computer Science 1977.** New York: R. R. Bowker; 1977. 1,294p.

A subset of the thirteenth edition of *American Men and Women of Science*, this one-volume reference presents brief biographical data on astronomers, astrophysicists, physicists, and so on. Data include birthplace, birth-date, major subject areas, educational and employment histories, special interests, current position, address, professional memberships, etc. An index by subject will help the user find names in specific subfields of astronomy and astrophysics.

R008.003 Codlin, E. M., ed. **Aslib Directory, Volume I: Information Sources in Science, Technology and Commerce.** 4th ed. London: Aslib; 1977. 634p.

A guide to 2,834 sources in Great Britain and Ireland ranging from libraries to societies to industrial firms. Arranged alphabetically, the entries are very brief: address, phone, Telex, subject coverage, special collections, and publications. Names of librarians and other contacts are omitted, but numerous cross-references from old or variant names of sources are interspersed throughout the text. There are more than 20 entries for astronomy, including the Royal Greenwich Observatory and the Institute of Astronomy Library of Cambridge University.

R008.004 **Directory of Physics & Astronomy Staff Members.** New York: American Institute of Physics. 1975/76- . Annual. (1979/80 ed., 389p.). Continues *Directory of Physics & Astronomy Faculties in North American Colleges and Universities.*

A listing of addresses and phone numbers for more than 28,000 scientists and over 2,900 academic and research institutions in North and Central America. The first part of the book is an alphabetic listing of institutions within each state, province, and country, giving names of staff and appropriate address data. The next three parts list federally funded research centers, government laboratories, and private laboratories and foundations. There is also a master list of all staff members from all entries in the book. Seven appendices present various statistical breakdowns of academic programs, degrees offered, research specialties, etc.

R008.005 Gall, J., comp. **Astronomical Directory.** Toronto: Gall Publications; 1978. 144p.

A geographical listing of more than 3,000 planetariums, amateur clubs, observatories, libraries, suppliers, publications, and individuals. A worldwide listing (62 countries) with an emphasis on the United States, the United Kingdom, and Canada, its entries include names, addresses, phone numbers, and very brief miscellaneous data such as size of planetarium, equipment sold, etc. Although the information is not very comprehensive or detailed, there is little else available which provides this type of data.

R008.006 **Graduate Programs in Physics, Astronomy and Related Fields.** New York: American Institute of Physics. 1976/77- . Annual. (1979/80 ed., 764p.).

Arranged by state and province, the entries in this guide describe advanced programs (master's and Ph.D.s) of study at institutions of higher education in the United States and Canada. Data includes basic information about the institution: size; names of faculty and their research interests; faculty publications during the past two to three years; size of graduate student body; requirements for admission; degree requirements; etc. A variety of appendices are included: an alphabetical listing of schools; statistics on degrees granted; research specialties; etc.

R008.007 **International Physics & Astronomy Directory, 1969-70.** New York: W. A. Benjamin; 1969. 802p.

Now outdated, this volume presents a variety of directory-type information, including lists of faculty members, addresses and phone numbers, laboratories, professional organizations, awards, meetings, books, and publishers.

R008.008 Kirby-Smith, Henry Tompkins. **U.S. Observatories: A Directory and Travel Guide.** New York: Van Nostrand Reinhold; 1976. 173p.

A listing of over 300 observatories, planetariums, and astronomical museums, including information on addresses and phone numbers, equipment, visiting hours, historical notes, and research projects. Detailed descriptions and histories of the 15 most prominent observatories are a major portion of the book.

R008.009 **List of Radio and Radar Astronomy Observatories.** Washington, DC: Committee on Radio Frequencies, National Academy of Sciences, National Academy of Engineering. 1970- . Irregular. 50p. (Apr 1979 ed.).

This pamphlet helps keep track of the world's radio and radar astronomy installations. Separated into U.S. and foreign institutions, the book contains (in tabular format) name, location, and altitude of the observatory; administration and address; sponsors; and descriptive data on the telescopes: type, size, height, sky coverage (degrees), collecting area, polarization, and remarks. Arranged by state and country, the work also includes tables of radio frequencies being monitored by the different observatories. The latter presents data on type of observation, frequency, bandwidth, etc.

R008.010 Palmer, Archie M., ed. **Research Centers Directory.** 6th ed. Detroit, MI: Gale Research; 1979. 1,121p.

A guide to university-related and independently operated nonprofit research centers, this volume contains 6,268 entries, 92 of which pertain to astronomy, including observatories (primarily), institutes, and planetaria. Similar in layout to *Directory of Special Libraries and Information Centers* (Gale), its entries list principal fields of research, equipment, staff, source of support, address, and where research results are published. Arranged by 15 major subject subdivisions, it includes name, subject, and institution indexes.

R008.011 **Scientific, Technical, and Related Societies of the United States.** 9th ed. Washington, DC: National Academy of Sciences; 1971. 213p.

A directory containing the following data for 531 societies: address, phone, chief officers, history and organization, purpose, membership (numbers and qualification), meetings, professional activities, and publications. Arranged alphabetically, the entries are about half a page long on the average. Astronomy has 11 entries in this edition, including the American Astronomical Society and seven scientific societies with astronomy as one aspect of their interest. The International Astronomical Union is not listed here since its home is in Europe.

R009 Dictionaries and Encyclopedias

R009.001 Bizony, M. T., ed. **The New Space Encyclopedia: A Guide to Astronomy and Space Exploration.** rev. ed. New York: E. P. Dutton; 1973. 326p.

This volume for the general reader presents brief and lengthy descriptions of a wide variety of topics, equally divided between astronomy and space. There are numerous illustrations, photographs and drawings, and a number of tables with numerical data. Selected topics include space missions, rockets, satellites, observatories, certain named celestial objects, etc.

R009.002 Fairbridge, Rhodes W., ed. **The Encyclopedia of Atmospheric Sciences and Astrogeology.** New York: Van Nostrand Reinhold; 1967. 1,200p. (Encyclopedia Earth Sci. Ser., II).

Although the emphasis in this volume is on meteorology, there are a substantial number of articles on astronomy and related topics. Entries, generally brief, concise, and well written, range from a single paragraph to several pages; bibliographic references are included. A comprehensive, though dated, volume.

R009.003 Hopkins, Jeanne. **Glossary of Astronomy and Astrophysics.** 2nd ed. Chicago: University of Chicago Press; 1980. 196p.

Approximately 2,600 terms and phrases are presented in this dictionary, including selected, related physics and mathematics terminology. Definitions are usually two to five lines long, although some are longer. Based primarily on terms found in *Astrophysical Journal*, the book also includes specific well-known celestial objects such as stars, radio sources, the planets, and asteroids. Aimed at the professional astronomer and physicist, it covers most terminology of importance to scientists, students, and science librarians. (1st ed.: 1976, 169p.).

R009.004 Illingworth, Valerie, ed. **The Macmillan Dictionary of Astronomy.** London: Macmillan; 1979. 378p.

Although this reference work concentrates on astronomy, it also includes a substantial number of entries on space exploration. Concise and clearly written, its 3,000 entries are primarily one or two paragraphs long, although some are lengthier. Coverage is comprehensive, with entries ranging from the constellations to nucleosynthesis. For all types of readers, this excellent work contains line drawings, graphs, and numerous cross-references. Also published as *The Facts on File Dictionary of Astronomy* (New York: Facts on File; 1979), and *The Anchor Dictionary of Astronomy* (pb) (Garden City, NY: Anchor Books/Doubleday; 1980).

R009.005 Marks, Robert W., ed. **The New Dictionary and Handbook of Aerospace: With Special Emphasis on the Moon and Lunar Flight.** New York: Praeger; 1969. 531p.

A one-volume encyclopedia with an emphasis on aerospace terminology, containing a fair number of astronomical entries. Quite a few physics terms are also included. Items range from one sentence to a page, with a single paragraph being most common. Concise, well written, with some black-and-white diagrams, the *Dictionary* includes appendices with numerical data, lists of space flights and satellites, and more.

R009.006 Mikhaĭlov, A. A., ed.; Melnikov, O. A., comps. **Anglo-russkiĭ astronomicheskiĭ slovar.** [English-Russian Astronomical Dictionary.] Moscow: Soviet Encyclopedia Publishing House; 1971. 504p.

This equivalency dictionary contains approximately 20,000 single terms and phrases in alphabetical order. There are a number of useful appendices: 1) common Western (not just English) astronomical abbreviations and acronyms and their Russian equivalents; 2) a list of named celestial objects such as stars, constellations, and comets, and their Russian equivalents; 3) a list of common Western astronomical journals, organizations, observatories, famous telescopes, star catalogs, and their Russian equivalents. There is also an index of Russian terminology to lead the reader to the English words and phrases.

R009.007 Mitton, S., ed. **The Cambridge Encyclopedia of Astronomy.** New York: Crown Publishers; 1977. 495p.

Not an encyclopedia in the traditional sense, that is an alphabetical list of many different topics, this book is divided into 23 chapters covering the various parts of our universe. Accompanied by a detailed index to easily locate specific topics, and beautifully illustrated, the volume is a complete and well-written work appropriate for any type of library and the home. There is a wealth of numeric and tabular data here, and therefore the book will be useful for reference work, as well as casual or serious study.

R009.008 Muller, Paul. **Concise Encyclopedia of Astronomy.** Glascow; London: William Collins; Chicago: Follett; 1968. 281p.

A nontechnical handbook for students and laypersons, this reference contains 850 entries of varying lengths on all aspects of the subject. The passages on astronomical topics cover basic definitions, historical notes, and theory. Besides the usual definitions, there is coverage of astronomers, star names, famous telescopes, units of measure, etc.

R009.009 Satterthwaite, Gilbert E. **Encyclopedia of Astronomy.** New York: St. Martin's Press; 1971. 537p.

A wide variety of information is presented in this comprehensive volume for the serious amateur, layperson, and non-astronomer scientist. Arranged alphabetically according to the most significant word in a phrase, the definitions range from one sentence to two or three pages, with the most frequent example somewhere in between. Among the topics covered are biographical data on famous astronomers, abbreviations, space-age terminology, observatories, etc. Black-and-white photographs and drawings are included.

R009.010 Teodorescu, Nicolae, coord. **Dicţionar Poliglot de Matematică, Mecanică şi Astronomie.** Bucharest: Editura Tehnică; 1978. 664p.

An equivalency dictionary of English, Romanian, German, French, and Russian terminology covering mathematics, mechanics, and astronomy, including nearly 9,600 words and phrases. Astronomy is very well represented with terms from all fields, including instrumentation, constellations, and some of the more recent terminology associated with non-visible astronomy. There are some omissions, but this book is very important, nonetheless, since the last book of this type (*Astronomical Dictionary: In Six Languages*, by J. Kleczek) was published in 1961.

R009.011 Tver, David F.; Motz, Lloyd; Hartmann, William K. **Dictionary of Astronomy, Space and Atmospheric Phenomena.** New York: Van Nostrand Reinhold; 1979. 281p.

This reference work contains more than 2,000 brief word and phrase entries, as well as 16 tables of data and numerous line drawings. Also included are selected mathematics and physics terminology useful for student, layperson, and nonspecialist.

R009.012 Weigert, A.; Zimmerman, H. **A Concise Encyclopedia of Astronomy.** 3rd ed. New York: American Elsevier; 1976, c1971. 548p.

Intended for the student and layperson, this book can also be used by the non-astronomer scientist to find a brief and clearly written explanation of an astronomical concept or description of a phenomenon or celestial object. A comprehensive volume, this reference also includes nine tables of numerical data, 16 photographic plates, and numerous black-and-white illustrations in the text. An elaborate system of cross-references within the text helps the reader easily find specific and related subjects.

The Anchor Dictionary of Astronomy. R009.004.

The Facts on File Dictionary of Astronomy. R009.004.

Illustrated Glossary for Solar and Solar-Terrestrial Physics. 080.004.

International Dictionary of Geophysics. 081.007.

R010 Handbooks and Manuals

R010.001 Allen, C. W. **Astrophysical Quantities.** 3rd ed. London: The Athlone Press, University of London; distr., Atlantic Highlands, NJ: Humanities Press; 1974, c1973. 310p.

A standard handbook containing basic numerical data for the astronomer, this reference work consists primarily of tabular information with references to the source of data. It is divided into 15 chapters: introduction; general constants and units; atoms; spectra; radiation; Earth; planets and satellites; interplanetary matter; Sun; normal stars; stars with special characteristics; star populations and the solar neighborhood; nebulae, sources, and interstellar space; clusters and galaxies; incidental tables.

R010.002 Burnham, Robert, Jr. **Burnham's Celestial Handbook.** rev. & enl. ed. New York: Dover Publications; 1978. 3v. 2,138p. 73 refs, 1879-1975.

Subtitled "An Observer's Guide to the Universe beyond the Solar System," this familiar handbook has been updated and expanded since its first edition in 1966. Volume 3 of the set is new; the others were part of the original handbook. Intended for amateur astronomers and college students, this handbook is a constellation by constellation catalog and detailed description of thousands of celestial objects visible in small- to medium-sized amateur instruments. A typescript text with many tables of data and black-and-white photographs, the set is a storehouse of information, an invaluable source.

R010.003 Lang, K. R. **Astrophysical Formulae: A Compendium for the Physicist and Astrophysicist.** 2nd ed. Berlin; New York: Springer-Verlag; 1980. 783p.

This standard handbook contains hundreds of basic formulae and their derivations, divided into five major sections: 1) continuum radiation; 2) monochromatic (line) radiation; 3) gas processes; 4) nuclear astrophysics and high energy particles; 5) astrometry and cosmology. Aimed at the scientist and graduate student, the book cites sources of reference for each formula and its derivation. The result is a staggering list of well over 2,000 references to the literature of the nineteenth and twentieth centuries.

R010.004 Robinson, J. H.; Muirden, J. **Astronomy Data Book.** 2nd ed. New York: John Wiley & Sons; 1979. 272p.

Primarily useful for amateur astronomers and students, this book of numerical and descriptive data could also be used by the science librarian to answer quick reference questions. Included are a glossary, a list of important historical dates, numerous tables of data, and individual chapters of basic information on a variety of topics. The latter includes the Sun, Moon, planets, comets, meteors, stars, galaxies, radio astronomy, telescopes, etc.

R010.005 Roth, G. D., ed. **Astronomy: A Handbook.** 2nd rev. ed. New York; Berlin: Springer-Verlag; 1975. 567p. 21 chapters. 442 refs, 1888-1974.

Aimed at advanced amateurs and students, this book may also be useful to astronomers. Covering a wide variety of subjects on a non-elementary level, and arranged primarily by type of phenomena to be observed, the volume is a collection of contributions by famous astronomers, professional and amateur. Topics: observing instruments; radio astronomy; effects of the atmosphere; spherical astronomy; sundials; applied mathematics for amateur astronomers; eclipses; photometry; etc. Translated and revised by Arthur Beer.

Handbook of Elemental Abundances in Meteorites. 105.020.

Handbook of Iron Meteorites. 105.006.

A Handbook of Quasistellar and BL Lacertae Objects. 141.011.

A Handbook of Radio Sources. 141.045.

R011 Texts, General Works, Miscellany

R011.001 Astronomical Society of Australia. Proceedings of the annual general meet-
C ing. **Proc. Astron. Soc. Australia** 1- , 1967- .

The proceedings usually include four to five short invited review papers and several dozen contributed papers, which are primarily abstracts.

R011.002 Astronomy Survey Committee. National Academy of Sciences. National Research Council. **Astronomy and Astrophysics for the 1970's.** Washington, DC: National Academy of Sciences; 1972. 2v. v.1: 136p. 6 chapters; v.2: 410p. 9 chapters.

A report of the NAS' Astronomy Survey Committee which outlines the present state of astronomy, identifies the most exciting new problem areas, and recommends a program for the United States for the decade. Priorities, including major new ground-based facilities and space science programs, are presented. Volume 1 is the general report, while volume 2 contains the reports of individual panels appointed to study specific areas such as optical, radio, infrared, space, and solar astronomy, etc.

R011.003 Avrett, Eugene H., ed. **Frontiers of Astrophysics.** Cambridge, MA: Harvard University Press; 1976. 554p. 12 chapters w/refs.

An intermediate text consisting of articles by specialists in different areas of astrophysics. Aimed at undergraduates and beginning graduate students, it covers cosmogony, solar research, stellar evolution, infrared astronomy, galaxies, cosmology, and much more.

R011.004 Billingham, John; Pešek, Rudolf, eds. **Communication with Extraterrestrial Intelligence.** Oxford, UK; New York: Pergamon Press; 1979. 225p. 17 papers. Also published as a special issue of *Acta Astronautica*, v.6, no. 1-2; 1979.

A collection of scientific papers dealing with the possibility and evolution of extraterrestrial intelligence and with the technology necessary to detect radio or other signals from distant civilizations. Search strategies and observing programs are considered.

R011.005 Brandt, John C.; Maran, Stephen P., eds. **The Astronomy and Space Science Reader.** San Francisco: W. H. Freeman; 1977. 371p. 12 chapters. 44 papers.

Intended to serve as supplemental reading for college astronomy and space science courses, this collection consists of articles and papers from *Natural History, Smithsonian, Scientific American, Science, Nature,* and *Astrophysical Journal.* The papers include selections on historical astronomy; instrumentation; asteroids, comets, meteors; Moon and planets; the Sun; stars; galaxies; space-age astronomy; etc.

R011.006 Caren, L. D.; Mallove, E. F.; Forward, R. L. A bibliography on interstellar communication. In: Ponnamperuma, Cyril; Cameron, A. G. W., eds. **Interstellar Communication: Scientific Perspectives.** Boston: Houghton Mifflin; 1974: 187-226. 658 refs, 1892-1974.

A listing without abstracts under five categories: probability of extrasolar intelligence; methods of communicating with extrasolar intelligence; philosophical, psychological, and sociological questions of interstellar communication and extrasolar intelligence; bibliographies, compendia, and books; miscellaneous. With an emphasis on journal and newspaper citations, the list includes both serious and popular items. Technical reports, conference proceedings, and books are also covered.

R011.007 C Castellani, V.; Gratton, L., eds. Proceedings of the Second European Regional Meeting in Astronomy. **Mem. Soc. Astron. Italiana** 45: 1-996; 1974. 141 papers.

Trieste, Italy: 2-5 Sep 1974. A report of current research on the following topics: general properties of the diffuse interstellar matter; condensation of interstellar matter and star formation; astrophysics of stars and galaxies; large European astronomical projects; and miscellaneous subjects.

R011.008 C Contopoulos, G., ed. **Highlights of Astronomy, Volume 3, as Presented at the XVth General Assembly and the Extraordinary General Assembly of the I.A.U. 1973.** Dordrecht, Holland; Boston: D. Reidel; 1974. 574p. 48 papers.

Sydney, Australia: 21-30 Aug 1973. Includes five invited discourses, six selected papers, and six joint discussions. Topics: the Sun; the early universe; galactic nuclei; x-ray sources; precession, planetary ephemerides and time scales; stellar infrared spectroscopy; kinematics and ages of stars near the Sun; origins of the Moon and satellites; Jovian radio bursts and pulsars; and the outer layers of novae and supernovae.

R011.009 C de Jager, C., ed. **Highlights of Astronomy, Volume 2, as Presented at the XIVth General Assembly of the I.A.U., 1970.** Dordrecht, Holland; Boston: D. Reidel; 1971. 793p. 82 papers.

Brighton, UK: 18-27 Aug 1970. Includes four invited discourses, six joint discussions, four invited papers, and five joint meetings. Topics: pulsars, spiral structure, exploration of the Moon, origin of the Earth and planets, helium in the universe, interstellar molecules, atomic data of importance for ultraviolet and x-ray astronomy, photoelectric observations of stellar occultations, cosmic rays and background radiation, and the absolute magnitudes of RR Lyrae stars.

R011.010 Fraknoi, Andrew. **Resource Book for the Teaching of Astronomy.** San Francisco: W. H. Freeman; 1977. 184p.

Subtitled "and Instructor's Guide for Ivan King's text *The Universe Unfolding*," this handy guide contains chapter summaries, discussion questions, and sample exam questions for each chapter of the text it accompanies. Its importance, however, comes from the very extensive reading lists for each subject, including bibliographies for the student and instructor, as well as lists of AV materials. There are hundreds of citations, primarily to books and journals, aimed at the science teacher and beginning student.

R011.011 Gingerich, Owen, ed. **Frontiers in Astronomy: Readings from Scientific American.** San Francisco: W. H. Freeman; 1970. 370p. 33 papers.

A collection of articles documenting some recent and exciting discoveries in the field: pulsars, quasars, 3° background radiation, intense infrared sources, etc. The papers are divided into eight chapters: Earth and Moon; Planetary System; Sun; Stellar Evolution; Milky Way; Galaxies; the New Astronomy; Cosmology. For the layperson and nonspecialist, this collection includes notes on the authors and lists of references.

R011.012 Gingerich, Owen, ed. **New Frontiers in Astronomy.** San Francisco: W. H. Freeman; 1975. 369p. 7 chapters. 31 papers.

A revised edition of an earlier collection of articles from *Scientific American* on some of the more important and exciting discoveries in recent years. Divided into seven parts, each with an introduction by the editor, the work looks at different aspects of the solar system; stellar evolution; the Milky Way; galaxies; and cosmology. Twelve of the papers appeared in the first edition (Freeman; 1970).

R011.013 Harwitt, Martin. **Astrophysical Concepts.** New York: John Wiley & Sons; 1973. 561p. 11 chapters. 285 refs, 1907-73.

An astrophysics text for seniors and graduate students providing broad coverage, with an emphasis on theory rather than on individual objects such as stars, etc. Topics: background; the cosmic distance scale; dynamics and masses of astronomical bodies; random processes; photons and fast particles; electromagnetic processes in space; quantum processes in astrophysics; stellar structure and evolution; cosmic dust and gas; etc.

R011.014 History of astronomy. **IAU Trans. Rep. Astron.** 17A(1): 187-189; 1979. - refs, 1976-78.

An overview of research activity worldwide with an emphasis on the previous three years, as reported by IAU Commission 41 (History of Astronomy). Additional reports: 14A: 481-489; 1970. 40 refs, 1967-69. 15A: 639-646; 1973. 72 refs, 1967-73. 16A(1): 202-205; 1976. 98 refs, 1971-75.

R011.015 International Astronomical Union. **IAU Transactions. Reports on Astronomy.** Dordrecht, Holland; Boston: D. Reidel. Triennial. (v.17, 1979).

An extremely important and useful source of information on recent research in astronomy and astrophysics. Each volume contains reports of the numerous IAU commissions which summarize research activity and highlight important results for the past three years. Since these reports often list key review papers, this title should be one of the first places a researcher should look when surveying a new field or reviewing a familiar one. Despite their potential usefulness, the reports vary from year to year in terms of completeness, number of references listed, length, and topics covered: the size and quality depend on the compiler of the report. In general, however, these reports are excellent and usually contain substantial numbers of references to recent literature, making them an invaluable bibliographic source. Selected commission reports appear in this work under a variety of topics.

R011.016 International review meeting on communication with extraterrestrial intelli-
C gence. **Astronaut. Acta** 18: 409-455; 1973. 5 papers.

Vienna, Austria: 8-15 Oct 1972. Held during the Twenty-third International Astronautical Congress of the IAF.

R011.017 Kaplan, S. A., ed. **Extraterrestrial Civilizations: Problems of Interstellar Communication.** Jerusalem: Israel Program for Scientific Translation; 1971. 265p. (NASA TT F-631; TT 70-50081). 6 chapters w/refs.

A non-elementary look at the various aspects of SETI and the related problems. Topics: the effect of space on the propagation of radio signals; the possibility of communication; methods of message decoding; rates of development of civilizations and forecasting; etc. Translated from the Russian: *Vnezemnye t͡sivilizat͡sii. Problemy mezhzvezdnoĭ svi͡azi* (Moscow: Nauka; 1969).

R011.018 Kruglak, H. Resource Letter EMAA-2: laboratory exercises for elementary astronomy. **American J. Phys.** 44: 828-833; 1976. 135 refs, 1953-75.

An annotated bibliography of materials for students from elementary school to college, this is an updated version of Resource Letter EMAA-1, by R. Berendzen and D. de Vorkin (*American J. Phys.* 41: 783-808; 1973). Entries include laboratory manuals; articles in periodicals; books; printed materials on planetariums; films and film loops; telescopes; atlases, ephemerides, handbooks, and other reference works; distributors of equipment and materials.

R011.019 Lang, Kenneth R.; Gingerich, Owen, eds. **A Sourcebook in Astronomy and Astrophysics, 1900-1975.** Cambridge, MA: Harvard University Press; 1979. 922p. (Source Books Hist. Sci.).

This collection of 132 key articles from the literature provides astronomers and laypersons with an overview of the important discoveries of the twentieth century. Relying heavily on professional journals for source material, the book presents most selections in abridged and annotated format, preceded by an introduction to provide perspective. Included also are citations to earlier important works not presented in the book. There are hundreds of references in the text to related materials. An important historical source.

R011.020 Mallove, E. F., et al. **A Bibliography on the Search for Extraterrestrial Intelligence.** Washington, DC: NASA; 1978. 132p. (NASA RP-1021). 1,488 refs.

An alphabetical listing of articles, books, and technical reports covering the literature through February 1977. Entries contain complete bibliographic information, but are not annotated. Both popular and technical sources are covered, in English and foreign languages, usually Russian. Selected topics include UFOs, philosophy, exobiology, book reviews, radio astronomy, etc. Twelve citations to related bibliographies are included as well.

R011.021 Mallove, E. F.; Forward, R. L. Bibliography of interstellar travel and communication. **J. Brit. Interplanet. Soc.** 27: 921-943; 1974 (pt.1). 28: 191-219; 1975 (pt.2). 28: 405-434; 1975 (pt.3).

An exhaustive listing of over 1,000 citations to journal articles, books, newspaper reports, etc., from both popular and technical sources. Covering the 1950s, 1960s, and up through the mid-1970s, this unannotated list covers interstellar transport and flight, extrasolar intelligence, philosophical and sociological questions, and more. International in scope, the bibliography supersedes previous efforts by the authors in 1967, 1971, and 1972.

R011.022 Mallove, E. F.; Forward, R. L.; Paprotny, Z. Bibliography of interstellar travel and communication. Updates. **J. Brit. Interplanet. Soc.** 29: 494-517; 1976 (Aug 1975 update, 312 refs). 31: 225-234; 1978 (Apr 1977 update, 485 refs).

See previous entry.

R011.023 Mikhailov, A. A., ed. **Physics of Stars and Stellar Systems.** Jerusalem: Israel Program for Scientific Translation; 1969. 718p. (NASA TT F-506; TT 68-50307). 22 chapters w/refs.

A lengthy, introductory text for advanced students written by a number of Soviet astronomers. The chapters/lectures are divided into six parts: 1) absolute stellar magnitudes and stellar masses; 2) binary stars; 3) variables and novae; 4) diffuse matter; 5) theory of stellar atmospheres and gaseous nebulae; 6) stellar systems. Both theoretical and observational data are presented. Translation by Z. Lerman of *Kurs astrofiziki i zvezdnoĭ astronomiĭ* (A Course in Astrophysics and Stellar Astronomy, v.2) (Moscow: Nauka; 1962). Volume 3 was translated and published in 1966; volume 1 has never been translated.

R011.024 Minnaert, M. G. J. **Practical Work in Elementary Astronomy.** Dordrecht, Holland: D. Reidel; distr., New York: Springer-Verlag; 1969. 247p. 5 chapters w/refs.

A collection of 40 classroom exercises for undergraduates and serious amateur astronomers. Included are both pencil-and-paper type problems and actual observational work. The exercises are divided into five categories: space and time, instruments (coordinate systems and telescopes); motions of celestial bodies (orbits and eclipses); planets and satellites; the Sun; the stars (magnitudes, spectra, orbits, binaries, variables, nebulae, etc.).

R011.025 Morrison, Philip; Billingham, John; Wolfe, John, eds. **The Search for Extraterrestrial Intelligence: SETI.** Washington, DC: NASA; 1977. 276p. (NASA SP-419).

A summary of findings from a series of workshops devoted to SETI. The book outlines major conclusions and reviews the various aspects of this area of research. Contains a selected bibliography of 14 key works.

R011.026 Müller, E. A., ed. **Highlights of Astronomy, Volume 4, as Presented at the XVIth General Assembly, 1976.** Dordrecht, Holland; Boston: D. Reidel; 1977. v.1: 370p.; v.2: 407p. 105 papers.

Grenoble, France: 24 Aug–2 Sep 1976. Proceedings include three invited discourses, seven joint discussions, and two joint meetings. Topics: infrared astronomy, planetary exploration and joint discussion on x-ray binaries, space missions, clusters of galaxies, space astrometry, galactic structure, stellar evolution, solar magnetic fields, spectral classification, and a joint meeting on heterogeneities of the stellar surfaces.

R011.027 Öpik, E. J. About dogma in science, and other recollections of an astronomer. **Annu. Rev. Astron. Astrophys.** 15: 1-17; 1977. 18 refs, 1921-76.

Philosophical treatise discussing the need for open minds and open lines of communication in science.

R011.028 **Priorities for Space Research, 1971-80.** Washington, DC: National Academy of Sciences; 1971. 147p. 29 refs, 1961-70.

The summary of the recommendations from the final report of a Study on Space Science and Earth Observations, sponsored by the Space Science Board of the National Research Council. Included are priorities and funding for space sciences studies to be conducted by NASA, including astronomy, lunar and planetary exploration, solar-terrestrial physics, life sciences, etc. The bibliography includes citations to previous similar reports.

R011.029 Schmadel, L. D. Ten years of "Astronomy and Astrophysics Abstracts." **Bull. Inf. Cent. Données Stellaires** 17: 2-11; 1979. 8 refs, 1755-1971.

A review of past achievements and future directions. The publication is described historically and with respect to content and intent. Statistics on the number of documents processed and sample entries are presented.

R011.030 Stephenson, F. Richard; Clark, David H. **Applications of Early Astronomical Records.** Bristol, UK: Adam Hilger; New York: Oxford University Press; 1978. 114p. (Monogr. Astron. Subj, 4). 4 chapters. 109 refs, 1695-1978.

A discussion of the use of pre-telescopic observations in modern astronomy. Topics: sources of historical astronomical records (Europe, Far East, Babylon, the Arab World); solar eclipses; new stars; sunspots and aurorae.

R011.031 Unsöld, Albrecht. **The New Cosmos.** 2nd rev. and enl. ed. New York; Berlin: Springer-Verlag; 1977. 451p. (Heidelberg Sci. Libr.). 3 chapters. 226 refs, 1892-1976.

An intermediate college-level survey text, for nonspecialists and students, on introductory astronomy. Also useful for certain laypersons, the book addresses three major topics: classical astronomy; astrophysics of individual stars and the Sun; and stellar systems (Milky Way; galaxies; cosmogony; cosmology). Translation by R. C. Smith of *Der neue Kosmos* (1974).

R011.032 Wayman, P. A., ed. **Highlights of Astronomy, Volume 5, as Presented at the**
C **XVII General Assembly of the IAU, 1979.** Dordrecht, Holland; Boston: D. Reidel; 1980. 117 papers.

Montreal, PQ, Canada: 14-23 Aug 1979. Three invited discourses, seven joint discussions, and three joint commission meetings. Topics include solar velocity fields, galactic nuclei, ultraviolet astronomy, intergalactic plasma, stellar instabilities, stellar atmospheres, high energy astrophysics, and more.

Exobiology: A Research Guide. 051.037.

Exobiology: The Search for Extraterrestrial Life. 051.011.

The historical quest for the nature of the spiral nebulae. *158.013.*

Key Problems in Physics and Astrophysics. 022.036.

Publications of the Exobiology Program. 051.027.

Publications of the Planetary Biology Program. 051.028.

R012 Files, Listings, Projects, Centers, Nomenclature

R012.001 **Astronomical Data Center Bulletin.** Greenbelt, MD: National Space Science Data Center/World Data Center A for Rockets and Satellites, NASA. v.1, no.1- (July 1980). Frequency unknown.

"A new publication designed to provide a vehicle for the dissemination of information about work in progress on astronomical catalogs." Also included in each issue is an updated list of astronomical catalogs available from the Astronomical Data Center at Goddard Space Flight Center (NASA), contributed papers on data projects, lists of errors for previously published data, etc. The list of catalogs in the July 1980 issue includes 246 items; this is a substantial update of the list mentioned in the Stellar Data Center *Information Bulletin* 16: 73-89; 1979 (see *R012.002*).

R012.002 **Bulletin d'Information du Centre de Données Stellaires.** Strasbourg, France: Centre de Données Stellaires, Observatoire de Strasbourg. no.1- , 1971- . Biannual. English/French.

Also known as the Stellar Data Center *Information Bulletin*, this observatory publication is *the* source of information on star catalogs and other listings of data for celestial objects. Entries run from one or two pages (most frequent) to several pages, and include reports of new catalogs, errata to existing catalogs, descriptions of new data compilations, descriptions of meetings on stellar and similar data, announcements of new data projects, short bibliographies on data-related topics, brief contributed papers, and much more. Includes a microfiche listing in each issue describing various catalogs and other listings available from the Stellar Data Center (over 230 catalogs in 1980).

R012.003 Cayrel, R., et al. **The Bibliographical Star Index.** Strasbourg, France: Centre de Données Stellaires. 1976- .

A cumulation of bibliographical references to individual stars obtained from approximately 32 journals in the field. Useful in obtaining all references on a given star which have appeared in the literature since 1950, the microfiche set is divided into three parts: 1) stars in HD number order; 2) all variable stars arranged by constellation and according to their order in the *General Catalogue of Variable Stars*, by Kukarkin, et al.; 3) all other stellar identifications in alphabetical order by catalog abbreviation, with most having DM numbers. Annual supplements were issued for 1973 through 1976, and a cumulative set for 1973-78 was published in 1980. The original set of seven fiche included references from 12 journals covering 1950-72. The total references included in the index through 1977 number more than 15,000.

R012.004 Cayrel, R.; Jung, J.; Valbousquet, A. The Bibliographical Star Index. **Bull. Inf. Cent. Données Stellaires** 6: 24-31; 1974.

A description of the early work on the *BSI*, a compilation of references from 1950 onward for any given star which has been mentioned in the literature. This paper presents criteria for inclusion, present status of the project, journals covered, etc. See also *R012.003*.

R012.005 **Directory of Astronomical Data Files.** Greenbelt, MD: National Space Science Data Center/World Data Center A for Rockets and Satellites; 1978. 122p.

The purpose of this compilation, prepared by the Data Task Force of the Interagency Coordination Committee for Astronomy (ICCA) in cooperation with the NSSDC, is to provide a listing which will enable a user to locate stellar and extragalactic data sources keyed along with sufficient descriptive information to permit assessment of the value of the files for use as well as the status and availability of the compilations. Contains information on 122 data files with indexes by contributor, title of data set, and name of object.

R012.006 Dixon, R. S.; Sonneborn, G., comps. **A Master List of Nonstellar Optical**
A **Astronomical Objects.** Columbus, OH: Ohio State University Press; 1980. 835p. 301 refs, 1784-1977.

Compendium of all known catalogs of nonstellar objects. Data for each object includes 1950 position, angular diameter, magnitude, and description. Approximately 185,000 listings from 270 catalogs are included.

R012.007 Elvove, S. **Astronomical Data in Machine Readable Form.** Washington, DC: U.S. Naval Observatory; 1974. 11p. U.S. Naval Obs. Circ., 146).

A catalog of catalogs, this pamphlet lists and describes numerical data available from the Naval Observatory on punched cards or magnetic tape. The data is primarily positional data on the Sun, Moon, planets, and stars, collected and compiled by the USNO. Also listed are some major star catalogs such as the *Cape Zone Catalogue*, Kukarkin's *General Catalogue of Variable Stars*, and the *Yale Observatory Catalogues*. This circular supersedes Circular no. 128, 1970.

R012.008 Gottlieb, D. M. Skymap: a new catalog of stellar data. **Astrophys. J. Suppl.**
A **Ser.** 38: 287-308; 1978. 30 refs, 1859-1977.

Description of a compilation of star data for all known stars brighter than 9.0 mag. (blue and visual). The article includes a description of methods used and sample lists, but the complete catalog itself is available only on magnetic tape.

R012.009 International Council of Scientific Unions. Panel on World Data Centres (Geophysical and Solar). **Fourth Consolidated Guide to International Data Exchange through the World Data Centres.** Washington, DC: Secretariat of the ICSU Panel on World Data Centres; 1979. 113p.

Originally established in 1957 as part of the International Geophysical Year (IGY), the World Data Centres (WDC) were to gather and make available the observational data obtained during the IGY program. The WDC system has been continued on a permanent basis to handle relevant geophysical and solar data worldwide. This book consists of six separate guides to the WDC data from around the world, including solar terrestrial physics, rockets and satellites, meteorology, oceanography, glaciology, and solid-earth physics. Each guide describes in detail the nature and extent of its data, and provides addresses of the appropriate WDC assigned to process the information. A useful source for scientists and librarians who deal with geophysical and related data.

R012.010 Jaschek, C. Data growth in astronomy. **Q. J. R. Astron. Soc.** 19: 269-276; 1978. 34 refs, 1910-77.

An analysis showing that growth rates vary from subject to subject and that there is generally no exponential growth. "A sizable fraction of all newly published data (28 percent) refers to objects already observed." Includes several tables illustrating various increases and decreases in output in various topical areas of the field.

R012.011 Jaschek, C. Information and catalogues. **IAU Symp.** 50: 275-284; 1973. 35 refs, 1927-71.

Brief review of the information flow in astronomy and how astronomers decide what catalogs, etc., to use. Explains the role of the Stellar Data Center in dealing with these problems.

R012.012 Jaschek, C.; Pecker, J.-C. A meeting on the nomenclature of astronomical objects. **Bull. Inf. Cent. Données Stellaires** 16: 57-70; 1979.

A brief discussion of a proposed meeting and the problem areas to be considered, followed by a list of 44 resolutions dealing with astronomical nomenclature presented at past IAU conferences. The text of each resolution and the meeting (date and place) are presented.

R012.013 C Jaschek, C.; Wilkins, G. A.; eds. **Compilation, Critical Evaluation and Distribution of Stellar Data.** Dordrecht, Holland; Boston: D. Reidel; 1977. 316p. (Astrophys. Space Sci. Libr., 64). (IAU Colloq., 35). 35 papers.

Strasbourg, France: 19-21 Aug 1976. Invited and contributed papers concerned with the management of data "common to astrometry, photometry, and spectrophotometry of stars and stellar systems." Besides the topics mentioned in the title, the major concern of the book is standards for the presentation of data, something greatly needed in astronomy.

R012.014 Lortet, M. C. Propositions pour un centre de données galactiques non stellaires. **Bull. Inf. Cent. Données Stellaires** 14: 79-104; 1978. In French.

A review of why it is necessary to establish a center for non-stellar galactic data. Past work, including bibliographies, is presented, as are suggestions for future work. Includes a list of 233 catalogs of non-stellar galactic objects, including radio sources, pulsars, H II regions; planetary nebulae, etc.

R012.015 Mermilliod, J.-C. Description of a code-numbering system for identifying stars in magnetic tape catalogues. **Bull. Inf. Cent. Données Stellaires** 14: 32-61; 1978. 12 refs, 1964-77.

A description of a system of transcription for computer use presently employed by the Geneva and Lausanne observatories. Not a new catalog of star names, this system was devised to prevent confusion where stars have several different catalog designations, many of which are unsuited for computer processing.

R012.016 **NSSDC Data Listing.** Greenbelt, MD: National Space Science Data Center/World Data Center A for Rockets and Satellites, NASA. 1979- . Annual.

This guide provides access to space science and supportive data from the NSSDC, including both satellite and ground-based data. The index also includes information on models, computer routines, and composite spacecraft data. The satellite data include spacecraft name; launch date; investigator name; experiment name; data set name; data form code; quantity of data; and time span for data. Non-satellite data encompass NSSDC ID; data type name; data contents name; data set name; form of data; quantity of data; time span of data.

R012.017 A Nagy, T. A.; Mead, J. **HD-SAO-DM Cross Index.** Greenbelt, MD: Goddard Space Flight Center; 1978. 802p. (NASA TM 79564). 8 refs, 1886-1973.

Corresponds to an SAO-HD-DM-GC machine-readable catalog from the Centre de Données Stellaires. Arranged by HD number, the index contains 180,411 entries. A machine-readable version of either the original (SAO-sorted) or the HD-sorted may be obtained from the authors.

R012.018 Nagy, T. A.; Mead, J. M.; Warren, W. H. Astronomical data center at NASA/GSFC. **Bull. Inf. Cent. Données Stellaires** 16: 73-89; 1979.

A description of the activities of the NASA data center at Goddard Space Flight Center, and an extensive list of catalogs available in machine-readable format. The latter information is presented as a photocopy of the January 5, 1979 "Status Report on Machine-Readable Astronomical Catalogues," which includes information on 118 catalogs and cross-indexes.

R012.019 Ochsenbein, F. Automated information retrieval at the Stellar Data Center. **Bull. Inf. Cent. Données Stellaires** 7: 7-37; 1974.

A very brief description, with several pages of sample printouts, of the types of requests for data that can be handled by the CDS. Included are retrieval of all designations for a given star; retrieval of available data (spectral, photometric, parallax, radial velocities) for a given star; retrieval of all stars meeting certain requirements (e.g., a given magnitude range).

R012.020 Ochsenbein, F. The Catalogue of Stellar Identifications: main features. **Bull. Inf. Cent. Données Stellaires** 15: 88-92; 1978.

A description of the various catalogs and data contained in the *CSI*, a CDS project which has merged many star catalogs into one file of 434,000 stars to eliminate duplication of data, to provide consistency in identifying individual stars, and to provide statistical data. This short article is the latest report on the project to appear in the *CDS Information Bulletin*. It lists the 34 star catalogs (including the SAO, Bright Star Catalogue, AGK2/3, Durchmusterung, etc.) in the *CSI* and the data (photometric, proper motion, coordinates, etc.) which can be extracted.

R012.021 Ochsenbein, F.; Spite, F. The Bibliographical Star Index. **Bull. Inf. Cent. Données Stellaires** 17: 19-20; 1979. 3 refs, 1964-77.

A brief description of the CDS microfiche index which assists scientists in quickly finding all information published about a star. Through 1977, the *BSI* included 15,000 references and 85,000 names of objects published in the literature after 1950. See also *R012.003*.

R012.022 Parsons, S. B.; Buta, N. S.; Bidelman, W. P. Machine-readable catalog of spectroscopic and bibliographic stellar data. **Bull. American Astron. Soc.** 11: 421; 1979. Abstract only.

A brief description of a database of "thousands of diverse articles and lists of data" for about 40,000 stars and nebulae. Coverage of the literature complements the *Bibliographical Star Index* (*R012.003*) and is "especially complete for observatory publications prior to 1962." Magnetic tape copies were to be available from the National Space Science Data Center at the end of 1979.

R012.023 Phelps, F. M., III. Machine readable M.I.T. Wavelength Tables (1979). **Bull. Inf. Cent. Données Stellaires** 17: 75-77; 1979.

A brief report announcing the availability on magnetic tape or printout of the second edition (1969) of the tables (see *022.058*).

R012.024 Teleki, G.; Sevarlic, B. A uniform system of designation of catalogues and surveys of star positions. **IAU Colloq.** 48: 483-488; 1979[?]. 1 ref, 1978.

A brief paper proposing the introduction of a uniform system of designation of catalogs and surveys (lists) of star positions. The system is proposed because of the lack of consistency or uniformity in the literature when star catalogs are cited.

R012.025 West, R. M. The southern sky surveys—a review of the ESO sky survey project. **ESO Bull.** 10: 25-40; 1974. no refs.

Information is given about the background of the project and the way in which it will be carried out. The various procedures are discussed, from the taking of the original plates to the publication of the resulting atlas.

R012.026 Wilkins, G. A., ed. Survey of astronomical data activities: Report of the Working Group on Numerical Data of IAU Commission 5. **Bull. Inf. Cent. Données Stellaires** 10: 17-27; 1976. 8 refs, no dates.

Included in this overview are a brief summary of the relevant activities of each IAU commission (39 in all); a listing of all data centers and major data projects reported to the

Working Group; and a short list of relevant publications. The list of data centers and projects is arranged by country and includes subjects and contact persons.

Analysis, processing and interpretation of geophysical data—a symposium. *081.002.*

A bibliographical catalogue of field RR Lyrae stars (magnetic tape). *123.004.*

Bibliographical index of planetary nebulae for the period 1965-1976. *135.002.*

International Information Bureau on Astronomical Ephemerides. Information Cards Nos. 78-93. *047.004.*

List of stellar catalogues used in the Bibliographic Star Index (*BSI*). *R013.011.*

Preliminary List of Star Catalogues 1963-76. R013.006.

Sources and Availability of IQSY Data. 085.016.

Star catalogs and files available at the Stellar Data Center. *R013.012.*

Stellar Data Center *Information Bulletin.* R012.002.

An x-ray astronomy bibliographic system. *142.007.*

R013 Star Catalogs (General)

R013.001 A Billaud, G.; Guallino, G.; Vigouroux, G. Catalogue général des étoiles observées à l'astrolabe (1957-1975). Corrections individuelles aux positions du FK4. [General catalogue of stars observed with astrolabes (1957-1975). Individual corrections to places of FK4 stars.] **Astron. Astrophys. Suppl. Ser.** 31: 159-165; 1978. 21 refs, 1956-76. In French.

Twenty catalogs of stars published from 1958 to 1975 give corrections to the positions of the FK4 catalog. Results are given for 1,139 stars.

R013.002 A Blackwell, K. C.; Buontempo, M. E. Second Greenwich catalogue of stars for 1950.0. **R. Obs. Ann.** 9; 1973. 66p. 15 refs, 1907-1973.

Observed right ascension of 375 FK3 stars September 29, 1937-September 5, 1940, reduced without proper motion to 1950.0. Part 1 contains the mean result for all observations of each star. Part 2 contains the separate results above and below pole for each of the 77 circumpolar stars contained in Part 1.

R013.003 A Derkach, K. N. Differential catalogue of right ascensions of 544 bright stars from FK4 for the observational epoch and equinox 1950.0 (zone -20° to +35°). **Vestn. Khar'kov. Univ.** 160: 42-55; 1977. 4 refs, 1973-76. In Russian.

R013.004 Eichhorn, H. **Astronomy of Star Positions.** New York: Frederick Ungar; 1974. 357p. 346 refs, 1857-1973.

A comprehensive guide to star catalogs, their data, construction, and use. Chapters include astronomical coordinate systems; acquisition of astronomical data; general discussion of star catalogs; compilation catalogs; systematic zone catalogs. Numerous annotated lists of catalogs in most chapters complement the excellent text of this reference book for students, astronomers, and librarians.

R013.005 Eichhorn, H. Star Catalogues. In: Mueller, Ivan I. **Spherical and Practical Astronomy as Applied to Geodesy.** New York: Frederick Ungar; 1969: 179-248. 69 refs, 1877-1967.

A review of the value, use, types, and constructions of "catalogues of accurate star positions." Also included are lists of important catalogs, a summary of AG-type catalogs, and a description of the Astrographic Catalogue (AC).

R013.006 Gliese, W. **Preliminary List of Star Catalogues 1963-76.** Heidelberg: Astronomisches Rechen-Institut; 1976. 27 leaves. 169 refs, 1963-76.

Compiled in preparation for the FK4, this list contains 142 catalogs of observed positions published since 1963. Information includes telescopes used, coordinates, number of stars, and magnitude range. Excluded are catalogs of photographically determined positions and catalogs observed with astrolabes. A supplement was issued in 1977.

R013.007 Halliwell, M. J. Possible nearby stars brighter than tenth magnitude.
A **Astrophys. J. Suppl. Ser.** 41: 173-190; 1979. 47 refs, 1952-78.

This compilation identifies and provides basic data for 436 stars which may be within 25 pc and which are missing from both the Gliese and Woolley et al. catalogs of nearby stars.

R013.008 Høg, E.; von der Heide, J. Perth 70. A catalogue of positions of 24900 stars.
A Derived from observations within the SRS-program during the years 1967 through 1970 with the Hamburg Meridian Circle at Perth Observatory. **Abh. Hamburger Sternw.** 9; 1976. 53 + 334p. 15 refs, 1928-76.

Part of the international effort on SRS (Southern Reference Stars). A total of 110,000 observations were distributed over 20,100 program stars of SRS about magnitude 9 and with $\delta < +5°$ and 4,800 bright stars with $\delta < +35°$, including the FK4.

R013.009 Marcus, E.; Rusu, I. **Bucharest Catalogues of Southern Reference Stars**
A **"SRS," Declination Zone -10° to +5° and Bright Stars "BS," Declination Zone -11° to +6°, Equinox 1950.0.** Bucharest: Central Institute of Physics, Bucharest Astronomical Observatory, Center for Astronomy and Space Sciences; 1979. 203p. 6 refs, 1969-77.

Catalogs resulting from differential visual observations at the meridian circle of the Bucharest Astronomical Observatory. SRS has 1,175 stars and BS has 625.

R013.010 Pierce, D. A. Star catalog corrections determined from observations of
A selected minor planets. **Astron. Pap.** 22: 207-360; 1978. 33 refs, 1928-72.

Provides corrections to the systematic errors in zodiacal regions of the Yale Zone and Boss General star catalogs determined from analyses of over 6,800 photographic observations of 15 selected minor planets.

R013.011 Spite, F.; Kirchner, S.; Lahmek, R. List of stellar catalogues used in the Bibliographic Star Index (BSI). **Bull. Inf. Cent. Données Stellaires** 15: 2-30; 1978.

A list of 128 items including name of catalog; code or alphanumeric designation used frequently in the literature; CDS catalog file number; type of objects; number of stars or celestial objects; bibliographical references and notes. A good quick-reference aid, this article also includes cross-references to alternate forms of catalog names.

R013.012 Star catalogs and files available at the Stellar Data Center. **Bull. Inf. Cent. Données Stellaires.**

A microfiche listing appearing in each issue of the *CDS Information Bulletin* beginning with issue no. 15, July 1978. Before that time it was printed in the *Information Bulletin* itself. The catalogs and listings are usually available in a variety of formats: microfiche, magnetic tape, punched cards, etc. The latest edition (July 1980, issue no. 19) contained 234 items in seven different categories: 1) astrometric data; 2) photometric data; 3) spectroscopic data; 4) cross identifications; 5) combined data; 6) miscellaneous data; 7) nonstellar objects. Each listing is cumulative and supersedes the previous edition. Each entry includes catalog designation; number of records; reference to the literature; contents; coordinates used; and remarks.

Astronomical Catalogues 1951-1975. R001.004.

Bibliographical Star Index. R012.003.

Bibliography of the catalogues of star positions. *041.017.*

The Catalogue of Stellar Identifications: main features. R012.020.

HD-SAO-DM Cross Index. R012.017.

Micrometer measures of double stars. 118.009.

APPLIED MATHEMATICS, PHYSICS

021 Mathematical Papers Related to Astronomy and Astrophysics, Computing, Data Processing

021.001 Bertiau, F. C.; Fierens, E. **Programmes for Pocket Calculators HP-67 and HP-97 in the Field of Theoretical and Observational Astronomy.** Leuven, Belgium: Leuven University Press; 1977. various paging. 6 chapters. no refs.

A collection of 20 program listings with examples of applications. Topics: coordinate transformation; star positions; heliocentric coordinates for sunspot observation; ephemerides; satellite launch; astrometric applications; photometric program; variable stars; perpetual calendar; etc.

021.002 Biskamp, D., ed. **Computing in Plasma Physics and Astrophysics.** Amsterdam;
C New York: North-Holland; 1976. 124p. 11 papers. Also in *Comput. Phys. Commun.*, 12: 1-124; 1976.

Second European Conference on Computational Physics. Garching, Germany: 27-30 Apr 1976. Review papers giving examples of numerical methods and computer applications used in solving problems in plasma physics and astrophysics. Topics: formation of protostars; numerical modeling of pulsar magnetospheres; convection in stars; computation of ideal MHD equilibria; etc.

021.003 Bracewell, R. N. Computer image processing. **Annu. Rev. Astron. Astrophys.** 17: 113-134; 1979. 88 refs, 1944-79.

A review of astronomical information processing, with emphasis on radio images, Fourier synthesis, restoration, speckle interferometry, and x-ray imaging.

021.004 Jones, Aubrey. **Mathematical Astronomy with a Pocket Calculator.** New York: John Wiley & Sons; London: David & Charles; 1978. 254p.

Aimed at the working astronomer, this volume describes methods of calculating commonly needed information for observational astronomy, using both algebraic and RPN calculators. Topics: time; precessional constants; proper motion; Sun, Moon, and comets; visual binary star orbits; cometary ephemerides; approximations; etc. For users of the RPN logic calculators (such as the HP-67), there is a lengthy appendix listing 57 complete applications programs.

021.005 Kugler, H. J. On programming languages for astronomical purposes. **Bull. Inf. Cent. Données Stellaires** 15: 31-38; 1978. 53 refs, 1967-78.

A discussion of the need for a new, astronomically oriented programming language (ASTROL) which combines the best and most useful aspects of existing languages. Also discussed are the need for portability and publicity of existing astronomical programs, and the advantages and disadvantages of existing programming languages used in astronomical applications.

021.006 Kurth, Rudolph. **Dimensional Analysis and Group Theory in Astrophysics.** Oxford, UK; New York: Pergamon; 1972. 235p. 6 chapters. 163 refs, 1619-1968.

A rigorous text which presents the applications of certain mathematical techniques (dimensional analysis and group theory) to certain astrophysical problems: the continuous spectra of stars; the mass-luminosity relation; the theory of stellar structure.

021.007 Meeus, Jean. **Astronomical Formulae for Calculators.** Hove, Belgium: Volkssterrenwacht Urania; 1979. 185p. (Monografieen over Astronomie en Astrofysica, 4). 39 chapters.

Primarily intended for advanced amateur astronomers, this guide presents a large number of formulae (not programs) useful for making frequently needed calculations related to observation. Selected topics: time calculations; planetary conjunction; precession; nutation; solar coordinates; parabolic motion; lunar phases; stellar magnitudes; etc. Most chapters have at least one example calculation for each major equation given.

Automation in optical astrophysics. *032.016.*

Group Theory and General Relativity. 066.006.

Mathematical Cosmology. 162.018.

Mathematical Cosmology and Extragalactic Astronomy. 162.032.

On-line computers for telescope control and data handling. *031.526.*

022 Physical Papers Related to Astronomy and Astrophysics

022.001 Adelman, S. J.; Snijders, M. A. J. A bibliography of atomic line identification lists. **Publ. Astron. Soc. Pacific** 86: 1018-1038; 1974. 434 refs, 1919-74.

"A bibliography of atomic-line-identification lists is presented to supplement the material contained in the *Ultraviolet and Revised Multiplet Tables* and in the finding list by Kelly and Palumbo (1973). The list covers the wavelength range from 911 Å to 8205 Å." Entries for each specific atomic line include author, journal title, year, volume, beginning page, wavelength range, and a designation of type of paper.

022.002 Adelman, S. J. A bibliography of atomic line identification lists II. August 1978 supplement. **Publ. Astron. Soc. Pacific** 90: 766-769; 1978. 145 refs, 1925-78.

References primarily since June 1974.

022.003 Adelman, S. J.; Shore, Steven N.; Nasson, Mark A. An astronomically oriented bibliography of atomic autoionization. **Publ. Astron. Soc. Pacific** 89: 780-791; 1977. 264 refs, 1936-77.

A bibliography (without abstracts) of interest to astronomers involved in or considering involvement in this field. Besides the citations and bibliographic data, there are alpha symbols indicating further the subject contents of each reference. The compilation is divided into five sections: 1) general reviews; 2) general theory; 3) applied theory (single and several elements); 4) spectroscopic analyses (single and several elements); 5) electron and ion impact analyses.

022.004 C Ahrens, L. H., ed. **Origin and Distribution of the Elements.** Oxford, UK: Pergamon; 1968. 1,178p. 89 papers.

Symposium on [title]. (1st). Paris, France: 8-11 May 1967. (Proceedings). Half of this large volume is devoted to the astronomical aspects of the subject: theories of origin; solar, stellar, and interstellar abundances; meteorites; planets, asteroids, comets, tektites.

022.005 C Ahrens, L. H., ed. **Origin and Distribution of the Elements.** Oxford, UK; New York: Pergamon; 1979. 909p. (Phys. Chem. Earth, 11). (Int. Ser. Earth Sci., 34). 79 papers. English/French.

Symposium on [title] (2nd). Paris, France: May 1977. (Proceedings). Papers reporting on recent research in geochemistry and cosmochemistry related to nucleosynthesis. Section I (Cosmochemistry) and Section II (Planetology) contain 15 papers on astronomical topics. Included are discussions of the evolution of the galaxy and origin of differentiated meteorites.

022.006 Audouze, J., ed. **CNO Isotopes in Astrophysics.** Dordrecht, Holland; Boston:
C D. Reidel; 1977. 195p. (Astrophys. Space Sci. Libr., 67). 20 papers.

Special IAU Session. Grenoble, France: 30 Aug 1976. (Proceedings). An overview of research on CNO isotopes in the following areas: solar system, stars, and the interstellar medium. Included are four pages on CNO isotope nucleosynthesis and chemical evolution.

022.007 Bely, O.; Van Regemorter, H. Excitation and ionization by electron impact. **Annu. Rev. Astron. Astrophys.** 8: 329-368; 1970. 228 refs, 1912-70.

A survey of the principles of different theoretical methods for calculating excitation and ionization cross sections. Also considered is the reliability and accuracy of the various approximations.

022.008 Bloor, D. Bibliography of far infrared spectroscopy. **Infrared Phys.** 10: 1-55; 1970. 1,819 refs, 1923-69.

Contains references to spectroscopic studies within the spectral region 50-2,000 microns (0.05-2 mm.) wavelength. A number of papers are included that deal with techniques and theoretical discussions applicable in this region. Papers are listed by year of publication, and then by first author's name in alphabetical order. A general subject index is included, and approximately 60 of the references are specifically atmospheric/astronomical.

022.009 Bulletin GRG, No. 37. Quantum physics and gravitation: list of publications. **Gen. Relativ. Gravitation** 9: 921-951; 1978. 473 refs, 1967-77.

An alphabetical listing by author of the world literature. Like other Bulletin GRGs, there is no subject access.

022.010 Bulletin GRG, No. 39. Solutions of the Einstein equation: list of publications. **Gen. Relativ. Gravitation** 10: 761-806; 1979. 682 refs, 1970-78.

An author-arranged, unannotated listing of references. There is no subject access to individual citations.

022.011 Burek, A. Crystals for astronomical x-ray spectroscopy. **Space Sci. Instrum.** 2: 53-104; 1976. 100 refs, 1922-75.

Review of those aspects of the diffraction of x-rays by crystals that influence spectrometer performance and affect the measurement of crystal diffraction properties. Single and double crystal parameters are discussed and compared, and crystal diffraction theory is explained. Various types of crystals used in x-ray astronomy are described.

022.012 Cameron, A. G. W., ed. **Cosmochemistry.** Dordrecht, Holland; Boston: D.
C Reidel; 1973. 173p. (Astrophys. Space Sci. Libr., 40). 9 papers; 18 abstracts. Also in *Space Sci. Rev.* 15: 1-119; 1973.

Symposium on Cosmochemistry. Cambridge, MA, USA: 14-16 Aug 1972. A collection of invited review papers dealing with the chemical composition of the universe. Topics: stellar and solar abundances; theories of nucleosynthesis; gas and dust in comets; chemistry of the solar nebula; the giant planets; etc.

022.013 Carson, T. R.; Roberts, M. J., eds. **Atoms and Molecules in Astrophysics.** Lon-
C don; New York: Academic; 1972. 367p. 14 papers.

Twelfth Session of the Scottish Universities Summer School in Physics (a NATO Advanced Study Institute). Stirling, UK: Aug 1971. The papers in this volume show how detailed observations of radiation from atoms and molecules are combined with basic theory to obtain the abundances, temperatures, and densities in stellar atmospheres and interstellar regions. Topics: spectra of gaseous nebulae; interstellar molecules; radio recombination lines; UV and x-ray spectroscopy; etc.

022.014 Chandra, N.; Joshi, S. K. Scattering of electrons by diatomic molecules. **Adv. Astron. Astrophys.** 7: 1-55; 1970. 139 refs, 1927-68.

Highly technical review of collision processes related to the study of electric discharges in gases, atmospheric physics, and astrophysics.

022.015 Chantry, G. W., ed. Submillimetre waves and their applications. **Infrared Phys.**
C 18: 375-927; 1978. 83 papers.

Third International Conference. Guildford, UK: 29 Mar—1 Apr 1978. Reports of recent research emphasizing physics, but also covering astronomy; atmospheric studies; dielectrics; detectors; etc.

022.016 Chretien, M.; Lipworth, E., eds. **Atomic Physics and Astrophysics.** New York;
C London: Gordon and Breach; 1973. 2v. v.1: 216p. 3 papers; v.2: 337p. 3 papers.

Twelfth Brandeis University Summer Institute in Theoretical Physics. Waltham, MA, USA: 16 June—25 July 1969. A collection of edited lectures for advanced graduate students aimed at showing the interactions of atomic physics and astrophysics. Topics: vol. 1: experiments with atomic hydrogen, quantum electrodynamics and the theory of the hydrogen atom, graphical methods in angular momentum theory; vol. 2: some aspects of the physics of gas lasers, experimental x-ray astronomy, atomic processes in astrophysics. Many references are included.

022.017 Cord, Marian S., et al. **Microwave Spectral Tables: Volume IV. Polyatomic Molecules without Internal Rotation.** Washington, DC: National Bureau of Standards, U.S. Department of Commerce; 1968. 418p. (NBS Monogr., 70). 1,032 refs, 1934-65.

A reference volume containing data on about 14,000 spectral lines of 166 polyatomic molecules. Data is the same type as in volume III of this series (see *022.076*).

022.018 Cord, Marian S.; Lojko, Matthew S.; Peterson, Jean D. **Microwave Spectral Tables: Volume V. Spectral Line Listing.** Washington, DC: National Bureau of Standards, U.S. Department of Commerce; 1968. 533p. (NBS Monogr., 70).

A compilation, in ascending magnitude of frequency, of spectral lines listed in volumes I, III, and IV of this series. The following data for each line is included: formula of the molecular isotropic series; volume and identification number; vibrational state; hyperfine quantum numbers; frequency and its accuracy.

022.019 de Cesare, L. Superconductivity in astrophysics. **Mem. Soc. Astron. Italiana** 44: 279-303; 1973. 66 refs, 1911-73.

Summary of superconductivity in general, followed by applications to large planets, white dwarfs, neutron stars, and gravitational fields.

022.020 Dixon, W. G. **Special Relativity: The Foundation of Macroscopic Physics.** Cambridge, UK; New York: Cambridge University Press; 1978. 261p. 5 chapters. 40 refs, 1887-1977.

An advanced text which "tries to show the unity of dynamics, thermodynamics and electromagnetism under the umbrella of special relativity." It further aims "to show that an understanding of the basic laws of macroscopic systems can be gained more easily within relativistic physics than within Newtonian physics." Topics: physics of space and time; affine spaces in mathematics and physics; foundations of dynamics; relativistic simple fluids; electrodynamics of polarizable fluids.

022.021 Dohnanyi, J. S. Particle dynamics. In: McDonnell, J. A. M., ed. **Cosmic Dust.** Chichester, UK: John Wiley & Sons; 1978: 527-605. 406 refs, 1923-76.

A review of the basic physics of motion useful in understanding the dynamics of cosmic dust, and a look at the dynamics of certain cosmic dust material. Topics: motion of bodies in gravitational fields (central force problem; orbits of interplanetary objects; perturbations); interaction with the interplanetary environment (solar radiation; influence of collisions; meteor streams; rotations).

022.022 C Earman, John; Glymour, Clark; Stachel, John, eds. **Foundations of Space-time Theories.** Minneapolis, MN: University of Minnesota Press; 1977. 459p. (Minn. Stud. Philos. Sci., VIII). 14 papers.

Conference on the Foundations of Space-time Theories. Minneapolis, MN, USA: 9-14 May 1974; Conference on Absolute and Relational Theories of Space and Space-time. North Andover, MA, USA: 3-5 Jun 1974. (Proceedings). Reports of research and essays on the various aspects of theoretical studies. Topics: philosophical prehistory of general relativity; indistinguishable space-times; time in general relativity; the casual theory of space-time; the curvature of physical space; simultaneity; absolute and relational theories of space and space-time; etc.

022.023 Edlen, B. The term analysis of atomic spectra. **Phys. Scr.** 7: 93-101; 1973. 224 refs, 1952-73.

A review "of the ground configurations of atoms in different stages of ionization through the periodic system," followed by a survey of "the present state of analysis for spectra of atoms and ions containing up to 28 electrons," i.e., those for cosmically abundant elements. Of interest to astronomers, this paper also includes a table of the analyses of atomic spectra of the specific atoms and ions covered, with references to the literature.

022.024 Ewing, G. E. Infrared spectroscopy. **Annu. Rev. Phys. Chem.** 23: 141-164; 1972. 188 refs, 1953-71.

Review of recent developments in tunable infrared sources and new techniques of non-linear absorption which provide very high resolution spectra.

022.025 Fechtig, H.; Grün, E.; Kissel, J. Laboratory simulation. In: McDonnell, J. A. M., ed. **Cosmic Dust.** Chichester, UK: John Wiley & Sons; 1978: 607-669. 95 refs, 1957-77.

A review of laboratory experiments used to infer particle properties from craters and other measured data. Topics: hypervelocity accelerators for simulation of micrometeoroids; high velocity impact processes; impact ionization detectors; penetration and impact phenomena; future developments.

022.026 Fernandez, A.; LeSqueren, A. M.; Lortet, M. C. **Recherches bibliographiques sur la detection de rais moleculaires dans les objets galactiques.** [Bibliography on Molecular Lines in Galactic Objects.] Strasbourg, France: Observatoire de Strasbourg; 1977. various paging. (Publ. Spec. CDS, 1). 410 refs, 1927-76.

Bibliography on the observations of the molecular lines of galactic objects, mainly those associated with regions of star formation, presented in the form of tables: A. references; B. objects; C. reference table of catalogs.

022.027 Fowler, W. A.; Caughlan, G. R.; Zimmerman, B. A. Thermonuclear reaction rates. II. **Annu. Rev. Astron. Astrophys.** 13: 69-112; 1975. 49 refs, 1957-75.

Update of an article by the same title and authors in *Annu. Rev. Astron. Astrophys.* 5: 525-570; 1967, wherein they presented the basic ideas underlying the determination of nuclear reaction rates under astrophysical conditions. Revision and updating of the charged particle reactions involving nuclei with $A \lesssim 30$.

022.028 Fuhr, J. R.; Miller, B. J.; Martin, G. A. **Bibliography on Atomic Transition Probabilities (1914 through October 1977).** Washington, DC: National Bureau of Standards, U.S. Department of Commerce; 1978. 270p. (NBS SP-505). Supersedes NBS SP-320 and SP-320 Supplements 1 and 2.

Approximately 2,400 references divided into four sections: 1) references of general interest; 2) references containing numerical data (arranged by element and ionization stage); 3) a listing of all references by year, then author, including all bibliographic information; 4) author index.

022.029 Fuhr, J. R.; Wiese, W. L.; Roszman, L. J. **Bibliography on Atomic Line Shapes and Shifts (1889 through March 1972).** Washington, DC: National Bureau of Standards; 1972. 154p. (NBS SP, 366). ~1,400 refs.

A listing of citations to journal articles, conference proceedings, report literature, etc. There are four parts: 1) papers of general interest; 2) papers which contain numerical data (theoretical and experimental) in order by element; 3) same as no. 2, but in chronological order; 4) an alphabetical list of all authors and reference numbers. Suppl. 1 (1974): 64p. 350 refs, 1972-73. Suppl. 2 (1975): 67p. 400 refs, 1973-75. Suppl. 3 (1978): 76p. 600 refs, 1975-78.

022.030 Fundamental spectroscopic data. **IAU Trans. Rep. Astron.** 17A(1): 37-71; 1979. 698 refs, 1974-79.

An overview of research activity worldwide with an emphasis on the previous three years, as reported by IAU Commission 14 (Fundamental Spectroscopic Data). Topics: wavelength standards; atomic transition probabilities; collision cross-sections and line broadening; structure of atomic spectra; molecular spectra. Additional reports: 14A: 125-140; 1970. 200 refs, 1965-70. 15A: 155-177; 1973. 449 refs, 1965-73. 16A(1): 31-60; 1976. 584 refs, 1961-76.

022.031 The GR8 symposium on singularities in general relativity. **Gen. Relativ.**
C **Gravitation** 10: 959-1069; 1979. 30 papers.

Waterloo, ON, Canada. 10 Aug 1977. The symposium was designed to give an overview of current understanding of the nature of space-time singularities in the classical (i.e., nonquantum) situation covered by standard general relativity.

022.032 Gabriel, A. H.; Jordan, C. Interpretation of spectral intensities from laboratory and astrophysical plasmas. **Case Stud. At. Collision Phys.** 2: 209-291; 1972. 137 refs, 1930-72.

A review of the present knowledge of various spectral line intensities arising from some simple types of ions. Topics: theoretical methods; collision rate experiments; lithium-like ions; beryllium-like ions; helium-like ions. Relationships to astrophysics are briefly explored.

022.033 Garstang, R. H. Iron in the Sun and stars. In: Wybourne, B. G., ed. **The Struc-**
C **ture of Matter.** Christchurch, New Zealand: University of Canterbury; 1972: 338-394. 168 refs, 1945-71.

Christchurch, New Zealand: 7-9 Jul 1971. A review covering laboratory studies of iron spectra and the detection of iron in the universe. Included are numerous tables listing ground states and ionization potentials for iron spectra, iron lines in the solar spectrum, representative forbidden lines of iron, etc. Topics: iron in the solar photosphere and corona; ultraviolet solar line identifications; abundances in stellar atmospheres; interpretation of abundances results; etc.

022.034 Garstang, R. H. Recent progress in laboratory astrophysics. **Astron. J.** 79: 1260-1268; 1974. 142 refs, 1965-74.

Presents a survey of recent work restricted to those topics falling within the scope of IAU Commission 14 (Fundamental Spectroscopic Data). There is no discussion of nuclear and particle physics or of macroscopic phenomena.

022.035 Gille, J. C. Methods of calculating infrared transfer—a review. In: Kuriyan,
C Jacob G., ed. **The UCLA International Conference on Radiation and Remote Probing of the Atmosphere.** North Hollywood, CA: Western Periodicals; 1974: 395-430. 59 refs, 1939-74.

Los Angeles, CA, USA: 28-30 Aug 1973. A summary of recent work (primarily since 1965) on techniques of calculating the transfer of infrared radiation in planetary atmospheres. Topics: detailed spectral investigations (line by line); band models; transmission through inhomogeneous atmosphere; flux and heating rate calculations.

022.036 Ginzburg, V. L. **Key Problems in Physics and Astrophysics.** 2nd rev. ed. Moscow: Mir; 1978. 167p. 278 refs, 1873-1978.

A collection of brief essays in macrophysics, microphysics, and astrophysics, aimed at university students. The purpose of the book is to provide an overview of some of the more exciting, interesting, and important issues available for study. Topics of an astrophysical nature include general relativity; gravitational waves; cosmology; quasars and galactic nuclei; neutron stars; pulsars; black holes; cosmic rays; etc. Translation by Oleg Glebov of *O fizike i astrofizike* (Moscow: Mir).

022.037 Ginzburg, V. L. **Theoretical Physics and Astrophysics.** Oxford, UK; New York: Pergamon; 1979. 457p. (Int. Ser. Nat. Philos., 99). 17 chapters. 326 refs, 1904-1978.

A graduate level text covering a wide variety of topics with an emphasis on physics. Selected topics include electrodynamics, synchrotron radiation, radiative transfer, particles, cosmic ray astrophysics, x-ray astronomy, and gamma-ray astronomy. Translated by D. ter Haar.

022.038 Ginzburg, V. L.; Syrovatskii, S. I. Developments in the theory of synchrotron radiation and its reabsorption. **Annu. Rev. Astron. Astrophys.** 7: 375-420; 1969. 47 refs, 1956-68.

Update of the author's article in *Annu. Rev. Astron. Astrophys.* 3: 297-350; 1965. Sections include the case of noncircular (spiral) particle motion, reabsorption by ultrarelativistic particles, and problems related to the theory.

022.039 Hagan, Lucy; Martin, W. C. **Bibliography on Atomic Energy Levels and Spectra: July 1968 through June 1971.** Washington, DC: National Bureau of Standards; 1972. 102p. (NBS SP, 363). ~1,100 refs.

With references classified by subject for individual atoms and atomic ions, this compilation continues C. E. Moore's *Bibliography on the Analyses of Optical Atomic Spectra* (see *022.060*). References contain data on energy levels, classified lines, wavelengths, Zeeman effect, Stark effect, hyperfine structure, isotope shift, ionization potentials, or theory which gives results for specific atoms or atomic ions. Includes an index to and reference numbers for spectra, and an author index. Supplement no. 1 (1977): 182p. 2,150 refs (July 1971—June 1975).

022.040 Hansen, C. J., ed. **Physics of Dense Matter.** Dordrecht, Holland; Boston: D.
C Reidel; 1974. 327p. (IAU Symp., 53). 22 papers, 1 abstract.

Boulder, CO, USA: 21-25 Aug 1972. Topics include a consideration of the nature of physical environments whose character is similar to, or more extreme than, nuclear matter; neutron stars, dense stars, white dwarfs, and the physics of these and similar objects.

022.041 Helig, K. Bibliography on experimental optical isotope shifts: 1918 through October 1976. **Spectrochim. Acta B.** 32B: 1-57; 1977. 666 refs, 1918-76.

A list of references pertaining to isotope shifts observed in atomic spectra. There are four parts to this compilation: A. references ordered by elements; B. references ordered by author; C. review articles; D. author index.

022.042 Henderson, S. T. **Daylight and Its Spectrum.** 2nd ed. New York: John Wiley & Sons; 1977. 349p. 19 chapters. 715 refs, 1624-1977.

A non-elementary, but very readable survey of the progress of the study of daylight from 544 B.C. to the present. Not a history book, but rather a text relying heavily on historical sources, this work covers a wide variety of topics, including spectroscopy, high altitude spectral measurements, the Sun and its radiation, less common forms of daylight, artificial daylight, etc.

022.043 Horský, J.; Novotný, J. Conservation laws in general relativity. **Czechoslovak J. Phys. B.** 19: 419-442; 1969. 114 refs, 1915-68.

Summary of the literature regarding attempts to transfer conservation laws to general relativity theory.

022.044 Hsu, D. K.; Smith, W. H. A review of spectroscopic information in the visible and ultraviolet region for diatomic molecules of astrophysical interest. **Spectrosc. Lett.** 10: 181-303; 1977. 247 refs, 1858-1977.

Spectroscopic information in the region between 950Å and 5000Å for 42 diatomic molecules is summarized from laboratory data and a general literature survey.

022.045 Hudson, R. D. Critical review of ultraviolet photoabsorption cross sections for molecules of astrophysical and aeronomic interest. **Rev. Geophys. Space Phys.** 9: 305-406; 1971. 187 refs, 1915-70.

A discussion of the relative merits of various experimental techniques is given, along with possible systematic and random errors that may be associated with them. The problems in data analysis associated with finite spectral bandwidths are reviewed, with special emphasis on the interpretation of published absorption cross-sections.

022.046 Hunt, G. E., et al., eds. Transport theory. **J. Quant. Spectrosc. Radiat. Transfer**
C 11: 511-1033; 1971. 38 papers.

Atlas Symposium No. 3: Proceedings of the Symposium on Interdisciplinary Applications of Transport Theory. Oxford, UK: 1-4 Sep 1970. The meeting was to cover transport problems in astrophysics, meteorology, and neutron transport. Of the 38 papers, 10 are reviews and 28 contributed.

022.047 International conference on infrared physics (CIRP). **Infrared Phys.** 16: 1-329;
C 1976. 54 papers and abstracts.

Zurich, Switzerland: 11-15 Aug 1975. Papers emphasize physics, not astronomy. Topics include: radiation; detectors; lasers; systems and measurements; far-infrared lasers; far-infrared detectors; far-infrared atmospheric transmission.

022.048 Jordan, C. Applications of atomic physics to astrophysical plasmas. In: Hanle, W.; Kleinpoppen, H., eds. **Progress in Atomic Spectroscopy** (Part B). New York; London: Plenum Press; 1979: 1453-1483. 91 refs, 1937-77.

A review of recent work on the interaction between atomic physics and astrophysics, in particular the "interpretation of the intensities of absorption and emission lines observed in astrophysical sources" using a wide variety of atomic data. Topics: energy levels and line identification (spectra of solar active regions and flares, other forbidden lines); atomic data for plasma diagnostics (oscillator strengths, collision cross sections).

022.049 Jordan, C., ed. Atomic data of importance for ultraviolet and x-ray astronomy.
C **Highlights Astron.** 2: 463-583; 1971. 17 papers.

Joint discussion at the Fourteenth General Assembly of the IAU. Organized with the aim of establishing the important requirements of ultraviolet and x-ray astronomy, of reviewing the state of knowledge of the relevant atomic data, and, from the interplay, perhaps of guiding and stimulating future developments.

022.050 Kelly, Raymond L.; Palumbo, Louis J. **Atomic and Ionic Emission Lines below 2000 Angstroms: Hydrogen through Krypton.** Washington, DC: U.S. Naval Research Laboratory; 1973. 992p. (NRL Rep., 7599). (AD-773 872/7GA). 374 refs, 1924-73.

A critical tabulation of 34,700 spectral lines obtained from the published literature through 1972. Intended primarily as an aid to physicists and astronomers who deal with the spectra of highly stripped atoms, it includes 36 elements and is divided into two sections: emission lines by spectrum and a finding list arranged by increasing wavelength. Entries for each element include ionization species, ground state term, ionization potential, and the best values for vacuum wavelength, intensity, and classification.

022.051 King, M. L., ed. **Catalog of Particles and Fields Data 1958-1965.** Greenbelt, MD: National Space Science Data Center, NASA; 1975. 142p. (NASA TM X-72574; N75-24211).

A listing by spacecraft name of NSSDC data sets (files) containing electric and magnetic field and particle information gathered by Earth-orbiting satellites. Entries include a description of the satellite and its goals, a description of the specific experiment to gather particules and fields data, and a description of the data set. The latter includes availability, quantity, time period, and a description of the data. Subject and other indexes are included.

022.052 King, M. L., ed. **Catalog of Particles and Fields Data 1966-1973.** Greenbelt, MD: National Space Science Data Center, NASA; 1975. 202p. (NASA TM X-72575; N75-24212).

See previous entry. Includes Apollo lunar surface data in addition to satellite information.

022.053 Király, P.; Sebestyén, Á.; Somogyi, A. J., eds. **Summer School on Particles and**
C **Fields in Space.** Budapest: Hungarian Academy of Science; 1974. 271p. 10 papers. English/Russian.

Balatonfüred, Hungary: 5-15 Jun 1973. A series of space science lectures for graduate students with an emphasis on galactic problems. Topics: formation of the elements; origin of high-energy particles and magnetic fields; x-rays, gamma rays, and neutrinos in the galaxy; plasmas; galactic dust; infrared radiation; neutron stars; black holes; etc.

022.054 Kleczek, Josip. **The Universe.** Dordrecht, Holland; Boston: D. Reidel; 1976. 259p. (Geophys. Astrophys. Monogr., 11). 5 chapters. no refs.

"This book is an attempt to interpret the structure and evolution of the universe in terms of elementary particles and of their interactions." Topics: elementary particles; particles and forces; agglomeration of particles; structures (e.g., atoms and nuclei, solar systems, galaxies); evolution. The latter includes evolutionary processes, the Big Bang, evolution of stars and galaxies, etc.

022.055 Landel, Robert F.; Rembaum, Alan, eds. **Chemistry in Space Research.** New York: American Elsevier; 1972. 653p. 11 papers.

Aimed at teachers, advanced students, and researchers, this text attempts to provide an introduction to and overview of recent chemical studies relevant to space research. Topics: planetary atmospheres; prebiological synthesis of organic compounds; carbonaceous meteorites; terrestrial and extraterrestrial stable organic molecules; chemical aspects of ablation; solid propellants; spacecraft sterilization; etc.

022.056 Lequeux, J. Intérèt astrophysique de la spectroscopie moléculaire Herzienne. [Astrophysical interests in r.f. molecular spectroscopy.] **Rev. Phys. Appl.** 6: 259-261; 1971. 69 refs, 1967-70. In French.

Bibliography of the molecular spectroscopy of OH, NH_3, H_20, and H_2CO.

022.057 Martin, W. C.; Zalubas, Romuald; Hagan, Lucy. **Atomic Energy Levels—The Rare Earth Elements.** Washington, DC: National Bureau of Standards; 1978. 411p. (NSRDS—NBS 60). 24 refs, 1932-77.

Energy level data for 66 atoms and atomic ions of the 15 rare earth elements compiled from published and unpublished material. A continuation of Moore's *Atomic Energy Levels* (see *022.061*), each element has its own chapter, including introductory and critical material, references, and the extensive numerical data for each configuration (odd and even parity) of the atom and ion.

022.058 Massachusetts Institute of Technology. **Wavelength Tables.** rev. ed. Cambridge, MA: MIT Press; 1969. 429p.

A reprinting and limited updating of the 1939 edition of this standard reference of about 110,000 spectral lines between 10,000Å and 2,000Å. About 600 new values for lines have been included in a separate table, while old values or erroneous entries have been lined out of the photographically reproduced pages of the first edition. Arranged in descending order of wavelength in angstroms, the book presents for each line: the element; arc, spark, or discharge tube intensity; and reference to the literature where located. The introduction discusses methods of obtaining data, error sources, intensity scales, precision, etc. See also *R012.023*.

022.059 Meggers, William F.; Corliss, Charles H.; Scribner, Bourdon F. **Tables of Spectral-Line Intensities.** 2nd ed. Washington, DC: National Bureau of Standards; 1975. 2v. pt.1: 403p.; pt.2: 228p. (NBS Monogr., 145). 37 refs, 1860-1961.

This reference set includes "the intensity, character, wavelength, spectrum, and energy levels of 39,000 lines between 2000Å and 9000Å observed in copper arcs containing 0.1 atomic percent of each of 70 elements." In addition, the source of reference for each element's wavelength and classification information is cited. This edition, which supersedes NBS Monograph 32, parts I and II, and its supplement (1962), contains 8,500 previously unclassified lines and improved wavelengths for 9,000 more. Part I is arranged by elements, while part II is arranged by wavelengths.

022.060 Moore, C. E. **Bibliography on the Analyses of Optical Atomic Spectra.** Washington, DC: National Bureau of Standards; 1969. 222p. (NBS SP, 306, Sec. 1-4).

A continuation of Moore's earlier work, *Atomic Energy Levels* (NBS Circ., 467) (see *022.061*). This listing covers references that deal with the outer structure of atoms as revealed by their spectra. Arranged by element, the citations contain no abstracts but do have key letters which indicate the item's contents or subject.

Sec. 1 (1968): 80p. ^{1}H-^{23}V.

Sec. 2 (1969): 57p. ^{24}Cr-^{41}Nb.

Sec. 3 (1969): 37p. ^{42}Mo-^{57}La & ^{72}Hf-^{89}Ac.

Sec. 4 (1969): 48p. ^{57}La-^{71}Lu & ^{89}Ac-^{99}Es.

022.061 Moore, Charlotte E. **Atomic Energy Levels: As Derived from the Analyses of Optical Spectra.** Washington, DC: National Bureau of Standards, U.S. Department of Commerce; 1971. 3v. v.1: 309p.; v.2: 230p.; v.3: 245p. (NSRDS-NBS 35).

Originally published in 1949, 1952, and 1958, this three-volume reprint set is a critical compilation of spectral data useful for physicists and astrophysicists. Elements are arranged in order of increasing atomic number, and include the following data for each entry: electron configuration; spectral designation; inner quantum number J; atomic energy level; interval; observed g-value. Also included are references to the literature, a brief review of past work on the element, and miscellaneous data.

vol. I: ^{1}H through ^{23}V.

vol. II: ^{24}Cr through ^{41}Nb.

vol. III: ^{42}Mo through ^{57}La and ^{72}Hf through ^{89}Ac.

022.062 Moore, Charlotte E., comp. **A Multiplet Table of Astrophysical Interest.** Washington, DC: National Bureau of Standards, U.S. Department of Commerce; 1972. 206p. 2v. in 1. (NSRDS-NBS 40). 410 refs, 1909-1944.

A handbook of spectral data useful for astrophysicists and physicists, being reprinted until a completely revised edition can be issued. Included are leading lines in 196 atomic spectra of 85 chemical elements listed in related groups called multiplets. In all there are 25,750 spectral lines between 2951 Å and 13164 Å. Data include estimated intensities, excitation potentials, multiplet designations, and references. There are two parts to the book: I.—the table of multiplets arranged by increasing ionization and atomic number; II.—a finding list arranged by increasing wavelength. A reprint of *Contributions* from the Princeton University Observatory, no. 20, 1945. Previously reprinted as *NBS Technical Note* no. 36, 1959.

022.063 Moore, Charlotte E. **Selected Tables of Atomic Spectra: Atomic Energy Levels and Multiplet Tables.** Washington, DC: National Bureau of Standards, U.S. Department of Commerce. 1965- . Irregular. (NSRDS-NBS 3 [sec. no.]).

A series of continuing updates to the data in *Atomic Energy Levels* (see *022.061*) and in two multiplet tables, *A Multiplet Table of Astrophysical Interest* (see *022.062*), and *An Ultraviolet Multiplet Table* (NBS Circular 488: 1950, 1952, and 1962). Issued so far: Section 1: (1965) Si II, Si III, Si IV. Section 2: (1967) Si I. Section 3: (1970) C I, C II, C III, C IV, C V, C VI. Section 4: (1971) N IV, N V, N VI, N VII. Section 5: (1975) N I, N III. Section 6: (1972) H I, D, T. Section 7: (1976) O I. Section 8: (1979) O VI, O VII, O VIII. Section 9: (1980) O V.

022.064 Mueller, Robert S.; Saxena, Surendra K. **Chemical Petrology: With Applications to the Terrestrial Planets and Meteorites.** New York; Berlin: Springer-Verlag; 1977. 394p. 15 chapters. 872 refs, 1877-1976.

An advanced text dealing with the physical chemistry of rocks and associated fluids, touching on aspects of mineralogy, geochemistry, geophysics, sedimentology, and cosmochemistry. Besides the basic Earth-based theory, the authors cover astronomical subjects such as meteorites (classification, mineralogy, origin, etc.); the terrestrial planets (origin and evolution, chemistry, petrology, comparative planetology); and chemical and petrogenetic processes in space. The book, though it does not emphasize astronomy, is important because the theory it presents is basic for the understanding of the growing fields of planetology, including studies of the Moon, Mars, Mercury, and Venus.

022.065 Muirhead, H. **The Special Theory of Relativity.** New York: John Wiley & Sons; 1973. 163p. 6 chapters w/refs.

An undergraduate text written from a high-energy physicist's point of view, i.e., emphasizing the principle of Lorentz invariance, the invariance of physical laws in different reference frames. Topics: historical notes; Einstein and relativity; applications of the Lorentz transformation; four-vectors; spin and special relativity; the Lorentz invariance of physical theories.

022.066 Nicholls, R. W. Transition probability data for molecules of astrophysical interest. **Annu. Rev. Astron. Astrophys.** 15: 197-234; 1977. 366 refs, 1927-77.

Review of transition probability data of molecular spectra, principally of diatomic molecular spectra viewed in the context of contemporary spectroscopic research.

022.067 C Oro, J., et al., eds. **Cosmochemical Evolution and the Origins of Life.** Dordrecht, Holland; Boston: D. Reidel; 1974. 2v. v.1: 523p. 41 papers; v.2: 334p. 38 papers.

Fourth International Conference on the Origin of Life and the First Meeting of the International Society for the Study of the Origin of Life. Barcelona, Spain: 25-28 June 1973. Volume 1: invited papers; volume 2: contributed papers. Included are review papers covering the necessary conditions for the origin of life in the universe, as well as reports on experiments (e.g., the Viking lander) to detect life elsewhere. A combination of chemistry, biology, astronomy, and technology, these two volumes are concerned with cosmochemistry, paleobiology, primordial organic chemistry, precellular organization, early biochemical evolution, exobiology, and more.

022.068 C Parker, L. The production of elementary particles by strong gravitational fields. In: Esposito, F. P.; Witten, L., eds. **Asymptotic Structure of Space-time.** New York: Plenum Press; 1977: 107-226. 168 refs, 1930-76.

Cincinnati, OH, USA: 14-16 Jun 1976. An introduction to and a survey of current research. Topics: quantized fields in curved space-time; particle creation by the expanding universe; particle creation by black holes; etc.

022.069 C Pomeroy, John H.; Hubbard, Norman J., eds. **The Soviet-American Conference on Cosmochemistry of the Moon and Planets.** Washington, DC: NASA; 1977. 2v. 929p. (NASA SP-370). 62 papers.

Moscow, USSR: 4-8 Jun 1974. Primarily reports of studies of lunar samples, these volumes consider the origin of the planets through lunar and planetary research. Topics: thermal history of the Moon; lunar gravitation and magnetism; differentiation of lunar and planetary material; cosmochemical hypotheses on the origin and evolution of the Moon and planets; etc. Also published in the Soviet Union as *Cosmochemistry of the Moon and Planets* (Moscow: Nauka; 1975). Abstracts of the papers were published in *The Moon* 11: 373-443; 1974.

022.070 Reader, Joseph, et al. **Wavelengths and Transition Probabilities for Atoms and Atomic Ions.** Washington, DC: U.S. Department of Commerce, National Bureau of Standards; 1980. 406p. (NSRDS-NBS, 68). 434 refs, 1916-79.

This reference work presents wavelengths for 47,000 spectral lines of atoms and atomic ions and transition probabilities for about 5,000 lines. Part I (Wavelengths) contains wavelengths of lines of neutral through quadruply ionized atoms in the range 40 to 40,000 Å. Part II (Transition Probabilities) contains transition probability data for atoms in various stages of ionization with emphasis on the neutral and singly ionized species. Both sections are arranged by element and subdivided according to ionization stage.

022.071 Rohlfs, K. **Lectures on Density Wave Theory.** Berlin; New York: Springer-Verlag; 1977. 184p. (Lecture Notes Phys., 69). 8 chapters w/refs.

Intended for graduate students and non-specialists, this book provides a simplified version of the spiral density wave theory used in galactic astronomy. Using the gas-dynamical approach, and covering linear theory, the work shows the role of the theory in galactic spiral structure. Topics: theory of spiral structure of galaxies; gas dynamical effects on density waves; and evolution and origin of density waves.

022.072 Rowan-Robinson, Michael. **Cosmic Landscape: Voyages back along the Photon's Track.** Oxford, UK: Oxford University Press; 1979. 149p. (Oxford Sci. Publ.). 8 chapters. no refs.

Aimed at students and intelligent laypersons, this compact volume succinctly describes in detail our solar system, galaxy, and the universe as they would appear in the various wavelengths of light: visible, radio, ultraviolet, x-ray, gamma-ray, infrared, and microwave. An excellent, highly readable book.

022.073 Sobolev, V. V. **Light Scattering in Planetary Atmospheres.** Oxford, UK; New York: Pergamon; 1975. 254p. (Int. Ser. Monogr. Nat. Philos., 76). 11 chapters w/refs.

An advanced text presenting the theory of radiative transfer for anistropic scattering. The first eight chapters deal with "the general problem of multiple scattering of light in an atmosphere consisting of plane-parallel layers illuminated by parallel radiation." The next two chapters cover the application of the theory to the determination of the physical characteristics of planetary atmospheres. Also covered is the theory of radiative transfer in spherical atmospheres. Translation by William M. Irvine of *Rasseyanie sveta v atmosferakh planet* (Moscow: Nauka; 1972).

022.074 Wacker, Paul F., et al. **Microwave Spectral Tables: Volume I. Diatomic Molecules.** Washington, DC: National Bureau of Standards, U.S. Department of Commerce; 1964. 144p. (NBS Monogr., 70). 305 refs, 1929-61.

A handbook containing numerical data "for about 1500 spectral lines of diatomic molecules observed by coherent radiation techniques." Data include measured frequencies, assigned molecular species, assigned quantum numbers, intensities, and various molecular data. References are given for all quantities.

022.075 Wacker, Paul F.; Pratto, Marlene R. **Microwave Spectral Tables: Volume II. Line Strengths of Asymmetric Rotors.** Washington, DC: National Bureau of Standards, U.S. Department of Commerce; 1964. 338p. (NBS Monogr., 70). 15 refs, 1929-63.

A reference databook containing "line strengths of asymmetric rotors ... tabulated as a function of Ray's asymmetry parameter κ for rotational quantum numbers J from 0 to 35."

022.076 Wacker, Paul F., et al. **Microwave Spectral Tables: Volume III. Polyatomic Molecules with Internal Rotation.** Washington, DC: National Bureau of Standards, U.S. Department of Commerce; 1969. 265p. (NBS Monogr., 70). 396 refs, 1932-63.

A reference volume containing data on about 9,000 spectral lines of 94 polyatomic molecules. Data include measured frequencies, assigned molecular species, assigned quantum numbers, rotational constants, dipole moments, and various coupling constants, as well as references to the literature for each entry.

022.077 Wainerdi, Richard E., ed. **Analytical Chemistry in Space.** Oxford, UK; New York: Pergamon; 1970. 275p. (Int. Ser. Monogr. Anal. Chem., 35). 7 chapters w/refs.

A collection of review papers concerned with the chemical constitution of the various elements of interplanetary space. Topics: composition and history of the solar system; space engineering consideration; solar system atmospheres; mass spectroscopy in solar system exploration; lunar and planetary surface analysis; etc.

022.078 Wickramasinghe, N. C. **Light Scattering Functions for Small Particles with Applications in Astronomy.** New York: John Wiley & Sons; London: Adam Hilger; 1973. 506p. 59 refs, 1909-1972.

Primarily a set of tables and curves for spherical and cylindrical particles useful in studies of light scattering, this handbook also includes a summary of the theory of light scattering by small particules and the relevant computational procedures essential to investigators in the field. Astrophysical background is presented, too, including evidence for interstellar dust and observed data on optical properties of dust.

Atlas for Objective Prism Spectra. 114.041.

Atomic and Molecular Physics and the Interstellar Matter. 131.004.

Atomic processes in astrophysical plasmas. *062.015.*

Chemical evolution and the origin of life—a comprehensive bibliography. *051.045.*

ASTRONOMICAL INSTRUMENTS AND TECHNIQUES

031 Astronomical Optics

031.001 Bernal, G. Enrique; Winsor, Harry V., eds. **Optics in Adverse Environments.**
C Bellingham, WA: Society of Photo-Optical Instrumentation Engineers; 1978. 180p. (SPIE Proc., 121). 21 papers.

San Diego, CA, USA: 25-26 Aug 1977. Brief papers on the many facets of optical systems used in difficult situations. At least three or four papers will be of interest to astronomers or engineers working on telescopes or similar systems. Topics: overview; space and underwater optics; optics for energy generation; power and industrial optics.

031.002 Crawford, D. L.; Meinel, A. B.; Stockton, Martha W., eds. **A Symposium on**
C **Support and Testing of Large Astronomical Mirrors.** Tucson, AZ: Kitt Peak National Observatory, University of Arizona; 1968. 252p. (Opt. Sci. Cent. Univ. Arizona Tech. Rep., 30). 29 papers.

Tucson, AZ, USA: 4-6 Dec 1966. Papers describing the practical problems of designing the optics for large reflecting telescopes. Presentations include both general remarks and descriptions of specific large mirror systems. Topics: theoretical aspects of mirror support; thermal effects; axial supports; radial supports; secondaries; the optical shop; testing; active optics.

031.003 de Jager, C.; Nieuwenhiujzen, H., eds. **Image Processing Techniques in Astron-**
C **omy.** Dordrecht, Holland; Boston: D. Reidel; 1975. 418p. (Astrophys. Space Sci. Libr., 54). 53 papers.

Utrecht, The Netherlands: 25-27 Mar 1975. (Proceedings). Review papers, reports of research, and descriptions of equipment and techniques. Topics: types of data (e.g., astrometric and photometric); acquisition and storage of information; processing hardware; software techniques; and applications to astronomical problems, including visual binaries observation, spectrophotometry, solar research, etc.

031.004 Horne, D. F. **Optical Production Technology.** London: Adam Hilger; distr., New York: Crane, Russak; 1972. 567p. 18 chapters. 166 refs, 1862-1971.

A comprehensive source for graduate and technical students, as well as managers and technicians in academia and industry. A detailed overview of techniques, materials, and equipment, this book presents all types of basic information along with chapters on applications, including one on large object glasses and mirrors in which the author describes the methods and problems of producing astronomical lenses and mirrors, especially the very large. Topics: grinding and polishing; optical tools; dioptric substances; lenses; prisms; non-spherical surfaces; testing; surface coating of glass; etc.

031.005 Lytle, John D.; Morrow, Howard, eds. **Stray-light Problems in Optical Sys-**
C **tems.** Bellingham, WA: Society of Photo-Optical Instrumentation Engineers; 1977. 182p. (SPIE Proc., 107). 22 papers.

Reston, VA, USA: 18-21 Apr 1977. Brief reports of research addressing stray light analysis and new, more sophisticated techniques of dealing with the problem. Four papers deal with astronomical applications. Topics: general aspects and phenomenology; software; instrumentation—design; instrumentation—experimental.

031.006 Malacara, D.; Cornejo, A.; Murty, M. V. R. K. Bibliography of various optical testing methods. **Appl. Opt.** 14: 1065-1080; 1975. 501 refs, 1853-1973.

Of interest to researchers in optical fabrication and testing, this list includes scholarly papers, review articles, and monographs. Selected subject categories include Newton, Fizeau, and Haidinger interferometers; lateral shearing interferometers; multiple reflection interferometers; Foucalt and wise tests; roughness measurements; etc.

031.007 Maréchal, A.; Courtès, G., eds. **Space Optics. [Optique spatiale.]** London;
C New York: Gordon & Breach; 1974. 389p. English/French.

Summer School. Marseille, France: 29 Jun–4 Jul 1970. (Proceedings). Headed by a basic preliminary lecture course (in French) in optics (refraction, lenses, etc.), this book addresses the following subjects: image formation; spectroscope; the space environment; and experiments that have been conducted. Equipment is discussed in terms of operation, design, and limitations.

031.008 **Optical Telescope Technology.** Washington, DC: NASA; 1970. 783p. (NASA
C SP-233). 83 papers.

Workshop. Huntsville, AL, USA: 29 Apr–1 May 1969. A collection of papers addressing the design and development of a large (orbiting) space telescope. Among the needed technical developments for this instrument which were discussed are improved diffraction gratings, lightweight optical mirrors, improved detectors, electronic imaging systems, etc. Topics: the use of space telescopes in astronomy; current technical status; mirrors and optical materials; optical design; structures; point and stabilization; instrumentation; operations and data handling; calibration, simulation, and testing.

031.009 Pacini, F.; Richter, W.; Wilson, R. N., eds. **Optical Telescopes of the Future.**
C Geneva: European Southern Observatory; 1978. 553p. 47 papers.

An ESO Conference. Geneva, Switzerland: 12-15 Dec 1977. A review of future needs and possible solutions covering the following topics: conventional large telescopes; incoherent arrays and multi-mirror telescopes; special techniques; coherent arrays and interferometers; image processing and live optics; astronomical implications.

031.010 Ryabova, N. V. Sectional active mirrors for telescopes. **Soviet J. Opt. Technol.** 42: 675-687; 1975. Transl. of *Opt.-Mekh. Prom.-St.* 42: 58-70; 1975. 58 refs, 1953-74.

Review of the design and application aspects of the use of controllable sectional mirrors for large terrestrial and space telescopes. Advantages and disadvantages of various configurations of mirrors are discussed.

031.011 Saito, Theodore T., ed. **Advances in Precision Machining of Optics.** Palos
C Verdes Estates, CA: Society of Photo-Optical Instrumentation Engineers; 1976. 159p. (SPIE Proc., 93). 22 papers.

San Diego, CA: 26-27 Aug 1976. Brief papers describing recent techniques and applications in the production of lenses and mirrors. Includes one paper on astronomical applications: "X-ray Telescope for Sounding Rocket-borne Observations," P. C. Agrawal, et al., pp. 113-18. Topics: unconventional optics; precision machines and design; analysis and applications of precision machined optics; applications and precision machines and services from industry.

031.012 Tanaka, S.-I., ed. ICO Conference on Optical Methods in Scientific and Indus-
C trial Measurements. **Japanese J. Appl. Phys.** 14: suppl. 14-1; 1975. 544p. 100 papers.

Tokyo, Japan: 26-30 Aug 1974. (Proceedings). A series of short, technical papers describing a multitude of equipment and techniques for performing optical measurements. Even though there is not much in the way of astronomical applications here, the work may be of interest to those who design, build, and use astronomical instrumentation. Topics: short pulses of light; spectroscopy; diffraction gratings; holography; speckle patterns; interferometry; etc.

031.013 Tescher, Andrew G., ed. **Efficient Transmission of Pictorial Information.** Palos
C Verdes Estates, CA: Society of Photo-Optical Instrumentation Engineers; 1975. 228p. (SPIE Proc., 66). 27 papers.

San Diego, CA, USA: 21-22 Aug 1975. Brief papers concerned with the improvement of sending and receiving of visual information over great distances. Topics: image representation; image coding; psychophysical considerations; channel effects; image quality; data management; display technology. Included is a paper describing an advanced imaging communication system for planetary exploration.

031.014 Thompson, B. J.; Shannon, R. R., eds. **Space Optics.** Washington, DC: Na-
C tional Academy of Sciences; 1974. 841p. 47 papers.

Ninth International Congress of the International Commission for Optics. Santa Monica, CA, USA: 9-13 Oct 1972. (Proceedings). An overview of techniques, theory, and equipment in the field, as well as reports of planned applications. There are no reports of results in this volume, which addresses space systems; ultraviolet instruments; infrared methods; communications and radiometry; thin films; image processing and holography; optical technology and methods; and instrumentation.

031.015 Weisskopf, M., ed. **Space Optics: Imaging X-ray Optics Workshop.** Bellingham,
C WA: Society of Photo-Optical Instrumentation Engineers; 1979. 294p. (SPIE Proc., 184). 36 papers.

Huntsville, AL, USA: 22-24 May 1979. A review of the design, construction, and use of x-ray telescopes and auxiliary equipment. The role of grazing incidence optics in x-ray astronomy is detailed, along with a description of NASA-launched satellites. Results are not addressed.

031.016 Wilkerson, Gary W.; Poindexter, Robert W., eds. **Effective Systems Integra-**
C **tion and Optical Design.** Palos Verdes Estates, CA: Society of Photo-Optical Instrumentation Engineers; 1975. 168p. (SPIE Proc., 54). 26 papers.

San Diego, CA, USA: 21-23 Aug 1974. Brief communications concerned with the integration of optics into various total systems, such as spacecraft, as well as the design of such optics. Various applications and systems are discussed, including those of astronomical interest: optics for satellites, orbiting telescopes, and the Viking lander camera.

Detection and Spectrometry of Faint Light. 034.012.

ESO/CERN Conference on Large Telescope Design. 032.019.

Large Space Telescope—A New Tool for Science. 032.503.

The Space Telescope. 032.508.

031.5 Methods of Observation and Reduction

031.501 Biraud, Y., ed. Proceedings of the symposium on the collection and analyses of
C astrophysical data. **Astron. Astrophys. Suppl. Ser.** 15: 321-534; 1974. 42 papers.

Charlottesville, VA, USA: 13-15 Nov 1972. Sponsored by the National Radio Astronomy Observatory, the IEEE Group on Aerospace and Electronic Systems, and URSI.

031.502 Boscarino, V. Richard, ed. **Astronomical Use of Television-Type Image Sensors.**
C Washington, DC: NASA; 1971. 217p. (NASA SP-256). 16 papers.

Princeton, NJ, USA: 20-21 May 1970. Reports of recent research and applications, including photometry and spectroscopy. Computer processing of image data is also covered.

031.503 C Burgess, D. D. Spectroscopy of laboratory plasmas. **Space Sci. Rev.** 13: 493-527; 1972. (IAU Colloq., 14). 174 refs, 1955-72.

A review of zeroth-order, first-order, and multi-order plasma spectroscopy. Topics: plasma sources; diagnostic techniques; spectroscopic effects of plasma waves; etc.

031.504 Carleton, N., ed. **Astrophysics; Part A: Optical and Infrared.** New York; London: Academic; 1974. 587p. (Meth. Exp. Phys., 12A). 14 chapters w/refs.

A collection of papers on observational methods and the physics involved, aimed at graduate students and nonspecialists. Each paper is written by an expert in the field; equipment per se is not emphasized. Topics: photomultipliers; observational technique and data reduction; reshaping and stabilization of astronomical images; two-dimensional electronic recording; x-ray and gamma-ray detection; polarization techniques; diffraction grating instruments; Fourier spectrometers; etc.

031.505 Clark, B. G. Information-processing systems in radio astronomy and astronomy. **Annu. Rev. Astron. Astrophys.** 8: 115-138; 1970. 68 refs, 1859-1969.

A summary of how astronomical data is gathered, stored, and processed for analysis, with an emphasis on radio astronomy systems.

031.506 Dainty, J. C. Stellar speckle interferometry. In: Dainty, J. C., ed. **Laser Speckle and Related Phenomena.** Berlin: Springer-Verlag; 1975: 255-280. (Top. Appl. Phys., 9). 41 refs, 1963-75.

A description of a "technique for obtaining diffraction-limited resolution of stellar objects despite the presence of the turbulent atmosphere which limits the resolution of conventional long-exposure pictures to approximately one arc second." Basic principles and theory are reviewed, and the transfer function and signal-to-noise ratio are discussed, as are numerical methods. Practical implementation, including data collection and processing and long baseline interferometry, is also considered.

031.507 C Davis, John; Tango, William J., eds. **High Angular Resolution Stellar Interferometry.** Sydney, Australia: Chatterton Astronomy Department, School of Physics, University of Sydney; 1979. various paging. (IAU Colloq., 50). 35 papers.

College Park, MD, USA: 30 Aug—1 Sep 1978. Invited and contributed papers covering recent research, both practical and theoretical. Topics: astronomical potential of high angular resolution stellar interferometry; two aperture interferometry in the visual and the infrared; astrometry; speckle interferometry; rotation-shearing interferometry; etc.

031.508 Dyce, R. B. Radar studies of the planets. In: Dollfus, A., ed. **Surfaces and Interiors of Planets and Satellites.** London: Academic; 1970: 140-168. 55 refs, 1958-68.

A review of techniques and results from studies of Venus, Mars, Mercury, and Jupiter. Background on techniques is presented in "Interpretation of Radar Observations," which covers radar albedo, scattering, surface differentiation techniques, and rotation determination.

031.509 Ford, W. K., Jr. Digital imaging techniques. **Annu. Rev. Astron. Astrophys.** 17: 189-212; 1979. 105 refs, 1966-78.

An overview of the techniques used in detecting and recording images of celestial objects in digital form. Provides a look at various types of digital systems, including television camera tubes, silicon arrays, hybrid systems, and digital image tubes. Serves as a continuation of the survey by Livingston (*Annu. Rev. Astron. Astrophys.* 11: 95-114; 1973).

031.510 Gabriel, A. H., ed. Ultraviolet and x-ray spectroscopy of astrophysical and
C laboratory plasmas. **Space Sci. Rev.** 13: 488-889; 1972. (IAU Colloq., 14). 47 papers.

Third Symposium. Utrecht, The Netherlands: 24-26 Aug 1971. Reports of research in the following areas: laboratory plasmas; excitation and ionization; stellar and solar UV radiation; the solar soft x-ray spectrum; the solar flare plasma; spectra of cosmic x-ray sources.

031.511 Garton, W. R. S., ed. UV and x-ray spectroscopy of astrophysical and labora-
C tory plasmas. **Astrophys. Space Sci.** 38: 167-190; 313-380; 1975. (IAU Colloq., 27). 10 papers.

Fourth Symposium. Cambridge, MA, USA: 9-11 Sep 1974. Reports of research presented at an IAU meeting. Topics: ultraviolet stellar spectrophotometry; hot gas in and between galaxies; solar UV emission; high temperature plasma; etc.

031.512 Gehrels, T. Photopolarimetry of planets and stars. **Vistas Astron.** 15: 113-129; 1973. 32 refs, 1946-71.

A review, completed in April 1971, of the wide range in polarization-wavelength dependence that is found in astronomy.

031.513 Glaspey, J. W.; Walker, G. A. H., eds. **Astronomical Observations with Tele-**
C **vision-Type Sensors.** Vancouver, BC: University of British Columbia; 1973. 440p. 33 papers.

Vancouver, BC, Canada: 15-17 May 1973. Review papers and contributed papers on low-light level television observational techniques in astronomy. Hardware design and use are discussed primarily, but various applications such as detection of supernovae, x-ray astronomy, and planetary exploration are also covered.

031.514 Golay, M. **Introduction to Astronomical Photometry.** Dordrecht, Holland; Boston: D. Reidel; 1974. 364p. (Astrophys. Space Sci. Libr., 41). 7 chapters. 310 refs, 1937-72.

Aimed at graduate students and nonspecialists, this advanced text emphasizes photometric measurements and resultant data, while de-emphasizing atmospheric extinction effects and equipment. Topics: general definitions and energy distribution for various spectral types; photometric measurements; effects of bandwidths and interstellar absorption; two-dimensional photometric representations of stars; multi-color and wide-band photometry; intermediate and narrow passband photometry; photometric parameters; photometry applied to various stellar objects.

031.515 Hack, M., ed. **High Resolution Spectrometry.** Trieste, Italy: Osservatorio
C Astronomico di Trieste; 1978. 712p. 59 papers.

Fourth International Colloquium on Astrophysics. Trieste, Italy: 3-7 Jul 1978. A series of papers addressing spectroscopy, stellar atmospheres, interstellar gas, instrumentation, and data processing of spectroscopic data. Observational data and theory are presented.

031.516 Hanel, R. A.; Kunde, V. G. Fourier spectroscopy in planetary research. **Space Sci. Rev.** 18: 201-256; 1975. 129 refs, 1927-74.

A review of the application of Fourier Transform Spectroscopy to planetary research. The survey includes FTS observations of the Sun, all of the planets except Uranus and Pluto, the Galilean satellites and Saturn's rings. Instrumentation and scientific results are considered.

031.517 Herzog, L. F. Mass spectroscopy in solar system exploration. In: Wainerdi, Richard E., ed. **Analytical Chemistry in Space.** Oxford, UK: Pergamon; 1970: 109-164. 125 refs, 1913-69.

The use of the mass spectroscope in making compositional analyses in space exploration is discussed, along with an explanation of its different forms, component parts, and theory. Also considered is the choice of a mass spectrometer for a particular solar system mission. Effects of the space environment are taken into account.

031.518 Huguenin, Maureen K.; McCord, Thomas B., eds. **Telescope Automation.**
C Springfield, VA: National Technical Information Service; 1975. 401p. (PB 260 660). 27 papers.

Cambridge, MA, USA: 29 Apr – 1 May 1975. (Proceedings). Conference papers and discussions reporting on a variety of computer-controlled telescopes and auxiliary instrumentation. Topics: computer software; automated radio telescopes; photometric telescopes; interactive data collection; microprocessors; space telescopes; etc. Certain specific telescopes and their control are covered in detail.

031.519 Ingrao, Hector C., ed. **New Techniques in Astronomy.** New York: Gordon and
C Breach; 1971. 446p. 47 papers.

A series of short papers from two Russian conferences held in 1961 and 1964. Topics: reflecting telescopes, automatic control systems, photometers, spectrometers, and astronomical photography. For the scientist, this volume provides historical perspective.

031.520 Lebeyrie, A. Stellar interferometry methods. **Annu. Rev. Astron. Astrophys.** 16: 77-102; 1978. 55 refs, 1873-1978.

A review of the main trends in the general evolution of stellar interferometry and their implications concerning observation. Concentration is on speckle and two-telescope interferometers, with some consideration of space interferometers.

031.521 Livingston, W. C. Image-tube systems. **Annu. Rev. Astron. Astrophys.** 11: 95-114; 1973. 77 refs, 1961-72.

A continuation of W. K. Ford's article: *Annu. Rev. Astron. Astrophys.* 6: 1-12; 1968. Consideration of image systems for stellar light levels as well as for solar and for spaceborne equipment. Looks critically at both operating and experimental systems.

031.522 McGee, J. D. Image tubes in astronomy. **Vistas Astron.** 15: 61-89; 1973. 27 refs, 1934-72.

Discussion of the use of photoelectronic image tubes for the purpose of focusing and directing light through efficient optical systems to a detector.

031.523 Meeks, M. L., ed. **Astrophysics; Part C: Radio Observations.** New York; London: Academic; 1976. 345p. (Meth. Exp. Phys, 12C). 16 chapters w/refs.

A companion to Part B (*033.012*), this collection of review papers examines additional aspects of radio astronomy equipment and methods, including single-antenna observations; interferometers and arrays; and computer programs. Topics: observations of small diameter sources; spectral line measurements; measurements of 21-cm. galactic hydrogen; theory of two-element interferometers; very long baseline interferometry; the fast Fourier transform; etc.

031.524 Nather, R. E. High-speed photometry. **Vistas Astron.** 15: 91-112; 1973. 64 refs, 1904-1971.

Discussion of the exploration at the University of Texas to define those areas of astronomy to which techniques of computer-directed observing can be most profitably employed.

031.525 Philip, A. G. D., ed. Problems of calibration of multicolor photometric systems. **Dudley Obs. Rep.** 14: 1-157; 1979. 29 papers.
C

Schenectady, NY, USA: 16-17 Mar 1979. Discussions of techniques and problems by photometrists, theorists involved in making stellar atmospheric models and synthetic spectra, and a few spectroscopists. Includes a microfiche of the transmission curves of filters, a list of standards in photoelectric systems, and the color indices of standard stars in the systems included in the conference. Methods discussed include the DDO Photometric System; the RGU system; and the Washington System.

031.526 Robinson, L. B. On-line computers for telescope control and data handling. **Annu. Rev. Astron. Astrophys.** 13: 165-185; 1975. **Lick Obs. Contr.** 404. 55 refs, 1964-74.

Characteristics of minicomputers that can be used for on-line applications for ground-based acquisition, data reduction, and telescope control are emphasized. Includes a table of small on-line computers in use.

031.527 Shore, B. W. Experimental techniques for the determination of fundamental spectroscopic data. **Q. J. R. Astron. Soc.** 12: 48-60; 1970. (IAU Colloq., 8). No papers: summary only.
C

London, UK: 1-4 Sep 1970. A summary of the meeting which included papers on astronomical needs, oscillator strengths, emission and absorption sources, line profiles, cross-sections, new discoveries and techniques. Papers were never formally published.

031.528 van Schooneveld, Cornelis, ed. **Image Formation from Coherence Functions in Astronomy.** Dordrecht, Holland; Boston: D. Reidel; 1979. 337p. (Astrophys. Space Sci. Libr., 76). (IAU Colloq., 49). 30 papers; 6 abstracts.
C

IAU Colloquium on the Formation of Images from Spatial Coherence Functions in Astronomy. Groningen, The Netherlands: 10-12 Aug 1978. Contributed papers covering the following topics: aperture synthesis methods; aperture synthesis with limited or no phase information; processing techniques and display methods; maximum entropy image reconstruction; optical interferometric methods; other image improvement methods; relevant inputs from other fields regarding image formation.

Computer image processing. *021.003.*

Conference on Ultraviolet and X-ray Spectroscopy of Astrophysical and Laboratory Plasmas. 062.003.

Image Processing Techniques in Astronomy. 031.003.

Instruments and techniques. *032.010.*

Modern Techniques in Astronomical Photography. 036.006.

Observational techniques (*X-ray Astronomy*). *032.008.*

Radio Astronomy: Instruments and Observations. 033.015.

Stellar photometry. 113.020.

032 Astronomical Instruments

032.001 Brown, R. Hanbury. **The Intensity Interferometer: Its Application to Astronomy.** London: Taylor & Francis; New York: Halstead Press, John Wiley & Sons; 1974. 184p. 12 chapters. 67 refs, 1915-74.

"A history of the interferometer together with a brief account of the theory, practice and application of this new instrument to a classical problem in astronomy—the measurement of the apparent angular diameter of stars." An advanced text for students and scientists, this volume covers three types of interferometers, data analysis, results, and future study.

032.002 Chaffee, F. H., Jr.; Schroeder, D. J. Astronomical applications of Echelle spectroscopy. **Annu. Rev. Astron. Astrophys.** 14: 23-42; 1976. 42 refs, 1949-76.

A review of general characteristics of Echelle grating and the astronomical applications for which Echelle spectrographs have been used at the Cassegrain focus.

032.003 Chase, Richard C.; Kuswa, Glenn W., eds. **X-ray Imaging.** Bellingham, WA:
C Society of Photo-Optical Instrumentation Engineers; 1977. 208p. (SPIE Proc., 106). 24 papers.

Reston, VA, USA: 18-21 Apr 1977. (Proceedings). Brief reports on methods and equipment used, with applications in laser fusion, astronomy, and medicine. Topics: overview presentations and detection with film; pinholes, slots, and coded apertures; reflection optics; special techniques and solid state detectors. Includes eight papers on astronomical topics, including an overview of x-ray astronomy from 1960 to 1980 by S. S. Murray (pp. 8-18).

032.004 Code, A. D. New generation of optical telescope systems. **Annu. Rev. Astron. Astrophys.** 11: 239-268; 1973. 46 refs, 1921-72.

A review of recent advances in optical instrumentation and the need therefor, including multimirror and optical telescope arrays along with large space telescopes.

032.005 Crawford, D. L., ed. **Instrumentation in Astronomy III.** Bellingham, WA:
C Society of Photo-Optical Instrumentation Engineers; 1979. 459p. (SPIE Proc., 172). 57 papers.

Tucson, AZ, USA: 29 Jan—1 Feb 1979. Covering instrumentation used to observe and gather data in the range of space ultraviolet to the far infrared, this book presents reports on state-of-the-art equipment and techniques used in astronomical research. Topics: telescopes; instrumentation I: optical; infrared and detectors; space instrumentation; instrumentation II: infrared and optical; data handling and optics.

032.006 Gillett, F. C.; Dereniak, E. L.; Joyce, R. R. Detectors for infrared astronomy. **Opt. Eng.** 16: 544-550; 1977. 86 refs, 1958-77.

A description of the inherent difficulties in making Earth-based infrared observations, and an explanation of the use of special instrumentations and techniques in this area. After outlining the background radiation problem, the paper describes photon and thermal infrared detectors.

032.007 Gursky, H. A survey of instruments and experiments for x-ray astronomy. **IAU Symp.** 37: 5-33; 1970. 23 refs, 1952-69.

Topics covered include x-ray detectors, modulation collimators, instrument sensitivity, measurements of source characteristics, description of the x-ray explorer, grazing incidence optics, and a description of telescope systems.

032.008 Gursky, H.; Schwartz, D. Observational techniques. In: Giacconi, Riccardo; Gursky, Herbert, eds. **X-ray Astronomy.** Dordrecht, Holland: D. Reidel; 1974: 25-98. 63 refs, 1935-73.

A technical article, reviewing equipment and methods used to gather x-ray data from above the atmosphere.

032.009 Infrared astronomy instrumentation. **Opt. Eng.** 16: 527-574; 1977. (special issue). 8 papers.

A collection of eight brief papers describing instruments on aircraft, space probes, satellites, balloons, and the ground. Both the design and use of these special telescopes and detectors are covered. Results are de-emphasized.

032.010 Instruments and techniques. **IAU Trans. Rep. Astron.** 17A(1): 25-36; 1979. - refs, 1976-79.

An overview of research activity worldwide, with an emphasis on the previous three years, as reported by IAU Commission 9 (Instruments and Techniques). Additional reports: 14A: 53-70; 1970. 11 refs, 1967-69. 15A: 43-73; 1973. 28 refs, 1967-72. 16A(1): 19-30; 1976. - refs, 1972-75.

032.011 Larmore, L.; Poindexter, R. W., eds. **Instrumentation in Astronomy.** Redondo
C Beach, CA: Society of Photo-Optical Instrumentation Engineers; 1972. 304p. (SPIE Proc., 28). 31 papers.

Tucson, AZ, USA: Mar 1972. Reports on new developments and planned projects utilizing innovative equipment primarily for satellite astronomy. A small selection of topics covered are the Orbiting Astronomical Observatory; large space telescopes; photographic analysis; image intensifier systems; far ultraviolet stellar spectrographs; television applications.

032.012 Menzel, D. H.; Larmore, L.; Crawford, D., eds. **Instrumentation in Astronomy**
C **II.** Palos Verdes Estates, CA: Society of Photo-Optical Instrumentation Engineers; 1974. 252p. (SPIE Proc., 44). 29 papers.

Tucson, AZ, USA: 4-6 Mar 1974. An update of the 1972 conference of the same name. The brief reports are divided into four groups: 1) telescopes in space; 2) instrumentation; 3) the Apollo Telescope Mount; 4) data processing and testing. Emphasis is on how equipment works and is constructed; problems in design and use are also addressed.

032.013 Meszaros, Stephen Paul. **World Atlas of Large Optical Telescopes.** Greenbelt, MD: Goddard Space Flight Center, NASA; 1979. 43p. (NASA TM-80246; N79-20944). 6 refs, 1960-79.

A listing by continent of the world's 100 largest (one meter and larger) optical telescopes. Included are maps showing locations of the instruments within each country, and the following data: country, city, observatory, size of mirror, type of instrument, and date operational. Just a list: no other information is provided.

032.014 Peterson, L. E. Instrumental technique in x-ray astronomy. **Annu. Rev. Astron. Astrophys.** 13: 423-509; 1975. 189 refs, 1947-75.

A review of nonfocusing, high-sensitivity counter techniques used in the detection of cosmic photons in the 0.20-300 keV range. Includes a discussion of the various known types of x-ray sources.

032.015 Reiz, A., ed. **ESO/SRC/CERN Conference on Research Programmes for the**
C **New Large Telescopes.** Geneva: ESO Telescope Project Division; 1974. 398p. 38 papers.

Geneva, Switzerland: 27-31 May 1974. (Proceedings). A review of various areas of interest to be studied by large (3.5-4 m.) reflectors in the coming decade. The papers describe current projects (e.g., southern globular clusters, infrared studies, QSOs) and discuss future work on such topics as the Magellanic Clouds and observational cosmology. Philosophy of telescope use and new instrumental capabilities are also addressed.

032.016 Seddon, H.; Smyth, M. J., eds. Automation in optical astrophysics. **Publ. R. Obs. Edinburgh.** 8: 1-211; 1971. (IAU Colloq., 11). 40 papers.
C

Edinburgh, UK: 12-14 Aug 1970. Papers describing current computer applications in four areas: control of telescopes, auxiliary telescope equipment, automated measuring equipment, and data handling and reduction.

032.017 Tescher, Andrew G., ed. **Applications of Digital Image Processing.** Bellingham, WA: Society of Photo-Optical Instrumentation Engineers; 1977. 310p. (SPIE Proc., 119). 38 papers.
C

San Diego, CA: 25-26 Aug 1977. Short papers dealing with digital procedures for improving final visual (analog) representation. Three papers address astronomical applications: Viking image processing; enhancement of solar corona and cometary details; general astronomical applications. Topics: space, surveillance, and guidance; image compression; hardware and implementation; theory and techniques. (Also: *Proceedings of the International Optical Computing Conference, v.2*).

032.018 van Speybroeck, L. Spectroscopic techniques in x-ray astronomy. **Space Sci. Rev.** 13: 845-869; 1972. 43 refs, 1952-71.

Review of how spectroscopic x-ray data is gathered and analyzed. Various types of equipment are described in some detail: proportional counters, scintillation detectors, Bragg crystal spectrometers, focusing instruments, focusing crystal spectrometers, grating spectrometers, and solid state detectors.

032.019 West, R. M., ed. **ESO/CERN Conference on Large Telescope Design.** Hamburg: European Southern Observatory; Geneva: CERN; 1971. 499p. 43 papers. English/French.
C

Geneva, Switzerland: 1-5 Mar 1971. Reports on large telescopes being planned and under construction, and papers covering specific aspects of the problems of designing and constructing large optical instruments. Covered are optical aspects, telescope mountings, and control and drive systems. Computer control of telescope operation is also discussed.

Advances in Precision Machining of Optics. 031.011.

Gamma ray astronomy. *061.045.*

Gamma-ray astrophysics. *061.036.*

Infrared spectroscopy. *022.024.*

Long-Wavelength Infrared. 034.017.

Low Light Level Devices for Science and Technology. 034.008.

Radio, Submillimeter, and X-ray Telescopes. 033.003.

Stray-light Problems in Optical Systems. 031.005.

Survey on new techniques for x-ray astronomy. *061.040.*

A Symposium on Support and Testing of Large Astronomical Mirrors. 031.002.

Telescope Automation. 031.518.

032.5 Space Instrumentation

032.501 Code, Arthur D., ed. **The Scientific Results from the Orbiting Astronomical**
C **Observatory (OAO-2).** Washington, DC: NASA; 1972. 590p. (NASA SP-310). 45 papers.

Amherst, MA, USA: 23-24 Aug 1971. A review and interpretation of data gathered by OAO-2 experiments. Topics: spacecraft and observations; solar system studies; interstellar matter; stellar observations; galactic and extragalactic systems.

032.502 Kuznetsov, S. N. Pribory dli͡a izmerenii͡a zari͡azhennykh chastit͡s v kosmicheskom prostranstve. [Instruments for the measurement of charged particles in space.] **Itogi Nauki Tekh. Ser. Issled Kosm. Prostranstva** 8: 70-105; 1976. 195 refs, 1963-75. In Russian.

"A general review of analysers, detectors and calorimeters for the registration of particles with energies 1 eV to 10^{13} eV." The emphasis is on instrumentation, not results. The bibliography includes both Russian and Western sources.

032.503 **Large Space Telescope—A New Tool for Science.** New York: American Institute
C of Aeronautics and Astronautics; 1974. 124p. 27 papers.

Washington, DC, USA: 30 Jan—1 Feb 1974. (Proceedings). A series of brief papers describing the history, concept, and use of the LST, to be launched in the early 1980s. Among the areas to be studied with the LST, described in this volume, are cosmology, galaxies, QSOs, infrared sources, the birth and death of stars, and the solar system. Half the book is devoted to brief overviews of various aspects of the technology involved in building and designing the LST.

032.504 Longair, M. S.; Warner, J. W., eds. **Scientific Research with the Space Tele-**
C **scope.** Marshall Space Flight Center, AL: NASA; 1980. 327p. (IAU Colloq., 54). (NASA CP-2111). 16 papers.

Princeton, NJ, USA: 8-11 Aug 1979. With an emphasis on future plans (the telescope had not yet been launched), these survey papers outline those areas of astronomy that would greatly benefit from observations above the atmosphere (with the Space Telescope) and what results can be expected. Selected topics include small solar system objects; ionized gaseous nebulae; globular clusters; central regions of galaxies and quasars; x-ray astronomy; etc.

032.505 Ness, N. F. Magnetometers for space research. **Space. Sci. Rev.** 11: 459-554; 1970. 158 refs, 1950-70.

A review of spacecraft instruments which have provided measurements of the magnetic fields of the Earth, Moon, Mars, Venus, and interplanetary measurements of the extended solar field obtained near the eliptic between the orbits of Mars and Venus.

032.506 Smyth, M. J. Infrared techniques. In: Mavridis, L. N., ed. **Stars and the Milky Way System.** Berlin: Springer-Verlag; 1974: 232-242. 59 refs, 1800-1972.

A brief review, aimed at the nonspecialist, outlining current developments and limitations for astronomical observations in the range 1 μm to 1 mm. Included are history of techniques, problems, detectors, instruments, telescopes, observatories, and space observations.

032.507 Sosnina, M. A. Opticheski sistemy kosmicheskikh teleskopov. [Optical systems for space telescopes.] **Itogi Nauki Tekh. Ser. Issled. Kosm. Prostranstva** 8: 7-69; 1976. 262 refs, 1961-75. In Russian.

A comprehensive review of the non-Soviet literature, with an emphasis on the NASA 20-year space astronomy program. Covers Apollo, Skylab, Orbiting Astronomical Observatory, and more. Includes many diagrams of space telescopes.

032.508 **The Space Telescope.** Washington, DC: NASA; 1976. 231p. (NASA SP-392). 43
C abstracts.

Twenty-first Annual Meeting of the American Astronautical Society. Denver, CO, USA: 26-28 Aug 1975. Summaries of selected papers (see *051.005*) describing the Space Telescope Mission and the equipment involved in carrying out the research program. Topics: telescope performance; instrument and detector development; mirror development; precision pointing and control systems; data management; maintainability, the future of space astronomy.

032.509 Van Der Hucht, K. A.; Vaiana, G., eds. **New Instrumentation for Space Astronomy.** Oxford, UK; New York: Pergamon; 1978. 339p. (Adv. Space Exploration,
C 1). 51 papers.

Twentieth Plenary Meeting of COSPAR. Tel Aviv, Israel: 7-18 Jun 1977. (Proceedings). A review of state-of-the-art equipment and techniques in ultraviolet and x-ray astronomy, emphasizing imaging and spectroscopy. Topics: cameras, detectors, telescopes, spectrometers, etc.

Advances in solar and cosmic x-ray astronomy: a survey of experimental techniques and observational results. 061.042.

Astronomy from a Space Platform. 051.020.

Infrared Detection Techniques for Space Research. 061.050.

Instrumentation in Astronomy. 032.005; 032.011; 032.012.

The interpretation of space observations of stars and interstellar matter. 131.045.

New results and techniques in space radio astronomy. *033.001.*

Optical Telescope Technology. 031.008.

Prospects in space astrometry. *041.008.*

Skylab Solar Workshop. 076.010.

Space Optics. 031.007; 031.014.

Space Optics: Imaging X-ray Optics Workshop. 031.015.

033 Radio Telescopes and Equipment

033.001 Alexander, J. K. New results and techniques in space radio astronomy. **IAU Symp.** 41: 401-418; 1971. 36 refs, 1957-70.

A brief summary of progress in low frequency space radio astronomy, emphasizing use of the Radio Astronomy Explorer (RAE) satellite.

033.002 Amitay, Noach; Galindo, Victor; Wu, Chen Pang. **Theory and Analysis of Phased Array Antennas.** New York: Wiley-Interscience; 1972. 443p. 9 chapters w/refs.

Aimed at engineers, physicists, and students working in radar, communications, and radio astronomy. Concerned with large-aperture antennas which are capable of high-speed scanning and simultaneous operation of multiple functions, the book provides an overview of the analysis and design of such systems and examines their electromagnetic properties. Selected subjects include array theory; the boundary value problem and solutions; arrays of rectangular waveguides; finite arrays; etc.

033.003 Basov, N. G., ed. **Radio, Submillimeter, and X-ray Telescopes.** New York; London: Consultants Bureau; 1976. 221p. (Proc. (Tr.) P. N. Lebedev Phys. Inst., 77). 18 papers.

A collection of advanced papers on the design and construction of special telescopes and related apparatus. Topics: an orbiting, reflecting x-ray telescope; mirror systems for x-ray telescopes; bandpass filters for the submillimeter range; polarizing devices for the submillimeter range; parabolic radio telescopes; fully steerable parabolic antennas; etc. Translation by E. U. Oldham of *Radioteleskopy, submilimetrovye i rentgenovskie teleskopy.*

033.004 Basov, N. G., ed. **Techniques and Methods of Radio-Astronomic Reception.** New York; London: Consultants Bureau; 1979. 150p. (Proc. (Tr.) P. N. Lebedev Phys. Inst., 93). 10 papers.

Papers reporting on recent Soviet advances in equipment development and on selected applications. Topics: computer applications to radio telescopes; structural design of movable parabolic radio telescope antennas; automation of the processing of pulsar observations; radio astronomical observations of the solar wind; etc. Translation of *Tekhnika i metody radio-astronomicheskogo priema.*

033.005 Christiansen, W. N.; Högbom, J. A. **Radiotelescopes.** Cambridge, UK: Cambridge University Press; 1969. 231p. (Cambridge Monogr. Phys.). 8 chapters. 107 refs, 1928-67.

A brief survey of the history of radio telescopes and a summary of the theory involved in design. Written by two designers of these instruments, the book does not deal with the results of observation. Instead, it covers theory; the steerable parabolic reflector; other types of filled aperture antennas; unfilled-aperture antennas; aperture synthesis; sensitivity. Not for amateurs, though they could understand the descriptive material, this work relies heavily on mathematics and physics in its presentation of telescope design and theory.

033.006 Findlay, J. W. Filled-aperture antennas for radio astronomy. **Annu. Rev. Astron. Astrophys.** 9: 271-292; 1971. 75 refs, 1960-71.

A review restricted to recent developments in the theory and practice of filled-aperture antennas. Several actual telescopes in this category are described. Gravitational distortions of antenna structures and atmospheric effects are reviewed.

033.007 Findlay, J. W., ed. Radio and radar astronomy. **Proc. IEEE** 61: 1169-1363; 1973. 35 papers.

Short, comprehensive review articles on radio and radar astronomy observations and instrumentation.

033.008 C Findlay, J. W., ed. Symposium on very long baseline interferometry. **Radio Sci.** 5: 1221-1292; 1970. 11 papers.

Charlottesville, VA, USA: 13-15 Apr 1970. The papers cover VLBI studies of Jupiter and galactic OH sources, as well as techniques and instrumentation.

033.009 Kislyakov, A. G. Radioastronomical investigations in the millimeter and submillimeter bands. **Soviet Phys. Usp.** 13: 495-521; 1971. Transl. of *Usp. Fiz. Nauk.* 101: 607-653; 1970. 328 refs, 1949-70.

A description of the present status and the near future of millimeter and submillimeter radio astronomy. Principal attention is paid to an analysis of the possibilities of land-based instruments with the aid of which most of the observations at 0.7-10 mm. were performed.

033.010 Love, A. W., ed. **Reflector Antennas.** New York: IEEE Press; distr., New York: John Wiley & Sons; 1978. 427p. (IEEE Press Sel. Repr. Ser.). 65 papers.

Of interest to radio astronomy engineers and other technicians, this volume presents selected papers primarily from the *IRE* and *IEEE Transactions on Antennas and Propagation*. The reprints are in nine categories: general interest papers; focal region fields: prime focus feed requirements; radiation pattern analysis of reflectors; cassegrain and dual reflector systems; polarization effects; offset or unsymmetrical reflectors; lateral feed displacement, scanning, and multiple beam formation; phase errors and tolerance theory; spherical reflectors.

033.011 C Mar, James W.; Liebowitz, Harold, eds. **Structures Technology for Large Radio and Radar Telescope Systems.** Cambridge, MA: MIT Press; 1969. 538p. 27 papers.

Cambridge, MA, USA: 18-20 Oct 1967. Detailed descriptions of the design and structure of existing and proposed instruments. Papers are primarily concerned with fixed and movable dish-type antennas.

033.012 Meeks, M. L., ed. **Astrophysics; Part B: Radio Telescopes.** New York; London: Academic; 1976. 309p. (Meth. Exp. Phys., 12B). 16 chapters w/refs.

A compendium of individual papers concerned with the design and use of radio telescopes and related apparatus, effects of the atmosphere on radio observations, and radiometers. Topics: radiometric measurements; astronomical antennas; antenna calibration; the ionosphere; absorption and emission by atmospheric gases; parametric and maser amplifiers; multichannel-filter spectrometers; etc.

033.013 Penzias, A. A.; Burrus, C. A. Millimeter-wavelength radio-astronomy techniques. **Annu. Rev. Astron. Astrophys.** 11: 51-72; 1973. 47 refs, 1939-73.

A review of coherent techniques being used to extend the short-wavelength boundaries of high-sensitivity radio astronomy. Includes a consideration of potentially useful new devices and techniques for the same application.

033.014 Rusch, W. V. T.; Potter, P. D. **Analysis of Reflector Antennas.** New York; London: Academic; 1970. 178p. (Electrical Sci. Ser. Monogr. Texts). 4 chapters. 112 refs, 1955-68.

Intended for engineers involved in the design and analysis of antennas, this text reviews some important developments in the field along with a presentation of basic principles. Of interest to radio astronomy engineers, too, the book covers equations of the electromagnetic field; performance analysis; and computer-aided analysis and design.

033.015 Skobel'tsyn, D. V., ed. **Radio Astronomy: Instruments and Observations.** New York: Consultants Bureau; 1971. 184p. (Proc. (Tr.) P. N. Lebedev Phys. Inst., 47). 20 papers.

A collection of "scientific research reports written in recent years at the Radio-Astronomy Laboratory of the P. N. Lebedev Physics Institute in the field of radio-astronomical apparatus and methods of observations." Topics: radiometers of high sensitivity (using masers and parametric amplifiers); meter-wave equipment; radio interferometers with radio relaying; radio telescope design; radio astronomy investigation of the solar wind; etc. Translation of *Radioastronomicheskie instrumenty i nabli͡udenii͡a*.

033.016 Skobel'tsyn, D. V., ed. **Wideband Cruciform Radio Telescope Research.** New York; London: Consultants Bureau; 1969. 175p. (Proc. (Tr.) P. N. Lebedev Phys. Inst., 38). 20 papers.

A collection of papers addressing both equipment and observations. Topics: the Lebedev cross radio telescope; wideband antenna arrays; resolution of correlation arrays; solar plasma observations; radiometer reliability; solar radio emission; etc. Translation of *Diapozonnyĭ krestoobraznyĭ radioteleskop*.

033.017 Steinberg, Bernard D. **Principles of Aperture and Array System Design; Including Random and Adaptive Arrays.** New York: John Wiley & Sons; 1976. 356p. 14 chapters w/refs.

A graduate-level text of interest also to designers of radio telescopes, providing an introduction to theory and design. Topics: scalar radiation theory; the radiation pattern; scanning and focusing; depth of field and aperture synthesis for the periodic array; non-conventional arrays; angular accuracy and resolution; adaptive beamforming and scanning; adaptive nulling; tolerance theory; and hard limiting.

033.018 Swenson, G. W., Jr. Synthetic-aperture radio telescopes. **Annu. Rev. Astron. Astrophys.** 7: 353-374; 1969. 35 refs, 1947-68.

Design and theory of correlation arrays operating in the supersynthesis mode are reviewed.

033.019 Swenson, G. W., Jr.; Mathur, N. C. The interferometer in radio astronomy. **Proc. IEEE** 56: 2114-2130; 1968. 55 refs, 1941-68.

History of radio interferometry in astronomy is discussed, and examples are given of the current uses of the technique in the precise measurement of source positions, in the measurement of the diameters of extremely small sources, and in the detailed mapping of complex sources.

Astrophysics; Part C: Radio Observations. 031.523

034 Auxiliary Instrumentation

034.001 C Advanced electronic systems for astronomy—1971 symposium. **Publ. Astron. Soc. Pacific** 84: 74-224; 1972. 25 papers.

Santa Cruz, CA, USA: 31 Aug—2 Sep 1971. Descriptions of specific instrumentation used in space and on the ground, e.g., photomultipliers, photometers, image tube scanners.

034.002 C Azzam, R. M. A.; Coffeen, David L., eds. **Optical Polarimetry.** Bellingham, WA: Society of Photo-Optical Instrumentation Engineers; 1977. 226p. (SPIE Proc., 112). 31 papers.

San Diego, CA, USA: 23-24 Aug 1977. A collection of short contributed papers on instrumentation and applications in the field. Six papers cover astronomical applications. Topics: instrumentation; surface and thin film ellipsometry; biological, chemical, and physical polarimetry; remote sensing.

034.003 Carruthers, G. R. Electronic imaging devices in astronomy. **Astrophys. Space Sci.** 14: 332-377; 1971. 90 refs, 1949-71.

A review of the types of devices that combine the best features of conventional photography and photoelectric photometry.

034.004 C Chincarini, Guido L.; Griboval, Paul J.; Smith, Harlan J., eds. **Electrography and Astronomical Applications.** Austin, TX: University of Texas Astronomy Department; 1974. 364p. 23 papers. English/French.

Austin, TX, USA: 11-12 Mar 1974. Papers reporting on recent research, results, techniques, and instrumentation. Topics: spectroscopy; photometry; far UV observations; studies of galaxies; etc.

034.005 Connes, P. Astronomical Fourier spectroscopy. **Annu. Rev. Astron. Astrophys.** 8: 209-230; 1970. 93 refs, 1954-70.

A summary of why, when, and how Fourier spectroscopy is used in astronomy. Emphasis is on techniques and experiments that have produced the greatest improvements over classical methods.

034.006 C Duchesne, M.; Lelievre, G., eds. **Astronomical Applications of Image Detectors with Linear Response.** Meudon, France: Observatoire de Paris-Meudon; 1977. various paging. (IAU Colloq., 40). 55 papers. English/French.

Meudon, France: 6-8 Sep 1976. A review of recent progress and research covering electronography, electronic output systems, and astronomical applications. Examples of the latter include the use of photoelectronic image detectors to study nebulae, double stars, and quasars. Applications in photometry and spectroscopy are also addressed.

034.007 C Franseen, R. E.; Schroder, D. K., eds. **Applications of Electronic Imaging Systems.** Bellingham, WA: Society of Photo-Optical Instrumentation Engineers; 1978. 187p. (SPIE Proc., 143). 23 papers.

Washington, DC, USA: 30-31 Mar 1978. A series of short presentations describing recent and current advances in the following areas: solid state visible imagers; image tubes; short wavelength imagers; infrared and other applications. Astronomy-related papers deal with detectors for the Space Telescope, electronic detectors in ground-based astronomy, and soft x-ray imaging experiments.

034.008 C Freeman, C., ed. **Low Light Level Devices for Science and Technology.** Palos Verdes Estates, CA: Society of Photo-Optical Instrumentation Engineers; 1976. 162p. (SPIE Proc., 78). 25 papers.

Reston, VA, USA: 22-23 Mar 1976. A description of instrumentation being used for public service, medical, general, and astronomical applications. There are six papers on the latter topic, covering imaging detectors, array photometers, and other devices. Emphasis is on the design and operation of equipment, not on the data to be collected.

034.009 C Laustsen, S.; Reiz, A., eds. **ESO/CERN Conference on Auxiliary Instrumentation for Large Telescopes.** Geneva: ESO Telescope Project Division, CERN; 1972. 525p. 49 papers. English/French.

Geneva, Switzerland: 2-5 May 1972. (Proceedings). Descriptions of auxiliary equipment presently in use or on the drawing board. The following areas are covered: grating spectrographs and spectrometers; interferometric spectrometers; electronographic photometry; and special instrumentation. Also includes reports on instrumentation developments and telescope projects at large observatories.

034.010 C McGee, J. D.; McMullan, D.; Kahan, E., eds. **Photo-Electronic Image Devices.** London; New York: Academic; 1972. 2v. 1,189p. (Adv. Electron. Electron Phys., 33A, 33B). 102 papers.

Fifth Symposium. London, UK: 13-17 Sep 1971. (Proceedings). An extensive compendium of papers describing innovative equipment and new techniques, including astronomical applications. Topics: electronography; image tubes; signal generating tubes; photocathodes and phosphors; electron optics; image tube assessment; applications in astronomy (11 papers); photon counting systems; applications in space research; low light-level systems; x-ray applications; high-speed photography; etc.

034.011 Meaburn, J. Astronomical spectrometers. **Astrophys. Space Sci.** 9: 206-297; 1970. 157 refs, 1911-70.

Review of the spectrometers that are being extensively applied to astronomical problems in the wavelength region from 13000 Å to 3300 Å.

034.012 Meaburn, John. **Detection and Spectrometry of Faint Light.** Dordrecht, Holland; Boston: D. Reidel; 1976. 265p. (Astrophys. Space Sci. Libr., 56). 11 chapters w/refs.

Although of interest to several branches of the physical sciences, the information presented in this advanced monograph will be of special interest to astronomers. The author presents descriptions of a variety of light detection equipment, illustrating their use (e.g., with telescopes) and limitations, and explaining the physics involved and the design criteria. Topics: principles of spectrometry; quantum detectors; diffraction gratings; filters; etc.

034.013 Morgan, B. L.; Airey, R. W.; McMullan, D., eds. **Photo-Electronic Image**
C **Devices.** London; New York: Academic; 1976. 2v. 1,074p. (Adv. Electron. Electron Phys., 40A, 40B). 86 papers.

Sixth Symposium. London, UK: 9-13 Sep 1974. (Proceedings). See entry *034.010*. Topics: image tubes and channel multipliers; camera tubes and targets; photocathodes and electron emission; electron optics; system assessment; electronography; digital detection; applications in astronomy (11 papers); etc.

034.014 Soifer, B. T.; Pipher, J. L. Instrumentation for infrared astronomy. **Annu. Rev. Astron. Astrophys.** 16: 335-369; 1978. 142 refs, 1958-78.

Overview of the variety of the very sensitive instrumentation now or soon to be available to the infrared observer, including a comparison of trade-offs between competing instruments.

034.015 Spiro, Irving J., ed. **Utilization of Infrared Detectors.** Bellingham, WA: Society
C of Photo-Optical Instrumentation Engineers; 1978. 147p. (SPIE Proc., 132). 17 papers.

Los Angeles, CA, USA: 16-18 Jan 1978. Short papers describing the construction and use of infrared devices in a number of scientific fields, including astronomy (six papers). Topics: materials and theory; testing and calibration; infrared astronomy; devices and techniques.

034.016 Van Bueren, H. G., ed. Workshop on coherent detection in optical and infrared
C astronomy. **Space Sci. Rev.** 17: 617-736; 1975. 15 papers.

Rhenen, The Netherlands: 25-26 Apr 1974. Four sections: principles and observational purposes, heterodyne spectroscopy, heterodyne interferometers, and instrumentation.

034.017 Wolfe, W. L., ed. **Long-Wavelength Infrared.** Palos Verdes Estates, CA:
C Society of Photo-Optical Instrumentation Engineers; 1975. 114p. (SPIE Proc., 67). 18 papers.

San Diego, CA, USA: 21-22 Aug 1975. The emphasis of this book is on new instrumentation methods, new detectors, lasers, and materials, for use in astronomy, solid-state physics, chemistry, and technology. Topics: instruments; components; astronomical measurements (five papers); calibration systems.

Applications of Digital Image Processing. 032.017.

Astronomical Observations with Television-Type Sensors. 031.513.

Astronomical Use of Television-Type Image Sensors. 031.502.

Proceedings of the International Optical Computing Conference, vol. 2. 032.017.

035 Clocks and Frequency Standards

[No entries].

036 Photographic Materials and Techniques

036.001 Allbright, G. S. Methods of hypersensitisation in astronomical photography—a state of the art review. **J. Photogr. Sci.** 24: 115-119; 1976. 64 refs, 1908-1975.

Available methods of hypersensitization are reviewed, and the most recently reported conditions for optimum hypersensitization are presented.

036.002 Eichhorn, Heinrich, ed. **Conference on Photographic Astrometric Technique.**
C Washington, DC: NASA; 1971. 267p. (NASA CR-1825). 27 papers.

Tampa, FL, USA: 18-20 Feb 1968. A series of papers describing recent results and observational techniques, including the use of automation. Topics: a proper motion survey; a semi-automatic measuring engine; progress in astrometric catalogs; long focus astrographics; photographic astrometry; etc.

036.003 Huang, T. S., ed. **Picture Processing and Digital Filtering.** New York; Berlin: Springer-Verlag; 1975. 289p. (Top. Appl. Phys., 6). 6 chapters. 245 refs, 1922-76.

Of interest to scientists and engineers involved in processing of signals (e.g., enhancement of lunar or Martian satellite photographs), this technical text is concerned specifically with digital processing of two-dimensional signals (i.e., pictures or images). Topics: 2-D transforms; 2-D nonrecursive filters; 2-D recursive filtering; image enhancement and restoration; noise considerations. No specific applications are mentioned.

036.004 Sim, M. Elisabeth. **Astronomical Photography.** Edinburgh: Royal Observatory; 1977. 63p. 512 refs, 1938-77.

Sponsored by the IAU Working Group on Photographic Problems, Commission 9, Astronomical Photography, this bibliography for scientists and laypersons is arranged under 12 major subjects in chronological order. Entries include bibliographic data, but no abstracts. Topics: general references (including books, reviews, sky surveys); materials and emulsions; storage; processing; measurement; image formation; structure of the developed image; information detection and storage; applications; photolab techniques and theory; color. Twenty review papers are listed, and there are cross-references for papers covering more than one topic or related subject.

036.005 Smith, A. G.; Hoag, A. A. Advances in astronomical photography at low light levels. **Annu. Rev. Astron. Astrophys.** 17: 43-71; 1979. 107 refs, 1890-1979.

Summary of current hypersensitization techniques, including chemical pre-exposure treatment, treatment with nitrogen and/or hydrogen, pre-flash, latensification, push development and intensification. Other topics include the effects of environment on photographic plates, evaluating photographic response, and future developments.

036.006 West, R. M.; Heudier, J. L., eds. **Modern Techniques in Astronomical Photog-**
C **raphy.** Geneva: European Southern Observatory; 1978. 304p. 32 papers.

ESO Workshop. Geneva, Switzerland: 16-18 May 1978. An overview of the topic, presenting details of methods in the following areas: hypersensitization, calibration, copying, and color work. Additional papers cover practical photographic work at the telescope and other photographic methods such as digital superposition of photographic plates.

POSITIONAL ASTRONOMY, CELESTIAL MECHANICS

041 Astrometry

041.001 Barbieri, C.; Bernacca, P. L., eds. **European Satellite Astrometry.** Padova,
C Italy: Tip. Antoniana; 1979. 303p. 30 papers.

Colloquium. Padova, Italy: 5-7 Jun 1978. (Proceedings). Papers primarily describing proposed and potential experiments to be carried out by the European Space Agency satellite, Hipparcos. Topics: space astrometry; ground-based astrometry; instrumentation; measurement of double stars; radio astrometry; fundamental reference systems; stellar kinematics; etc.

041.002 Counselman, C. C., III. Radio astrometry. **Annu. Rev. Astron. Astrophys.** 14: 197-214; 1976. 57 refs, 1966-76.

Discussion of the present state and directions of growth of this technique concerned with accurate measurement of positions.

041.003 Fricke, W. Fundamental systems of positions and proper motions. **Annu. Rev. Astron. Astrophys.** 10: 101-128; 1972. 121 refs, 1785-1972.

Review of the present status of astrometry and proper motion measurement.

041.004 Gliese, Wilhelm; Murray, C. Andrew; Tucker, R. H., eds. **New Problems in**
C **Astrometry.** Dordrecht, Holland: D. Reidel; 1974. 333p. (IAU Symp., 61). 36 papers, 29 abstracts.

Perth, Australia: 13-17 Aug 1973. Sessions included reference systems, current and future projects for Southern Hemisphere reference systems, radio astrometry, astrometry with large telescopes, proper motions and galactic problems, astronomical refraction problems, and astrometric techniques.

041.005 Ianna, Philip A., ed. **Proceedings of the Fourth Astrometric Conference.** Char-
C lottesville, VA: University of Virginia Press; 1971. 305p. (Publ. Leander McCormick Obs., 16). 12 papers.

Charlottesville, VA, USA: 8-10 Oct 1969. Devoted solely to parallaxes, in particular annual trigonometric parallaxes, this volume also covers astrometry for the spectroscopist, astrometric accuracy of photographs, parallax catalogs, and statistical parallaxes. Various programs for determining parallaxes are also discussed.

041.006 Kołaczek, B.; Weiffenbach, G., eds. **On Reference Coordinate Systems for**
C **Earth Dynamics.** Warsaw: Zakładzie Graficznym Politechniki Warszawskiej; 1975. 478p. (IAU Colloq., 26). 35 papers.

Torun, Poland: 26-31 Aug 1974. Papers considering the need for and the problems associated with an optimum reference coordinate system for Earth dynamics. Reports are arranged according to the subjects of the four working groups of IAU Commission no. 4: requirements for reference coordinate systems for Earth dynamics; conceptual definitions for reference coordinate systems; practical realization of the reference coordinate system; relations between different reference systems.

041.007 Kovalenko, N. D. Catalogue of the positions of 587 FKSZ stars compiled from
A observations with the meridian circle at the Kiev Observatory in 1972-1975. **Vestn. Kiev. Univ.** 20: 88-100; 1978. no refs. In Russian.

041.008 Lacroute, P. Prospects in space astrometry. **Highlights Astron.** 4(1): 345-370;
C 1977. 4 papers.

Joint meeting of commissions 8 and 24 at the XVI General Assembly of the IAU. Brief review of ESA plans, the Large Space Telescope, and general principles and future prospects for space astrometry.

041.009 Mackie, J. B. **The Elements of Astronomy for Surveyors.** 8th ed. London: Charles Griffin; 1978. 308p. 14 chapters w/refs.

First published in 1918, this how-to-do-it handbook presents the usual topics and fundamental concepts, providing numerous examples based on actual observations. A major change in this edition is that there are no longer explanations of the use of mathematical tables once so common and necessary for surveyors; this is due to the recent advent of advanced, hand-held calculators. Topics: celestial and astronomical coordinates; time; timing of observations; determination of altitude, azimuth, longitude, and local time; etc.

041.010 McNally, D. **Positional Astronomy.** New York: John Wiley & Sons; 1974[?]. 375p. 12 chapters. no refs.

A beginning undergraduate text on astrometry. Topics include, but are not limited to, systems of astronomical coordinates; time; refraction, parallax and aberration; precession and nutation; eclipses and occultations; orbital motion.

041.011 Morrison, L. V. Catalogue of observations of occultations of stars by the Moon
A for the years 1943-1971. **R. Greenwich Obs. Bull.** 183: 1-14 + 5 microfiche; 1978. 10 refs, 1927-78.

Fiche include collected times of 59,193 occultations, coordinates for the 2,418 telescopes used for making the observations, descriptions of the 652 telescopes used in professional observations to make the observations, and the names of the 2,730 observers.

041.012 Nemiro, A. A., et al. Catalogue of absolute right ascensions of 1023 bright and
A faint fundamental stars of the Northern Hemisphere. **Tr. Glav. Astron. Obs. Pulkovo,** ser. 2. 82: 4-52, 1977. 22 refs, 1934-76. In Russian.

Absolute right ascensions of 505 bright (FK4) and 518 faint (FKSZ) fundamental stars.

041.013 Nguyen, T. D.; Battrick, B. T., eds. **Space Astrometry.** Noordwijk, The Nether-
C lands: ESRO Scientific and Technical Information Branch, ESTEC; 1975. 137p. 18 papers. English/French.

Frascati, Italy: 22-23 Oct 1974. (Proceedings). Papers resulting from an ESRO meeting to discuss the feasibility and potential benefits of a space (craft) astrometry mission. Additional topics covered are the fundamental reference system; motions in the solar system; and stellar astrometry.

041.014 Positional astronomy. **IAU Trans. Rep. Astron.** 17A(2): 1-10; 1979. + refs, 1976-79.

An overview of research activity worldwide, with an emphasis on the previous three years, as reported by IAU Commission 8 (Positional Astronomy). Additional reports: 14A: 39-51; 1970. + refs, 1967-69. 15A: 29-41; 1973. + refs, 1970-72. 16A(2): 1-12; 1976. + refs, 1973-76.

041.015 Prochazka, F. V.; Tucker, R. H., eds. **Modern Astrometry.** Vienna: Institute of
C Astronomy, University Observatory; 1979[?]. 603p. (IAU Colloq., 48). 75 papers and abstracts.

Vienna, Austria: 12-14 Sep 1978. Reports of recent research and selected review papers on the subject of stellar positions. The papers are arranged by session topics: astrophysical astrometry; astrometry of radio sources; new developments in ground-based astrometry; impact of new techniques in astrometry.

041.016 Rybka, P. Pozycyjne katalogi gwiazd. [Positional catalogues of stars.] Parts I and II. **Postepy Astron.** 22: 81-107; 1974. 43 refs, 1872-1971. In Polish.

Part I describes basic problems with positional star catalogs and Part II gives a chronological description of fundamental catalogs.

041.017 Ševarlić, B. M.; Teleki, G.; Szádeczky-Kardoss, G. Bibliography of the catalogues of star positions. In: Sevarlic, B. M.; Teleki, G., eds. **Epitome Fundamentorum Astronomiae. Pars I. Catalogues of Star Positions.** Belgrade: University of Beograd; 1978: 69-272. (Publ. Dep. Astron., Univ. Beograd, 7).

"A list of 2,087 catalogues and lists of star positions from the earliest times up to our days." Arranged by equinoxes of the star positions in them, the references in this impressive list include type of catalog (observational, fundamental, etc.), number of stars, type of coordinates used, abbreviated name of list or catalog, bibliographic citation, and notes. There are three indexes which facilitate the use of this work: abbreviated designation (title), observatory, and author.

041.018 Teleki, G. Fundamental astrometry—its present state and future prospects. In: Ševarlic, B. M.; Teleki, G., eds. **Epitome Fundamentorum Astronomiae. Pars I. Catalogues of Star Positions.** Belgrade: University of Beograd; 1978: 31-67. (Publ. Dep. Astron., Univ. Beograd, 7). 143 refs, 1830-1977.

Review of the field, with an analysis of present results and accuracy. Also included is a look at the sources of observational errors, characteristics of the various types of star catalogs, astronomical constants, and proper motions.

041.019 Tengström, E.; Teleki, G., eds. **Refractional Influences in Astrometry and**
C **Geodesy.** Dordrecht, Holland; Boston: D. Reidel; 1979. 394p. (IAU Symp., 89). 35 papers.

Uppsala, Sweden: 1-5 Aug 1978. Joint meeting of the IAU and IAG, cosponsored by IAG and IUGG. Topics include astronomical refraction, environmental systematics, refraction corrections, meteorological measurements, geodetic measurements, photogrammetry, and so on.

041.020 Van Herk, G. Report on IAU Colloquium No. 20, "Meridian Astronomy." In:
C Gliese, W.; Murray, C. A.; Tucker, R. H., eds. **New Problems in Astrometry.** Dordrecht, Holland: D. Reidel; 1974: 7-21. (IAU Symp., 61). no refs.

Copenhagen, Denmark: 25-28 Sep 1972 (IAU Colloquium No. 20). A summary of the meeting for which no proceedings were published. Topics covered were instrumentation, reference systems, various investigations, and resolutions adopted.

041.021 Westerhout, G., ed. Space astrometry. **Highlights Astron.** 5: 777-807; 1980. 6
C papers.

Joint commission meeting at the XVII General Assembly of the IAU. Various space telescopes and needs are discussed.

041.022 Zuev, N. G. Catalogue of right ascensions of 1355 bright stars in the declination
A zone from +30° to +90°. **Vestn. Khar'kov. Univ.** 176: 40-70; 1978. no refs. In Russian.

Astronomy of Star Positions. R013.004.

Conference on Photographic Astrometric Technique. 036.002.

The nearby stars. *111.014.*

Positions of radio stars. *116.009.*

042 Celestial Mechanics, Figures of Celestial Bodies

042.001 Abalakin, V. K. The development of theoretical astronomy in the USSR. **Vistas Astron.** 19: 163-177; 1975. 172 refs, 1798-1971.

Describes work in celestial mechanics in pre-revolutionary Russia and the USSR.

042.002 Celestial mechanics. **IAU Trans. Rep. Astron.** 17A(1): 15-23; 1979. + refs, 1976-78.

An overview of research activity worldwide, with an emphasis on the previous three years, as reported by IAU Commission 7 (Celestial Mechanics). Additional reports: 14A: 19-37; 1970. 759 refs, 1964-70. 15A: 19-27; 1973. + refs, 1970-73. 16A(1): 9-17; 1976. + refs, 1973-75.

042.003 Cohen, C. J.; Hubbard, E. C.; Oesterwinter, C. Elements of the outer planets for one million years. **Astron. Pap.** 22: 1-92; 1973. 14 refs, 1895-1971.

Coordinates of the outer planets were computed by numerical integration for one million years centered at the present time. Results are presented in the form of element plots, and a comparison is made with general theory.

042.004 Duncombe, R. L.; Seidelmann, P. K.; Klepczynski, W. J. Dynamical astronomy of the solar system. **Annu. Rev. Astron. Astrophys.** 11: 135-154; 1973. 218 refs, 1873-1973.

Review of celestial mechanics through the study of applications of computers as well as the use of optical, radar, and laser observations for the improvements of orbits and the determination of astronomical constants.

042.005 Fitzpatrick, Philip M. **Principles of Celestial Mechanics.** New York; London: Academic; 1970. 405p. 14 chapters. 54 refs, 1889-1970.

An advanced, introductory text requiring a knowledge of calculus, differential equations, vector analysis, mechanics, and complex variables. Primarily an overview of basic theory, with less emphasis on applications, the book covers the usual topics: one- and two-body problems; elliptical motion; astronomical coordinates; Lagrange planetary equations; mechanics; Hamilton-Jacobi Theory; artificial satellites. Includes exercises.

042.006 C Franz, O. G.; Pişmiş, P. Observational parameters and dynamical evolution of multiple stars. **Rev. Mexicana Astron. Astrofis.** 3: May 1977 (special issue). 216p. (IAU Colloq., 33). 38 papers.

Oaxtepec, Mexico: 13-16 Oct 1975. Reports of research and review papers examining the celestial mechanics of multiple star systems. Included are considerations of evolution and various observational aspects (e.g., photometry) of these stars.

042.007 C Giacaglia, G. E. O., ed. **Periodic Orbits, Stability and Resonances.** Dordrecht, Holland: D. Reidel; 1970. 530p. 44 papers.

São Paulo, Brazil: 4-12 Sep 1969. (Proceedings). Papers reviewing the periodic orbits problems of celestial mechanics, presenting new methods of applied mathematics in solving these problems.

042.008 Hagihara, Y. **Celestial Mechanics.** Cambridge, MA: MIT Press; 1970, 1972 [v.1, 2]; Tokyo: Japan Society for the Promotion of Science, 1974-76 [v.3-5]. 5v. in 9. v.I: 689p.; v.II, pt. 1 & 2: 919p.; v.III, pt. 1 & 2: 1160p.; v.IV, pt. 1 & 2: 1243p.; v.V, pt. 1 & 2: 1560p., all w/refs.

The purpose of this classic set is "to recapitulate the results of the whole field of celestial mechanics and the associated branches of science during the last hundred years."

042.009 Kovalevsky, J., ed. Precession, planetary ephemerides and time scales.
C **Highlights Astron.** 3: 207-232; 1974. 4 papers.

Joint discussion at the fifteenth General Assembly of the IAU. Topics: changes in conventional values of precession; calculation of nutation; planetary ephemerides; astronomical units, constants, and time scales.

042.010 Kovalevsky, Jean. **Introduction to Celestial Mechanics.** Dordrecht, Holland: D. Reidel; distr., New York: Springer-Verlag; 1967. 126p. (Astrophys. Space Sci. Libr., 7). 7 chapters. 8 refs, 1923-64.

An advanced text primarily concerned with the classical methods useful in calculating the trajectory of a body in space. The main chapter of the book provides a detailed solution of the problem of motion of an artificial satellite in the Earth's gravitational field; Von Zeipel and Brouwer methods are employed. Topics: principles of celestial mechanics; the two-body problem; systems of canonical equations; perturbation theory; lunar theory and motion of satellites; planetary theory.

042.011 Kozai, Y., ed. **The Stability of the Solar System and of Small Stellar Systems.**
C Dordrecht, Holland; Boston: D. Reidel; 1974. 313p. (IAU Symp., 62). 27 papers, 14 abstracts.

Copernicus Symposium I. Warsaw, Poland: 5-8 Sep 1973. Sponsored by Commissions 7 (Celestial Mechanics), 4 (Ephemerides), 37 (Star Clusters and Associations), and by IUTAM.

042.012 Lecar, M., ed. **Gravitational N-Body Problem.** Dordrecht, Holland: D. Reidel;
C 1972. 441p. (Astrophys. Space Sci. Libr., 31). (IAU Colloq., 10). 39 papers.

Cambridge, UK: 12-15 Aug 1970. Reports of recent developments in celestial mechanics. Papers cover analytic treatments and numerical experiments for both collisional and collisionless systems, as well as plasma physics applications and methods of computer simulation of the gravitational n-body problem. Fifteen additional papers from this meeting appeared in *Astrophys. Space Sci.* 14: 3-178; 1971.

042.013 Nacozy, Paul E.; Ferraz-Mello, Sylvio, eds. **Natural and Artificial Satellite**
C **Motion.** Austin, TX: University of Texas Press; 1979. 434p. 39 papers.

Austin, TX, USA: 5-10 Dec 1977. (Proceedings). An attempt to synthesize and summarize the many diverse aspects of the study of satellite motion. Topics: observations of natural satellites; theories of motion; orbit stability and evolution; analytical artificial satellite theory; reports on specific man-made satellites.

042.014 Parry, W. E. **The Many-Body Problem.** Oxford, UK; New York: Oxford University Press; 1973. 217p. (Oxford Stud. Phys.). 9 chapters. 153 refs, 1926-72.

A basic graduate text and a handbook for researchers addressing a classic problem of celestial mechanics, that of presenting various methods of solution. Topics: second quantization; perturbation theory; Green functions and correlation functions; diagrammation perturbation theory; the equation of motion method; magnetism; linear response and transport processes; many-body systems at zero temperature; the variational principle and pair-wave approximation.

042.015 Proceedings of the fifth conference on mathematical methods in celestial
C mechanics—Part I, II. **Celestial Mech.** 14: 2-149, 286-392; 1976. 28 papers.

Oberwolfach, Fed. Rep. Germany: 24-30 Aug 1975. Proceedings divided into three parts: foundations of celestial mechanics, numerical methods, and perturbation methods.

042.016 Roy, A. E. **Orbital Motion.** New York: John Wiley & Sons; 1978. 489p. 15 chapters w/refs.

An advanced but basic text introducing university students to the many aspects of celestial mechanics and astrodynamics, requiring background in calculus and vector methods. Topics: coordinate and time-keeping systems; data reduction; the two-body problem; the many-body problem; perturbations; artificial satellite dynamics; dynamics of binary stars; etc. Includes problem sets and more than 200 references.

042.017 C Seidelmann, P. K., ed. Copernicus ... and modern dynamical astronomy. **Celestial Mech.** 9: 295-363; 1974. 6 papers.

Part of the annual meeting of the AAAS in December 1972. The colloquium was part of the program marking the 500th anniversary of the birth of Copernicus. Invited speakers reviewed the various aspects of dynamical astronomy.

042.018 Siegel, C. L.; Moser, J. K. **Lectures on Celestial Mechanics.** New York: Springer-Verlag; 1971. 290p. (Grundlehren math. Wiss. Einzeldarstellungen, Band 187). 3 chapters. 84 refs, 1767-1969.

A presentation of ideas and results from "the study of solutions to differential equations in the large, in which of course applications to Hamiltonian systems and in particular the equations of motion for the three-body problem occupy an important place." Topics: the three-body problem; periodic solutions; stability. An advanced text; not a bibliographic work.

042.019 Sternberg, Shlomo. **Celestial Mechanics.** New York: W. A. Benjamin; 1969. 2v. pt.1: 158p.; pt.2: 304p. (Math. Lecture Note Ser.). 4 chapters. 91 refs, 1878-1968.

An advanced text based on lecture notes for a course at Harvard in 1968. The book centers on recent developments, primarily the ideas of A. N. Kolmogoroff, later developed by V. I. Arnold and J. Moser. Includes a brief historical background of celestial mechanics. Emphasis is on the solution of the three-body problem.

042.020 C Stiefel, E., ed. Conference on celestial mechanics. **Celestial Mech.** 2: 272-447; 1970. 22 papers.

Oberwolfach, Germany: 17-23 Aug 1969.

042.021 Stiefel, E. L.; Scheifele, G. **Linear and Regular Celestial Mechanics.** New York; Berlin: Springer-Verlag; 1971. 301p. (Grundlehren math. Wiss. Einzeldarstellungen, Band 174). 11 chapters. 38 refs, 1765-1970.

A rigorous treatise using linear differential equations to describe the pure two-body motion of celestial mechanics. These linear differential equations are said to be "everywhere regular." Topics: regularized theory; Kepler motion; the initial value problem; typical perturbations; refined numerical methods; canonical theory; geometry of the KS-transformation.

042.022 C Szebehely, V.; Tapley, B. D., eds. **Long-Time Predictions in Dynamics.** Dordrecht, Holland; Boston: D. Reidel; 1976. 358p. (NATO Adv. Stud. Inst. Ser., Ser. C, Math. Phys. Sci., 26). 23 papers.

NATO Advanced Study Institute. Cortina d'Ampezzo, Italy: 3-16 Aug 1975. (Proceedings). A series of rigorous lectures addressing problems of interest to mathematicians and celestial mechanicians. Topics: fundamentals (general concepts); numerical and statistical analysis; three- and many-body problems; dynamics in astronomy.

042.023 C Szebehely, V. G., ed. **Instabilities in Dynamical Systems: Applications to Celestial Mechanics.** Dordrecht, Holland; Boston: D. Reidel; 1979. 314p. (NATO Adv. Stud. Inst. Ser., Ser. C, 47). 20 papers.

NATO Advanced Study Institute. Cortina d'Ampezzo, Italy: 30 Jul—12 Aug 1978. A rigorous volume addressing the problems of instability in classical and celestial mechanics. Topics: fundamental considerations of stability; aspects of numerical analysis and statistical mechanics; stability of planetary systems; the problem of three bodies.

042.024 Szebehely, Victor, ed. **Dynamics of Planets and Satellites and Theories of Their**
C **Motion.** Dordrecht, Holland; Boston: D. Reidel; 1978. 375p. (Astrophys. Space Sci. Libr., 72). (IAU Colloq., 41). 28 papers; 14 abstracts. English/French.

Cambridge, UK: 17-29 Aug 1976. Papers reviewing current knowledge and progress in celestial mechanics covering planetary theory and analytical methods; lunar theory and minor planet motions; numerical and other techniques; satellites of Jupiter and Saturn, and artificial satellites; gravitational problems of three or more bodies.

042.025 Tapley, B. D.; Szebehely, V., eds. **Recent Advances in Dynamical Astronomy.**
C Dordrecht, Holland; Boston: D. Reidel; 1973. 468p. (Astrophys. Space Sci. Libr., 39). 37 papers.

NATO Advanced Study Institute. Cortina d'Ampezzo, Italy: 9-21 Aug 1972. (Proceedings). Papers summarizing recent celestial mechanics research on the following topics: regularization; three-body problem; n-body problem and stellar dynamics; theory of general perturbations; solar system and orbital resonances; trajectory determination and the motion of rigid bodies. Emphasis is on fundamental mathematical and astronomical aspects, not applications.

Planetary Satellites. 091.008.

043 Astronomical Constants

043.001 Emerson, B.; Wilkins, G. A., eds. The IAU system of astronomical constants.
C **Celestial Mech.** 4: 128-280; 1971. (IAU Colloq., 9). 16 papers.

Heidelberg, West Germany: 12-14 Aug 1970. A substantial number of numerical data are presented in this conference issue devoted primarily to the consideration of the fundamental constants of precession and nutation, and the system of planetary masses.

043.002 Seidelmann, P. K. Numerical values of the constants of the Joint Report of the Working Groups of IAU Commission 4. **Celestial Mech.** 16: 165-177; 1977. 59 refs, 1931-78.

The numerical values underlying the Joint Report of the Working Groups of IAU Commission 4 on Precession, Planetary Ephemerides, Units and Time Scales are summarized and additional explanations and references provided.

Nutation and the Earth's Rotation. 044.003.

044 Time, Rotation of the Earth

044.001 Enslin, H.; Proverbio, E., eds. **Time Determination, Dissemination and Synchronization.**
C Cagliari, Italy: 3T Edizioni Anasatatiche; 1975. 291p. 19 papers.

Second Cagliari International Meeting. Cagliari, Italy: 1974. (Proceedings). Reports of recent research, experimental results, and theoretical studies. Papers are divided into the following subjects: atomic time scales and fundamentals; time synchronization; time services and standard laboratories; astronomical fundamentals and dynamical time; rotational time.

044.002 Essen, L. The measurement of time. **Vistas Astron.** 11: 45-67; 1969. 43 refs, 1919-66.

A review of astronomical time scales and measurement, along with a discussion of the caesium atomic clock and its importance in defining an accurate time unit.

044.003 Fedorov, E. P.; Smith, M. L.; Bender, P. L. **Nutation and the Earth's Rotation.** Dordrecht, Holland; Boston: D. Reidel, 1980. 266p. (IAU Symp., 78). 20 papers, 18 abstracts.
C

Kiev, USSR: 23-28 May 1977. Topics: the specifications of nutation in the IAU system of astronomical constants; determination of forced nutation and nearly diurnal free polar motion; expected use of lunar ranging data and long baseline interferometers for precise measurement; models of the internal constitution of the Earth as the basis of a new theory of nutation; the effect of the ocean and liquid core on the rotation of the Earth; and the interaction between Earth tides and nutation.

044.004 McCarthy, Dennis D.; Pilkington, John D. H., eds. **Time and the Earth's Rotation.** Dordrecht, Holland; Boston: D. Reidel; 1979. 332p. (IAU Symp., 82). 44 papers, 11 abstracts.
C

San Fernando, Spain: 8-12 May 1978. Discussions of modern research in the field of the rotation of the Earth, with particular emphasis on the role of new observational techniques. Topics include time, polar motion, reference systems, conventional radio interferometry and VLBI, Doppler satellite methods, satellite laser ranging, lunar laser ranging, and geophysical research concerning the Earth's rotation.

044.005 Melchior, Paul; Yumi, Shigeru, eds. **Rotation of the Earth.** Dordrecht, Holland: D. Reidel; 1971. 244p. (IAU Symp., 48). 40 papers, 13 abstracts.
C

Morioka, Japan: 9-15 May 1971. Overview of the topic, with discussions of research methods, Chandler wobble, polar motion, rotational accelerations and variations, core models, and tidal friction.

044.006 Rotation of the Earth. **IAU Trans. Rep. Astron.** 17A(1): 123-132; 1979. 63 refs, 1971-78.

An overview of research activity worldwide, with an emphasis on the previous three years, as reported by IAU Commission 19 (Rotation of the Earth). Lunar laser range measurements were reported on in volume 15 (cited below). Additional reports: 14A: 177-185; 1970. no refs. 15A: 215-224; 1973. 54 refs, 1962-72. 16A(1): 115-116; 1976. 75 refs, 1973-75.

045 Latitude Determination, Polar Motion

045.001 Cáceres, O., ed. **The Problem of the Variation of the Geographical Coordinates in the Southern Hemisphere.** La Plata, Argentina: Astronomical Observatory, National University of La Plata; 1972. 124p. (IAU Colloq., 1). 15 papers.
C

La Plata, Argentina: 4-5 Nov 1968. Papers discussing the title topic and the future organization of the observations of latitude and time.

The Elements of Astronomy for Surveyors. 041.009.

Kinetic models of the solar and polar winds. *074.019.*

Nutation and the Earth's Rotation. 044.003.

046 Astronomical Geodesy, Satellite Geodesy, Navigation

046.001 Anderson, J. D. Planetary geodesy. **Rev. Geophys. Space Phys.** 13(3): 274-275, 292-293; 1975. 42 refs, 1971-74.

Review covering the preceding four years. Discussion is restricted to the traditional subjects of geometrical and physical geodesy, with particular emphasis on the size and shape of the planets and on their masses and gravity fields.

046.002 Bomford, G. **Geodesy.** 3rd ed. Oxford, UK: Clarendon Press; 1971. 731p. 7 chapters. 531 refs, 1743-1969.

This standard, advanced text emphasizes principles rather than equipment and techniques, although the latter is addressed. This much-expanded latest edition includes most of the previous subjects but has added emphasis to recent developments such as the use of computers and artificial satellites in geodetic work. Topics: triangulation, traverse, and trilateration (field work); computation of triangulation, traverse, and trilateration; heights above sea level; geodetic astronomy; geometrical use of artificial satellites; gravity and seismic surveys; physical geodesy.

046.003 Ewing, Clair E.; Mitchell, Michael M. **Introduction to Geodesy.** New York: American Elsevier; 1970. 304p. 11 chapters w/refs.

An intermediate-level academic text emphasizing principles, not techniques or methods. Topics: triangulation; the ellipsoid of revolution and related computations; geodetic astronomy; coordinate systems; electronic surveying; gravity; leveling; satellites; adjustment computations.

046.004 Groten, Erwin. **Geodesy and the Earth's Gravitational Field. Vol. 1. Principles and Conventional Methods.** Bonn: Ferd. Dümmlers Verlag; 1979. 409p. 5 chapters. 145 refs, 1911-77.

A graduate text concentrating on basic theory and classical geodetic methods. Also useful for geologists, geophysicists, and surveyors, the book covers basic principles; celestial reference systems; concepts of astronomical positioning and orientation; geometrical geodesy; dynamic satellite geodesy; and physical geodesy.

046.005 C Henriksen, Soren W.; Mancini, Armando; Chovitz, Bernard H., eds. **The Use of Artificial Satellites for Geodesy.** Washington, DC: American Geophysical Union; 1972. 298p. (Geophys. Monogr., 15). 38 papers.

Third International Symposium. Washington, DC, USA: 15-17 Apr 1971. The papers in this volume emphasize a critical analysis of past results and a look to the future for new data-gathering methods to improve geodetic knowledge. Topics: geometric geodesy theory and results; physical geodesy theory and results; instrumentation and environment; extra-terrestrial geodesy; data management.

046.006 Kovalevsky, J.; Barlier, F. Géodésie par satellites. [Geodesy by satellites.] **Ciel Terre** 87: 50-107; 1971. 79 refs, 1897-1970.

Reviews the various means of studying the geometrical shape of the Earth. Covers geodesy by satellite in more detail, explaining the advantages to geodetic measurements unencumbered by atmospheric and spatial interference.

046.007 Lambeck, K. Methods and geophysical applications of satellite geodesy. **Rep. Prog. Phys.** 42: 547-628; 1979. 194 refs, 1921-79.

Satellite tracking methods, geodetic procedures, and results are described, with emphasis on the geophysical significance of the results.

046.008 Mueller, Ivan I. **Spherical and Practical Astronomy as Applied to Geodesy.** New York: Frederick Ungar; 1969. 615p. 12 chapters w/refs.

A text and reference work for the scientist and student covering "the art and science of determining the precise location of a terrestrial point from measurements on natural celestial bodies." Subjects covered include the celestial sphere and its coordinate systems; variations in celestial coordinates (e.g., nutation and precession); time systems (e.g., sidereal, universal, ephemeris); star catalogs; instrumentation; determination of astronomic latitude, longitude, and time; solar eclipses and occultations.

046.009 Wright, Frances W. **Celestial Navigation.** Cambridge, MD: Cornell Maritime Press; 1969. 137p. 11 chapters. 22 refs, 1942-67.

A basic how-to book for any reader wishing to learn how to navigate by the stars. Including many examples, diagrams, and tables, the volume covers coordinate systems and other astronomy necessary for celestial navigation; identification of the north and south celestial poles; use of the sextant; latitude determination; different types of solar time; etc. Useful for the novice only.

Refractional Influences in Astrometry and Geodesy. 041.019.

047 Ephemerides, Almanacs, Calendars, Chronology

047.001 **The Astronomical Almanac for the Year 1981.** Washington, DC: U.S. Government Printing Office; London: Her Majesty's Stationery Office; 1980- . 544p. (1981 ed.). Annual. Supersedes and combines *The American Ephemeris and Nautical Almanac* (U.S.A.: 1852-1979) and *The Astronomical Ephemeris* (U.K.: 1960-79).

A standard reference book found in all observatory libraries and telescope domes. Includes a variety of numerical data on various astronomical phenomena, time-scales, coordinate systems, and positions and motions of the Sun, Moon, and planets. Additional information is presented: lists of radio and optical observatories, a glossary, miscellaneous tables, and an index.

047.002 Ephemerides. **IAU Trans. Rep. Astron.** 15A: 1-10; 1973. 27 refs, 1969-73.

An overview of research activity worldwide, with an emphasis on the previous three years, as reported by IAU Commission 4 (Ephemerides).

047.003 Goldstine, Herman H. **New and Full Moons 1001 B.C. to A.D. 1651.** Phila-
A delphia, PA: American Philosophical Society; 1973. 221p. (American Philos. Soc. Mem., 94).

A computer-produced list containing 32,800 new and full Moon data including date, time (for an observer in Babylon), and longitude. Primary use is for historians who need this data. Considered [by Goldstine] to be a supplement to Bryant Tuckerman's *Planetary, Lunar, and Solar Positions 601 B.C. to A.D. 1* (American Philos. Soc. Mem., 56; 1962), and *Planetary, Lunar, and Solar Positions A.D. 2 to A.D. 1649* (American Philos. Soc. Mem., 59; 1964).

047.004 International Information Bureau on Astronomical Ephemerides. Information Cards Nos. 78-93. **Bull. Bur. Int. Inf. Ephemerides Astron.** Paris: BIIEA; 1973.

These particular cards contain notes on data files available at the Astronomisches Rechen-Institut, Heidelberg. These "data files ... are primarily produced or collected in connection with its work on the fundamental catalogue." The data consists primarily of positional star catalogs in machine-readable form (cards and tape). Examples are the General Catalogue (GC), Fundamentalkatalog (FK3), etc. Fifteen catalogs are described, including type, contents, number of records, bibliographic reference, source, and date of issue.

047.005 Meeus, Jean. **Phases of the Moon 1801-2010.** Brussels: Vereniging voor Sterrenkunde; 1976. 33p. (Vereniging Voor Sterrenkunde Mem., 2).

A set of tables in three parts. The first consists of only the dates of full moons, new moons, first and last quarters for the period 1801-1950. Part 2 consists of dates and times (instants) with an accuracy of 0.1 minute of time. Part 3 gives tables for calculating the approximate time of any lunar phase between 1900 B.C. and 2999 A.D.

047.006 Seidelmann, P. K. The ephemerides: past, present and future. **IAU Symp.** 81: 99-114; 1979. 46 refs, 1884-1978.

Review of the efforts towards improving agreement on accurate astronomical constants, planetary, lunar, and satellite theories, and cooperative methods of printing the annual ephemerides in different languages.

047.007 U.S. Nautical Almanac Office. **Sunrise and Sunset Tables for Key Cities and**
A **Weather Stations in the United States.** Detroit, MI: Gale Research; 1977. unpaged. 369 tables.

Usable in any twentieth-century year, this collection of data for 369 cities is accurate to one to two minutes. It is more convenient than the tables in *The Astronomical Almanac* (*047.001*) because the latter requires that one know latitude and longitude and it is necessary to interpolate to obtain the correct times.

The American Ephemeris and Nautical Almanac. 047.001.

The Astronomical Ephemeris. 047.001.

SPACE RESEARCH

051 Extraterrestrial Research, Spaceflight Related to Astronomy and Astrophysics

051.001 Adler, I.; Trombka, J. I. **Geochemical Exploration of the Moon and Planets.** New York: Springer-Verlag; 1970. 243p. (Phys. Chem. Space, 3). 6 chapters w/refs.

A review of recent research and future programs, using manned and unmanned spacecraft for data collection. Topics: introduction (overview of Mars, Venus, Moon, outer planets); instruments used for compositional exploration; instruments and techniques under development; Apollo surface missions; data processing and analysis; orbital and surface exploration after early Apollo.

051.002 Astronomical observations from outside the terrestrial atmosphere. **IAU Trans. Rep. Astron.** 17A(3): 165-201; 1979. 258 refs, 1970-79.

An overview of research activity worldwide with an emphasis on the previous three years, as reported by IAU Commission 44 (Astronomical Observations from outside the Terrestrial Atmosphere). Topics: solar astronomy; interplanetary medium; UV astronomy; x-ray sources; x- and gamma-ray background radiation; radio astronomy. Additional reports: 14A: 525-546; 1970. 248 refs, 1962-70. 15A: 671-695; 1973. 47 refs, 1969-72. 16A(3): 195-220; 1976. 94 refs, 1973-76.

051.003 Bowhill, S. A.; Jaffe, L. D.; Rycroft, M. J., eds. **Space Research XII.** Berlin:
C Akademie-Verlag; 1972. 1,815p. 2v. v.1: 88 papers; v.2: 103 papers.

Open Meetings of Working Groups of the Fourteenth Plenary Meeting of COSPAR, Seattle, WA, USA: 21 Jun—2 Jul 1971; Symposium on Total Solar Eclipse of 7 March 1970, Seattle, WA: 18-19 & 21 Jun 1971; Symposium on Dynamics of the Thermosphere and Ionosphere above 120 km, Seattle, WA: 24-26 Jun 1971; Symposium on High Angular Resolution Astronomical Observations from Space, Seattle, WA: 28-30 Jun & 1 Jul 1971. (Proceedings). A series of papers reporting on physical and life sciences and space research. The reports and reviews are arranged under 11 topics: the Moon; Venus and Mars; cosmic dust; Earth observed from space; Earth's neutral atmosphere; dynamics of the thermosphere and ionosphere above 120 km.; ionosphere; polar ionosphere and precipitation of low energy charged particles; solar terrestrial relationships; astronomy; high angular resolution astronomical observations from space.

051.004 Bruzek, Anton; Pilkuhn, Hartmut, eds. **Lectures on Space Physics.** Düsseldorf: Bertelsmann Universitätsverlag; 1973. 2v. v.1: 353p. 17 papers; v.2: 347p. 12 papers.

A series of lectures on four topics: cosmic rays (v.1); space biophysics (v.1); Sun and interplanetary medium (v.2); relativistic astrophysics (v.2). Aimed at the advanced student and nonspecialist, the papers contain a great number of references to the literature.

051.005 Bursnall, W. J.; Morgenthaler, G. W.; Simonson, G. E., eds. **Space Shuttle**
C **Missions of the 80's.** San Diego, CA: American Astronautical Society; distr., San Diego, CA: Univelt; 1977. 1,308p. 2v. (Adv. Astronaut. Sci., 32, pt. 1-2). 159 papers.

Twenty-first Annual Meeting of the American Astronautical Society. Denver, CO, USA: 26-28 Aug 1976. (Proceedings). Contributed papers discussing the many potential scientific uses of the Space Shuttle and Spacelab. Both experiments and instrumentation are reviewed. Fifty-nine of the papers are concerned with astronomical research, in particular the plans for the orbiting Large Space Telescope (LST).

051.006 Courtes, G. Optical systems for UV space researches. **IAU Symp.** 41: 273-301; 1971. 36 refs, 1905-1971.

General review of the best ways to adapt the UV optical designs to the peculiar conditions of guidance of space vehicles.

051.007 Donahue, T. M.; Smith, P. A.; Thomas, L., eds. **Space Research X.** Amster-
C dam: North-Holland; 1970. 1,049p. 126 papers.

Proceedings of Open Meetings of Working Groups of the Twelfth Plenary Meeting of COSPAR. Prague, Czechoslovakia: 11-24 May 1969; Symposium on Thermospheric Properties Concerning Temperatures and Dynamics with Special Application to H and He, Prague, Czechoslovakia: 12-14 May 1969. A collection of papers in the interdisciplinary field of space science, covering the following topics: satellite tracking and its applications; meteorology; stratosphere and mesosphere; cosmic dust and related studies; thermosphere and exosphere; ionosphere; solar proton experiments; magnetosphere; radiations; Moon and planets.

051.008 Doyle, Robert O., ed. **A Long-Range Program in Space Astronomy.** Washington, DC: NASA; 1969. 305p. (NASA SP-213). 7 chapters. 9 refs, 1962-68.

Subtitled "Position Paper of the Astronomy Missions Board," this volume presents recommendations to NASA for future astronomical research via spacecraft. The majority of the book consists of reports of subdisciplinary panels which include proposed programs of study. Examples are x-ray and gamma-ray astronomy; optical and infrared space astronomy; solar and planetary space astronomy; etc. Also addressed are the role of man in space astronomy; the relation to a ground-based program; and other subjects.

051.009 Enzmann, Robert Duncan, ed. **Use for Space Systems for Planetary Geology**
C **and Geophysics.** Washington, DC: American Astronautical Society; 1968. 605p. (AAS Sci. Technol. Ser., 17). 32 papers; 16 abstracts.

AAS Symposium. Boston, MA, USA: 25-27 May 1967. (Proceedings). Review papers, reports of research, and proposed projects on space geology and geophysics, with an emphasis on the required hardware. Topics: geophysical environments; instrumentation requirements; vehicular and communications requirements; data and sample collection processing and dissemination; systems integration and space mission planning.

051.010 Fourth international symposium on the origin of life. **Origins of Life** 5: 1-505;
C 1974. 42 papers.

Barcelona, Spain: Jun 1973. Sections include cosmochemistry, paleobiology, primordial organic chemistry, precellular organization, early biochemical evolution, and exobiology.

051.011 Freundlich, Martin M.; Wagner, Bernard M., eds. **Exobiology: The Search for**
C **Extraterrestrial Life.** Washington, DC: American Astronautical Society; 1969. 172p. (AAS Sci. Technol. Ser., 19). 7 papers.

An AAS/AAAS Symposium. New York, NY, USA: 30 Dec 1967. (Proceedings). Typed transcripts of conference papers reviewing current knowledge on the subject of extraterrestrial life. Topics: solar system exobiology; life detection techniques; interstellar communication methods; sociological aspects of exobiology; etc.

051.012 Grard, R. J. L., ed. **Photon and Particle Interactions with Surfaces in Space.**
C Dordrecht, Holland; Boston: D. Reidel; 1973. 577p. (Astrophys. Space Sci. Libr., 37). 39 papers.

Sixth ESLAB Symposium. Noordwijk, The Netherlands: 26-29 Sep 1972. (Proceedings). Papers reviewing research into the effects of "photon and particle fluxes on the electronic properties of the surface of a space probe," and evaluating how much the output of a scientific instrument will be affected by such phenomena. How to counteract the effects of interactions with celestial objects is also considered. Emphasis is on the solar wind and magnetospheric particles.

051.013 Haymes, Robert C. **Introduction to Space Science.** New York: John Wiley & Sons; 1971. 556p. (Space Sci. Text Ser.). 16 chapters. 54 refs, 1928-69.

A text aimed at seniors and graduate students knowing basic physics and differential equations. Since space science is an interdisciplinary field, many topics are covered: celestial coordinates and time; celestial mechanics; Earth's atmosphere; planetary atmospheres; aurora and airglow; planetary interiors; planetary magnetism; stellar structure and evolution; cosmic rays; cosmology; radio astronomy; exotic astronomy; Van Allen radiation; comets, meteors, and the interplanetary medium.

051.014 Hong, Frances; Pulliam, Jean. **Scientific Publications of the Bioscience Programs Division. Volume 3: Exobiology.** 2nd ed. Washington, DC: George Washington University Medical Center; 1968. 154p. (NASA CR-103224; N69-36436). 715 refs, 1960-67.

A bibliography (without abstracts) of journals, conference proceedings, and books arranged by year and by author within each year. There are author and permuted title indexes, as well as lists of senior authors and their addresses, and 129 journals publishing exobiology material.

051.015 Katterfel'd, G. N., ed., Suslov, A. K., comp. **Bibliography on Problems of Astrobiology.** Washington, DC: NASA; 1969. 54p. (NASA TT-F-12201; N69-25197). 255 refs, 1896-1964.

A review article on life on other worlds with a comprehensive bibliography of Russian and Western sources. Translation of *Literatura po problemam astrobiologii.*

051.016 Kondratyev, K. Ya.; Rycroft, M. J.; Sagan, C., eds. **Space Research XI.** Berlin:
C Akademie-Verlag; 1971. 2v. 1,415p. v.1: 86 papers; v.2: 87 papers.

Open Meetings of Working Groups of the Thirteenth Plenary Meeting of COSPAR. Leningrad, USSR: 20-29 May 1970; Symposium on Remote Sounding of the Atmosphere, Leningrad, USSR: 22 & 25-26 May 1970. (Proceedings). A series of papers reporting on physical and life sciences and space research. The reports and reviews are organized into nine sections: the Moon; planets; cosmic dust; solid Earth physics; remote sounding of the atmosphere; neutral atmosphere; ionosphere; solar terrestrial relationships; astronomical measurements. Of special interest are reports of the first manned lunar landings, including a paper by Neil Armstrong.

051.017 Lacroute, P. Prospects in space astrometry. **Highlights Astron.** 4(1): 345-370;
C 1977. 4 papers.

Joint meeting of commissions 8 and 24 at the Sixteenth General Assembly of the IAU. Topics: European Space Agency plans; the Large Space Telescope Astrometric Instrument; general principles; future prospects.

051.018 Lazarev, A. I. Optical studies from space. **Soviet J. Opt. Technol.** 44: 566-573; 1977. 60 refs, 1970-77.

Review of some of the major results of optical studies from manned Soviet spacecraft and unmanned interplanetary spacecraft. Covered are visual observations on early manned flights, vertical-ray structure of the emission radiation of the Earth's upper atmosphere;

spectrophotometry and multizonal photography of the Earth and its atmosphere, astrophysical studies, and planetary studies.

051.019 Looney, John J. **Bibliography of Space Books and Articles from Non-Aerospace Journals 1955-1977.** Washington, DC: NASA; 1979. 243p. (HHR-51). ~3,800 refs.

Although this large, unannotated bibliography is decidedly slanted towards aerospace sources, there are a large number of entries (over 300) on space science which include many references to astronomical applications. Entries are arranged by author under subject categories and contain author, title of book or journal, journal article title, year, volume, and pagination.

051.020 Morgenthaler, G. W.; Greyber, H. D., eds. **Astronomy from a Space Platform.**
C Tarzana, CA: American Astronautical Society; 1972. 398p. (AAS Sci. Technol. Ser., 28). 21 papers.

Philadelphia, PA, USA: 27-28 Dec 1971. A series of papers describing future space astronomy activity and the benefits to be derived, with an emphasis on the equipment and techniques to be used. Topics: planetary and solar astronomy; stellar and galactic astronomy; large space telescopes; etc.

051.021 Morgenthaler, George W.; Hollstein, Manfred, eds. **Space Shuttle and Spacelab**
C **Utilization.** San Diego, CA: American Astronautical Society; distr., San Diego: Univelt; 1978. 820p. 2v. (Adv. Astronaut. Sci., 37, pt. 1-2). 51 papers.

Twenty-fourth AAS Annual Meeting & Sixteenth Goddard Memorial Symposium. Washington, DC, USA: 8-10 Mar 1978. Subtitled "Near-Term and Long-Term Benefits for Mankind," this two-volume set explores the many applications, both planned and potential, of the Space Shuttle and Spacelab. Included are transportation; satellite launching; Earth resources; astronomy; biology; etc. Equipment is emphasized, and astronomical applications are discussed: x-ray, UV, and infrared astronomy.

051.022 Papagiannis, Michael D. **Space Physics and Space Astronomy.** New York: Gordon and Breach; 1972. 293p. 8 chapters w/refs.

A text for beginning graduate students, advanced undergrads, and nonspecialists, providing a nonelementary introduction to the two fields, with more emphasis on space physics. Topics: planetary atmospheres; ionosphere; magnetosphere; the active Sun; interplanetary space; Earth-Sun interactions; x-ray, ultraviolet, infrared, and gamma-ray astronomy; space radio astronomy; etc.

051.023 Pecker, Jean-Claude. **Experimental Astronomy.** Dordrecht, Holland: D. Reidel; distr., New York: Springer-Verlag; 1970. 105p.(Astrophys. Space Sci. Libr., 18). 6 chapters. 39 refs, 1955-69.

A brief introduction to astronomy aided by spacecraft, covering a variety of topics: experimental astronomy; the use of artificial satellites; initiation in astronautics; direct exploration of the Moon and planets; the plurality of inhabited worlds. For the scientist, student, and advanced layperson.

051.024 Pecker, Jean-Claude. **Space Observatories.** Dordrecht, Holland: D. Reidel; distr., New York: Springer-Verlag; 1970. 120p. (Astrophys. Space Sci. Libr., 21). 13 chapters. 44 refs, 1957-69.

A brief introduction to the advantages of space astronomy over Earth-based studies, and an overview of the information to be gained and areas of study. The first part of the book discusses why and how the atmosphere hinders astronomy on Earth. The latter part covers the solar corona; galaxies and faint objects; ultraviolet astronomy; x-ray and gamma ray studies.

051.025 **Physics and Astrophysics from Spacelab.** Noordwijk, The Netherlands: ESTEC,
C ESA Scientific and Technical Publications Branch; 1977. 155p. (ESA SP. 132). 7 papers.

Trieste, Italy: 6-11 Sep 1976. After an introductory paper on the Space Shuttle and Spacelab programs, this book describes astronomical research programs which will be carried out on Spacelab, including infrared, x-ray, cosmic ray, and ultraviolet studies.

051.026 Ponnamperuma, C., ed. **Chemical Evolution of the Giant Planets.** New York:
C Academic; 1976. 240p. 16 papers.

Colloquium. College Park, MD, USA: Oct 1974. Following the theme of the possibility of life on the giant planets and their satellites, this volume reviews the necessary conditions for life and discusses how it might be detected. The composition and evolution of planetary atmospheres are a major topic of discussion here, along with biological considerations and details of planetary exploration by spacecraft.

051.027 **Publications of the Exobiology Program for [yr.]: A Special Bibliography.** Washington, DC: NASA; 1980- . Annual.

A continuation of *Publications of the Planetary Biology Program* (*051.028*). Additional topics include extraterrestrial life and planetary protection.

1980 Pleasant, L. G.; DeVincenzi, D. L., comp. Oct 1980. 34p. (NASA TM-82182; N80-34106). 146 refs.

051.028 **Publications of the Planetary Biology Program for [yr.]: A Special Bibliography.** Washington, DC: NASA. 1976-79. Annual. Continued by *Publications of the Exobiology Program for [yr.]: A Special Bibliography.*

A bibliography concerned with "the origin, evolution, and distribution of life in the universe," this list includes primarily journal articles, proceedings, papers, and annual reviews. The Planetary Biology Program (NASA) and the bibliographies addressed the following areas: chemical evolution; organic geochemistry; life detection; biological adaptation; bioinstrumentation; planetary environments; origin of life. Each bibliography includes complete citations but not abstracts. A complete list of publications follows:

1975 Souza, K. A.; Young, R. S., comp. Jul 1976. 21p. (NASA TM-X-74313; N76-26874). 131 refs.

1976 Bradley, F. D.; Young, R. S., comp. 1977. 37p. (NASA TM-75017; N78-25773). 182 refs.

1977 Pleasant, L. G.; Young, R. S., comp. Apr 1979. 33p. (NASA TM-80338; N79-21777). 177 refs.

1978 Pleasant, L. G.; Young, R. S., comp. Oct 1979. 34p. (NASA TM-80745; N79-33854). 187 refs.

051.029 Rycroft, M. J., ed. **Space Research XV.** Berlin: Akademie-Verlag; 1975. 737p.
C 93 papers.

Proceedings of Open Meetings of Working Groups on Physical Sciences of the Seventeenth Plenary Meeting of COSPAR. Sąo Paulo, Brazil: Jun 1974. Contributed and review papers on recent studies conducted in the Earth's atmosphere and beyond, with emphasis on atmospheric studies. Papers are arranged in eight major topics: geodesy; remote sensing; upper atmosphere; ionosphere; magnetosphere; cosmic dust; Moon; astronomy. Specific subjects covered in the latter include solar studies; ultraviolet, x-ray, and high-energy astronomy; and the interstellar medium.

051.030 Rycroft, M. J., ed. **Space Research XVI.** Berlin: Akademie-Verlag; 1976.
C 1,077p. 160 papers.

Open Meetings of Working Groups on Physical Sciences of the Eighteenth Plenary Meeting of COSPAR, Varna, Bulgaria: 29 May—7 Jun 1975; COSPAR Symposium and Workshop on Results from Coordinated Upper Atmosphere Measurement Programs, Varna, Bulgaria: 29-31 May 1975. (Proceedings). A compendium of papers reviewing the use of space technology in astronomical and atmospheric research. Topics: remote sensing; satellite geodesy; Earth's neutral atmosphere; thermosphere and ionosphere; magnetosphere; interplanetary medium; Sun; astronomy; Skylab; Moon; planets.

051.031 Rycroft, M. J., ed. **Space Research, Vol. XVII.** Oxford, UK; New York: Pergamon; 1977. 860p. 128 papers.
C

Open Meetings of Working Groups on Physical Sciences of the Nineteenth Plenary Meeting of COSPAR. Philadelphia, PA, USA: 8-19 Jun 1976; COSPAR/IAGA Symposium on Minor Constituents and Excited Species, Philadelphia, PA: 9-10 Jun 1976. (Proceedings). Reviews and reports of research on physical, space, and atmospheric sciences from around the world. Topics: remote sensing of the Earth; satellite orbits and tracking; stratosphere and mesosphere; upper atmosphere; thermosphere; ionosphere; solar-terrestrial relations; cosmic dust; the planets; astronomy; balloon research; Space Shuttle.

051.032 Rycroft, M. J., ed. **Space Research, Vol. XVIII.** Oxford, UK; New York: Pergamon; 1978. 543p. 112 papers.
C

Open Meetings of the Working Groups on Physical Sciences of the Twentieth Plenary Meeting of COSPAR. Tel Aviv, Israel: 7-18 Jun 1977. (Proceedings). A compendium of invited and contributed papers addressing recent scientific studies. The volume is arranged topically, covering the following areas: remote sensing of the Earth's environment; atmospheric response to solar and geomagnetic activity; the thermosphere; the ionosphere; the magnetosphere; the Sun and interplanetary medium; cosmic dust; Moon and planets; astronomy; materials science under micro-gravity conditions.

051.033 Rycroft, M. J., ed. **Space Research, Vol. XIX.** Oxford, UK; New York: Pergamon; 1979. 615p. 105 papers.
C

Open Meetings of the Working Groups on Physical Sciences of the Twenty-first Plenary Meeting of COSPAR. Innsbruck, Austria, 29 May—10 Jun 1978. (Proceedings). A collection of papers reporting on physical, atmospheric, and space sciences research. Topics: remote sensing; the middle atmosphere; the thermosphere; the ionosphere; the magnetosphere; the Sun and the interplanetary medium; cosmic dust; materials science in space.

051.034 Rycroft, M. J., ed. **Space Research, Vol. XX.** Oxford, UK; New York: Pergamon; 1980. 285p. 50 papers.
C

Open Meetings of the Working Groups on Physical Sciences of the Twenty-second Session of COSPAR. Bangalore, India: 29 May—9 Jun 1979. Contributed papers reporting on recent studies on the Earth's neutral atmosphere; the Earth's plasma envelope; planetary science; and astronomy. Selected topics: ionospheric chemistry; atmospheric radiation; Venus; the Sun; cosmic ray studies; etc.

051.035 Rycroft, M. J.; Reasenberg, R. D., eds. **Space Research XIV.** Berlin: Akademie-Verlag; 1974. 800p. 126 papers.
C

Open Meetings of Working Groups of the Sixteenth Plenary Meeting of COSPAR. Constance, Germany, 23 May—5 Jun 1973; Symposium on Noctilucent Clouds and Interplanetary Dust, Constance, Germany, 24-26 May 1973. (Proceedings). A collection of contributed and invited papers, including reviews, discussing the use of space techniques to study the Earth, solar system, and beyond. Topics: satellite geodesy; the Earth's neutral atmosphere; ionosphere; magnetosphere; Sun; astronomy; comets; Moon; Venus; noctilucent clouds and interplanetary dust.

051.036 Rycroft, M. J.; Runcorn, S. K., eds. **Space Research XIII.** Berlin: Akademie-Verlag; 1973. 2v. 1,198p. 166 papers.
C

Open Meetings of Working Groups on Physical Sciences of the Fifteenth Plenary Meeting of COSPAR. Madrid, Spain: 10-24 May 1972. (Proceedings). Papers reporting on current research and recent experimental results. The papers are arranged under 10 categories: tracking of artificial Earth satellites; remote sensing of the Earth; the neutral atmosphere; dynamics of the thermosphere and ionosphere; the ionosphere; the magnetosphere; the Sun; astronomical measurements; the Moon; cosmic dust.

051.037 Sable, Martin H. **Exobiology: A Research Guide.** Brighton, MI: Green Oak Press; 1978. 324p. 3,832 refs, 1648-1975.

There are no abstracts in this bibliography, which includes references to books, journal and newspaper articles, films, TV shows, and more. Topics: UFOs, SETI, exobiology, and selected astronomical subjects related to exobiology.

051.038 Schwartz, Alan W., ed. **Theory and Experiment in Exobiology, Vol. 1.** Groningen, The Netherlands: Wolters-Noordhoff Publishing; 1971. 160p. 7 papers.

A collection of review papers on "the study of the origin, distribution, and properties of life in the universe." Topics: the role of ionizing radiation in primordial organic synthesis; catalysis; the origin of biological phosphates; exobiology of porphyrins; analysis of lunar samples for carbon compounds; chemistry and photochemistry of the Jovian atmosphere; exobiology and planetary exploration.

051.039 Singer, S. Fred, ed. **Manned Laboratories in Space.** Dordrecht, Holland: D.
C Reidel; distr., New York: Springer-Verlag; 1969. 133p. (Astrophys. Space Sci. Libr., 16). 9 papers.

Second International Orbital Laboratory Symposium. New York, NY, USA: 18 Oct 1968. A brief overview of possible applications of orbiting space stations, including meteorology, geoscience investigations, ocean exploration, and astronomy. Also covered are engineering problems of space stations, an orbital laboratory, and manned flight to other planets.

051.040 Space Sciences Symposium. **Indian J. Radio Space Phys.** 6: no. 3; 1977. 13
C papers.

Trivandrum, India: 18-21 Jan 1977. Brief communications on recent research in India. Topics: cosmic rays; solar particle effects; travelling ionospheric disturbances; etc.

051.041 Space Sciences Symposium. **Indian J. Radio Space Phys.** 7: 171-214; 1978. 9
C papers.

Waltair, India: 9-12 Jan 1978. Brief reports of recent research in India. Topics: effects of solar activity on cosmic ray intensity and geomagnetic field variation; radio meteor rates; interplanetary magnetic field structure; high energy cosmic ray particles approaching solar minimum; etc.

051.042 Tross, Carl H., ed. **Future Space Activities.** Tarzana, CA: American Astro-
C nautical Society; distr., San Diego: Univelt; 1976. 169p. (AAS Sci. Technol. Ser., 40). 9 papers.

Thirteenth Goddard Memorial Symposium. Washington, DC, USA: 11 Apr 1975. (Proceedings). A brief look at coming activities in space research, including astronomical applications. Topics: Mars landers; Spacelab; Space Shuttle; automated planetary missions; etc.

051.043 Vernov, S. N.; Kocharov, G. E., eds. **Winter School on Space Physics.** 6th.
C Jerusalem: Israel Program for Scientific Translation; 1971. 2v. pt. 1: 332p. 37 papers; pt. 2: 192p. 24 papers.

Apatity, USSR: 18 Mar–1 Apr 1969. (Proceedings). Basic lectures on the following topics: pt. 1: the universe, galaxies, and stars; x-rays, gamma rays, electrons and neutrinos; solar-terrestrial relations; pt. 2: comets, meteorites and the Moon; cosmic rays; methodological problems.

051.044 Weete, J. D., ed. Conference on space biology. **Space Life Sci.** 4: 1-220; 1973.
C 14 papers.

Houston, TX, USA: 11-12 Oct 1971. The primary objective was to bring together active investigators in the field of exobiology for the presentation of data and exchange of concepts as they relate to the applications of the biological and biochemical sciences to problems in space. Topics included the chemical evolution of life, the detection of extraterrestrial life, and the influence of various environmental parameters on the growth of plant, animal, and microbial systems.

051.045 West, M. W.; Ponnamperuma, C. Chemical evolution and the origin of life—a comprehensive bibliography. **Space Life Sci.** 2: 225-295; 1970. 1,588 refs, 1864-1969.

A comprehensive guide to the literature, with the focus on experimental and theoretical material, excluding exobiology, biological evolution, and geochemistry. Includes a subject index. Supplements have been published as follows: *Space Life Sci.* 3: 293-304; 1972. *Space Life Sci.* 4: 309-328; 1973. *Origins of Life* 5: 507-527; 1974. *Origins of Life* 6: 285-300; 1975. *Origins of Life* 7: 75-85; 1976. *Origins of Life* 9: 67-74; 1978.

051.046 White, R. Stephen. **Space Physics.** New York; London: Gordon and Breach; 1970. 318p. 6 chapters w/refs.

A basic text for students, engineers, and nonspecialists covering a variety of subjects including experimental facts, important discoveries, and ideas. Topics: the Earth's radiation belts; Earth's atmosphere; Earth's ionosphere; the Sun; interplanetary space (including solar wind); Earth's magnetosphere (including the aurora); etc.

051.047 Zuckerberg, Harry, ed. **Exploitation of Space for Experimental Research.**
C Tarzana, CA: American Astronautical Society; 1968. 351p. (Adv. Astronaut. Sci., 24). 17 papers.

Fourteenth Annual AAS Meeting. Dedham, MA, USA: 13-15 May 1968. (Proceedings). Contributed papers centering on possible spacecraft experiments and the necessary instrumentation in astronomy, atmospheric sciences, and space science. Topics: experimental research; planetary phenomena research; astronomical investigations (x-ray; above the atmosphere; space telescopes); systems for space exploration.

Astrophysics and Space Science: An International Journal of Cosmic Physics. R005.007.

Bibliography of Soviet solar system space research. *091.005.*

A Catalog of NASA Special Publications. R001.001.

Catalog of Particles and Fields Data 1958-1965. 022.051.

Catalog of Particles and Fields Data 1966-1973. 022.052.

Cometary Missions. 102.001.

Comets: Scientific Data and Missions. 102.010.

A discussion on solar studies with special reference to space observations. *080.027.*

Exploration of the Planetary System. 091.075.

History of Aeronautics and Astronautics: A Preliminary Bibliography. R001.005.

Mass spectroscopy in solar system exploration. 031.517.

Microparticle studies by space instrumentation. 131.038.

The New Dictionary and Handbook of Aerospace. R009.005.

The New Space Encyclopedia. R009.001.

Priorities for Space Research, 1971-80. R011.028.

Proceedings of the JOSO workshop: future solar optical observations—needs and constraints. *080.017.*

Scientific Research with the Space Telescope. 032.504.

The Scientific Results from the Orbiting Astronomical Observatory (OAO-2). 032.501.

Space Astrometry. 041.013; 041.021.

Space Physics: The Study of Plasmas in Space. 062.002.

Space relativity. *066.050.*

Summer School on Particles and Fields in Space. 022.053.

052 Astrodynamics, Navigation of Space Vehicles

052.001 Bate, Roger R.; Mueller, Donald D.; White, Jerry E. **Fundamentals of Astrodynamics.** New York: Dover Publications; London: Constable; 1971. 455p. 9 chapters w/refs.

An undergraduate text reviewing basic celestial mechanics and its application (astrodynamics) to the motions of artificial satellites. Topics: two-body orbital mechanics; orbit determination from observations; basic orbital maneuvers; position and velocity as a function of time; orbit determination from two positions and time; lunar and interplanetary trajectories; perturbations; etc.

052.002 Herrick, Samuel. **Astrodynamics.** London; New York: Van Nostrand Reinhold; 1971, 1972. 2v. v.1: 540p. 141 refs, 1778-1971; v.2: 348p. 241 refs, 1743-1971. 19 chapters.

A classic in the field, this two-volume work for specialists and nonspecialists alike provides a comprehensive overview of the many aspects of the celestial mechanics of man-made satellites and spacecraft. A highly rigorous work, it covers basic celestial mechanics; conic sections; coordinate systems; the two-body problem; orbit determination and design; perturbation theory; and much more.

Natural and Artificial Satellite Motion. 042.013.

Orbital Motion. 042.016.

053 Lunar and Planetary Probes and Satellites

053.001 Dollfus, A., ed. **Moon and Planets II.** Amsterdam: North-Holland; 1968. 196p.
C 13 papers.

Tenth Plenary Meeting of COSPAR: A Session of the Joint Open Meeting of Working Groups I, II and V. London, UK: 26-27 Jul 1967. A collection of papers reporting primarily on early spacecraft exploration. Included are results from Zond-3, Luna 13, Surveyor I & III, and Lunar Orbiter. The Moon, Mars, and Venus are discussed.

053.002 Donahue, T. M. Planetary exploration: accomplishments and goals. **Rev. Geophys. Space Phys.** 9: 437-443; 1971. 30 refs, 1967-70.

A short review of observations from space vehicles of Mars and Venus, with attention to practical returns of comparative studies of them and Earth.

053.003 Kondrat'ev, K. Ya. Solar system planets (Mars). **Itogi Nauki Tekh. Ser. Issled. Kosm. Prostranstva** 10: 5-186; 1977. 424 refs. In Russian.

Recent spacecraft studies results are reviewed. Topics: Martian relief and properties of the upper crustal layer; constitution and structural parameters of the atmosphere; atmospheric dust and clouds; general circulation of the atmosphere; upper atmosphere; results of investigations of the atmosphere and Martian surface by Vikings 1 and 2.

053.004 Lukashevich, N. L. Results of recent investigations of Mars and Venus. **Itogi Nauki Tekh. Ser. Issled. Kosm. Prostranstva** 6: 7-64; 1975. 213 refs. In Russian.

A review of recently published results obtained by U.S. and Soviet space probes. The reviews and references are in separate sections for each planet. Russian and Western sources are included in the bibliographies.

053.005 Rasool, I., et al. Rationale for NASA planetary exploration program. **IAU Symp.** 65: 549-561; 1974. no refs.

NASA programs of future solar system exploration for 1973-1990 are presented.

053.006 Runcorn, S. K., ed. Space missions to the Moon and planets. **Highlights Astron.**
C 4(1): 173-241; 1977. 6 papers.

Joint discussion no. 3 of the Sixteenth General Assembly of the IAU. Commissions 16 and 17. Topics: Mercury's magnetic field; lunar magnetism; magnetospheres of Jupiter, Saturn, and Uranus; Venus' surface; cratering of terrestrial planets; etc.

053.007 Strangway, D. W., ed. Geophysical and geochemical exploration of the Moon
C and planets. **Moon** 9: 5-245; 1974. 18 papers, 13 abstracts.

Houston, TX, USA: 10-12 Jan 1973. The purpose was to bring together a variety of disciplines having an interest in the results of the Apollo series of missions as well as other planetary missions which had geophysics or geochemistry as part of their objectives.

053.008 Tucker, R. B. **Viking Lander Imaging Investigation: Picture Catalog of Primary**
A **Mission; Experiment Data Record.** Washington, DC: NASA; 1978. 558p. (NASA RP 1007). 21 refs, 1975-77.

Provides a general reference for the imaging data from the Viking Lander Primary Mission. All of the images returned by the two landers during the primary phase of the Viking Mission are presented. Includes a section on terminology to assist with the interpretation of the listings and the image presentation.

A Catalog of NASA Special Publications. R001.001.

Experimental Astronomy. 051.023.

Geochemical Exploration of the Moon and Planets. 051.001.

Lunar orbital science. *094.001.*

Lunar science: the Apollo legacy. *094.005.*

The magnetic fields of Mercury, Mars, and Moon. *091.045.*

Space Missions to Comets. 102.017.

Surface history of Mercury: a review. *092.004.*

Use for Space Systems for Planetary Geology and Geophysics. 051.009.

054 Artificial Earth Satellites

[No entries; cross reference only].

European Satellite Astrometry. 041.001.

THEORETICAL ASTROPHYSICS

061 General Aspects (Nucleosynthesis, Neutrino Astronomy, etc.)

061.001 Allen, David A. **Infrared: The New Astronomy.** New York: John Wiley & Sons; 1975. 228p. 11 chapters. 288 refs, 1800-1975.

A comprehensive review for the nonspecialist and intelligent layperson. The first book of its kind, it provides a history of the development of infrared astronomy, as well as a look at techniques and prospects for future study. A large portion of the book is spent reviewing infrared emission from a variety of sources: the solar system; cool stars; hot stars; young celestial objects; hidden sources.

061.002 Arnett, W. D. Explosive nucleosynthesis in stars. **Annu. Rev. Astron. Astrophys.** 11: 73-94; 1973. 143 refs, 1957-73.

A review of recent quantitative investigations of explosive nucleosynthesis in stars.

061.003 Arnett, W. D., et al., eds. **Nucleosynthesis.** New York; London: Gordon and
C Breach; 1968. 273p. 24 papers.

Seventh Interdisciplinary Meeting on Topics in Space Physics. Greenbelt, MD, USA: 25-26 Jan 1965. A discussion of the problems of the formation of elements in stars. Topics: abundances of various elements, observational evidence, aspects of stellar evolution, physical processes, role of supernovae, etc.

061.004 Arnould, M. Etat actuel de la theorie de la nucleosynthese. I. La formation des elements jusqu'au groupe du fer. [Actual state of the theory of nucleosynthesis. I. Formation of elements as far as the Fe group.] **Ciel Terre** 88: 233-272; 1972. 169 refs, 1950-72. In French.

Review of the theory of nucleosynthesis and its possible association with the Big Bang, galaxy and stellar formation, and the chemical evolution of the galaxy.

061.005 **Astrophysics and Gravitation.** Bruxelles: Editions de l'Universite de Bruxelles;
C 1974. 494p. 12 papers.

Sixteenth Solvay Conference on Physics. Brussels, Belgium: Sep 1973. A collection of review papers covering pulsars, black holes, quasars, x-ray sources (binary and compact galactic), neutron stars, etc.

061.006 Audouze, Jean; Vauclair, Sylvie. **An Introduction to Nuclear Astrophysics: The Formation and the Evolution of Matter in the Universe.** Dordrecht, Holland; Boston: D. Reidel; 1980. 167p. (Geophys. Astrophys. Monogr., 18). 9 chapters w/refs.

A textbook based on lectures to advanced students, also suitable for educated laypersons, on the topic of nucleosynthesis. Topics: evolution of matter in the universe; chemical composition of the observable universe; thermonuclear reactions and nuclear reactions in stellar interiors; explosive nucleosynthesis in stars; formation of heavy and light elements; chemical evolution of galaxies; etc. An enlarged and updated translation of *L'Astrophysique nucléaire* (Paris: les Presses Universitaires de France; 1972).

061.007 Baity, W. A.; Peterson, L. E., eds. **X-ray Astronomy.** Oxford, UK; New York:
C Pergamon; 1979. 558p. 85 papers.

Symposium of the Twenty-first Plenary Meeting of COSPAR. Innsbruck, Austria: 29 May—10 Jun 1978. (Proceedings). Topics covered include source positions; identification and properties; low-luminosity sources and transients; binary x-ray emitting systems; diffuse galactic phenomena; interstellar medium; supernovae; extragalactic compact and extended sources; transient x- and γ-ray cosmic sources. Includes a catalog and bibliography of galactic x-ray sources.

061.008 Barkat, Z. Neutrino processes in stellar interiors. **Annu. Rev. Astron. Astrophys.** 13: 45-68; 1975. 126 refs, 1941-75.
Discussion of the role of neutrinos in different relevant epochs of stellar evolution.

061.009 Baschek, B. Abundance anomalies in early-type stars. In: Baschek, B.; Kegel, W. H.; Traving, G., eds. **Problems in Stellar Atmospheres and Envelopes.** New York: Springer-Verlag; 1975: 101-148. 190 refs, 1942-75.
A summary "of the quantitative abundance data for some selected groups of peculiar early-type stars," discussed in terms of current ideas of stellar evolution and hypotheses on the Ap anomalies phenomenon.

061.010 Beckman, J. E.; Moorwood, A. F. M. Infrared astronomy. **Rep. Prog. Phys.** 42: 87-157; 1979. 402 refs, 1943-78.
Summarizes progress in understanding of the interaction of stars with the interstellar medium brought about by observations in the infrared spectral range.

061.011 Bernacca, P. L.; Renzini, A., eds. Ultraviolet stellar astronomy. **Mem. Soc.**
C **Astron. Italiana** 47: 311-632; 1976. 15 papers.
Fifth Course of the Advanced School of Astronomy, "E. Majorana" Centre for Scientific Culture. Erice, Sicily: 26 May—8 Jun 1975. Topics include data obtained using space telescopes, a description of future projects, and an overview of some of the celestial bodies studied by UV methods.

061.012 Biémont, E.; Grevesse, N. f-values and abundances of the elements in the Sun and stars. **Phys. Scr.** 16: 39-47; 1977. 71 refs, 1957-77.
Survey of atomic and molecular f-values presently needed for solar abundance determinations, with stellar ones being treated more briefly.

061.013 Blecha, A.; Maeder, A., eds. **Extragalactic High Energy Astrophysics.** Sauverny,
C Switzerland: Geneva Observatory; 1979. 247p. 3 papers.
Ninth Advanced Course of the Swiss Society of Astronomy and Astrophysics. Saas-Fee, Switzerland: 26-31 Mar 1979. Edited transcripts of three graduate-level lectures: 1) Radiogalaxies and the origin of the high-energy activity in galactic nuclei (F. Pacini, pp. 1-72); 2) Phenomenology of extragalactic high-energy sources (C. E. Ryter, pp. 73-152); 3) Lectures on high energy extragalactic astrophysics (P. A. Strittmatter, pp. 153-247).

061.014 Bless, R. C.; Code, A. D. Ultraviolet astronomy. **Annu. Rev. Astron. Astrophys.** 10: 197-226; 1972. 65 refs, 1964-72.
Outline of recent developments in ultraviolet astronomy, especially with the OAO-2 satellite data.

061.015 Borghesi, A.; Bussoletti, E.; Blanco, A., eds. Proceedings of the conference
C on infrared astronomy. **Mem. Soc. Astron. Italiana** 49: 1-296; 1978. 19 papers.
Lecce, Italy: 22-24 Sep 1977. A review of current research and theory.

061.016 Borgman, J. Infrared astronomy. In: Mavridis, L. N., ed. **Stars and the Milky Way System.** Berlin: Springer-Verlag; 1974: 188-208. 78 refs, 1947-73.

A review paper addressing selected topics: techniques and practical problems; distribution of IR sources; radiation from stars; other galactic sources such as H II regions, planetary nebulae and the galactic center; extragalactic sources.

061.017 Boyd, R. L. F. The Bakerian lecture 1978. Cosmic exploration by x-rays. **Proc. R. Soc. London, Ser. A.** 366: 1-21; 1979. 60 refs, 1930-79.
Review of x-ray astronomy since the conception and employment of reflexion optics.

061.018 Bradt, H.; Giacconi, R., eds. **X- and Gamma-ray Astronomy.** Dordrecht, Hol-
C land; Boston: D. Reidel; 1973. 323p. (IAU Symp., 55). 23 papers.
Madrid, Spain: 11-13 May 1972. Cosponsored by COSPAR. Topics: galactic sources, theoretical models for compact sources, extragalactic sources, and the interstellar medium and soft x-ray background.

061.019 Brancazio, P. J.; Cameron, A. G. W., eds. **Infrared Astronomy.** New York:
C Gordon and Breach; 1968. 248p. 18 papers.
Greenbelt, MD, USA: 1-2 Apr 1966. A series of short pieces describing data from ground-based and rocket observations, and reviewing current understanding of various aspects of the field. Emphasis is on the observation of stars and the planets.

061.020 Brecher, K.; Setti, G., eds. **High Energy Astrophysics and Its Relation to Ele-**
C **mentary Particle Physics.** Cambridge, MA: MIT Press; 1974. 591p. 13 papers.
NATO Advanced Study Institute. Erice, Sicily: 16 Jun—6 Jul 1972. Lectures reviewing the observations, theories, and problems of high energy astrophysics which might lend themselves to new physical interpretations. Topics: quasistellar sources; cosmology; x-ray astronomy; weak and strong interactions; etc.

061.021 Canuto, V. Equation of state at ultra-high densities. Part 1. **Annu. Rev. Astron. Astrophys.** 12: 167-214; 1974. 50 refs, 1949-74.
Primarily interested in the physics involved in the construction of an equation of state, which is the fundamental ingredient for deriving parameters like mass, radius, moment of inertia, pulsation properties, and so on. Part 1 is concerned with densities within the range $10^6 \leqq \rho \leqq 5 \times 10^{14}$ g cm^{-3}.

061.022 Canuto, V. Equation of state at ultra-high densities. Part 2. **Annu. Rev. Astron. Astrophys.** 13: 335-380; 1975. 78 refs, 1934-75.
Review of the work done so far in the region $\rho \geqq 2 \times 10^{14}$ g cm^{-3}. Begins with a review of the work concerning the appearance of hyperons at densities higher than nuclear density and their influence on the equation of state. This is followed by a review of the work concerning the possible existence of a solid neutron core.

061.023 Cavallo, G. Recent advances in short-time-constant astrophysics. **Nuovo Cimento, Riv.** 3: 205-232; 1973. 141 refs, 1939-73.
Broad view of a large number of recent advances in the areas of periodic phenomena and sporadic phenomena.

061.024 Chretien, M.; Deser, S.; Goldstein, J., eds. **Astrophysics and General Relativity.**
C New York: Gordon and Breach; 1969, 1971. 2v. v. 1: 300p., 4 papers; v. 2: 383p. 5 papers.
Eleventh Brandeis Summer Institute in Theoretical Physics. Waltham, MA, USA: 17 Jun—26 Jul 1968. A series of lectures for graduate students on the following topics: statistical mechanics of stellar systems; physics of the interstellar and intergalactic medium; gravitational collapse; extragalactic radio astronomy; observational data on the extremes of stellar evolution; physics of cosmic x-ray, gamma-ray, and particle sources; cosmogonic processes; density wave theory of spiral structure; kinetic theory and cosmology.

061.025 Clark, G. The present state of gamma-ray astronomy. **IAU Symp.** 41: 3-13; 1971. 30 refs, 1948-70.

The current state is reviewed in terms of the most recent experimental results that define, in either measured fluxes or upper limits, the cosmic photon spectrum in the energy range from 0.5 MeV to 10^{16} eV. Methods of research developed during the last 10 years are discussed, and specific results are cited.

061.026 Collins, George W., II. **The Virial Theorem in Stellar Astrophysics.** Tucson, AZ: Pachart Publishing; 1978. 135p. (Astron. Astrophys. Ser., 7). 4 chapters w/refs.

A highly technical work providing an introduction to the subject. Topics: development of the virial theorem; contemporary aspects of the virial theorem; the variational form of the virial theorem; some applications of the virial theorem. The latter chapter includes applications to stability of white dwarfs and neutron stars.

061.027 Colloque astrophysique des hautes energie. [Colloquium on high energy astro-
C physics.] **J. Physique** 36: C5-137—C5-199; 1975. 15 papers and abstracts.

Dijon, France: 30 Jun—4 Jul 1975.

061.028 Culhane, J. L. Extragalactic x-ray astronomy. **Q. J. R. Astron. Soc.** 19: 1-37; 1978. 108 refs, 1958-77.

Description of the classification and identification of extragalactic x-ray sources. Extended sources associated with clusters of galaxies and the compact sources located in the nuclei of active galaxies are treated in some detail.

061.029 Culhane, J. L. X-ray astronomy. **Vistas Astron.** 19: 1-67; 1975. 283 refs, 1939-75.

A review covering the production of x-radiation and its interaction with matter, need for special instrumentation, study of solar x-rays, the x-ray sources in our galaxy and the universe as a whole, and the diffuse x-ray background.

061.030 Dailey, Carroll; Johnson, Wendell, eds. **HEAO Science Symposium.** Hunts-
C ville, AL: NASA; 1979. 459p. (NASA CP-2113). 25 papers.

Marshall Space Flight Center, AL, USA: 8-9 May 1979. Reports of results obtained from the High Energy Astronomy Observatories, with an emphasis on x-ray astronomy studies. Papers cover data from HEAO-1 principal investigator programs (including a report on the progress of the HEAO A-1 Catalog); results from the HEAO-1 guest observer program; results from HEAO-2 non-imaging and imaging instrumentation.

061.031 Danziger, I. J. The cosmic abundance of helium. **Annu. Rev. Astron. Astrophys.** 8: 161-178; 1970. 136 refs, 1938-70.

An overview of observation and theory leading to the present knowledge of the extent of helium in the universe.

061.032 Davis, R., ed. Ultraviolet astronomy—new results from recent space
C experiments. **Highlights Astron.** 5: 227-329; 1980. 11 papers.

Topics include UV observations of x-ray sources, planetary nebulae, hot subluminous stars, cool stars, peculiar stars, novae, interstellar clouds, normal galaxies, Seyfert galaxies, along with some theoretical considerations.

061.033 de Graaf, T. Neutrinos in the universe. **Vistas Astron.** 15: 161-181; 1973. 68 refs, 1922-70.

Review of research relating to the relations between the properties of the neutrino and the behavior of the universe as a whole.

061.034 A discussion on cosmic x-ray astronomy. **Proc. R. Soc. London, Ser. A.** 313:
C 299-402; 1969. 8 papers.

Held 27 Nov 1968. A consideration of ground-based observations, theoretical aspects, physical characteristics, stellar x-ray sources, current developments, future requirements, and a specific source, Sco X-1.

061.035 Fazio, Giovanni G., ed. **Infrared and Submillimeter Astronomy.** Dordrecht,
C Holland; Boston: D. Reidel; 1977. 226p. (Astrophys. Space Sci. Libr., 63). 15 papers.

Philadelphia, PA, USA: 8-10 Jun 1976. (Proceedings). Reports of recent results obtained by observations from aircraft, balloons, rockets, satellites, and space probes. Topics: the Sun; the solar system; the galactic center; galactic and extragalactic sources; the cosmic background radiation; instrumentation; observational techniques.

061.036 Fichtel, C. E. Gamma-ray astrophysics. **Space Sci. Rev.** 20: 191-234; 1977. 159 refs, 1948-77.

Review of instrumentation used to detect gamma-rays, followed by a discussion of the various types of gamma-ray emitters.

061.037 Gallino, R.; Masani, A. Galactic neutrino sources and experimental neutrino astronomy. **Nuovo Cimento, Riv.** 6: 495-528; 1976. 101 refs, 1950-76.

Review of experimental neutrino astronomy and the theoretical expectations of cosmic $\nu-\bar{\nu}$ bursts.

061.038 Gal'per, A. M.; Kirillov-Ugryunov, V. G.; Luchkov, B. I. Observational gamma ray astronomy. **Soviet Phys. Usp.** 17: 186-198; 1974. Transl. of *Usp. Fiz. Nauk.* 112: 491-515; 1974. 128 refs, 1964-74.

A historical review of the observational data on detection of cosmic gamma rays.

061.039 Gal'per, A. M., et al. The study of cosmic γ rays. **Soviet Phys. Usp.** 14: 630-654; 1972. Transl. of *Usp. Fiz. Nauk.* 105: 209-250; 1971. 299 refs, 1953-71.

This review systematizes the methods, experimental data, and theoretical studies of cosmic gamma rays.

061.040 Giacconi, R. Survey on new techniques for x-ray astronomy. **IAU Symp.** 41: 104-133; 1971. 35 refs, 1948-71.

Review of the status of knowledge of x-ray astronomy in the eight years since the discovery of cosmic x-ray sources. Summary of the instrumentation is given from the point of view of the developments and improvements being carried out.

061.041 High energy astrophysics. **IAU Trans. Rep. Astron.** 17A(3): 223-228; 1979. 56 refs, 1975-79.

An overview of research activity worldwide, with an emphasis on the previous three years, as reported by IAU Commission 48 (High Energy Astrophysics). Topics: x-ray astronomy; gamma-ray astronomy; cosmic rays; particle astronomy; etc. Additional reports: 15A: 737-755; 1973. 165 refs, 1968-73. 16A(3): 221-222; 1976. 31 refs, 1973-75.

061.042 Hoover, R. B.; Thomas, R. J.; Underwood, J. H. Advances in solar and cosmic x-ray astronomy: a survey of experimental techniques and observational results. **Adv. Space Sci. Technol.** 11: 1-214; 1972. 736 refs, 1937-72.

Review of x-ray observations and the discoveries therefrom, particularly with satellite observatories. Appendix includes review of instrumentation.

061.043 Huber, Martin C. E.; Nussbaumer, Harry, eds. **Atomic and Molecular**
C **Processes in Astrophysics.** Sauverny, Switzerland: Geneva Observatory; 1975. 308p. 3 papers.

Fifth Advanced Course of the Swiss Society of Astronomy and Astrophysics. Valais, Switzerland: 17-22 Mar 1975. This volume contains three lectures of interest to graduate students: 1) Molecular processes in interstellar clouds (A. Dalgarno, pp. 1-98); 2) Introduction to collision theory and to some astrophysical applications (F. Masnou-Seeuws, pp. 99-184); 3) The contribution of laboratory measurements to the interpretation of astronomical spectra (R. W. P. McWhirter, pp. 185-308).

061.044 Kharadze, E. K., ed. **Stars and Galaxies from Observational Points of View.**
C Tbilisi, USSR: Izdatel'stvo Metsniereba; 1976. 534p. 36 papers. English w/Russian abstracts.

Third European Astronomical Meeting. Tbilisi, USSR: 1-5 Jul 1975. (Proceedings). A wide variety of topics are covered in this volume, which concentrates on reports of research in Europe. Topics: spectral classification; stellar and galactic evolution; supernovae; galactic x-ray sources; missing mass in galaxies; etc.

061.045 Kniffen, D. Gamma ray astronomy. **IEEE Trans. Nucl. Sci.** 22: 45-53; 1975. 63 refs, 1948-75.

A survey of instruments developed for research, along with a brief summary of observational results, including studies of galactic gamma ray emission, the diffuse gamma radiation, and localized gamma ray sources.

061.046 Kuchowicz, B. **The Cosmic ν.** Warsaw: Nuclear Energy Information Center; 1972. 130p. (NEIC-RR-47). 7 chapters. 618 refs, 1965-72.

A detailed review of all the major developments in cosmic neutrino research, from astrophysical to cosmic-ray and gravitational aspects. Topics: solar neutrinos; neutrinos, Universal Fermi Interaction, and stellar evolution; extrasolar sources of neutrino radiation; cosmological neutrinos; neutrinos from cosmic radiation; various theoretical developments.

061.047 Labuhn, F.; Lüst, R., eds. **New Techniques in Space Astronomy.** Dordrecht,
C Holland; Boston: D. Reidel; 1971. 419p. (IAU Symp., 41). 33 papers, 32 abstracts.

Munich, Germany: 10-14 Aug 1970. Topics include gamma-ray astronomy, x-ray astronomy, UV astronomy, optical systems, detecting systems, calibration, and radio astronomy.

061.048 Lebovitz, N. R.; Reid, W. H.; Vandervoort, P. O., eds. **Theoretical Principles**
C **in Astrophysics and Relativity.** Chicago: University of Chicago Press; 1978. 258p. 10 papers.

Chicago, IL, USA: 27-29 May 1975. A conference volume dedicated to S. Chandrasekhar containing reviews of important general astrophysical and relativistic problems which may be interconnected. Sample topics are stellar structure, rotation of stars and stellar systems, radiative transfer, relativistic astrophysics, and singularities of space-time.

061.049 McDonald, Frank B.; Fichtel, Carl E., eds. **High Energy Particles and Quanta in Astrophysics.** Cambridge, MA: MIT Press; 1974. 476p. 9 chapters w/refs.

An advanced text covering the many subfields of high energy astrophysics, written by specialists in the field. Topics: cosmic ray nuclei; cosmic electrons; solar modulation; solar particles; radio astronomy; galactic x-ray sources; gamma-ray astronomy; etc.

061.050 Manno, V.; Ring, J., eds. **Infrared Detection Techniques for Space Research.**
C Dordrecht, Holland: D. Reidel; 1972. 344p. (Astrophys. Space Sci. Libr., 30). 35 papers.

Fifth ESLAB/ESRIN Symposium. Noordwijk, The Netherlands: 8-11 Jun 1971. (Proceedings). Reviewing the present and future methods of infrared detection, this volume examines existing instrumentation and considers the improvements necessary to make satellite observations. After a paper reviewing the results of infrared space astronomy through 1971, the book is divided into five parts: systems for space research;

detectors; cryogenics; filters; interferometers. The use of liquid helium to cool detectors is discussed at some length.

061.051 Massey, H.; Boyd, R. L. F.; Willmore, A. P., eds. Some recent results in x-ray
C astronomy. **Proc. R. Soc. London, Ser. A.** 366: 277-489; 1979. 14 papers.

Discussion held 25-26 Apr 1978. The main purpose was to review recent results in x-ray astronomy in the United Kingdom and to relate these to their background in current astronomy.

061.052 Massey, H.; Ring, J., eds. A discussion on infrared astronomy. **Philos. Trans.**
C **R. Soc. London, Ser. A.** 264: 107-320; 1969. 27 papers.

London, UK: 1-2 May 1966. Covers Moon and planets, stars, nebulae and extragalactic objects, and infrared techniques.

061.053 Massey, H.; Wilson, R., eds. A discussion on astronomy in the ultraviolet.
C **Philos. Trans. R. Soc. London, Ser. A.** 279: 297-485; 1975. 23 papers.

Meeting held 9-10 Apr 1974. Abstracts and short papers dealing with results from satellites and terrestrial observations of sources in the ultraviolet.

061.054 Mavridis, L. N., ed. **Stars and the Milky Way System.** Berlin: Springer-Verlag;
C 1974. 368p. 39 papers.

First European Astronomical Meeting. Athens, Greece: 4-9 Sep 1972. (Proceedings, v.2). Reports of research by European astronomers. Topics: variable stars; binary stars; stellar kinematics; interstellar matter; galactic center; chemical evolution of the galaxy; infrared astronomy; instruments; celestial mechanics; galactic dynamics.

061.055 Neugebauer, G.; Becklin, E.; Hyland, A. R. Infrared sources of radiation. **Annu. Rev. Astron. Astrophys.** 9: 67-102; 1971. 169 refs, 1928-71.

A review of research including infrared excesses in stars, infrared associated with selected galactic objects, infrared emission from the galactic nucleus, and extragalactic sources.

061.056 Ögelman, H.; Wayland, J. R., eds. **Introduction to Experimental Techniques of**
C **High-Energy Astrophysics.** Washington, DC: NASA; 1970. 255p. (NASA SP-243). 6 papers.

Edited lectures from a graduate course at the University of Maryland in 1969. Topics: measurements of non-relativistic charged particles of extra-terrestrial origins; x-ray detectors; high-energy photon detectors; extensive air showers; radio-astronomical observations of high-energy particles.

061.057 Ögelman, H.; Wayland, J. R., eds. **Lectures in High Energy Astrophysics.**
C Washington, DC: NASA; 1969. 165p. (NASA SP-199). 8 papers.

An expanded and edited version of lecture notes accompanying an introductory graduate-level course at the University of Maryland in 1968. Topics: cosmic rays (origin, propagation, photons, electrons); relativistic stochastic processes; solar cosmic rays; interplanetary medium (dust, gas, and magnetic fields).

061.058 Pagel, B. E. J. Stellar and solar abundances. In: Cameron, A. G. W., ed. **Cosmochemistry.** Dordrecht, Holland: D. Reidel; 1973: 1-21. 190 refs, 1942-73.

A review of recent spectroscopic results with reference to "a) standard abundance distribution; b) abundance peculiarities related to stellar evolution; c) population effects that may be related to the evolution of our galaxy." Includes an informative table of spectral data for solar and solar system abundances of 63 elements.

061.059 Reeves, H. On the origin of the light elements. **Annu. Rev. Astron. Astrophys.** 12: 437-469; 1974. 131 refs, 1938-74.

Discussion of the origin of all stable nuclei between hydrogen and carbon. Advances the theory of nucleosynthesis in the Big Bang for the origin of the lighter elements and spallation induced by galactic cosmic rays in interstellar space for the heavier ones.

061.060 Reeves, H.; Meyer, J.-P. Cosmic-ray nucleosynthesis and the infall rate of extragalactic matter in the solar neighborhood. **Astrophys. J.** 226: 613-631; 1978. 163 refs, 1959-78.

An up-to-date discussion of the galactic abundances of lithium, beryllium, and boron.

061.061 Schramm, D. N. Nucleo-cosmochronology. **Annu. Rev. Astron. Astrophys.** 12: 383-406; 1974. 86 refs, 1929-74.

Discussion of the use of the observed or implied relative abundance of radioactive nuclei to determine the time scales for the nucleosynthesis of these nuclei in order to derive time scales for the age of the galaxy.

061.062 Schramm, D. N.; Arnett, W. D., eds. **Explosive Nucleosynthesis.** Austin, TX:
C University of Texas Press; 1973. 301p. 22 papers.

Austin, TX, USA: 2-3 Apr 1973. (Proceedings). A summary of the processes involved in the formation of atomic nuclei in supernovae. Topics: abundances and chemical evolution of galaxies; explosive processes; aspects of pre-carbon burning; the carbon detonation model; massive stars, deuterium and gamma rays; nuclear reaction rates. Tables of data, including abundances of nuclei, are presented.

061.063 Schramm, D. N.; Wagoner, R. V. Element production in the early universe. **Annu. Rev. Nucl. Sci.** 27: 37-74; 1977. 143 refs, 1922-77.

A review to explore the nature and consequences of the process of Big Bang nucleosynthesis.

061.064 Setti, G.; Fazio, G. G., eds. **Infrared Astronomy.** Dordrecht, Holland; Boston:
C D. Reidel; 1978. 353p. (NATO Adv. Stud. Inst. Ser., Ser. C., 38). 16 papers.

Fourth International School of Astrophysics. (Proceedings). Erice, Sicily: 9-20 Jul 1977. A series of basic lectures covering infrared research and related topics. Two basic themes are addressed: 1) galactic infrared sources and their role in star formation; the nature of the interstellar medium and galactic structure; 2) the interpretation of infrared, optical, and radio observations of extra-galactic sources and their role in the origin and structure of the universe.

061.065 Shen, B. S. P.; Merker, M., eds. **Spallation Nuclear Reactions and Their Appli-**
C **cations.** Dordrecht, Holland; Boston: D. Reidel; 1976. 235p. (Astrophys. Space Sci. Libr., 59). 10 papers.

Philadelphia, PA, USA: May 1975. Revised and expanded invited papers dealing with astrophysics and radiotherapy applications. Selected topics include experimental results (review paper covering 1967-75); spallation reactions; calculations; cosmic ray applications; radioactive isotopes on the Moon; etc.

061.066 Sobolev, V. V. **Course in Theoretical Astrophysics.** Washington, DC: NASA; 1967. 493p. (NASA TT F-531; N70-15814). 8 chapters w/refs.

An advanced, rigorous text based on 20 years of lectures at the University of Leningrad. Topics: stellar photospheres; stellar and solar atmospheres; planetary atmospheres; gaseous nebulae; variable stars; interstellar space; internal structure of stars; etc. Translation of *Kurs teoreticheskoĭ astrofiziki* (Moscow: Nauka; 1967).

061.067 Strugalski, Z. Podstawy eksperymentalne astronomii gamma. [Experimental bases of gamma astronomy]. **Postepy Astron.** 26: 73-93; 1978. 87 refs, 1941-75. In Polish.

Survey of the experimental basis of gamma-ray astronomy. Detectors, counters, and telescopes used are presented.

061.068 Swihart, Thomas L. **Astrophysics and Stellar Astronomy.** New York: John Wiley & Sons; 1968. 299p. (Space Sci. Text Ser.). 4 chapters. 90 refs, 1937-67.

An intermediate text for university science majors with a knowledge of physics and calculus. Topics: basic physics of stellar processes; stellar positions and magnitudes; binaries and variables; astrophysics (spectra, continuous radiation, stellar interiors, stellar evolution, etc.); galaxies and cosmology.

061.069 Tananbaum, H.; Tucker, W. H. Compact x-ray sources. In: Giacconi, Riccardo; Gursky, Herbert, eds. **X-ray Astronomy.** Dordrecht, Holland: D. Reidel; 1974: 207-266. 149 refs, 1944-74.

A review of x-ray sources which are typically members of binary star systems. Besides a general look at theory, the authors present detailed accounts of observations of 10 such objects or systems.

061.070 Tayler, R. J. **The Origin of the Elements.** London: Wykeham Publications; distr., New York: Springer-Verlag; 1972. 169p. (Wykeham Sci. Ser., 23). 9 chapters. 16 refs, no dates.

A brief text for upper-level science undergraduates, providing an introduction to the evolution of the chemical composition of the universe. Topics: chemical composition of the stars, nebulae, and the Sun; elemental abundances in the solar system and cosmic rays; etc. Also presented are the equations and nuclear reactions involved in the formation of the elements in stars.

061.071 Trombka, J., et al. Gamma-ray astrophysics: a new look at the universe. **Science** 202: 933-938; 1979. 58 refs, 1971-77.

A good, brief review of what has been and can be learned using gamma-ray astronomy.

061.072 Truran, J. W. Theories of nucleosynthesis. In: Cameron, A. G. W., ed. **Cosmochemistry.** Dordrecht, Holland: D. Reidel; 1973: 23-49. 116 refs, 1939-73.

A review of current theories, including cosmological nucleosynthesis; nucleosynthesis in supermassive stars; nucleosynthesis in stellar evolution; and nova and supernova element formation. Each process is summarized, with an emphasis on the various elements which are by-products of each type of nucleosynthesis.

061.073 Tyson, J. A.; Giffard, R. P. Gravitational-wave astronomy. **Annu. Rev. Astron. Astrophys.** 16: 521-554; 1978. 123 refs, 1959-78.

An examination of the present observational situation plus a brief review of present calculations of theoretical models for gravitational wave emission.

061.074 Usov, V. V. Galactic and intergalactic astronomy (high energy astrophysics). **Itogi Nauki Tekh. Ser. Issled. Kosm. Prostranstva** 9: 1-159; 1977. 318 refs. In Russian.

A comprehensive review of recent studies in three areas: pulsars; x-ray point sources; and cosmic gamma-ray bursts.

061.075 Webber, W. R. X-ray astronomy—1968 vintage. **Proc. Astron. Soc. Australia** 1: 160-164; 1968. 23 refs, 1963-68.

A review of x-ray astronomy, giving details of the 22 confirmed x-ray sources. Observations, theories, and sources are summarized.

061.076 Weekes, Trevor C. **High-Energy Astrophysics.** London: Chapman and Hall; 1969. 209p. 12 chapters. 136 refs, 1939-69.

A brief survey for the nonspecialist and advanced layperson covering three types of high-energy topics: high energies relative to the rest mass of the object (supernovae, radio galaxies, QSOs); individual quanta possessing high energies (cosmic rays, x-rays, gamma rays); possible large cosmic energy densities (neutrinos, microwaves). Astrophysical theories for all these phenomena are presented without too much detail or mathematics.

061.077 Weiss, N. O. The dynamo problem. **Q. J. R. Astron. Soc.** 12: 432-446; 1971. 79 refs, 1919-71.

Review of work on the symmetry properties necessary for a motion to maintain a planetary or stellar magnetic field. Includes a summary of observations.

061.078 Westerlund, Bengt E., ed. **Stars and Star Systems.** Dordrecht, Holland; Boston:
C D. Reidel; 1979. 264p. (Astrophys. Space Sci. Libr., 75). 17 papers.

Fourth European Regional Meeting in Astronomy. Uppsala, Sweden: 7-12 Aug 1978. (Proceedings). A variety of papers on galaxies, high-energy astrophysics, stars, interstellar processes, and instrumentation.

061.079 Weymann, R. J., et al. **Lecture Notes on Introductory Theoretical Astrophysics.**
C Tucson, AZ: Pachart Publishing; 1976. 230p. (Astron. Astrophys. Ser.). 6 papers. 146 refs, 1924-77.

Papers based on lecture notes from a graduate course at the University of Arizona. Topics: radiative transfer and stellar atmospheres; stellar interiors; gaseous nebulae; relativity; synchrotron spectra; physics and astronomy of celestial x-ray sources.

061.080 Wilson, R.; Boksenberg, A. Ultraviolet astronomy. **Annu. Rev. Astron. Astrophys.** 7: 421-472; 1969. 218 refs, 1929-68.

Discussion of the types of objects being observed in the ultraviolet, of observational techniques, and of some important data being gathered.

Astronomical observations from outside the terrestrial atmosphere. *051.002.*

Astrophysical Concepts. R011.013.

Astrophysics. 065.034.

Atomic Physics and Astrophysics. 022.016.

Atoms and Molecules in Astrophysics. 022.013.

CNO Isotopes in Astrophysics. 022.006.

Conference on solar system plasmas and x-ray astronomy. *091.013.*

Detectors for infrared astronomy. *032.006.*

Gamma-Ray Astronomy: Nuclear Transition Region. 142.501.

Gamma-ray Astrophysics. 142.507.

General-relativistic astrophysics. *066.045.*

The impact of ultraviolet observations on spectral classification. 114.022.

Infrared techniques. *032.506.*

062 Hydrodynamics, Magnetohydrodynamics, Plasma

062.001 C Axford, W. I., ed. Very hot plasmas in circumstellar, interstellar, and intergalactic space. **Highlights Astron.** 5: 331-428; 1980. 10 papers.

Topics include solar plasma, plasma in the solar system, the interstellar medium, supernova remnants, intergalactic medium, clusters of galaxies, and more.

062.002 Boyd, R. L. F. **Space Physics: The Study of Plasmas in Space.** London: Clarendon Press, Oxford University Press; 1974. 100p. (Oxford Phys. Ser., 7). 4 chapters. 21 refs, 1931-73.

An intermediate undergraduate text which examines four different regions of cosmic plasma, discussing their basic properties, and showing how space techniques have and are being used in their study. Topics: ionospheres; magnetospheres; the solar atmosphere; cosmic x-ray astronomy.

062.003 C **Conference on Ultraviolet and X-ray Spectroscopy of Astrophysical and Laboratory Plasmas.** 5th. London: Imperial College, Blackett Laboratory; 1977[?]. unpaged. (IAU Colloq., 43). Abstracts only.

London, UK: 4-7 Jul 1977. Program and abstracts only. Papers never published.

062.004 Cowling, T. G. **Magnetohydrodynamics.** London, UK: Adam Hilger; distr., New York: Crane, Russak & Co.; 1976. 136p. (Monogr. Astron. Subj.). 6 chapters. 97 refs, 1919-74.

A text discussing basic phenomena and theory, as well as current research efforts. Primarily concerned with continuum theory, it considers particle aspects "only as indicating the possible generalizations, as well as limitations, of the continuum approach." Topics: general principles; magnetohydrostatics; waves and oscillations; dynamics instability and convection; dynamo theories; rarefied plasmas.

062.005 Egeland, A.; Holter, Ø.; Omholt, A., eds. **Cosmical Geophysics.** Oslo: Universitetsforlaget; 1973. 360p. (Scandinavian Univ. Books). 23 chapters w/refs.

An advanced text for seniors and graduate students covering a wide variety of subjects concerned with "a magnetized plasma of solar origin interacting with the Earth's magnetic field and atmosphere." Each chapter is written by experts in the field. Topics: solar radiation and particle emission; the geomagnetic field; the neutral atmosphere; solar wind; the magnetosphere; auroral particles; auroral morphology; cosmic rays; etc.

062.006 C Fälthammar, C.-G. Plasma phenomena in astrophysics. In: Davenport, P. A., ed. **Phenomena in Ionized Gases.** Oxford, UK: Donald Parsons; 1971: 1-35. 88 refs, 1950-71.

Oxford, UK: 13-18 Sep 1971. A review article describing the cosmical plasma, its distribution and classification, ionospheric processes, wave phenomena in the magnetosphere, and more. The most familiar and spectacular types of cosmical plasma, the aurorae and magnetic storms, are also described, as is the interaction of interplanetary plasma (solar wind) with the Earth's magnetosphere.

062.007 Greenstadt, E. W.; Fredricks, R. W. Shock systems in collisionless space plasmas. In: Lanzerotti, Louis J.; Kennel, Charles F.; Parker, E. N., eds. **Solar System Plasma Physics. Volume III.** Amsterdam: North-Holland; 1979: 3-43. 104 refs, 1950-78.

An overview of current "knowledge of shock structure from a plasma physics point of view, based on direct observations in space." The paper includes a description of the Earth's bow shock, planetary shocks, and astrogenic shocks, as well as shock theory and methods of observation.

062.008 Griem, Hans R. **Spectral Line Broadening by Plasmas.** New York; London: Academic; 1974. 408p. (Pure Appl. Phys., 39). 4 chapters. 358 refs, 1906-1974.

An advanced text dealing with theory, experiment, and application. The latter includes two topics of interest to astrophysicists: stellar atmospheres and radio-frequency lines. Seven appendices present a variety of numerical data including Stark profiles for hydrogen and ionized helium lines, and Stark broadening parameters and profiles for isolated neutral atom lines.

062.009 C Habing, H. J. **Interstellar Gas Dynamics.** Dordrecht, Holland: D. Reidel; New York: Springer-Verlag; 1970. 388p. (IAU Symp., 39). 15 papers.

Sixth Symposium on Cosmical Gas Dynamics. Yalta, The Crimea, USSR: 8-18 Sep 1969. Papers cover the description and general theory of the interstellar medium and the interaction of stars and the interstellar medium. Cosponsored by IUTAM (International Union of Theoretical and Applied Mechanics).

062.010 C Hultqvist, Bengt; Stenflo, Lennart, eds. **Physics of the Hot Plasma in the Magnetosphere.** New York: Plenum Press; 1975. 369p. 16 papers.

Thirtieth Nobel Symposium. Kiruna, Sweden: 2-4 Apr 1975. (Proceedings). With an emphasis on considering unsolved problems, the following topics are addressed: entry of the hot plasma into the magnetosphere; distribution within the magnetosphere and large scale dynamical processes; acceleration/heating and instabilities/turbulence; interaction with the ionosphere.

062.011 The interplanetary plasma and the heliosphere. **IAU Trans. Rep. Astron.** 17A(1): 199-213; 1979. 102 refs, 1970-79.

An overview of research activity worldwide, with an emphasis on the previous three years, as reported by IAU Commission 49 (The Interplanetary Plasma and the Heliosphere). Additional report: 16A(1): 175-187; 1976. 58 refs, 1963-75.

062.012 Kaplan, S. A.; Tsytovich, V. N. **Plasma Astrophysics.** Oxford, UK; New York: Pergamon Press; 1973. 299p. (Int. Ser. Monogr. Nat. Philos., 59). 4 chapters. 296 refs, 1946-72.

A rigorous look at the physical processes in various types of astrophysical plasmas. Chapters: 1) the physics of plasma turbulence; 2) sporadic radio-emission of the Sun; 3) galactic nuclei, radio-galaxies, quasars; 4) pulsar emission. Includes tables of averaged probabilities for plasma processes.

062.013 Lanzerotti, Louis J.; Kennel, Charles F.; Parker, E. N., eds. **Solar System Plasma Physics. Volume III. Solar System Plasma Processes.** Amsterdam; New York: North-Holland; 1979. 371p. 10 papers.

A collection of review papers on recent research and current knowledge of the field. Topics: shock systems in collisionless space plasmas; magnetic field reconnection; hydromagnetic waves; magnetospheric plasma waves; etc.

062.014 Mestel, L.; Weiss, N. O. **Magnetohydrodynamics.** Sauverny, Switzerland: Geneva Observatory; 1974. 248p. 3 papers.

Fourth Advanced Course of the Swiss Society of Astronomy and Astrophysics. Saas-Fee, Switzerland: 1-6 Apr 1974. Presents three edited lectures of interest to advanced students: 1) Introduction to magnetohydrodynamics (N. O. Weiss, pp. 1-36); 2) Applications to magnetic stars, cosmical gas dynamics, and pulsars (L. Mestel, pp. 37-182); 3) Dynamo maintenance of magnetic fields in stars (N. O. Weiss, pp. 183-248).

062.015 Myerscough, V. P.; Peach, G. Atomic processes in astrophysical plasmas. **Case Stud. At. Collision Phys.** 2: 293-397; 1972. 343 refs, 1919-71.

A review of stellar spectral analysis and the interpretation of the resultant data in terms of physical processes in stellar atmospheres. Predicted and observed spectra are compared for different types of stars. Topics: stellar atmospheres; the continuous spectrum; the line spectrum; collisional excitation and ionization; interpretation of observations; problems for further study.

062.016 Palmadesso, Peter J.; Papadopoulos, Konstantinos, eds. **Wave Instabilities in**
C **Space Plasmas.** Dordrecht, Holland; Boston: D. Reidel; 1979. 309p. (Astrophys. Space Sci. Libr., 74). 20 papers.

Symposium from the XIXth URSI General Assembly. Helsinki, Finland: 31 Jul—8 Aug 1978. (Proceedings). Review papers and reports of observations covering 1) natural noise in space (including ELF and VLF waves); 2) plasma turbulence; 3) nonlinear effects; 4) ionospheric f-region irregularities.

062.017 Parker, E. N. **Cosmical Magnetic Fields: Their Origin and Their Activity.** Oxford, UK: Clarendon Press, Oxford University Press; 1979. 841p. 23 chapters w/refs.

Guided by the observed behavior of magnetic fields in space (especially of the Earth and Sun), this advanced text presents the study of physics of large-scale magnetic fields in the fluids of high electrical conductivity. Topics: magnetic field stress and energy; magnetic equilibrium; the isolated and twisted flux tubes; the isolation and concentration of magnetic flux tubes; submerged magnetic fields; the generation of large-scale fields in turbulent fluids; magnetic fields of planets, the Sun, stars, and the galaxy; etc.

062.018 Plasmas and magnetohydrodynamics in astrophysics. **IAU Trans. Rep. Astron.** 15A: 669-670; 1973. 44 refs, 1969-72.

An overview of research activity worldwide, with an emphasis on the previous three years, as reported by IAU Commission 43 (Plasmas and Magnetohydrodynamics in Astrophysics). Additional report: 14A: 515-524; 1970. 12 refs, 1966-70.

062.019 Schindler, K., ed. **Cosmic Plasma Physics.** New York; London: Plenum Press;
C 1972. 369p. 45 papers.

Conference on Cosmic Plasma Physics. Frascati, Italy: 20-24 Sep 1971. (Proceedings). An overview of the evidence for astronomical plasmas, as well as a review of the physical processes involved. Topics: planetary environments; solar wind; solar physics; stellar and interstellar plasmas; pulsars; general theory; shock waves and turbulence; cosmic rays.

062.020 Shawhan, S. D. Magnetospheric plasma waves. In: Lanzerotti, Louis J.; Kennel, Charles F.; Parker, E. N., eds. **Solar System Plasma Physics. Volume III.** Amsterdam: North-Holland; 1979: 211-270. 112 refs, 1961-77.

A review and interpretation of plasma wave observations made in the Earth's magnetospheres and cosmic plasma systems in general. Also included are tables of physical characteristics and properties of various types of plasma waves.

062.021 Tsytovich, V. N. Interaction of fast particles with waves in cosmic magnetoactive plasma. **Annu. Rev. Astron. Astrophys.** 11: 363-386; 1973. 76 refs, 1937-73.

A review of cosmic plasma physics with an emphasis on the recent expansion of theoretical concepts.

062.022 Wentzel, D. G.; Tidman, D. A., eds. **Plasma Instabilities in Astrophysics.** New
C York: Gordon and Breach; 1969. 417p. 19 papers; 45 abstracts.

Monterey Peninsula, CA, USA: 14-17 Oct 1968. (Proceedings). A series of papers on plasma physics and astrophysics designed to promote mutual interest and cooperation between the two fields. The invited papers on astrophysical topics (solar activity, Earth environment, and cosmic radiation) attempt to emphasize the plasma aspects of their studies.

Applications of atomic physics to astrophysical plasmas. *022.048.*

Interplanetary fields and plasmas. *074.036.*

Interpretation of spectral intensities from laboratory and astrophysical plasmas. *022.032.*

Observations of the interplanetary plasma. 106.001.

On relativistic magnetohydrodynamic processes in the fields of black holes. *066.037.*

Processes in collapsing interstellar clouds. *131.030.*

Solar System Plasma Physics. Volume I. Solar and Solar Wind Plasma Physics. *074.026.*

Spectroscopy of laboratory plasmas. *031.503.*

Ultraviolet and x-ray spectroscopy of astrophysical and laboratory plasmas. *031.510.*

UV and x-ray spectroscopy of astrophysical and laboratory plasmas. *031.511.*

063 Radiative Transfer, Scattering

063.001 Athay, R. Grant. **Radiation Transport in Spectral Lines.** Dordrecht, Holland: D. Reidel; 1972. 263p. (Geophys. Astrophys. Monogr., 1). 8 chapters w/refs.

An advanced monograph concerned with some particular applications of the theory of radiation transfer "to the flow of line photons through the outer layer of a star or some other tenuous media." Based on a large amount of numerical data generated by computer, the book covers the line source function; the two-level case (one line and two or more lines); line profiles; total line intensities; the line blanketing effect; numerical methods.

063.002 Avrett, E. H. Solution of non-LTE transfer problems. **J. Quant. Spectrosc. Radiat. Transfer** 11: 511-529; 1971. 99 refs, 1950-71.

Deals with computational methods for the solution of non-LTE transfer problems; that is, for the solution of coupled equations of radiative transfer and statistical equilibrium for atomic number densities that depart in general from the thermodynamic-equilibrium values that correspond to the local temperature and density. Emphasis is on progress in the field from 1967 to 1971.

063.003 Cook, A. H. **Celestial Masers.** Cambridge, UK: Cambridge University Press; 1977. 135p. 6 chapters. 156 refs, 1929-76.

An advanced treatise which reviews both what is and what is not known. Following an introduction, the author covers molecular structure and spectra, observations, theory of amplification by stimulated emission, pumping schemes, analysis and interpretation of maser radiation. Results of observations of hydroxyl and water sources are presented.

063.004 Gough, D. O. Stellar convection. **IAU Colloq.** 38: 349-363; 1977. 131 refs, 1938-76.

A review of research primarily covering 1972-76. Topics: attempts to model thermal convection; penetration and overshooting; subcritical convection; rotation and magnetic fields; time dependent convection. Reprinted from *IAU Trans. Rep. Astron.* 16A(2): 169-176; 1976.

063.005 Hummer, D. G. Line formation in expanding atmospheres. **IAU Symp.** 70: 281-312; 1976. 87 refs, 1945-76.

A review of current understanding of radiative transfer and line formation processes in expanding atmospheres, and the successes and limitations of current computational techniques.

063.006 Hummer, D. G.; Rybicki, G. The formation of spectral lines. **Annu. Rev. Astron. Astrophys.** 9: 237-270; 1971. 102 refs, 1931-72.

Review addressed to the nonspecialist intended to provide an introduction to the physics and phenomenology of the radiation field in spectral lines. Omits all discussion of direct astrophysical application.

063.007 Rybicki, George B.; Lightman, Alan P. **Radiative Processes in Astrophysics.** New York: John Wiley & Sons; 1979. 382p. 11 chapters w/refs.

With an emphasis on physics rather than formulae, this advanced text for seniors, graduate students, and scientists provides a broad overview of the field. Topics: radiative transfer; basic theory of radiation fields; radiation from moving charges; relativistic covariance and kinematics; Bremsstrahlung; synchrotron radiation; Compton scattering; plasma effects; atomic structure; radiative transitions; molecular structure.

063.008 Spiegel, E. A.; Zahn, J. P., eds. **Problems of Stellar Convection.** Berlin:
C Springer-Verlag; 1977. 363p. (IAU Colloq., 38). (Lecture Notes Phys., 71). 27 papers.

Nice, France: 16-20 Aug 1976. Contributed papers cover the following aspects of the field: mixing-length theory; linear theory; observational aspects; numerical solutions; rotation and magnetic fields; penetration, waves, turbulence; special topics.

063.009 Tucker, Wallace H. **Radiation Processes in Astrophysics.** Cambridge, MA: MIT Press; 1975. 311p. 8 chapters. 158 refs, 1912-75.

A graduate text covering basic formulae for classical radiation processes and for quantum radiation processes; cyclotron and synchrotron radiation; electron scattering; Bremsstrahlung and collision losses; radiative recombination; photoelectric effect; emission and absorption lines.

The composition of planetary atmospheres. *091.041.*

Line and continuum problems in gaseous nebulae. *134.002.*

064 Stellar Atmospheres, Stellar Envelopes, Mass Loss, Accretion

064.001 Baschek, B.; Kegel, W. H.; Traving, G., eds. **Problems in Stellar Atmospheres and Envelopes.** New York: Springer-Verlag; 1975. 375p. 11 papers.
A collection of review papers on current research and issues. Topics: solar energy flux; model atmospheres and heavy element abundances; helium stars; early-type stars; white dwarfs; cosmic masers; radio emission from stellar and circumstellar atmospheres; etc.

064.002 Bessell, M. S. Some applications of model atmospheres. **Proc. Astron. Soc. Australia** 2: 230-235, 1974. 66 refs, 1960-74.
A review of model stellar atmosphere applications, covering models of white dwarfs; hot stars; Ap, Bp, and Am stars; cool stars; synthetic spectra and colors.

064.003 Carbon, D. F. Model atmospheres for intermediate- and late-type stars. **Annu. Rev. Astron. Astrophys.** 17: 513-549; 1979. 157 refs, 1935-79.
Examination of the current status of the ability to construct model atmospheres for intermediate- and late-type stars. Topics include line blanketing, departures from local thermodynamic equilibrium (LTE), convection in stellar atmospheres, and extended atmospheres.

064.004 Cayrel, R., ed. Stellar atmospheres as indicator and factor of stellar evolution.
C **Highlights Astron.** 4(2): 99-218; 1977. 11 papers.
Joint discussion no. 5 of the Sixteenth General Assembly of the IAU.

064.005 Cayrel, R.; Steinberg, M., eds. **Physique des mouvements dans les atmo-**
C **sphères stellaires.** [Physics of motion in stellar atmospheres.] Paris: Editions du CNRS; 1976. 478p. (Colloq. Int. Cent. Natl. Rech. Sci., 250). 32 papers. English with French abstracts.
Nice, France: 1-5 Sep 1975. Reviews covering hydrodynamic instabilities, non-linear dynamics, solar studies and their use in hydrodynamics, stellar winds, and stellar observations.

064.006 Conti, P. S. Mass loss in early-type stars. **Annu. Rev. Astron. Astrophys.** 16: 371-392; 1978. 120 refs, 1929-78.
Discussion of the phenomenon of mass loss in early-type stars by stellar winds, including the effects in close binaries.

064.007 Conti, Peter S.; de Loore, Camiel W. H., eds. **Mass Loss and Evolution of O-**
C **type Stars.** Dordrecht, Holland; Boston: D. Reidel; 1979. 501p. (IAU Symp., 83). 13 papers, 54 abstracts.
Vancouver Island, BC, Canada: 5-9 Jun 1978. The purpose of the symposium was to bring together observers and theoreticians to discuss stellar winds and mass loss rates and their effects on evolutions of O-type stars. The first three sessions dealt with outlining the existing data on mass-loss rates. Further sessions dealt with theoretical considerations of evolution of O-type stars when mass-loss is taken into account.

064.008 Deutsch, A. J. Mass loss from stars: a review. In: Hack, M., ed. **Mass Loss from Stars.** Dordrecht, Holland: D. Reidel; 1969: 1-14. 110 refs, 1946-69.
A brief summary of the evidence supporting the existence of less conspicuous mass-loss processes in stars other than those in supernovae, ordinary novae, and planetary novae. Both single star and binary star mass loss are described.

064.009 Gingerich, O., ed. **Theory and Observation of Normal Stellar Atmospheres.**
C Cambridge, MA: MIT Press; 1969. 472p. 56 papers.

Third Harvard-Smithsonian Conference on Stellar Atmospheres. Cambridge, MA, USA: 8-11 Apr 1968. (Proceedings). The extent to which quantitative spectral classification can be interpreted by "classical" model atmospheres is the theme of this conference volume, which considers both the theory and spectra of normal stars. Topics: quantitative spectral classification; properties of synthetic spectra; comparison of synthetic spectra with real spectra. Included is a computer-produced grid of model stellar atmospheres.

064.010 Godoli, G. Stellar activity of the solar type. Observational aspects. **IAU Symp.** 71: 421-446; 1976. 221 refs, 1913-76.

Summary of the latest results in the study of the manifestations of stellar activity determined by a mechanism of the same kind as that producing solar activity, such as the interplay of solar poloidal magnetic fields, differential rotation, and convection.

064.011 Gray, D. F.; Linsky, J. L., eds. **Stellar Turbulence.** Berlin: Springer-Verlag;
C 1980. 308p. (IAU Colloq., 51). (Lecture Notes Phys., 114). 21 papers.

London, ON, Canada: 27-30 Aug 1979. Invited papers and abstracts covering both stellar and solar topics related to turbulence. Topics: physical origins of turbulence (convection, rotation, and oscillations); observed properties; conceptualizations of turbulence; and effects of turbulence on chromospheres, coronae, and mass loss.

064.012 Gray, David F. **The Observation and Analysis of Stellar Photospheres.** New York; London: John Wiley & Sons; 1976. 471p. 18 chapters w/refs.

An advanced text for graduate students and researchers covering basic physics, observational techniques, and data interpretation. Theory is discussed, but not overemphasized. Topics: spectroscopic tools; detectors; radiative and convective energy transport; the model photosphere; measurement of spectral lines, stellar temperatures and radii, and photospheric pressure; stellar rotation; turbulence in stellar photospheres; etc.

064.013 Groth, H. G.; Wellman, P., eds. **Spectrum Formation in Stars with Steady-**
C **State Extended Atmospheres.** Washington, DC: National Bureau of Standards; 1970. 332p. (IAU Colloq., 2). (NBS SP-332). 20 papers.

Munich, Germany: 16-19 Apr 1969. A collection of papers discussing current problems and research in extended stellar atmospheres. Besides a consideration of problems, the papers deal with theoretical methods for handling non-LTE problems; chromospheres and coronae of stars. Specific topics include model atmospheres; line formation; spectroscopy; etc.

064.014 Hack, M., ed. **Mass Loss from Stars.** Dordrecht, Holland: D. Reidel; 1969.
C 345p. (Astrophys. Space Sci. Libr., 13). 44 papers.

Second Trieste Colloquium on Astrophysics. Trieste, Italy: 12-17 Sep 1968. (Proceedings). Reports of research concerning stars which gradually and continually lose matter in their lifetimes, as opposed to the dramatic, quick losses experienced by ordinary novae, supernovae, and planetary nebulae. Observational evidence and theoretical work are presented for both single star and binary situations. There is also a section dealing with mass loss from unstable stars such as Nova Delphini 1967 and CH Cygni, a variable star.

064.015 Holzer, T. E.; Axford, W. I. The theory of stellar winds and related flows. **Annu. Rev. Astron. Astrophys.** 8: 31-60; 1970. 47 refs, 1942-69.

Review of the general theory of steady, radial, spherically symmetric flow of an ideal gas and how it can be applied to several astronomical situations: stellar winds, galactic winds, comet winds, the polar wind, and the classical accretion problem.

064.016 Joint conference on the physics of stellar atmospheres. **Proc. Astron. Soc.**
C **Australia** 1: 355-390; 1970. 16 papers.

University of Queensland, Australia: 27-29 May 1970. Cosponsored by the Astronomical Society of Australia and the Australian Institute of Physics.

064.017 Lyubimkov, L. S. Model atmospheres for normal stars (a survey from 1965 to 1973). **Astrophysics** 11: 462-483; 1975. Transl. of *Astrofizika* 11: 703-733; 1975. 83 refs, 1958-74.

Basic data for O-M type stars is considered, and the influence of blanketing effect, convection, departures from LTE, and abundance anomaly on atmospheric structure and emergent radiation is discussed.

064.018 Marlborough, J. M. Models for the circumstellar envelopes of Be stars. **IAU Symp.** 70: 336-370; 1976. 93 refs, 1867-1975.

A survey of the theoretical attempts to determine the structure of the circumstellar matter around Be stars. The general equations describing the structure and dynamics of Be stars are given.

064.019 Mihalas, D.; Pagel, B.; Souffrin, B. **Théorie des atmosphères stellaires.** [Theory
C of the Stellar Atmospheres.] Sauverny, Switzerland: Observatoire de Geneve; 1971. 312p. 3 papers. In English.

First Advanced Course of the Swiss Society of Astronomy and Astrophysics. Saas-Fee, Switzerland: 28 Mar—3 Apr 1971. Three advanced lectures with references on the following topics: 1) Theoretical analysis of stellar spectra (D. Mihalis, pp. 1-156); 2) Determination of stellar abundances (B. Pagel, pp. 157-237); 3) Convection (B. Souffrin, pp. 238-312).

064.020 Mihalas, D. H., ed. Physics of the chromosphere—corona-wind complex and
C mass loss in stellar atmospheres. **Highlights Astron.** 5: 521-613; 1980. 9 papers.

Joint discussion at the XVIIth General Assembly of the IAU. The topic is presented with regard to hot stars, late-type stars, x-ray binaries, and so on.

064.021 Mihalas, Dimitri. **Stellar Atmospheres.** 2nd ed. San Francisco: W. H. Freeman; 1978. 632p. (Ser. Books Astron. Astrophys.). 15 chapters. 694 refs, 1905-1977.

A graduate text emphasizing the theory of radiative transfer, along with a thorough discussion of the classical stellar atmospheres problem: atmospheres in hydrostatic, radiative, and steady-state statistical equilibrium. Topics: the radiation fields; the equation of transfer; the gray atmosphere; absorption cross-sections; equations of statistical equilibrim; model atmospheres; the line absorption profile; non-LTE transfer; line formation; radiative transfer; stellar winds; etc.

064.022 Novotny, Eva. **Introduction to Stellar Atmospheres and Interiors.** New York; London: Oxford University Press; 1973. 543p. 10 chapters. 360 refs, 1922-71.

An intermediate-level university text. Topics: observational data; radiative transfer; properties of gas; model atmospheres; line absorption in stellar atmospheres; stellar interior equations; stellar evolution; calculation of model atmospheres and interiors.

064.023 Pecker, J. C. The use of model atmospheres for temperature-gravity calibration. **IAU Symp.** 54: 173-221; 1973. 135 refs, 1931-73.

A review of the general principles of the use of model atmospheres in stellar calibration, providing a few specific examples of applications.

064.024 Skobel'tsyn, D. V., ed. **Methods in Stellar Atmosphere and Interplanetary Research.** New York; London: Consultants Bureau; 1974. 202p. (Proc. [Tr.] P. N. Lebedev Phys. Inst., 62). 23 papers.

A selection of short papers describing Soviet research. Topics: solar wind; radiative acceleration of gas in stellar atmospheres; pulsars; interstellar plasmas; radiometry; radio telescope techniques; etc. Translation of *Atmosfery zvezd i mezhplanetnaia plazma. Tekhnika radioastronomicheskogo priema.*

064.025 Slettebak, A. **Be and Shell Stars.** Dordrecht, Holland; Boston: D. Reidel; 1976.
C 465p. (IAU Symp., 70). 37 papers, 11 abstracts.

Merrill-McLaughlin Memorial Symposium. Bass River, MA, USA: 15-18 Sep 1975. Topics include observations of Be stars, Be stars as rotating stars, observational techniques, line formation in expanding atmospheres, models, and single versus binary stars.

064.026 Slettebak, A. The Be stars. **Space Sci. Rev.** 23: 541-580; 1979. 191 refs, 1866-1979.

Models for the envelopes of Be stars are reviewed, followed by a discussion of the evolution of Be stars.

064.027 Swihart, Thomas L. **Basic Physics of Stellar Atmospheres.** Tucson, AZ: Pachart Publishing; 1971. 86p. (Intermediate Short Texts Astrophys.). 4 chapters. 11 refs, 1950-70.

Aimed at the nonspecialist, this advanced volume provides an overview of the theory of stellar atmospheres. Topics: radiative transfer; equation of transfer; the gray atmosphere; the non-gray atmosphere; model atmospheres; line formation (absorption and emission, line broadening, profiles).

064.028 Theory of stellar atmospheres. **IAU Trans. Rep. Astron.** 17A(2): 193-210; 1979. 263 refs, 1970-79.

An overview of research activity worldwide, with an emphasis on the previous three years, as reported by IAU Commission 36 (Theory of Stellar Atmospheres). Additional reports: 14A: 425-436; 1970. - refs, 1967-69. 15A: 537-569; 1973. 660 refs, 1968-73. 16A(2): 199-205; 1976. 249 refs, 1970-79.

064.029 Vardya, M. S. Atmospheres of very late-type stars. **Annu. Rev. Astron. Astrophys.** 8: 87-114; 1970. 196 refs, 1916-70.

Concentrating on stars of spectral classes M, S, and C, this review looks at their gross properties obtained by spectral analysis. Model atmospheres are presented and discussed at length.

The Interaction of Variable Stars with Their Environment. 122.015.

Mass Loss and Evolution in Close Binaries. 117.006.

065 Stellar Structure and Evolution

065.001 Bodenheimer, P. Stellar evolution toward the main sequence. **Rep. Prog. Phys.** 35: 1-54; 1972. Also: **Lick Obs. Contr.** 358. 215 refs, 1919-72.

Review of stellar evolution from just after formation in interstellar clouds to the point of stabilization on the main sequence.

065.002 Bouvier, P.; Maeder, A., eds. **Advanced Stages in Stellar Evolution.** Sauverny, Switzerland: Geneva Observ.; 1977. 363p. 3 papers.

Seventh Advanced Course of the Swiss Society of Astronomy and Astrophysics. Saas-Fee, Switzerland: 28 Mar—2 Apr 1977. Contains the text of three lectures: 1) Stellar structure and evolution with emphasis on the evolution of intermediate mass stars (I. Iben, pp. 1-148; 226 refs, 1923-77); 2) The evolution of population II stars and mass loss and stellar evolution (A. Renzini, pp. 149-283; 230 refs, 1955-77); 3) Nucleosynthesis and the later stages of stellar evolution (D. N. Schramm, pp. 284-363; 85 refs, 1931-77).

065.003 Bulletin GRG, no. 38. Stars (including black holes): list of publications. **Gen. Relativ. Gravitation** 10: 149-179; 1979. 463 refs, 1968-78.

An alphabetical listing by author of the world literature, with an emphasis on journal articles and conference papers.

065.004 Burki, G. Formation of open clusters. **IAU Symp.** 85: 169-190; 1980. 102 refs, 1953-79.

Overview of star formation including mechanisms of cluster formation, birth of clusters from molecular clouds, the fragmentation process and the initial mass function, and properties of open clusters with star formation.

065.005 Carson, T. R. Stellar opacity. **Annu. Rev. Astron. Astrophys.** 14: 95-117; 1976. 91 refs, 1926-76.

Review of the calculations of stellar opacity with attention to the equation of state, nuclear reaction rates, convective energy transport, and the opacity of stellar material.

065.006 Cayrel de Strobel, G.; Delplace, A. M., eds. **Age des Etoiles.** Paris: Observa-
C toire de Paris-Meudon; 1972[?]. various paging. (IAU Colloq., 17). 53 papers.

Paris, France: 18-22 Sep 1972. Papers on the study and determination of stellar ages pertaining to the following topics: age from location on the H-R diagram; age and kinematical properties; nucleosynthesis and chronology; age from spectroscopic studies.

065.007 Chiu, H.-Y.; Muriel, A., eds. **Stellar Evolution.** Cambridge, MA: MIT Press;
C 1972. 812p. 21 papers.

Third Summer Institute for Astronomy and Astrophysics. Stony Brook, NY, USA: 18 Jun – 16 Jul 1969. A series of basic lectures describing the life of a star from birth to death, aimed at beginning graduate students and nonspecialists. Topics: normal stellar evolution and main sequence stars; variable stars; white dwarfs; novae; pulsars; stellar opacity; stellar magnetism and rotation; and nuclear burning shells.

065.008 Clayton, Donald C. **Principles of Stellar Evolution and Nucleosynthesis.** New York: McGraw-Hill; 1968. 612p. 7 chapters w/refs.

An introductory text for advanced students on the closely allied subjects in the title. Topics: basic physics and characteristics of stars; stellar interiors (thermodynamic state and energy transport); thermonuclear reaction rates; major nuclear burning stages in stellar evolution; calculation of stellar structure; synthesis of the heavy elements.

065.009 Cox, A. N., ed. Stellar instabilities. **Highlights Astron.** 5: 429-519; 1980. 24
C papers.

Topics include shell flashes, solar pulsations, variability, white dwarfs, supergiants, Cepheids, red giants, and so on.

065.010 Cox, J. P. Nonradial oscillations of stars: theories and observations. **Annu. Rev. Astron. Astrophys.** 14: 247-273; 1976. 178 refs, 1938-76.

Primarily concerned with a review of more general kinds of stellar oscillations than the purely radial, spherically symmetric type that have been considered extensively in the literature.

065.011 Cox, J. P. Stellar oscillations, stellar stability and application to variable stars. **Mem. Soc. R. Sci. Liège** 6e ser. 8: 129-159; 1975. 157 refs, 1918-74.

A general review of the present status of pulsation theory, including linear, nonlinear, adiabatic, non-adiabatic, radial and nonradial oscillations.

065.012 **Évolution stellaire avant la sequence principale.** [Pre-main sequence stellar evol-
C ution.] Liège, Belgium: Societe Royale des Sciences de Liège; 1970. 377p. (Mem. Soc. R. Sci. Liège. Collect. 8°, 5 ser., t. 19, fasc. unique). 38 papers. In English.

Sixteenth International Colloquium on Astrophysics. Liège, Belgium: 30 Jun—2 Jul 1969. Contributed papers addressing pre-main sequence stellar evolution. Topics: 1) theory of star formation, relevant properties of the interstellar medium, early phases of contraction; 2) stellar evolution from the phases of dynamical collapse to the contraction towards the main sequence; 3) effects of rotation and magnetic fields; 4) nucleosynthesis; 5) observational data.

065.013 Faulkner, J. The abundance of helium in stellar interiors. **Highlights Astron.** 2: 269-287; 1971. Also: **Lick Obs. Contr.** 333. 73 refs, 1955-70.

Review of the theoretical implications deduced from stellar structure. Concentration is on Population II stars.

065.014 Freedman, D. Z.; Schramm, D. N.; Tubbs, D. L. The weak neutral current and its effects in stellar collapse. **Annu. Rev. Nucl. Sci.** 27: 167-207; 1977. 119 refs, 1931-77.

Included is a discussion of current understanding of the role of the neutral current in weak interactions in relation to gravitational collapse and explosion.

065.015 Hack, M., ed. **Colloquium on Supergiant Stars.** Opicina, Tip. Villaggio del
C fanciullo, [1972?]. 371p. 45 papers.

Third Colloquium on Astrophysics. Trieste, Italy: 6-8 Sep 1971. (Proceedings). Reports of observations, physical characteristics, evolution, spectroscopy, models, etc.

065.016 Hack, M.; Swings, J. P., eds. Observational evidence of the heterogeneities of
C the stellar surfaces. **Highlights Astron.** 4(2): 373-407; 1977. 6 papers.

Topics: heterogeneity of the solar atmosphere; heterogeneity of surfaces of magnetic Ap stars; starspots; etc.

065.017 Hansen, C. J. Secular stability: application to stellar structure and evolution. **Annu. Rev. Astron. Astrophys.** 16: 15-32; 1978. 80 refs, 1926-77.

A review concentrating on purely radial instabilities and evolutionary effects that take place on time scales roughly comparable to t_{KH}. For the most part, it concentrates on phenomena that operate on thermal time scales and ignore dynamical and nuclear evolutionary effects.

065.018 Iben, I., Jr. Post main sequence evolution of single stars. **Annu. Rev. Astron. Astrophys.** 12: 215-256; 1974. 259 refs, 1939-74.

A sequel to the author's 1967 article (*Annu. Rev. Astron. Astrophys.* 5: 571-626; 1967). Discussion is limited to the evolution of single stars in quasi-static phases.

065.019 Kourganoff, V. **Introduction to the Physics of Stellar Interiors.** Dordrecht, Holland; Boston: D. Reidel; 1973. 115p. (Astrophys. Space Sci. Libr., 34). 5 chapters. 143 refs, 1935-71.

"An introduction to the theory of internal structure and of thermonuclear reactions," this brief treatise is aimed at graduate students and nonspecialists. The emphasis is on a clarification of the principal physical concepts used in formulating the theory. Topics: mechanical equilibrium; determination of internal structure by density distribution; energy equilibrium and nuclear reactions; evolutionary models. Translation by J. R. Lesh of *Introduction à la physique des interieurs stellaires* (Paris: Dumond; 1969).

065.020 Ledoux, P. Non-radial oscillations. **IAU Symp.** 59: 135-173; 1974. 105 refs, 1863-1974.

The problem of the adiabatic non-radial oscillations of spherical stars is reviewed and results recalled for a variety of models.

065.021 Ledoux, P. Stellar stability. In: Leibovitz, N. R.; Reid, W. H.; Vandervoort, P. O., eds. **Theoretical Principles in Astrophysics and Relativity.** Chicago: University of Chicago Press; 1978: 15-58. 263 refs, 1928-76.

A review of recent studies limited to cases of "purely hydrostatic stars most of the time in thermal balance." Both radial and nonradial perturbations are considered; dynamical, vibrational, and secular stability are detailed.

065.022 C Ledoux, P.; Noels, A.; Rodgers, A. **Stellar Instability and Evolution.** Dordrecht, Holland; Boston: D. Reidel; 1974. 198p. (IAU Symp., 59). 7 papers, 29 abstracts.

Mount Stromlo, Canberra, Australia: 16-18 Aug 1974. Topics include a survey of the problems in radial stellar instability and its relation to stellar evolution; pulsation and the young disc population; large mass stars, stability in supergiants, critical masses; halo and old disc populations; eruptive and explosive variables; and instability mechanisms.

065.023 C Luyten, W. J., ed. **White Dwarfs.** Dordrecht, Holland: D. Reidel; New York: Springer-Verlag; 1971. 164p. (IAU Symp., 42). 16 papers, 6 abstracts.

St. Andrews, Fife, UK: 11-13 Aug 1970. Topics include red subluminous stars, parallaxes, temperatures, spectra, origin, and models of white dwarfs.

065.024 Massevitch, A. G.; Tutukov, A. V. Evolution of stars with $M \geq 8 M_{\odot}$. **IAU Symp.** 66: 73-92; 1974. 87 refs, 1941-73.

Review of research in the field of evolution of massive stars during the last three years.

065.025 Meadows, A. J. **Stellar Evolution.** 2nd ed. Oxford, UK; New York: Pergamon Press; 1978. 171p. (Pergamon Int. Libr. Sci. Technol. Eng. Soc. Stud.). 9 chapters. 4 refs, 1968-74.

A nontechnical introduction for the nonspecialist, student, and intelligent layperson. The usual topics are covered.

065.026 Mullan, D. J. Convection in A stars: convection suppressed in magnetic A stars? **Irish Astron. J.** 11: 32-82; 1973. 210 refs, 1906-1971.

Discussion of how convection in A stars is expected to be affected by the presence of magnetic fields. The paper is divided into two parts: part 1 contains a nontechnical account of how magnetic fields in sunspots and stars can suppress convection; part 2 is a more theoretical account.

065.027 C Ögelman, H.; Rothschild, R., eds. **Recognition of Compact Astrophysical Objects.** Washington, DC: NASA; 1977. 194p. (NASA SP-421). 8 chapters w/refs.

Edited versions of lectures presented at a graduate course at the University of Maryland in 1975. Topics: stellar evolution from main sequence to white dwarf, neutron star, or black hole; pulsars; binary stars and compact x-ray sources; x-ray signatures of compact objects; gamma rays from compact objects; temporal analysis of x-ray data; pulsar theory; etc.

065.028 Paczynski, B. Evolution of stars with $M \leq 8 M_{\odot}$. **IAU Symp.** 66: 62-69; 1974. 59 refs, 1955-74.

Discussion of the late stages of evolution of stars that develop degenerate carbon-oxygen cores.

065.029 Payne-Gaposchkin, Cecilia. **Stars and Clusters.** Cambridge, MA: Harvard University Press; 1979. 262p. (Harvard Books Astron.). 16 chapters. no refs.

An excellent overview for students and serious laypersons on the evolution of stars. Clearly written and profusely illustrated, the book covers stellar birth, growth, and death for all types of stars, both main sequence and otherwise. Topics: the H-R diagram;

pre-and post-main sequence stages; red giants; Cepheid variables; open and globular clusters; short period variables; novae; supernovae; white dwarfs; etc.

065.030 Preston, G. W. The chemically peculiar stars of the upper main sequence. **Annu. Rev. Astron. Astrophys.** 12: 257-277; 1974. 136 refs, 1933-74.

Summarizes hypotheses advanced to account for the various CP stars and those data needed to construct and test explanations of CP stars.

065.031 Reddish, V. C. **The Physics of Stellar Interiors: An Introduction.** New York: Crane, Russak & Co.; Edinburgh: Edinburgh University Press; 1974. 107p. 7 chapters. no refs.

A brief intermediate university text, based on lecture notes from a course given at Case Western Reserve University in 1969, covering stable stars, explosive events, and nuclear synthesis. Topics: main sequence stars; red giants; pre-main sequence stars; white dwarfs, neutron stars, and pulsars; supernovae; supermassive objects.

065.032 Reddish, V. C. **Stellar Formation.** Oxford, UK; New York: Pergamon Press; 1978. 28p. (Int. Ser. Nat. Philos., 97). 13 chapters. 485 refs, 1928-75.

An advanced text reviewing work in the field, with an emphasis on recent results such as cool clouds in regions of star formation. Divided into chapters under data (a review of observations and associated information) and theory, the book covers locations of star formation; frequency distribution of stellar masses; rate of formation; necessary environment; cloud structure; evolution of galaxies; etc.

065.033 Rolfs, C.; Trautvetter, H. P. Experimental nuclear astrophysics. **Annu. Rev. Nucl. Sci.** 28: 115-159; 1978. 99 refs, 1931-78.

Restricted to charged-particle nucleosynthesis processes, the review concentrates on recent nuclear experiments. Topics include charged-particle reaction rates under stellar conditions, experimental procedures and some practical considerations, hydrogen burning, helium burning, carbon and oxygen burning, and aspects of nuclear astrophysics.

065.034 Rose, William K. **Astrophysics.** New York: Holt, Rinehart & Winston; 1973. 287p. 13 chapters. 137 refs, 1953-72.

An advanced introduction for nonspecialists to the various aspects of the field, with an emphasis on stellar evolution. Topics: matter and energy in space; stellar theory; star formation and protostars; red giants; variable stars; planetary nebulae; novae and galactic x-ray sources; white dwarfs; supernovae and the formation of the heavy elements; neutron stars; galaxies; cosmology.

065.035 Sherwood, V. E.; Plaut, L., eds. **Variable Stars and Stellar Evolution.**
C Dordrecht, Holland; Boston: D. Reidel; 1975. 614p. (IAU Symp., 67). 53 papers, 32 abstracts.

Moscow, USSR: 29 Jul–4 Aug 1974. Topics include flare stars, T Tauri stars, R Coronae Borealis stars, Cepheids, δ Scuti stars, Wolf-Rayet stars, novae, U Geminorum stars, variable x-ray sources, RR Lyrae and W Virginis stars, and variable QSOs and nuclei of galaxies.

065.036 Shklovskii, Iosif S. **Stars: Their Birth, Life, and Death.** San Francisco: W. H. Freeman; 1978. 442p. 24 chapters. 265 refs, 1929-79.

An intermediate-level university text, also useful for educated laypersons and scientists. Topics: stellar evolution; sources of energy; stellar models; types of stars; unstable stars (e.g., supernovae); neutron stars and pulsars; x-ray stars; black holes; etc. Revised translation by R. B. Rodman of *Zvezdy, ikh rozhdenie, shizn' i smert* (Moscow: Nauka; 1975).

065.037 Spiegel, E. A. Convection in stars: I. basic Boussinesq convection. **Annu. Rev. Astron. Astrophys.** 9: 323-352; 1971. 114 refs, 1937-71.

Review of the problem of convective transfer in the laboratory situation, with a focus on approaches that seem to have a direct bearing on the problems of stellar convection.

065.038 Spiegel, E. A. Convection in stars: II. special effects. **Annu. Rev. Astron. Astrophys.** 10: 261-304; 1972. 233 refs, 1883-1972.

Review of stability theory, thermosolutal convection, followed by a discussion of some thermodynamic and dynamic effects that complicate stellar convection and Boussinesq models.

065.039 Stellar constitution. **IAU Trans. Rep. Astron.** 17A(2): 173-192; 1979. 78 refs, 1962-78.

An overview of research activity worldwide, with an emphasis on the previous three years, as reported by IAU Commission 35 (Stellar Constitution). Topics: stellar interiors; stellar stability; stellar oscillations; evolution of close binaries; neutrino astrophysics; stellar convection; neutron stars; white dwarfs; etc. Additional reports: 14A: 405-423; 1970. 173 refs, 1950-70. 15A: 507-535; 1973. + refs, 1967-73. 16A(2): 161-188; 1976. - refs, 1973-75.

065.040 **La structure interne des étoiles.** Sauverny, Switzerland: Observatoire de Genève;
C 1970. 419p. 4 papers. English/French.

XIe cours de perfectionnement de l'Association Vaudoise des Chercheurs en Physique. Saas-Fee, Switzerland: 24-29 Mar 1969. This volume contains four lectures, each with an extensive number of references: 1) An introduction to the internal structure of stars (in French); 2) Oscillations and stellar stability (in French); 3) Stellar evolution according to numerical models (in English); 4) Nuclear reactions in stellar surfaces and their relations with stellar evolution (in English).

065.041 Swihart, Thomas L. **Physics of Stellar Interiors.** Tucson, AZ: Pachart Publishing; 1972. 119p. (Intermediate Short Texts Astrophys.). (Astron. Astrophys. Ser.). 6 chapters w/refs. 15 refs, 1954-72.

A companion to the author's *Basic Physics of Stellar Atmospheres* (Tucson: Pachart; 1971), this book provides an introduction for the nonspecialist with the aim of providing an understanding of stellar structure. Topics: radiation theory; gas in thermodynamic equilibrium; polytropic models; stellar energies and energy transport; structure and evolution of the stars.

065.042 Tayler, R. J. **The Stars: Their Structure and Evolution.** London: Wykeham Publications; distr., New York: Springer-Verlag; 1970. 201p. (Wykeham Sci. Ser.). 9 chapters. no refs.

A brief introductory text for college students and educated laypersons. Topics: observational properties of stars; equations of stellar structure; stellar interiors; main sequence stars; various aspects of stellar evolution; post-main sequence, advanced stages, and final stages.

065.043 Tayler, R. J. Stellar evolution. **Rep. Prog. Phys.** 31: 167-224; 1968. 267 refs, 1917-67.

Discussion of the results of recent theoretical calculations and comparison of these results with observation.

065.044 Tayler, R. J.; Hesser, J. E., eds. **Late Stages of Stellar Evolution.** Dordrecht,
C Holland; Boston: D. Reidel; 1974. 269p. (IAU Symp., 66). 16 papers, 46 abstracts.

Warsaw, Poland: 10-12 Sep 1973. Topics include nuclear reactions and neutrinos, mass loss, circulation and mixing, presupernovae, novae, nebulae, and supernovae.

065.045 Underhill, Anne B. **The Early Type Stars.** Dordrecht, Holland: D. Reidel; 1966. 282p. (Astrophys. Space Sci. Libr., 6). 16 chapters. 611 refs, 1867-1965.

An advanced monograph describing the characteristics, structure, and evolution of a small group of hot stars in the O, B, A, and early F classes. Topics: spectral classification systems; O and B star luminosities; distribution of O and B stars; spectroscopic binaries; Wolf-Rayet stars; Be stars, Shell stars, and O stars; Beta Canis Majoris stars; etc. Photometric, spectrophotometric, and wavelength studies are discussed.

065.046 Unno, Wasaburo, et al. **Nonradial Oscillations of Stars.** Tokyo: University of Tokyo Press; 1979. 323p. 6 chapters. 296 refs, 1863-1979.

Restricted mainly to discussion of the linear theory of nonradial oscillations of spherically symmetric stars, this advanced text reviews recent research and summarizes fundamental concepts. The book covers the two theoretical aspects: the variation of oscillations in wavelength and frequency, and the source and control of these oscillations within the stars themselves.

065.047 Wheeler, J. C. The final evolution of stars. **Mem. Soc. Astron. Italiana** 49: 349-373; 1978. 85 refs, 1967-79.

Review of both theory and observations of evolution of single stars, white dwarfs, pulsars, supernova remnants, optical supernovae, and compact binary systems.

Abundance anomalies in early-type stars. *061.009.*

Astrophysics and Stellar Astronomy. 061.068.

The Evolution of Galaxies and Stellar Populations. 151.038.

Evolution of rotating stars. *116.002.*

The Hertzsprung-Russell Diagram and stellar ages. *115.003.*

Introduction to Stellar Atmospheres and Interiors. 064.022.

Lithium and Beryllium in stars. *114.050.*

An observational approach to stellar evolution. I. Flare stars and related objects. *123.003.*

Observational evidence for atmospheric chemical composition peculiarities relevant to stellar evolution. *114.039.*

The Origin of the Elements. 061.070.

Stellar atmospheres as indicator and factor of stellar evolution. *064.004.*

Stellar Chromospheres. 114.026.

Stellar Turbulence. 064.011.

---

066 Relativistic Astrophysics, Gravitation Theory, Background Radiation, Black Holes

---

066.001 Adler, Ronald; Bazin, Maurice; Schiffer, Menahem. **Introduction to General Relativity.** 2nd ed. New York: McGraw-Hill; 1975. 549p. (Int. Ser. Pure Appl. Phys.). 15 chapters w/refs.

A graduate text attempting to show the close interaction of mathematical and physical ideas, and to express the importance and beauty of the laws of general relativity. Topics: tensor algebra; tensor analysis; tensors in physics; the gravitational field equations; the Schwarzschild solution; descriptive cosmic astronomy; cosmological models; the role of relativity in stellar structure and gravitational collapse; etc.

066.002 Atwater, H. A. **Introduction to General Relativity.** Oxford, UK; New York: Pergamon Press; 1974. 221p. (Int. Ser. Monogr. Nat. Philos., 63). 9 chapters. no refs.

A text for graduate students or seniors with knowledge of classical mechanics, matrix algebra, and Maxwell's equations. Topics: tensor analysis; equations of motion of free particles; solutions of field equations; electromagnetism; gravitational fields and waves; relativity in cosmology; unified theories and quantized theories; etc.

066.003 Bergmann, P. G.; Fenyves, E. J.; Motz, L., eds. **Texas Symposium on Relativistic Astrophysics.** 7th. New York: New York Academy of Sciences; 1975. 500p. (Ann. NY Acad. Sci., 262). 53 papers.
C

Dallas, TX, USA: 16-20 Dec 1974. Review papers on theoretical and observational results as well as reports of recent discoveries and research. Topics: abundances of the elements; supernovae; black holes; gravitational waves; neutrino astrophysics; relativity; quasars; cosmology; etc.

066.004 Bertotti, B., ed. **Experimental Gravitation.** London; New York: Academic; 1974. 574p. 22 papers.
C

International School of Physics "Enrico Fermi," Course LVI. Varenna, Italy: 17-29 Jul 1972. (Proceedings). A review of various experiments, primarily astronomically oriented, which could help verify the theory of relativistic gravity. Topics: theoretical tools of experimental gravitation; radio tracking; solar oblateness and relativity; low-temperature technology; the gyroscope experiment; etc.

066.005 **Black Holes 1970-74.** London: Institution of Electrical Engineers; 1974. 83p. (INSPEC Bibliogr. Ser., 1). 304 refs, 1969-74.

A bibliography of citations appearing in *Physics Abstracts* in 1970-74, covering papers with 1969-74 imprints. Entries include authors, affiliation, sources, volume, pagination, date, abstract, and number of references. Citations are arranged topically, including the following subjects: general and review papers; general theories; stability, vibrations, and perturbations; black holes in x-ray and binary star systems; key papers; etc.

066.006 Carmeli, Moshe. **Group Theory and General Relativity.** New York: McGraw-Hill; 1977. 392p. (Int. Ser. Pure Appl. Phys.). 12 chapters. ~1,000 refs, 1890-1976.

An advanced text for graduate students on the applications of group theory to general relativity, this book is subtitled "representations of the Lorentz group and their applications to the gravitational field." Topics: the Rotation Group; the Lorentz Group; principle series of representations of SL(2,C); complete series of representations of SL(2,C); general relativity theory; spinors of general relativity; analysis of the gravitational field; exact solutions of the gravitational field equations; the Bondi-Metzner-Sachs Group; etc.

066.007 Carmeli, Moshe; Fickler, Stuart I.; Witten, Louis, eds. **Relativity.** New York: Plenum Press; 1970. 381p. 17 papers.
C

Relativity Conference in the Midwest. Cincinnati, OH, USA: 2-6 Jun 1969. (Proceedings). Papers primarily describing current theoretical work. Topics: quantum gravitation models; general relativity; gravitational radiation experiments; Schwarzschild singularity; superspace; etc.

066.008 Chandrasekhar, S. The black hole in astrophysics: the origin of the concept and its role. **Contemp. Phys.** 15: 1-24; 1974. 36 refs, 1926-73.

Basic review of the theoretical possibility of the existence of black holes through a consideration of stellar evolution. Partial condensation of the author's Halley Lecture, 1972.

066.009 DeWitt, C.; DeWitt, B. S., eds. **Black Holes.** New York: Gordon and Breach;
C 1973. 552p. 7 papers.

Twenty-third Session, Les Houches Summer School. Les Houches, France: Aug 1972. Aimed at the scientist and advanced student, this volume contains seven edited lectures outlining both theoretical arguments and observational evidence for black holes. Highlighted by a chapter on x-ray observations pointing to Cygnus X-1 as a likely black hole, and a detailed overview of the astrophysics involved, the book provides a rigorous introduction to the field. Translation of *Les astres occlus.*

066.010 DeWitt-Morette, Cecile M., ed. **Gravitational Radiation and Gravitational Col-**
C **lapse.** Dordrecht, Holland; Boston: D. Reidel; 1974. 223p. (IAU Symp., 64). 11 papers, 38 abstracts.

Copernicus Symposium I. Warsaw, Poland: 5-8 Sep 1973. Sessions included gravitational radiation, stability and collapse, and accretion of matter and x-ray sources.

066.011 Doroshkevich, A. G.; Sunyaev, R. A.; Zel'dovich, Ya. B. The formation of galaxies in Friedmannian universes. **IAU Symp.** 63: 213-225; 1974. 59 refs, 1929-73.

Short review of the theory of formation of galaxies and clusters of galaxies within the framework of the non-linear theory of gravitational instability.

066.012 Drever, R. W. P. Gravitational wave astronomy. **Q. J. R. Astron. Soc.** 18: 9-27; 1977. 68 refs, 1960-76.

Introductory review dealing with experimental techniques used in searches for gravitational radiation.

066.013 Eardley, D. M.; Press, W. H. Astrophysical processes near black holes. **Annu. Rev. Astron. Astrophys.** 13: 381-422; 1975. 196 refs, 1798-1975.

Discussion of the main astrophysical processes relevant to black holes of various possible masses. Includes description of gravitational collapse of rotating, nonrotating, and supermassive stars as well as galactic nuclei.

066.014 Ehlers, J. Survey of general relativity theory. In: Israel, Werner, ed. **Relativity, Astrophysics and Cosmology.** Dordrecht, Holland: D. Reidel; 1973: 1-125. 124 refs, 1923-73.

An in-depth, advanced overview of Einstein's theory of gravitation, emphasizing "the general structure of the theory and the connection between physical and mathematical concepts and ideas." Only brief mention is made of astronomical applications.

066.015 Ehlers, J.; Perry, J. J.; Walker, M., eds. **Texas Symposium on Relativistic**
C **Astrophysics.** 9th. New York: New York Academy of Sciences; 1980. 599p. (Ann. NY Acad. Sci., 336). 38 papers.

Munich, Germany: 14-19 Dec 1978. Reports of research, short review papers, and workshop summaries are included in this volume, which covers the following subjects: extragalactic sources; cosmology and galaxy formation; gamma-ray astronomy; gravitation theory and fundamental interactions; supernovae and collapse; neutron stars, pulsars, and supernova remnants; x-ray astronomy.

066.016 Giacconi, R.; Ruffini, R., eds. **Physics and Astrophysics of Neutron Stars and**
C **Black Holes.** Amsterdam: North Holland; New York: Elsevier; 1978. 871p. 37 papers.

International School of Physics "Enrico Fermi" (Course 65). Varenna on Lake Como, Italy: 14-26 Jul 1975. (Proceedings). A series of lectures aimed at reviewing "the progress made in recent years in our understanding of neutron stars and black holes in light of the extensive experimental knowledge acquired from binary x-ray sources." Experimental results and data and theoretical topics are presented.

066.017 C Halpern, Leopold, ed. **On the Measurement of Cosmological Variations of the Gravitational Constant.** Gainesville, FL: University Presses of Florida; 1978. 116p. (Monogr. Publ. Demand: Imprint Ser.). 6 papers.

Workshop. Tallahassee, FL, USA: 12-14 Nov 1975. (Proceedings). Papers reporting on current investigations including "The Large Numbers Hypothesis and the Cosmological Variation of the Gravitational Constant," by P. A. M. Dirac.

066.018 Hawking, S. W.; Israel, W. **General Relativity: An Einstein Centenary Survey.** Cambridge, UK: Cambridge University Press; 1979. 919p. 16 papers. ~2,000 refs.

A collection of advanced papers commemorating Einstein's birth, this book surveys the current state of research, covering gravitation theory and experiment; gravitational-radiation experiments; global structure of spacetimes; black hole astrophysics; Big Bang cosmology; quantum gravity; etc.

066.019 C Hegyi, D. J., ed. **Texas Symposium on Relativistic Astrophysics.** 6th. New York: New York Academy of Sciences; 1973. 364p. (Ann. NY Acad. Sci., 224). 39 papers.

New York, NY, USA: 18-22 Dec 1972. Short review papers and reports of research. Selected topics include quasars, x-ray astronomy, gravitational radiation, pulsars, neutron stars, black holes, solar oblateness, and interstellar deuterium.

066.020 Hohler, G.; Niekisch, E. A., eds. **Astrophysics.** Berlin; New York: Springer-Verlag; 1973. 115p. (Springer Tracts Mod. Phys., 69). 2 papers.

Contains two review papers: "On the Properties of Matter in Neutron Stars," by G. Borner, pp. 1-67, 134 refs, 1926-73; and "Black Holes: The Outside Story," by J. Stewart and M. Walker, pp. 69-115, 83 refs, 1934-73.

066.021 C Isham, C. J.; Penrose, R.; Sciama, D. W., eds. **Quantum Gravity.** Oxford, UK: Clarendon Press; 1975. 605p. 9 papers.

An Oxford Symposium. Chilton, UK: 15-16 Feb 1974. Rigorous papers which attempt to relate the separate and different fields of general relativity and quantum mechanics. Three papers address astronomical topics: 1) quantum cosmological models; 2) particle creation by black holes; 3) is physics legislated by cosmogony?

066.022 C Israel, Werner, ed. **Relativity, Astrophysics and Cosmology.** Dordrecht, Holland: D. Reidel; 1973. 329p. (Astrophys. Space Sci. Libr., 38). 6 papers.

Summer School. Banff, AB, Canada: 14-26 Aug 1972. (Proceedings). An introduction to the theory of relativity with some astronomical applications. Aimed at graduate students, the lectures also cover black holes, gravity, and differential geometry.

066.023 Kilmister, C. W. **General Theory of Relativity.** Oxford, UK; New York: Pergamon Press; 1973. 365p. (Commonwealth Int. Libr. Sel. Readings Phys.). 14 chapters w/refs.

An introduction to general relativity followed by 11 classical or otherwise important papers from the past on specific topics. The introduction reviews the principle of equivalence, the beginnings of general relativity, and recent developments. The additional selections (by Einstein, Bondi, Reimann, and others) provide insight into the foundations and theory of gravitation, relativity, and related subjects. For the graduate student and scientist.

066.024 Kuchowicz, B. **Nuclear and Relativistic Astrophysics and Nuclidic Cosmochemistry: 1963-67.** Warsaw: Nuclear Energy Information Center. 4v. I: 1968. 366p. (NEIC-RR-34); II: 1969. 304p. (NEIC-RR-37); III: 1969. 265p. (NEIC-RR-38); IV: 1971. 209p. (NEIC-RR-45). 5,690 refs, 1963-67.

A comprehensive literature survey of the various subfields of the topics in the title (v. I), extensive lists of references (v. II & III), and an index to volumes II and III (v. IV). The latter contains indexes to authors, editors, books, theses, reports, conferences, lectures, languages, etc. The bibliographic volumes' entries contain author, affiliation (if known), title, sources, volume, pagination, year, and index numbers where cited (e.g., *Physics Abstracts, Chemical Abstracts,* etc.) up to six or seven numbers. Entries are arranged by author for each individual year. The survey volume (I) covers 30 separate topics.

066.025 C Lévy, M.; Deser, S., eds. **Recent Developments in Gravitation.** New York: Plenum Press; 1979. 596p. (NATO Adv. Stud. Inst. Ser., Ser. B., Phys., 44). 16 papers.

1978 Cargese Summer Institute. Cargese, Corsica: 10-29 Jul 1978. (Proceedings). Advanced papers discussing classical relativity, quantum gravity, and supergravity. Includes a review paper by B. Bertotti (pp. 3-22) on the present experimental status of the theory of gravitation.

066.026 C Longair, M. S., ed. **Confrontation of Cosmological Theories with Observational Data.** Dordrecht, Holland; Boston: D. Reidel; 1974. 382p. (IAU Symp., 63). 31 papers, 10 abstracts.

Second Copernicus Symposium. Cracow, Poland: 10-12 Sep 1973. Topics include the contemporary structure and dynamics of the universe, relic radiation, origin of structure in the expanding universe, structure of singularities, matter-antimatter universes and physical processes near the singularity.

066.027 Longair, M. S.; Sunyaev, R. A. The universal electromagnetic background radiation. **Soviet Phys. Usp.** 14: 569-599; 1972. Transl. of *Usp. Fiz. Nauk* 105: 41-96; 1971. 222 refs, 1962-71.

Discussion of the properties of the principal sources of radiation, existing observations, and the problems of accounting for the background emission.

066.028 Misner, Charles W.; Thorne, Kip S.; Wheeler, John Archibald. **Gravitation.** San Francisco: W. H. Freeman; 1973. 1,279p. 44 chapters. ~800 refs.

An advanced monograph on gravitation physics, this book covers a wide variety of physical, mathematical, and astronomical topics. With an overall theme of geometrodynamics (Einstein's general relativity), the book covers spacetime physics; physics in flat spacetime; mathematics of curved spacetime; Einstein's geometric theory of gravity; relativistic stars; the universe; gravitational collapse and black holes; gravitational waves; experimental tests of general relativity; etc.

066.029 Narlikar, Jayant V. **Lectures on General Relativity and Cosmology.** London: Macmillan; 1979, c1978. 279p. 18 chapters. 115 refs, 1687-1977.

An advanced text aimed at graduate students and theoretical physicists. A variety of subjects are presented, including theories of gravitation; spacetime curvature and symmetries; Einstein's equations of gravitation; the Schwarzschild solution; experimental tests of general relativity; relativistic astrophysics; black holes; observational tests in cosmology; advanced topics; etc.

066.030 C **Ondes et radiations gravitationnelles.** [Waves and Gravitational Radiation.] Paris: Editions du Centre National de la Recherche Scientifique; 1974. 392p. (Colloq. Int. Cent. Nat. Rech. Sci., 20). 38 papers. French/English.

Paris, France: 18-22 Jun 1973. Reports of current research in the detection of gravitational waves and radiation.

066.031 Papagiannis, M. D., ed. **Texas Symposium on Relativistic Astrophysics.** 8th.
C New York: New York Academy of Sciences; 1977. 689p. (Ann. NY Acad. Sci., 302). 53 papers.

Boston, MA, USA: 13-17 Dec 1976. Review papers on theoretical and observational results, and reports of recent discoveries and research. Topics: close binaries; supernovae; x-ray bursters; globular clusters; cosmology; gravitational theories; quasars; galaxies; etc.

066.032 Papapetrou, A. **Lectures on General Relativity.** Dordrecht, Holland; Boston: D. Reidel; 1974. 203p. 12 chapters. no refs.

An advanced text assuming prior knowledge of the fundamentals of special relativity. Topics: tensor calculus; covariant differentiation; Riemannian geometry; the gravitational field; the Schwarzschild solution; other exact solutions; weak gravitational fields; variational principle; Einstein-Maxwell equations; equations of motion in general relativity; gravitational radiation; the cosmological problem. Includes student exercises.

066.033 Pisarev, A. F. Problems of emission and reception of gravitational waves. **Soviet J. Part. Nucl.** 6: 98-117; 1976. Transl. of *Fiz. Elem. Chastits At. Yadra.* 6: 244-291; 1975. 192 refs, 1937-74.

Presentation, in concise form, of the contemporary physical picture and mathematical description of the radiation processes, of the propagation and interaction of gravitational waves with matter and with an electromagnetic field, and a discussion of the published results of experimental searches for gravitational radiation and an analysis of new ideas in the field.

066.034 Press, W. H.; Thorne, K. S. Gravitational-wave astronomy. **Annu. Rev. Astron. Astrophys.** 10: 335-374; 1972. 146 refs, 1933-72.

Review of gravitational-wave astronomy, including properties, generation, astrophysical sources, and gravitational wave receivers.

066.035 Rees, Martin; Ruffini, Remo; Wheeler, John Archibald. **Black Holes, Gravitational Waves and Cosmology: An Introduction to Current Research.** New York: Gordon and Breach; 1974. 321p. (Top. Astrophys. Space Phys., 10). 19 chapters. 459 refs, 1879-1972.

A graduate text providing an introduction to the many aspects of relativistic astrophysics and cosmology. Topics: physics of superdense stars; pulsars; black holes; supernovae; QSOs; gravitational radiation; the expanding universe; the microwave background radiation; models of the universe; etc. Includes selected reprints of 13 articles from the literature as an appendix.

066.036 Rindler, Wolfgang. **Essential Relativity: Special, General, and Cosmological.** 2nd ed. New York; Berlin: Springer-Verlag; 1977. 284p. (Texts Monogr. Phys.). 9 chapters w/refs.

A text for advanced undergraduates with equal treatment given to the three types of relativity. Topics: absolute space; Einsteinian kinematics; Einsteinian optics; relativistic particle mechanics; electrodynamics; cosmology; Kruskal space; the plane gravitational wave; linearized general relativity. The first edition (Van Nostrand Reinhold) was issued in 1969.

066.037 Ruffini, R. On relativistic magnetohydrodynamic processes in the fields of black
C holes. In: Ruffini, Remo, ed. **Marcel Grossman Meeting on General Relativity.** 1st. Amsterdam; New York: North-Holland; 1977: 349-392. 130 refs, 1922-77.

Trieste, Italy: 7-12 Jul 1975. (Proceedings). A rigorous summary of the following topics: observation of neutron stars and potential black holes; black hole physics; structure of a black hole under relativistic astrophysical conditions; the basic equations of relativistic magnetohydrodynamics; analytic solution for the magnetohydrodynamics equations; Klein paradox and black holes; x-ray sources and gravitational radiation.

066.038 Ruffini, R. Physics outside the horizon of a black hole. In: Giacconi, R.; Ruffini, R., eds. **Physics and Astrophysics of Neutron Stars and Black Holes.** Amsterdam: North-Holland; 1978: 287-355. 204 refs, 1910-77.

A review of research related to constructing models of black holes and to proving their existence. Topics: the Kerr-Newman black hole; stability against small perturbations; radiation processes; mass formula; accretion; the Klein paradox; etc.

066.039 Sachs, R. K., ed. **General Relativity and Cosmology.** New York; London: Academic; 1971. 387p. 15 papers.
C

International School of Physics "Enrico Fermi." Course 47. Varenna on Lake Como, Italy: 30 Jun—12 Jul 1969. (Proceedings). A volume which will help bridge "the gap between the standard textbooks and the flood of research articles on these topics." There are five major topics: general relativity and kinetic theory; spacetime structure from a global viewpoint; relativistic cosmology; astrophysical cosmology; relativistic stars, black holes, and gravitational waves.

066.040 Sears, Francis W.; Brehme, Robert W. **Introduction to the Theory of Relativity.** Reading, MA: Addison-Wesley; 1968. 216p. (Addison-Wesley Ser. Phys.). 11 chapters. no refs.

A brief text for junior and senior college physics students emphasizing the use of the Lorentz transformation in its presentation. Topics: the Galilean transformation; the Lorentz transformation; time; space; accelerated frames of reference; four-vectors; energy and momentum; curved spacetime and gravitation.

066.041 Sexl, Roman; Sexl, Hannelore. **White Dwarfs—Black Holes: An Introduction to Relativistic Astrophysics.** New York; London: Academic; 1979. 203p. 10 chapters. 32 refs, 1966-77.

A brief, intermediate text for university students, describing some of the more important and interesting aspects of the field. Topics: general relativity; curved spacetime; stars and planets; pulsars; gravitational collapse; black holes; gravitational waves; cosmology; cosmogony; etc. Translation by Patrick P. Weidhaas of *Weisse Zwerge—Schwarze Locher* (Hamburg: Rowohlt Taschenbuch Verlag; 1975).

066.042 Shaviv, G.; Rosen, J., eds. **General Relativity and Gravitation.** New York: Halsted Press, John Wiley & Sons; Jerusalem: Israel Universities Press; 1975. 344p. 14 papers.
C

Seventh International Conference (GR 7). Tel Aviv, Israel: 23-28 Jun 1974. (Proceedings). A summary of current ideas and theory, also covering cosmology and high-energy astrophysics. The conference summary by J. A. Wheeler contains a bibliography of 182 references, covering 1915-75.

066.043 Smarr, Larry L., ed. **Sources of Gravitational Radiation.** Cambridge, UK: Cambridge University Press; 1979. 505p. 25 papers.
C

Battelle Seattle Workshop. Seattle, WA, USA: 24 Jul—4 Aug 1978. (Proceedings). Edited, advanced lectures reviewing the four major aspects of the subject: astrophysics; theoretical relativity; numerical relativity; and experimental relativity. Papers are arranged under four categories: gravitational wave detectors; theoretical foundations; black holes; neutron stars. The book's general theme is the possible sources of gravitational radiation and how they might be detected.

066.044 Taylor, J. G. **Special Relativity.** Oxford, UK: Clarendon Press; London: Oxford University Press; 1975. 106p. (Oxford Phys. Ser., 10). 8 chapters w/refs.

A brief, intermediate-level university text. Topics: speed of light; measuring time and distance; the Lorentz transformation; relativistic kinematics; relativistic dynamics; structure of special relativity; extensions of special relativity.

066.045 Thorne, K. S. General-relativistic astrophysics. In: Leibovitz, N. R.; Reid, W. H.; Vandervoort, P. O., eds. **Theoretical Principles in Astrophysics and Relativity.** Chicago: University of Chicago Press; 1978: 149-216. 254 refs, 1916-77.

An extensive review of the effects of general relativity on astrophysical systems such as stars, globular clusters, galactic nuclei, black holes. In particular, the author discusses in great detail gravitational wave astronomy in three parts: 1) gravitational waves that bathe the Earth; 2) problems and prospects for detection of the predicted waves; 3) the theory of small perturbations of relativistic stars and black holes, and the gravitational waves they generate.

066.046 **Unified Field Theory: General Relativity—Gravitation Theory—Cosmology; Volume 1.** Alexandria, VA: Defense Documentation Center; 1969. 204p. 159 refs, 1961-68.

A bibliography of unclassified documents in the DDC collection on the general relativity-gravitation theory-cosmology aspects of "current research for mathematical solutions and theoretical explanations of Einstein's equations of Unified Field Theory. Arranged in accession number (AD) order, there are author, report number, and title indexes, as well as listings of documents under five subject headings: cosmology; gravitational collapse; gravitational fields; gravitational radiation; gravitational redshift.

066.047 van der Laan, H., ed. Extragalactic high energy astrophysics. **Highlights**
C **Astron.** 5: 619-774; 1980. 21 papers.

Joint meeting at the XVII General Assembly of the IAU. Topics include active nuclei and Seyferts, radio galaxies and quasars, clusters, and cosmology.

066.048 Wagoner, R. V. Physics of massive objects. **Annu. Rev. Astron. Astrophys.** 7: 553-576; 1969. 91 refs, 1926-69.

Review of the theory and evolution of objects with a mass greater than 1,000 solar masses. The emphasis is on the equilibrium stages of evolution.

066.049 Weinberg, Steven. **Gravitation and Cosmology: Principles and Applications of the General Theory of Relativity.** New York: John Wiley & Sons; 1972. 657p. 16 chapters w/refs.

An advanced text providing "a comprehensive picture of the experimental tests of general relativity and observational cosmology," using a nontraditional, non-geometrical approach. Topics: history; special relativity; the principle of equivalence; tensor analysis; curvature; effects of gravitation; Einstein's field equations; gravitational radiation; stellar equilibrium and collapse; cosmology; etc.

066.050 Wrigley, W., ed. Space relativity. **Acta Astronaut.** 5: 3-130; 1978. 13 papers.
C Second and Third Space Relativity Symposia. Lisbon, Portugal: 21-27 Sep 1975; Anaheim, CA, USA: 10-16 Oct 1976. Reports of recent studies in space on both general and special relativistic effects, covering theory, instrumentation, and procedures.

066.051 Zakharov, V. D. **Gravitational Waves in Einstein's Theory.** Jerusalem: Israel Program for Scientific Translation; New York: John Wiley & Sons; 1973. 183p. 13 chapters. 465 refs, 1916-70.

"A review of work on gravitational waves in general relativity. Its central theme is the exposition of rigorous approaches to the problem, principally the definitions and criteria for distinguishing wavelike gravitational fields from other solutions of Einstein's equations." A highly technical work, it contains only one chapter devoted to the detection of gravitational waves and a review of results so far.

066.052 Zel'dovich, Ya. B.; Novikov, I. D. **Relativistic Astrophysics, Volume 1: Stars and Relativity.** Chicago: University of Chicago Press; 1971. 522p. 14 chapters. ~800 refs, 1890-1970.

A text applying the theory of general relativity to various astrophysical problems and processes. Topics: the theory of gravitation; the equation of state of matter; relativistic stages of evolution of cosmic objects (stars, star clusters, QSOs, etc.). Also discussed are physical processes in the vicinities of relativistic objects and a comparison with observations. This English edition was revised and enlarged by the authors; it was translated by E. Arlock and edited by K. S. Thorne and W. D. Arnett.

Astrophysics and General Relativity. 061.024.

Astrophysics and Gravitation. 061.005.

Black holes and neutron stars: evolution of binary systems. *066.508.*

Bulletin GRG, no. 38. Stars (including black holes). *065.003.*

The chemical evolution of the galaxies. *151.040.*

The cosmic microwave background radiation. *162.035.*

Galaxies and Relativistic Astrophysics. 151.006.

Neutron Stars, Black Holes and Binary X-Ray Sources. 066.504.

The physics of gravitationally collapsed objects. *066.510.*

Principles of Cosmology and Gravitation. 162.004.

Quantum physics and gravitation: list of publications. *022.009.*

Solutions of the Einstein equation: list of publications. *022.010.*

Special Relativity: The Foundation of Macroscopic Physics. 022.020.

The Special Theory of Relativity. 022.065.

Stars and Relativity. 066.052.

Stars: Their Birth, Life, and Death. 065.036.

Theoretical Principles in Astrophysics and Relativity. 061.048.

066.5 Neutron Stars

066.501 Baym, G.; Pethick, C. Physics of neutron stars. **Annu. Rev. Astron. Astrophys.** 17: 415-443; 1979. 184 refs, 1964-79.

Includes a review of recent developments; a summary of the physical characteristics of neutron stars; and a discussion of the equation of state, neutron star models, and nonequilibrium processes.

066.502 Cameron, A. G. W. Neutron stars. **Annu. Rev. Astron. Astrophys.** 8: 179-208; 1970. 92 refs, 1932-70.

A review of theoretical work done from 1966 to 1970, with an emphasis on stellar models. Topics include the atmosphere, crust, interiors, vibration, magnetic fields, and rotation.

066.503 Cameron, A. G. W.; Canuto, V. Neutron stars: general review. In: **Astrophysics and Gravitation.** Bruxelles: Editions de l'Universite Bruxelles; 1974: 221-278. 113 refs, 1932-74.

A review of the physics of the interiors of neutron stars, covering the surface, central region, solid core, ultra-high density regime, equation of state and models, and magnetic structure. Evolution of these stars is also detailed.

066.504 Gursky, Herbert; Ruffini, Remo, eds. **Neutron Stars, Black Holes and Binary**
C **X-Ray Sources.** Dordrecht, Holland; Boston: D. Reidel; 1975. 441p. (Astrophys. Space Sci. Libr., 48). 28 papers.

Articles based on a session of the February 1974 annual meeting of the American Association for the Advancement of Science held in San Francisco. Review papers, reprints of key papers on collapsed objects, and papers on current research on gravitationally collapsed stars. Includes a list of 92 galactic x-ray sources.

066.505 Hartle, J. B. Bounds on the mass and moment of inertia of nonrotating neutron stars. **Phys. Rep.** 46: 201-247; 1978. 94 refs, 1916-78.

Concerned chiefly with optimum bounds—bounds for which there is at least one configuration consistent with the assumptions on the matter for which the boundary inequality becomes an equality—within the context of Einstein's general relativity.

066.506 Irvine, J. M. **Neutron Stars.** Oxford, UK: Clarendon Press, Oxford University Press; 1978. 157p. (Oxford Stud. Phys.). 6 chapters w/refs.

A brief, advanced look at the physical properties, especially internal structure, of neutron stars. Topics: introduction, pulsars, neutron star temperatures, exterior of neutron stars, neutron star structure, neutron star equations of state.

066.507 Källman, C. G. Neutron stars: dynamics and structure. **Fundam. Cosmic Phys.** 4: 167-210; 1979. 109 refs, 1916-78.

Discussion of some of the dynamical properties of dense, relativistic stars and their implications for neutron star structure, with emphasis on the observable parameters.

066.508 Kraft, R. P. Black holes and neutron stars: evolution of binary systems. In: Gursky, Herbert; Ruffini, Remo, eds. **Neutron Stars, Black Holes and Binary X-Ray Sources.** Dordrecht, Holland: D. Reidel; 1975: 235-255. 75 refs, 1944-74.

A review of work which tries to prove the existence of neutron stars and black holes in double star systems. Observation and theory are both presented.

066.509 Pandharipande, V. R.; Pines, D.; Smith, R. A. Neutron star structure: theory, observation, and speculation. **Astrophys. J.** 208: 550-566; 1976. 81 refs, 1949-76.

The broad physical aspects of the neutron-neutron interaction are reviewed, and an examination is made of the extent to which the equation of state of neutron star matter is influenced by the various phase transitions which have been proposed for the high-density regime.

066.510 Ruffini, R. The physics of gravitationally collapsed objects. In: Gursky, Herbert; Ruffini, Remo, eds. **Neutron Stars, Black Holes and Binary X-Ray Sources.** Dordrecht, Holland: D. Reidel; 1975: 59-118. 142 refs, 1929-75.

A technical review which considers neutron stars, black holes, gravitational radiation detectors, regularly pulsating and bursting binary x-ray sources, gravitational collapse, and other topics.

066.511 Sheldon, E. The structure and physics of neutron stars. **Nukleonika** 23: 1091-
C 1139; 1978. 199 refs.

Eleventh Masurian School in Nuclear Physics. Mikolajki, Poland: 30 Aug–2 Sep 1978. (Proceedings). A review of physical characteristics with emphasis on the physics of matter under extreme conditions. Certain pulsars (e.g., those in the Crab Nebula and Vela supernova remnant) are discussed in detail, with spectral data being considered at some length.

Black holes: the outside story. *066.020.*

On the properties of matter in neutron stars. *066.020.*

Physics and Astrophysics of Neutron Stars and Black Holes. 066.016.

Physics of Dense Matter. 022.040.

SUN

071 Photosphere, Spectrum

071.001 Gomez, M. T. Lo spettro infrarosso solare: lo spettro continuo. [The infrared solar spectrum: the continuous spectrum.] **Mem. Soc. Astron. Italiana** 48: 617-646; 1977. 60 refs, 1879-1975. In Italian.

A study of infrared solar radiation with historical outline and references to the present state of the art.

071.002 Hauge, Ø.; Engvold, O. Compilation of solar abundance data. **Inst. Theor. Astrophys., Blindern-Oslo, Rep.** 49; 1977. 23p. 167 refs, 1929-77.

Solar abundance data of 67 elements are tabulated from the literature. Supersedes reports no. 31 (1970) and no. 39 (1974).

071.003 Linsky, J. L. The solar output and variability in the broader context of stellar activity. In: White, Oran R., ed. **The Solar Output and Its Variation.** Boulder, CO: Colorado Associated Universities Press; 1977: 477-515. 231 refs, 1934-76.

A review of various solar-type phenomena, diagnostics for identifying and describing this phenomena, and evidence (if any) for these phenomena in cool stars. The phenomena considered are non-radiative heating in the upper photosphere; chromospheres; transition regions; cornae and winds; magnetic fields and stellar cycles; and flares, starspots, and plages. An appendix describes various chromospheric diagnostics such as CaII, MgII, and HI.

071.004 Oertel, G. K.; Epstein, G. L. Current solar spectroscopic research. **Appl. Spectrosc. Rev.** 10: 139-200; 1975. 260 refs, 1817-1975.

Review addressed to the laboratory spectroscopist, with emphasis on observations made from space vehicles. Topics include solar physics, spectroscopic applications, requirements for solar instrumentation, and satellite experiments.

071.005 Stix, M., ed. Large-scale velocity fields on the Sun. **Highlights Astron.** 5: 71-
C 127; 1980. 4 papers.

Joint discussion at the XVII General Assembly of the IAU. Topics include global oscillations, circulation, differential rotation, and H-alpha patrols.

071.006 Swensson, J. W., et al. The solar spectrum from λ 7498 to λ 12016. A table of
A measures and identifications. **Mem. Soc. R. Sci. Liège.** 5e ser. 5 (special volume): 7-449; 1970. 148 refs, 1868-1969.

A list of more than 10,000 spectral lines, with accurate wavelengths, and the central depression of the lines given in a scale ranging between 1 and 200.

071.007 Walker, A. B. C., Jr. The coronal x-spectrum: problems and prospects. **Space Sci. Rev.** 13: 672-730; 1972. 223 refs, 1948-72.

Review of high resolution observations of the spectrum of the non-flaring sun.

071.008 White, Oran R., ed. **The Solar Output and Its Variation.** Boulder, CO: Colorado
C Associated Universities Press; 1977. 526p. 35 papers.

Solar Output Workshop. Boulder, CO, USA: 26-28 Apr 1976. A comprehensive look at the various types of radiation emitted by the Sun, along with a wealth of quantitative data resulting from observations. Aimed at scientists in climatology, aeronomy, stellar physics, planetary physics, and solar physics, the book covers solar variability, integrated solar flux, solar spectrum, stellar variability, etc.

Iron in the Sun and stars. *022.033*.

Stellar and solar abundances. *061.058*.

072 Sunspots, Faculae, Activity Cycles, Solar Patrol

072.001 Bartholomew, C. F. The discovery of the solar granulation. **Q. J. R. Astron. Soc.** 17: 263-289, 1976. 121 refs, 1801-1971.
Historical survey of the attempts to determine the physical constitution of the Sun.

072.002 Bruzek, A. Properties of solar active regions. In: Dyer, E. R., ed. **Solar-Terrestrial Physics/1970.** Dordrecht, Holland: D. Reidel; 1972: pt. I, 49-60. 108 refs, 1939-72.
A brief review of four aspects of the solar active (magnetic) regions: 1) appearance of new active regions; 2) strong field phenomena (e.g., spots); 3) moderate field phenomena (e.g., plages, coronal condensations); and 4) flares.

072.003 C Bumba, V.; Kleczek, J., eds. **Basic Mechanisms of Solar Activity.** Dordrecht, Holland; Boston: D. Reidel; 1976. 481p. (IAU Symp., 71). 30 papers, 12 abstracts.
Prague, Czechoslovakia: 25-29 Aug 1975. Stresses theoretical problems; does not deal with short-lived phenomena of the solar activity or with individual active regions. Sections include basic observed parameters of the solar cycle; solar convection and differential rotation; dynamo theory and magnetic dissipation; and stellar activity of the solar type.

072.004 Castelli, J. P.; Guidice, D. A. Impact of current solar radio patrol observations. **Vistas Astron.** 19: 355-384; 1976. 53 refs, 1961-75.
Review of solar radio astronomy and where such research may be going.

072.005 Kopecky, M. Babcock's theory of the 22-year solar cycle and the latitude drift of the sunspot zone. **Adv. Astron. Astrophys.** 7: 57-81; 1970. 21 refs, 1944-69.
Review of theories dealing with solar activity periodicity, with emphasis on the work of Babcock and those who modified his theory.

072.006 Kuklin, G. V. Cyclical and secular variations of solar activity. **IAU Symp.** 71: 147-190; 1976. 263 refs, 1901-1975.
Review of the wealth and diversity of the forms of solar activity phenomena and the various indices used to summarize them.

072.007 Michalitsanos, A. G. Recent theoretical interpretations of the solar five minute period oscillation. **Earth Extraterr. Sci.** 2: 125-138; 1973. 80 refs, 1945-73.
Consideration of current models of the solar five-minute-period oscillations.

072.008 Newkirk, G., Jr. Solar activity as relevant to solar planetary relationships. **EOS Trans. American Geophys. Union** 52: IUGG 485-490; 1971. 175 refs, 1961-71.
Review of recent developments in the areas of the general activity cycle, active regions, the role of magnetic fields, transient events, and solar-geomagnetic relations.

072.009 Parker, E. N. The enigma of solar activity. **IAU Symp.** 71: 3-16; 1976. 114 refs, 1894-1976.

Discussion of outstanding puzzles in the understanding of solar activity. Topics include solar convection, circulation and the dynamo, sunspots and intense flux tubes, and flares, x-rays, and eruptions.

072.010 Schröter, E. H. On magnetic fields in sunspots and active regions. **IAU Symp.** 43: 167-180; 1971. 93 refs, 1960-71.

Review of the topic for the past four years. Consideration of thermodynamical structure of sunspots, the average magnetic field, fine structure of the sunspot magnetic field, the magnetic field in pores, and so on.

072.011 **Study of the Solar Cycle from Space.** Washington, DC: NASA; 1980. 368p.
C (NASA CP-2098). 28 papers.

Wellesley, MA, USA: 14-15 Jun 1979. Papers and discussions reviewing possible future research on the solar interior and solar cycle. Topics: dynamics of the solar interior and solar dynamo; the corona and heliosphere; solar interior structure; global surface dynamics; x-ray corona; solar cycle effect on the solar wind; etc.

072.012 Vrabec, D. Streaming magnetic features near sunspots. **IAU Symp.** 56: 201-231; 1974. 68 refs, 1959-73.

Review of small moving magnetic features near sunspots and a discussion of magnetic flux outflow and inflow.

Large-scale velocity fields on the Sun. *071.005*.

073 Chromosphere, Flares, Prominences

073.001 Athay, R. Grant, ed. **Chromospheric Fine Structure.** Dordrecht; Boston: D.
C Reidel; 1974. 310p. (IAU Symp., 56). 36 papers, 23 abstracts.

Surfer's Paradise, Queensland, Australia: 3-7 Sep 1973. Topics: the quiet chromosphere; limb and disk phenomena; the upper chromosphere; the chromosphere in active regions; evolution of chromospheric fine structures; and energy balance, heat transfer, and heating mechanisms in chromospheric fine structures.

073.002 Athay, R. Grant. **The Solar Chromosphere and Corona: Quiet Sun.** Dordrecht, Holland; Boston: D. Reidel; 1976. 504p. (Astrophys. Space Sci. Libr., 53). 10 chapters w/refs.

An advanced text in the field of solar physics. Topics: structural features of the chromosphere and corona; macroscopic motions; magnetic fields; spectroscopy; empirical models of the chromosphere and corona; chromospheric structure; energy and momentum balance; wave generation and heating.

073.003 Beckers, J. M. Solar spicules. **Annu. Rev. Astron. Astrophys.** 10: 73-100; 1972. 174 refs, 1877-1971.

Updates earlier reviews of spicules by including observational and derived results from recent observations in the optical, far ultraviolet, and radio regions of the solar spectrum.

073.004 Bonnet, R.-M.; Delache, P., eds. **The Energy Balance and Hydrodynamics of**
C **the Solar Chromosphere and Corona.** Clermont-Ferrand, France: G. de Bussac; 1977. 504p. (IAU Colloq., 36). 13 papers.

Nice, France: 6-10 Sep 1976. (Proceedings). Primarily a review of ground-based data along with early results from Skylab and satellites, this volume also contains transcripts of discussions related to the papers presented. Topics: solar non-magnetic hydrodynamics; influence of magnetic fields; important physical problems; and energy balance in the non-active and active Sun.

073.005 Bray, R. J.; Loughhead, R. E. **The Solar Chromosphere.** London: Chapman and Hall; 1974. 384p. (Int. Astrophys. Ser.). 7 chapters w/refs.

An overview with an emphasis on recent research. Aimed at scientists and graduate students, the book contains hundreds of references to the literature. Topics: historical introduction; spicules and other fine structures at the solar limb; the morphology and dynamics of the quiet chromosphere; physical conditions in the quiet chromosphere; fine structure of the active chromosphere; wave propagation and dissipation; theories of heating and the origin of spicules.

073.006 Bruzek, A. Solar flares. In: Wentzel, D. G.; Tidman, D. A., eds. **Plasma Instabilities in Astrophysics.** New York: Gordon and Breach; 1969: 71-90. 107 refs, 1941-69.

A review of the characteristics, development, and effects of solar flares, the latter being various types of emission due to flare components involving plasma of different types. The origin and storage of flare energy, and changes in the magnetic field are also covered.

073.007 de Feiter, L. D. Solar flares as sources of energetic particles. **Space Sci. Rev.** 16: 3-43; 1974. 151 refs, 1910-74.

Review of solar flare phenomena with special emphasis on the production of suprathermal particles and their solar effects.

073.008 de Jager, C.; Švestka, Z., eds. **Solar Flares and Space Research.** Amsterdam:
C North-Holland; 1969. 419p. 38 papers.

Eleventh Plenary Meeting of COSPAR. Tokyo, Japan: 9-11 May 1968. (Proceedings). In addition to four basic review papers on solar flares, magnetic fields, and flare-associated optical phenomena, there are 34 other papers describing both observational and theoretical work. These include EUV, X- and gamma-ray observation; flare-associated radio and particle events; theories and laboratory experiments; and forecasting solar activity.

073.009 Dryer, M. Interplanetary shock waves generated by solar flares. **Space Sci. Rev.** 15: 403-468; 1974. 202 refs, 1950-73.

Review of recent observational and theoretical studies of interplanetary shock waves associated with solar flares. An attempt is made to outline their genesis, life, and demise.

073.010 Frazier, E. N. Motions of chromospheric fine structures. **IAU Symp.** 56: 97-135; 1974. 75 refs, 1954-73.

Concerned with the measurement of the effects of chromospheric motions and the diagnosis of those motions themselves over approximately the past 10 years.

073.011 Jensen, E.; Maltby, P.; Orrall, F. Q., eds. **Physics of Solar Prominences.** Blin-
C dern, Norway: Institute of Theoretical Physics; 1979. 375p. (IAU Colloq., 44). 44 papers.

Oslo, Norway: 14-18 Aug 1978. Review papers, reports of research, and discussions. Topics: spectra; magnetic fields; filaments; models; quiescent prominences; coronal loops; x-ray emission; origin and physical characteristics.

073.012 Kane, S. R. Impulsive (flash) phase of solar flares: hard x-ray, microwave, EUV and optical observations. **IAU Symp.** 57: 105-141; 1974. 165 refs, 1947-74.

Recent observations are briefly reviewed in order to deduce the characteristics of the impulsive phase phenomenon in small solar flares, particularly from the point of view of the acceleration of electrons and their role in producing the various impulsive phase emissions.

073.013 Kleczek, J.; Leroy, J.-L.; Orrall, F. Q. A general bibliography of solar prominence research 1880-1970. **Publ. Czech. Acad. Sci. Astron. Inst.** no. 53; 1972. 1,366 refs, 1880-1971.

The emphasis is on results and data analysis in this listing of journal articles and other sources. Instrumentation is not covered. Topics: statistical properties; morphology; motions; relation to other solar phenomena; radiation; spectrum; theoretical interpretations; models; specific types of prominences. Citations are arranged by author, and there is a classified subject index.

073.014 Křivský, L. Solar proton flares and their prediction. **Publ. Czech. Acad. Sci. Astron. Inst.** 52: 1-121; 1977. 250 refs, 1946-75.

A review of papers on the topic issued between 1946 and 1976, primarily by the Astronomical Institute of the Czechoslovak Academy of Sciences. In addition, a catalog of solar proton flares for the 30-year period is included.

073.015 Lanzerotti, L. J. Observations of solar particle propagation. In: Page, D. E., ed. **Correlated Interplanetary and Magnetospheric Observations.** Dordrecht, Holland: D. Reidel; 1974: 345-379. (Astrophys. Space Sci. Libr., 42). 125 refs, 1962-73.

A review of propagation characteristics from impulsive flare events between 1969 and 1973. The paper is limited to a discussion of the "observations of impulsive solar flare events in the theoretical context of the complete particle transport equation, including the convection and energy loss terms."

073.016 McGuire, James P., ed. **A Solar Flare Bibliography for 1973 through 1975.** Marshall Space Flight Center, AL: NASA; 1976. 87p. (NASA TM-X-73312; N76-26131). 185 refs, 1973-75.

An annotated listing containing three sections: 1) results from Skylab; 2) results from ground-based data during the Skylab time period; 3) other pertinent publications. Entries contain the usual bibliographic data and are arranged by author within each section. There is no subject index. Entries are taken from scientific journals, primarily *Solar Physics.*

073.017 Pallavicini, R. Observational aspects of flare phenomena. **Mem. Soc. Astron. Italiana** 48: 161-196; 1977. 105 refs, 1949-75. In Italian.

A review of the observations of electromagnetic and particle emission from solar flares, with special emphasis on the relationships between characteristic features of flares in different spectral bands.

073.018 **The Proton Flare Project (The July 1966 Event).** Cambridge, MA: MIT Press; 1969. 511p. (Ann. Int. Years Quiet Sun, 3). 60 papers.

Short reports of research on the study of a solar flare (July 7, 1966) producing high-energy particles, or a proton flare. Topics: the evolution of the proton-flare active region; the proton flare itself; the coronal effects of the ejected particles; the particle propagation in space; the associated phenomena on Earth; etc. (see also *080.037*).

073.019 Rosenberg, H. Solar radio observations and interpretations. **Philos. Trans. R. Soc. London, Ser. A.** 281: 461-471; 1976. 83 refs, 1946-75.

Summary of work in recent years on observations of solar flares with particular emphasis on the unsolved problems and on all those aspects that lead to more understanding of particle acceleration processes.

073.020 Rust, D. M. Solar flares. In: Parker, E. N.; Kennel, Charles F.; Lanzerotti, Louis J., eds. **Solar System Plasma Physics. Volume I.** Amsterdam: North-Holland; 1979: 51-98. 145 refs, 1859-1979.

A review of the phenomenon, its physical characteristics and effects, as well as certain observational data, such as that obtained by Skylab. Topics: flares and plasma physics; the preflare state; energy release; flare models.

073.021 Sakurai, K. Solar flare emissions and geophysical disturbances. **Indian J. Radio Space Phys.** 3: 289-312; 1974. 190 refs, 1950-74.

A review of the nature of solar flares, their development, and geophysical effects. Topics: sudden ionospheric disturbances (SIDs); polar cap absorption (PCA); wave and particle emissions; flare-generated interplanetary disturbances; forecasting solar flares; etc.

073.022 Smolkov, G. Ya.; Bashkirtsev, V. S. Magnetic fields in solar prominences. In: Sýkora, J., ed. **Solar Activity and Solar-Terrestrial Relations.** Bratislava, Czechoslovakia: VEDA; 1976: 175-194. 320 refs, 1913-75.

A review covering magnetic fields in both the solar corona and prominences. Covered are the results of magnetographic and spectrographic measurements of the magnetic fields, as well as data on the middle corona resulting from studies of the prominences.

073.023 Sturrock, Peter A., ed. **Solar Flares.** Boulder, CO: Associated Universities
C Press; 1980. 513p. 9 chapters w/refs.

A monograph from Skylab Solar Workshop II, Boulder, CO, USA: 1976-77. A review of results from observations made on Skylab using solar instruments known as the Apollo Telescope Mount (ATM). Topics: the preflare state; primary energy release; energetic particles in solar flares; impulsive phase of solar flares; the chromosphere and transition region; mass ejections; thermal x-ray flare plasma; flare models.

073.024 Švestka, Z., ed. Flare build-up study. **Sol. Phys.** 47: 1-432; 1976. Also:
C Dordrecht, Holland; Boston: D. Reidel; 1977. 42 papers.

Falmouth, Cape Cod, MA, USA: 8-11 Sep 1975. Topics include a review of present knowledge, ATM observations, a general discussion, and plans for the future. The purpose was to bring together solar and magnetospheric physicists to discuss problems which might be common both to the solar flares and the Earth magnetosphere.

073.025 Švestka, Z., ed. Proceedings of the meeting "How can flares be understood?"
C **Sol. Phys.** 53: 217-301; 1977. 17 papers.

Grenoble, France: 27 Aug 1976. The discussion is divided into two parts: magnetic configurations and instabilities in flares and location of the primary flare site and energy transfer in flares.

073.026 Švestka, Z. Spectra of solar flares. **Annu. Rev. Astron. Astrophys.** 10: 1-24; 1972. 106 refs, 1954-71.

Discussion of the flare spectrum in the region accessible to ground-based optical solar spectrographs. Topics include general features of the spectrum, Balmer line broadening, electron density, optical thickness, hydrogen density, filamentary structure, helium lines, H and K lines, metallic lines, line asymmetry, and continuous emission.

073.027 Švestka, Z.; Simon, P., eds. **Catalog of Solar Particle Events, 1955-69.** Dordrecht, Holland; Boston: D. Reidel; 1975. 428p. (Astrophys. Space Sci. Libr., 49). 679 refs, 1955-74.

A compendium of observational data in three parts: 1) the catalog; 2) list of flares identified as sources of particle events; 3) maps of selected active (solar) regions which were sources of particle events. The catalog data includes the kind of observation or the name of the spacecraft, kind of particles and recorded energy range, onset time, time of maximum particle flux, duration, magnitude, and references. Two very lengthy lists of references are included.

073.028 Švestka, Zdeněk. **Solar Flares.** Dordrecht, Holland; Boston: D. Reidel; 1976. 399p. (Geophys. Astrophys. Monogr., 8). 6 chapters. 1,148 refs, 1859-1975.

An advanced monograph reviewing all aspects of the nature, physics, and effects of the phenomena, with emphasis on emissions. Topics: introduction; the low-temperature flare (observations, spectra, models); the high-temperature flare (thermal and non-thermal phenomena); flare-associated optical phenomena; particle emission; flare models (including energy source, storage, and release).

073.029 Sweet, P. A. Mechanisms of solar flares. **Annu. Rev. Astron. Astrophys.** 7: 149-176; 1969. 89 refs, 1859-1968.

Critical account of theories of the origin of flares.

073.030 Tandberg-Hanssen, E. Physics of solar flares. **Earth Extraterr. Sci.** 2: 89-98; 1973. 101 refs, 1933-73.

Discusses flare phenomena in the framework of the preexisting magnetic fields, and distinguishes three flare components: plage, prominence, and sunspot.

073.031 Tandberg-Hanssen, Einar. **Solar Prominences.** Dordrecht, Holland; Boston: D. Reidel; 1974. 155p. (Geophys. Astrophys. Monogr., 12). 7 chapters. 456 refs, 1733-1973.

An advanced, introductory text for graduate students and nonspecialists. Topics: history; observational data; models; formation of prominences; stability; interaction of prominences with centers of activity (e.g., sunspots); prominences as part of the corona.

073.032 Withbroe, G. L.; Noyes, R. W. Mass and energy flow in the solar chromosphere and corona. **Annu. Rev. Astron. Astrophys.** 15: 363-387; 1977. 133 refs, 1961-77.

Discussion of the results of investigations done in the hope of developing a physical model to account for the physical conditions in the outer atmosphere of the Sun which can be applied to the study of chromospheres and coronae of other stars.

The enigma of solar activity. *072.009.*

Properties of solar active regions. *072.002.*

X rays from the Sun. *076.008.*

074 Corona, Solar Wind

074.001 Axford, W. I. The interaction of the solar wind with the interstellar medium. In: Sonett, C. P.; Coleman, P. J., Jr.; Wilcox, J. M., eds. **Solar Wind.** Washington, DC: NASA; 1972: 609-660. 327 refs, 1951-72.

An extensive review of work in the field, first describing the properties of the interstellar medium near the Sun and beyond the Earth's orbit. Interactions with the following solar wind components are discussed at great length: ionized interstellar gas, the interplanetary magnetic field, and neutral interstellar gas.

074.002 Barnes, A. Hydromagnetic waves and turbulence in the solar wind. In: Parker, E. N.; Kennel, Charles F.; Lanzerotti, Louis J., eds. **Solar System Plasma Physics. Volume I.** Amsterdam: North-Holland; 1979: 249-319. 236 refs, 1946-79.

A review of "the current state of knowledge of interplanetary fluctuations, their origins and their effects on the wind," and a discussion of some astrophysical problems for which solar-wind hydromagnetic-wave studies have relevance.

074.003 Brandt, J. C.; Mendis, D. A. The interaction of the solar wind with comets. In: Kennel, Charles F.; Lanzerotti, Louis J.; Parker, E. N., eds. **Solar System Plasma Physics. Volume II.** Amsterdam: North-Holland; 1979: 253-292. 94 refs, 1836-1978.

An overview of current knowledge and past research. Topics: continuum description of the interaction; cometary plasma; generation of tail streamers; gross tail structure; fine structure in the plasma tail; etc.

074.004 Brandt, John C. **Introduction to the Solar Wind.** San Francisco: W. H. Freeman; 1970. 199p. (Ser. Books Astron. Astrophys.). 7 chapters w/refs.

A text for advanced students and nonspecialists. Topics: historical introduction; solar physics summary; theory; ground-based methods of observation; space observations; effects on solar system; implications for astrophysics (origin of solar wind; solar and stellar rotation; stellar winds; etc.).

074.005 Cuperman, S. Theoretical contributions to solar wind research—a review. In: Shea, M. A.; Smart, D. F.; Wu, S. T., eds. **Study of Travelling Interplanetary Phenomena/1977.** Dordrecht, Holland: D. Reidel; 1977: 165-194. 87 refs, 1951-78.

A description of work done since 1958, primarily the development and refinement of E. N. Parker's macroscopic model with an emphasis on steady-state, spherically symmetric flow. A brief discussion of the generalization of Parker's initial model to microscopic (kinetic) aspects is also included in the paper.

074.006 Dobrowolny, M.; Moreno, G. Plasma kinetics in the solar wind. **Space Sci. Rev.** 20: 577-620; 1977. 101 refs, 1953-76.

Review of observations and theoretical ideas concerning the role of kinetic processes in the solar wind.

074.007 Dryer, M.; Tandberg-Hanssen, E., eds. **Solar and Interplanetary Dynamics.**
C Dordrecht, Holland; Boston: D. Reidel; 1980. 558p. (IAU Symp., 91). 38 papers, 42 abstracts.

Cambridge, MA, USA: 27-31 Aug 1979. Sections include the life history of coronal structures and fields, coronal and interplanetary responses to long-time scale phenomena, solar transient phenomena affecting the corona and interplanetary medium, coronal and interplanetary responses to short-time scale phenomena, and future directions.

074.008 Fisk, L. A. The interactions of energetic particles with the solar wind. In: Parker, E. N.; Kennel, Charles F.; Lanzerotti, Louis J., eds. **Solar System Plasma Physics. Volume I.** Amsterdam: North-Holland; 1979: 177-247. 142 refs, 1935-79.

A review of current and recent studies with an emphasis on theories of interaction. Topics: theory of solar modulation of galactic cosmic rays; theories for the propagation of energetic particles in the solar wind; and theories for their acceleration.

074.009 Harvey, J. W.; Sheeley, N. R., Jr. Coronal holes and solar magnetic fields. **Space Sci. Rev.** 23: 139-158; 1979. 95 refs, 1889-1978.

Review of observations of coronal holes and their associated photospheric magnetic fields made from 1972 to 1978.

074.010 Hollweg, J. V. Some physical processes in the solar wind. **Rev. Geophys. Space Phys.** 16: 689-720; 1978. 257 refs, 1953-78.

Summary of some of the relevant processes and a new model of the solar wind flow that incorporates these processes.

074.011 Hollweg, J. V. Waves and instabilities in the solar wind. **Rev. Geophys. Space Phys.** 13: 263-289; 1975. 156 refs, 1956-75.

Review article concentrating on those aspects that are likely to play important roles in influencing the dynamic and thermodynamic states of the general solar wind expansion.

074.012 Holzer, T. E. The solar wind and related astrophysical phenomena. In: Parker, E. N.; Kennel, Charles F.; Lanzerotti, Louis J., eds. **Solar System Plasma Physics. Volume I.** Amsterdam: North-Holland; 1979; 101-176. 203 refs, 1952-79.

A review of our current understanding of the large-scale dynamics of the solar wind, and the relationship between physical processes in the solar wind and in various astrophysical plasmas. Most of the discussion of the solar wind is presented in the context of recent studies of coronal holes.

074.013 Hundhausen, A. J. Composition and dynamics of the solar wind plasma. **Rev. Geophys. Space Phys.** 8: 729-811; 1970. 160 refs, 1961-70.

Review of the relationship between solar and solar wind properties. Concentration on observations and theoretical interpretations relating solar wind chemical composition and dynamical properties to solar properties and dynamical processes.

074.014 Hundhausen, A. J. **Coronal Expansion and Solar Wind.** New York: Springer-Verlag; 1972. 238p. (Phys. Chem. Space, 5). 7 chapters. 337 refs, 1919-72.

A review of solar wind observations and coronal expansion models from the past decade. Aimed at the graduate student and nonspecialist, the book attempts to interpret "observed solar wind phenomena as the effects of basic physical processes occurring in the coronal and interplanetary plasma and as the natural manifestations of solar properties and structure." Topics: identification and classification of important solar wind phenomena; dynamics of a structureless coronal expansion; chemical composition of coronal and interplanetary plasma; high speed plasma streams; etc.

074.015 Hundhausen, A. J. Streams, sectors, and solar magnetism. In: Eddy, John A., ed. **The New Solar Physics.** Boulder, CO: Westview Press; 1978: 59-133. 110 refs, 1858-1977.

A review of the relationship between coronal holes and the interplanetary stream-sector structure, emphasizing results obtained by Skylab research. The paper includes a considerable number of illustrations showing coronal hole/solar wind activity. Originally published as chapter 7 of *Coronal Holes and High Speed Wind Streams* (*074.040*).

074.016 Jokipii, J. R. Propagation of cosmic rays in the solar wind. **Rev. Geophys. Space Phys.** 9: 27-87; 1971. 28 refs, 1935-70.

A coherent exposition of the modern statistical theory of the transport of fast-charged particles in the solar wind. Observations are discussed only as they illustrate the phenomena under discussion.

074.017 Kavanagh, L. D., Jr.; Schardt, A. W.; Roelof, E. C. Solar wind and solar energetic particles: properties and interactions. **Rev. Geophys. Space Phys.** 8: 389-460; 1970. 271 refs, 1955-70.

Review of the findings since 1966 about the properties of the interplanetary magnetic field, the solar wind plasma, solar wind interactions with the Earth, the Moon, Mars, and Venus; and the properties and propagation characteristics of energetic solar flare particles.

074.018 Lee, M. A.; Lerche, I. Waves and irregularities in the solar wind. **Rev. Geophys. Space Phys.** 12: 671-687; 1974. 50 refs, 1940-73.

Review for the nonspecialist who wants a simple physical picture.

074.019 Lemaire, J.; Scherer, M. Kinetic models of the solar and polar winds. **Rev. Geophys. Space Phys.** 11: 427-468; 1973. 125 refs, 1922-73.

Discussion of the application of the kinetic theory to the collisionless regions of the polar and solar winds.

074.020 Livshits, M. A. Physical characteristics of sources of x-rays in the solar corona. **Itogi Nauki Tekh. Ser. Astron.** 9: 111-166; 1974. 289 refs, 1930-73. In Russian.
Presents a systematic survey and analysis of phenomena in x-ray emitting solar coronal plasma, in light of recent observations. The bibliography contains 243 items in English.

074.021 Macris, C. J., ed. **Physics of the Solar Corona.** Dordrecht, Holland: D. Reidel;
C 1971. 345p. (Astrophys. Space Sci. Libr., 27). 22 papers.
NATO Advanced Study Institute. Athens, Greece: 6-17 Sep 1970. (Proceedings). A review of current research and theory. Topics: atomic processes; magnetohydrodynamics and plasma physics; magnetic fields; coronal events; active regions and flare-associated events; x-ray spectroscopy; radio emission; solar bursts; etc.

074.022 Ness, Norman F., ed. **Solar-Wind Interaction with the Planets Mercury, Venus,**
C **and Mars.** Washington, DC: NASA; 1976. 170p. (NASA SP-397). 14 papers.
Seminar. Moscow, USSR: 17-21 Nov 1975. Review of current knowledge, based primarily on results from Venera, Mariner, and Mars spacecraft missions. Topics: Martian magnetic field and magnetosphere; planetary ionospheres; Mercury's magnetic field; solar wind flow around terrestrial planets; etc.

074.023 Newkirk, G., Jr., ed. **Coronal Disturbances.** Dordrecht, Holland; Boston: D.
C Reidel; 1974. 508p. (IAU Symp., 57). 20 papers, 45 abstracts.
Surfer's Paradise, Queensland, Australia: 7-11 Sep 1973. Topics include magnetic structure responsible for coronal disturbances; the flash phase of solar flares; shock waves and plasma ejection; acceleration, containment, and emission of high-energy flare particles; and reports on special observations.

074.024 Parker, E. N. Recent developments in theory of solar wind. **Rev. Geophys. Space Phys.** 9: 825-835; 1971. 76 refs, 1953-71.
Current problems and developments in the theory of the large-scale expansion of the solar corona. Brief review, with good basic list of other reviews.

074.025 Parker, E. N. Solar wind interaction with the geomagnetic field. **Rev. Geophys.** 7: 3-10; 1969. 29 refs, 1958-69.
Points out the various major effects involved directly in the interaction, and assesses the degree to which the effects are, and are not, understood.

074.026 Parker, E. N.; Kennel, Charles F.; Lanzerotti, Louis J., eds. **Solar System Plasma Physics. Volume I. Solar and Solar Wind Plasma Physics.** Amsterdam; New York: North-Holland; 1979. 344p. 6 papers.
A collection of review papers on recent research and current knowledge in the field. Topics: solar physics; solar flares; solar wind and related phenomena; interaction of energetic particles with the solar wind; hydromagnetic waves and turbulence in the solar wind; kinetic processes in the solar wind.

074.027 Russell, C. T. The interaction of the solar wind with Mars, Venus and Mercury. In: Kennel, Charles F.; Lanzerotti, Louis J.; Parker, E. N., eds. **Solar System Plasma Physics. Volume II.** Amsterdam: North-Holland; 1979: 207-252. 102 refs, 1963-78.
A review of past studies along with a comparison of data and theory for each planet. Drawing heavily on spacecraft-gathered data, especially from Soviet missions, the paper also attempts to define a magnetic hierarchy for the terrestrial planets. Future areas of study are also presented.

074.028 Russell, C. T. The solar wind and magnetospheric dynamics. In: Page, D. E., ed. **Correlated Interplanetary and Magnetospheric Observations.** Dordrecht,

Holland: D. Reidel; 1974: 3-47. (Astrophys. Space Sci. Libr., 42). 157 refs, 1930-74.

A review of the effects of the solar wind on the Earth's magnetosphere, including a history of such studies. The author examines evidence for magnetospheric dynamics, the effects of the interplanetary field, and the various dynamical changes that occur. Geomagnetic variations and substorms are also considered.

074.029 Russell, C. T., ed. **Solar Wind Three.** Los Angeles: Institute of Geophysics and
C Planetary Physics, UCLA; 1974. 487p. 54 papers.

Pacific Grove, CA, USA: Mar 1974. (Proceedings). A combination of review papers and reports on research covering the following subjects: solar abundances; history and evolution; structure and dynamics of the solar corona; macroscopic properties; cosmic rays; stellar winds; spatial gradients; microscopic properties; solar wind interactions.

074.030 **Solar Wind, Volume I. Report Bibliography, 1953-Aug 1968.** Alexandria, VA: Defense Documentation Center; 1969. 151p. (DDC-TAS-68-79-Vol-1; AD-684400; N69-28296). 123 refs, 1953-68.

A computer-produced list arranged by AD number of references from the DDC collection. Entries include bibliographic data, descriptors, and abstracts. Indexes include personal and corporate author, subject, and report number.

074.031 Sonett, C. P.; Coleman, P. J., Jr.; Wilcox, J. M., eds. **Solar Wind.** Washing-
C ton, DC: NASA; 1972. 717p. (NASA SP-308). 53 papers.

Pacific Grove, CA, USA: 21-26 Mar 1971. (Proceedings). With an emphasis on the solar wind as an astrophysical phenomenon, rather than its effects on the solar system, this book presents a combination of papers and discussions addressing both theory and observations. Topics: magnetic fields; solar wind plasma; angular momentum of the solar wind; shock waves; radio observations; etc.

074.032 Spreiter, J. R.; Alksne, A. Y. Solar-wind flow past objects in the solar system. **Annu. Rev. Fluid Mech.** 2: 313-354; 1970. 129 refs, 1931-70.

Review of the movement of the solar wind and its interaction with the Earth, Moon, Mars, and Venus, from a fluid mechanics point of view. Primarily based on spacecraft observations.

074.033 Summers, D. Fluid models of the solar wind. **J. Inst. Math. Appl.** 22: 71-87; 1978. 74 refs, 1896-1976.

Analysis of the fluid dynamical models that have been constructed to describe the solar wind since Parker's (1958) first model. Brief historical development is sketched, and an account is given of the gross properties of the solar wind medium.

074.034 Tandon, J. N., ed. **Solar Radiations and the Earth.** Delhi, India: Hindustan
C Publishing; 1973. 342p. 15 papers.

Delhi, India: 21-26 Jun 1971. (Proceedings). A combination of review papers and reports of research in India centering on the solar wind and solar flares, as well as the interactions of these phenomena with the Earth. Other topics include solar microwave bursts, interplanetary shockwaves, magnetospheric convection, etc.

074.035 Vaiana, G. S.; Rosner, R. Recent advances in coronal physics. **Annu. Rev. Astron. Astrophys.** 16: 393-428; 1978. 170 refs, 1946-78.

Summary of the present state of understanding of coronae, focusing particularly on the interplay between data interpretation and theoretical model construction.

074.036 Vasyliunas, V. M. Interplanetary fields and plasmas. **EOS Trans. American Geophys. Union** 52: IUGG 490-498; 1971. 274 refs, 1967-70.

Review of solar wind research from 1967 to 1970, including consideration of plasma, magnetic and electric fields, and interaction with the Moon and planets.

074.037 Wild, J. P.; Smerd, S. F. Radio bursts from the solar corona. **Annu. Rev. Astron. Astrophys.** 10: 159-196; 1972. 114 refs, 1951-72.

Review of radio emission from traveling disturbances in the corona and the role they play in the flare phenomena.

074.038 Wilson, Robert M. **Results of Coronal Hole Research: An Overview.** Marshall Space Flight Center, AL: NASA; 1976. 38p. (NASA TM-X-73317; N76-27151). 98 refs, 1931-76.

An overview of the past 10 years, with an emphasis on 1970-76, of both observational and theoretical results. Skylab data is included.

074.039 Wilson, Robert M. **Results of Coronal Hole Research: An Update.** Marshall Space Flight Center, AL: NASA; 1979. 96p. (NASA TM-78237; N79-32146). 253 refs, 1974-79.

A continuation of the previous entry, with an emphasis on results from 1976 to 1979. The 98-entry bibliography cited in the earlier paper is included here, in addition to the 253 new references.

074.040 Zirker, J. B., ed. **Coronal Holes and High Speed Wind Streams.** Boulder, CO:
C Colorado Associated Universities Press; 1977. 454p. 9 chapters w/refs.

A monograph based on the Skylab Solar Workshop I. Boulder, CO, USA: 1975-76. A review of current knowledge and data obtained from the Skylab solar observations and other sources during the mission. Coronal holes are discussed with regard to temporal behavior, solar magnetic fields, the chromospheric and transition layers, the Sun's interior, etc. Contains hundreds of references to the literature.

074.041 Zombeck, M. V., et al. An atlas of soft x-ray images of the solar corona from
A Skylab. **Astrophys. J. Suppl. Ser.** 38: 69-85; 1978. 219 refs, 1971-78.

The atlas contains 104 plates of x-ray images of the Sun obtained by the American Science and Engineering Company x-ray telescope on the Apollo telescope mount of Skylab, launched in May 1973. A complete bibliography (through February 1978) of scientific papers that have made extensive use of the database is also presented, along with a list of catalogs and atlases which provide complementary or supplementary data.

The Energy Balance and Hydrodynamics of the Solar Chromosphere and Corona. 073.004.

Hydromagnetic waves in interplanetary space. *106.005.*

The interplanetary plasma and the heliosphere. *062.011.*

Mass and energy flow in the solar chromosphere and corona. *073.032.*

The Solar Chromosphere and Corona: Quiet Sun. 073.002.

075 Magnetic Fields

075.001 Beckers, J. M. The measurement of solar magnetic fields. **IAU Symp.** 43: 3-23; 1971. 53 refs, 1913-70.

The different methods which have been used, or which may be used in the future, to measure solar magnetic fields are described and discussed.

075.002 Bumba, V. Large-scale solar magnetic fields. **IAU Symp.** 71: 47-67; 1976. 58 refs, 1955-74.

The characteristics of the large-scale distribution of the solar magnetic fields on the basis of a series of solar magnetic charts covering more than 15 years of observations are given.

075.003 Deubner, F. L., ed. The small scale structure of solar magnetic fields. **Highlights Astron.** 4(2): 219-275; 1977. 7 papers.
C

Joint discussion no. 6 of the Sixteenth General Assembly of the IAU. Topics: observations; theory; line profiles of faculae and pores; magnetic flux tubes; etc.

075.004 Howard, R. Large-scale solar magnetic fields. **Annu. Rev. Astron. Astrophys.** 15: 153-173; 1977. 69 refs, 1908-1977.

Review of two decades of magnetograph observations, showing that solar activity is the result of the motions of magnetic fields in and around the Sun.

075.005 Howard, R., ed. **Solar Magnetic Fields.** Dordrecht, Holland: D. Reidel; 1971. 782p. (IAU Symp., 43). 88 papers, 6 abstracts.
C

Paris, France: 31 Aug—4 Sep 1970. Topics include instrumentation, interpretation of magnetograph results, observations of sunspot and active region magnetic fields, observations of magnetic fields associated with flares, theories of small-scale magnetic fields, optical and radio observations of large-scale magnetic fields, polar fields and the magnetic activity cycles, and theories of large-scale fields.

075.006 Newkirk, G., Jr. Large scale solar magnetic fields and their consequences. **IAU Symp.** 43: 547-568; 1970. 88 refs, 1889-1971.

The general properties of large-scale solar magnetic fields are reviewed.

075.007 Parker, E. N. The origin of solar activity. **Annu. Rev. Astron. Astrophys.** 15: 45-68; 1977. 142 refs, 1894-1977.

Review of the basic physical problems and principles of solar activity, concentrating on the general question of why magnetic fields produce activity.

075.008 Parker, E. N. The origin of solar magnetic fields. **Annu. Rev. Astron. Astrophys.** 8: 1-30; 1970. 142 refs, 1907-1969.

Review concentrating on the origin and behavior of the general magnetic field of the Sun.

075.009 Stenflo, J. O. The measurement of solar magnetic fields. **Rep. Prog. Phys.** 41: 865-907; 1978. 203 refs, 1908-1978.

Deals primarily with how the Sun's magnetic field is measured, plus a brief review of the present observational pictures of the magnetic field.

075.010 Stenflo, J. O. Small-scale solar magnetic fields. **IAU Symp.** 71: 69-99; 1976. 132 refs, 1903-1975.

Review of the observed properties of small-scale solar magnetic fields, along with a discussion of the implications of the small-scale magnetic structure.

075.011 Svalgaard, L.; Wilcox, J. M. A view of solar magnetic fields, the solar corona, and the solar wind in three dimensions. **Annu. Rev. Astron. Astrophys.** 16: 429-443; 1978. 36 refs, 1889-1978.

Discussion of the information currently available about the dominance of the role of large-scale magnetic fields in determining the global structure of the solar wind.

075.012 Weiss, N. O. Small scale solar magnetic fields: theory. **Highlights Astron.** 4(2): 241-250; 1977. 60 refs, 1958-76.

Review of the principal problems raised by the detection of filamentary magnetic fields. Includes a discussion of the interaction of magnetic fields with convection in the Sun.

Coronal holes and solar magnetic fields. *074.009.*

Magnetic fields in solar prominences. *073.022.*

On magnetic fields in sunspots and active regions. *072.010.*

076 UV, X, Gamma Radiation

076.001 Acton, L. W.; Culhane, J. L., eds. Proceedings of the symposium on the tech-
C niques of solar and cosmic x-ray spectroscopy. **Space Sci. Instrum.** 2: 1-378; 1976. 24 papers.

Holmbury, UK: 22-23 May 1975. Selected conference papers emphasizing the design and use of equipment used both in Earth-based and space x-ray spectroscopy.

076.002 Brown, J. C. The interpretation of hard and soft x-rays from solar flares. **Philos. Trans. R. Soc. London, Ser. A.** 281: 473-490; 1976. 78 refs, 1952-75.

Review of the present status of observations of hard x-ray bursts in terms of the light they shed on alternative source models and on general characteristics of electron acceleration in flares. Observational characteristics of soft x-ray flares are cursorily reviewed.

076.003 Brown, J. C. The interpretation of spectra, polarization, and directivity of solar hard x-rays. **IAU Symp.** 68: 245-282; 1975. 102 refs, 1959-75.

Critical review of observations of burst characteristics, followed by a discussion of the problem of analytic and numerical inversion of the x-ray spectrum to give the flare electron spectrum.

076.004 Culhane, J. L.; Acton, L. W. The solar x-ray spectrum. **Annu. Rev. Astron. Astrophys.** 12: 359-381; 1974. 110 refs, 1931-74.

Structure of the solar corona, spatial distribution of the x-ray emission, and the observed features of both active region and flare x-ray spectra are discussed. Techniques for x-ray observations and x-ray emission processes are detailed.

076.005 de Feiter, L. D. Solar flare x-ray measurements and their relation to microwave bursts. **IAU Symp.** 68: 283-297; 1975. 78 refs, 1959-75.

A review of the available observational material of solar hard x-ray bursts, their interpretation in terms of a model of the source region, and their relation with other flash-phase phenomena, in particular the impulsive microwave bursts.

076.006 Kane, S. R., ed. **Solar Gamma-, X-, and EUV Radiation.** Dordrecht, Holland;
C Boston: D. Reidel; 1975. 439p. (IAU Symp., 68). 25 papers, 15 abstracts.

Buenos Aires, Argentina: 11-14 Jun 1974. Co-sponsored by COSPAR. Topics include general solar activity, coronal holes and bright points, active regions, and solar flares.

076.007 Korchak, A. A. On the origin of solar flare x-rays. **Sol. Phys.** 18: 284-304; 1971. 53 refs, 1939-71.

The origin of x-ray solar bursts is investigated on the basis of the theoretical model developed by Syrovatskii, with a brief review of other models.

076.008 Neupert, W. M. X rays from the Sun. **Annu. Rev. Astron. Astrophys.** 7: 121-148; 1969. 146 refs, 1935-68.

A description of above-the-atmosphere observations and a review of data and its interpretation. Emphasis is on solar flares and resultant x-ray emission. Instrumentation is also discussed.

076.009 Ramaty, R.; Stone, R. G., eds. **High Energy Phenomena on the Sun.** Washing-
C ton, DC: NASA; 1973. 641p. (NASA SP-342; GSFC X-693-73-193). 50 papers.

Greenbelt, MD, USA: 28-30 Sep 1972. A review of observational data and theory on the following topics: flares; microwaves and hard x-rays; ultraviolet and soft x-ray emissions; nuclear reactions in flares; energetic particles; magnetic fields; radio emissions in the corona and interplanetary space.

076.010 Righini, Guglielmo, ed. **Skylab Solar Workshop.** Arcetri, Italy: Osservatorio
C Astrofisico di Arcetri; 1975. 210p. (Oss. Mem. Oss. Astrofis. Arcetri, 104). 22 papers.

Preliminary results from the S-054 x-ray telescope and the correlated ground-based observations (Florence, Italy: 21-22 Mar 1974). Coverage includes "the major instrumental and theoretical problems present in the data reduction and interpretation of x-ray photographs of the solar corona." Ground-based optical and radio observations obtained during the Skylab mission were also presented and used for comparison with space data.

076.011 Vaiana, G.; Tucker, W. Solar x-ray emission. In: Giacconi, Riccardo; Gursky, Herbert, eds. **X-ray Astronomy.** Dordrecht, Holland: D. Reidel; 1974; 169-205. 47 refs, 1956-74.

An overview of observations and theory regarding x-ray emission from various parts of the Sun. Includes many x-ray photographs illustrating the different types of emission, as well as a description of the various parts of the Sun and the associated physical processes: convection zone, photosphere, corona, chromosphere. The roles of sunspots, plages, prominences, and flares are discussed as well.

076.012 Walker, A. B. C., Jr. Interpretation of the x-ray spectra of solar active regions. **IAU Symp.** 68: 73-100; 1975. 92 refs, 1962-75.

Review of recent analytical studies of the coronal x-ray spectrum below 25Å.

076.013 Wilson, Robert M.; Reynolds, John M.; Fields, Stanley A. **A Solar X-ray Astronomy Summary and Bibliography.** Huntsville, AL: Marshall Space Flight Center, NASA; 1970. 143p. (NASA TM-X-53991; N70-39828). 153 refs, 1951-69.

An annotated bibliography of articles and papers arranged alphabetically by author. Each entry has complete bibliographic data and an extremely detailed and lengthy abstract/annotation. The articles are followed by an author index and brief subject index. There are also four extensive tables of data on rocket, balloon, and satellite experiments and observed solar spectral lines (1 to 100 Å).

The coronal x-spectrum: problems and prospects. *071.007.*

International Conference on X-rays in Space (Cosmic, Solar and Auroral X-Rays). 142.043.

Physical characteristics of sources of x-rays in the solar corona. *074.020.*

Ultraviolet studies of the solar atmosphere. *080.029.*

077 Radio, Infrared Radiation

077.001 Barletti, R.; Pampaloni, P. The quiet sun emission at mm wavelengths. **Mem. Soc. Astron. Italiana** 43: 547-566; 1972. 73 refs, 1949-72.
Observational results and models are emphasized in this review, which addresses the spectrum at the disk center and the center-limb distribution. A table summarizing solar brightness temperatures at various wavelengths is included.

077.002 Elgaroy, E. O. **Solar Noise Storms.** Oxford, UK; New York: Pergamon Press; 1976. 363p. (Int. Ser. Nat. Philos., 90). 13 chapters. 300 refs, 1946-75.
An advanced text dealing with only one aspect of solar radio astronomy, that of noise storms. The author reviews recent observations, interpretations, and theories in order to provide groundwork for further study. Specific subjects include, but are not limited to, relations between noise storms and optically visible solar features; spectrum of noise storms; polarization of noise storm emission; directivity; coronal scattering of radiation; etc.

077.003 Fomichev, V. V.; Chertok, I. M. Fine structure of solar radio bursts at meter wavelengths: a survey. **Radiophys. Quantum Electron** 20: 869-898; 1977. Transl. of *Izv. VUZ Radiofiz.* 20: 1255-1301; 1977. 140 refs, 1960-77.
Covered are noise storms and bursts of types I, II, III, and IV. Each type of burst is analyzed, origins suggested, and characteristics described.

077.004 Fürst, E. The quiet sun at cm- and mm-wavelengths. **IAU Symp.** 86: 25-39; 1980. 103 refs, 1946-80.
Review covering the brightness temperature at the center of the disk, center-to-limb variation of the solar brightness and the radio radius, and interpretation of the brightness variation and the radius.

077.005 Krüger, Albrecht. **Physics of Solar Continuum Radio Bursts.** Berlin: Akademie-Verlag; 1972. 206p. 5 chapters. 479 refs, 1912-70.
An advanced text "describing (a) the main observational background, (b) the theoretical basis of continuum radio burst physics, and (c) a synthesis between observations and theory yielding the results and problems of the interpretation of the flare-burst radiations." Emissions covered are Bremsstrahlung, gyroemission, synchrotron radiation, Čerenkov radiation, and particle energy. Includes a data table of over 300 major type IV events during 1956-1969.

077.006 Kundu, M. R.; Gergely, T. E., eds. **Radio Physics of the Sun.** Dordrecht, Holland; Boston: D. Reidel; 1980. 475p. (IAU Symp., 86). 31 papers, 36 abstracts.
C
College Park, MD, USA: 7-10 Aug 1979. Topics: the quiet Sun; active regions; solar bursts at centimeter, meter, and meter-decameter wavelengths; decimeter and low frequency observations of solar bursts; and radio, white light, and x-ray observations of solar bursts.

077.007 Lin, R. P., ed. Proceedings of the workshop on mechanisms for solar type III radio bursts. **Sol. Phys.** 46: 433-544; 1976. 12 papers, 7 abstracts.
C
Berkeley, CA, USA: 8-9 May 1975.

077.008 Noci, G. Problemi moderni di radioastronomia solare. [Modern problems of solar radio astronomy.] **Nuovo Cimento, Suppl.** 6: 835-856; 1968. 175 refs, 1944-67. In Italian.
Review concerned with radio emission of the quiet Sun and emission connected with radio bursts.

077.009 Smith, D. F. Type III solar radio bursts. **Adv. Astron. Astrophys.** 7: 147-226; 1970. 100 refs, 1946-69.

A review which includes the consideration of non-linear effects in a plasma as well as the setting up of a model of a type III burst for study.

077.010 Stewart, R. T. Ground-based observations of type III bursts. **IAU Symp.** 57: 161-181; 1974. 91 refs, 1950-74.

Review of observations over the past 20 years, with emphasis on recent high spatial resolution observations.

Radio bursts from the solar corona. *074.037.*

Radio Emission of the Sun and Planets. 141.080.

Solar radio observations and interpretations. *073.019.*

Wideband Cruciform Radio Telescope Research. 033.016.

078 Cosmic Radiation

078.001 Dorman, L. I.; Miroshnichenko, L. I. **Solar Cosmic Rays.** New Delhi: Indian National Scientific Documentation Centre; 1976. 582p. (NASA TT-F-674; TT 70-57262). 5 chapters. 682 refs, 1947-68.

Focusing on the problems of generation and propagation of solar cosmic rays, this book presents, summarizes, and analyzes voluminous experimental data gathered during 1942-67. The possibility of forecasting solar flares and other areas of future study are considered. Topics: solar cosmic rays as a source of information on electromagnetic state of the solar system; geophysical effects; energy spectrum; nuclear composition and time variations of solar cosmic rays; propagation; etc. Translation of *Solnechnye kosmicheskie luchi* (Moscow: Nauka; 1968).

078.002 McCracken, K. G.; Rao, U. R. Solar cosmic ray phenomena. **Space Sci. Rev.** 11: 155-233; 1970. 176 refs, 1953-70.

Review of the several types of solar cosmic ray phenomena including the properties of both prompt and delayed events.

078.003 Sakurai, K. Energetic particles from the Sun. **Astrophys. Space Sci.** 28: 375-519; 1974. 616 refs, 1908-1973.

Discussion of solar cosmic ray phenomena and related topics from the solar physical point of view. Considers basic physics of the solar atmosphere and solar flare phenomena in some detail.

078.004 Sakurai, Kunitomo. **Physics of Solar Cosmic Rays.** Tokyo: University of Tokyo Press; 1974. 428p. 10 chapters. 1,186 refs, 1900-1971.

"Deals with the physics of solar cosmic rays and related subjects such as the interplanetary plasma and magnetic fields, the solar atmosphere and associated phenomena." As background to explaining the problem, the author also discusses in detail the basic physics of the solar atmosphere, active phenomena, and interplanetary space. Topics: generation of solar cosmic rays; mechanism of solar flares; acceleration mechanism; propagation of solar cosmic rays; origin theory; etc.

078.005 Shcherbina-Samoǐlova, I. S. Cosmic rays of solar origin. **Itogi Nauki Tekh. Ser. Issled. Kosm. Prostranstva** 12: 7-184; 1978. 355 refs. In Russian.

A review of recent studies by Soviet and Western scientists.

079 Solar Eclipses

079.001 Houtgast, J., ed. The international symposium on the 1970 solar eclipse. **Sol.**
C **Phys.** 21: 259-495; 1971. 29 papers.

Seattle, WA, USA: 18-21 Jun 1971. Sponsored by COSPAR, IAGA, IAU, and URSI. Papers in this collection have to do with the physics of the Sun. Other papers from the conference on aeronomy and ionospheric physics were published in the *J. Atmos. Terr. Phys.*

Solar Eclipses and the Ionosphere. 083.001.

080 Atmosphere, Figure, Internal Constitution, Neutrinos, Rotation, Solar Physics (General), etc.

080.001 Bahcall, J. N. Solar neutrinos. In: de Boer, J.; Mang, H. J., eds. **International Conference on Nuclear Physics, Vol. II.** Amsterdam: North-Holland; New York: American Elsevier; 1973: 682-716. 173 refs, 1946-73.

A review of the discrepancy between calculations and observations of the phenomena. Cross-sections, solar models, and methods of detection are discussed.

080.002 Bahcall, J. N. Solar neutrinos: theory. In: Frenkel, A.; Marx, G., eds. **Neutrino**
C **'72.** Budapest: OMKDT-Technoinform; 1972: 29-75. 235 refs, 1946-72.

Europhysics Conference. Balatonfüred, Hungary: 11-17 Jun 1972. (Proceedings). A review which addresses, among other things, the discrepancy between calculation and observation of the solar neutrino phenomenon. Other subjects include cross-sections, solar models, and observational implications. The question of whether or not solar neutrinos reach the Earth is also considered.

080.003 Bahcall, J. N.; Sears, R. L. Solar neutrinos. **Annu. Rev. Astron. Astrophys.** 10: 25-44; 1972. 152 refs, 1946-72.

Review of the literature on solar neutrinos, with a discussion of the discrepancy between calculation and observation, nuclear energy generation in stars, and stellar evolution.

080.004 Bruzek, Anton; Durrant, Christopher J., eds. **Illustrated Glossary for Solar and Solar-Terrestrial Physics.** Dordrecht, Holland; Boston: D. Reidel; 1977. 204p. (Astrophys. Space Sci. Libr., 69).

Divided into 14 major subject groupings, this book for scientists and advanced students contains 255 entries, most of them one-quarter to one-half page in length. Taking the form of brief encyclopedic-type articles, many entries are accompanied by photographs or diagrams of the phenomenon in question. Most entries also have one to five references to basic sources in the astronomical literature. Topics: solar interior; non-spot magnetic fields; solar corona; spots and faculae; solar wind and interplanetary medium; solar-terrestrial physics; etc.

080.005 Chinderi, C.; Landini, M.; Righini, A., eds. Proceedings of the first European
C Solar Meeting. **Oss. Mem. Oss. Astrofis. Arcetri** 105: 1-143; 1975. 16 papers and abstracts.

Florence, Italy: 25-27 Feb 1975. Proceedings include four topical review papers: 1) Mass, energy, and momentum transport in the solar atmosphere; 2) Photospheric and chromospheric magnetic structures; 3) Coronal magnetic structure; and 4) Onset of instabilities.

080.006 Cram, L. E. Mass, energy and momentum transport in the solar atmosphere. **Oss. Mem. Oss. Astrofis. Arcetri** 105: 13-38; 1975. 118 refs, 1958-75.

The emphasis in this review is on those classes of atmospheric phenomena which are of fundamental relevance to the study of general stellar atmospheric dynamics. Topics include theoretical aspects of solar atmospheric dynamics, spherically symmetric solar models, and departures from spherical symmetry.

080.007 de Jager, C.; Kuperus, M.; Rosenberg, H. Physics of the solar atmosphere. **Philos. Trans. R. Soc. London, Ser. A.** 281: 415-426; 1976. 60 refs, 1926-75.

Summary of recent research on the physics of the quiet solar atmosphere and active regions. Includes solar rotation, velocity fields and waves, magnetic field concentration, the transition region, coronal magnetic field structure, and prominences.

080.008 Dicke, R. H. Internal rotation of the Sun. **Annu. Rev. Astron. Astrophys.** 8: 297-328; 1970. 79 refs, 1895-1970.

The theory of a rapidly rotating solar core is supported by observational evidence, especially the oblateness of the Sun.

080.009 Durney, B. R. On the theories of solar rotation. **IAU Symp.** 71: 243-295; 1976. 169 refs, 1877-1976.

A critical review of some of the main theories of solar rotation.

080.010 Eddy, John A., ed. **The New Solar Physics.** Boulder, CO: Westview Press; 1978. 214p. (AAAS Sel. Symp., 17). 5 papers.

Published for the American Association for the Advancement of Science, this volume is a collection of papers summarizing recent developments in the field, including historical and arboreal evidence for a changing Sun; solar neutrinos; streams, sectors, and solar magnetism; and seismic sounding. The papers are taken from a number of AAAS annual conferences.

080.011 Eddy, John A.; Ise, Rein (ed.). **A New Sun: The Solar Results from Skylab.** Washington, DC: NASA; 1979. 198p. (NASA SP-402). 8 chapters. no refs.

Primarily a collection of beautiful color photographs, this volume also provides an excellent summary of the solar observations made by Skylab astronauts in the first-ever above-the-atmosphere studies. Useful for students, laypersons, and nonspecialists, this volume covers solar structure; history of observations; solar telescopes on Skylab; and the results of the mission. The latter includes the quiet Sun (optical, x-ray, and U-V observations; coronal holes; prominences; and the outer corona) and the active Sun (active regions; flares; prominences; and coronal transients).

080.012 Engvold, O.; Hauge, Ø. Elemental abundances, isotope ratios and molecular compounds in the solar atmosphere. **Inst. Theor. Astrophys., Blindern-Oslo, Rep.** 39; 1974. 25p. 144 refs, 1929-74.

Solar abundances of 67 chemical elements are listed, and solar isotope ratios are tabulated. Upper limits for 5 elements are listed. A separate list is given of 21 molecules identified in the solar atmosphere, plus 15 possibly found.

080.013 C Gebbie, K. B., ed. **The Menzel Symposium on Solar Physics, Atomic Spectra, and Gaseous Nebulae.** Washington, DC: National Bureau of Standards; 1971. 203p. (NBS SP., 353). 14 papers.

Cambridge, MA, USA: 8-9 Apr 1971. (Proceedings). A series of short review papers on the topics in the title. Selected subjects include experimental studies of atomic spectra and transition probabilities; solar instrumentation; the corona; planetary nebulae; radio recombination lines.

080.014 Gibson, E. G. Description of solar structure and processes. **Rev. Geophys. Space Phys.** 10: 395-461; 1972. 18 refs, 1951-70.

A general introduction to solar structure. The Sun is first viewed as a spherically symmetric steady-state system, and the energy generated in the core is traced as it flows outward.

080.015 Gibson, Edward G. **The Quiet Sun.** Washington, DC: NASA; 1973. 330p. (NASA SP-303). 6 chapters w/refs.

A graduate-level solar physics text covering all aspects of the subject: background; solar structure and processes; interior; photosphere; chromosphere; and corona. Specific topics include, but are not limited to, emissions; spectroscopy; energy sources; solar wind; flares; composition; etc. Contains a substantial number of references to the literature.

080.016 Gilman, P. A. Solar rotation. **Annu. Rev. Astron. Astrophys.** 12: 47-70; 1974. 73 refs, 1939-74.

Discussion of rotation and its influence on other solar phenomena and basic physical theory. Defines solar rotation, gives explanations of rotation, and describes outstanding problems in both observations and theory.

080.017 Godoli, G.; Noci, G.; Righini, A., eds. Proceedings of the JOSO workshop:
C future solar optical observations—needs and constraints. **Oss. Mem. Oss. Astrofis. Arcetri** 106: 1-317; 1979. 25 papers.

Florence, Italy: 7-10 Nov 1978. Includes sessions on instrumentation, solar observations from space, and scientific problems.

080.018 Hill, H. A. Seismic sounding of the Sun. In: Eddy, John A., ed. **The New Solar Physics.** Boulder, CO: Westview Press; 1978: 135-214. 106 refs, 1908-1977.

A review presenting "recent observational data which suggest the existence of global oscillations of the Sun that are of sufficient magnitude as to be observable." Observational techniques are discussed and compared to previous methods that were unsuccessful. Evidence is presented in the form of data, which is reviewed and interpreted.

080.019 Hirshberg, J. Helium abundance of the Sun. **Rev. Geophys. Space Phys.** 11: 115-131; 1973. 47 refs, 1955-72.

Critical review of the four methods that have been used to estimate the ratio of helium to hydrogen (solar neutrino flux, spectral intensity of helium lines in prominence and the chromosphere, elemental abundance of solar cosmic rays, and variations of solar wind He/H). Concludes that the solar abundance of He remains uncertain to within a factor of 2 or 3.

080.020 Howard, R. The rotation of the sun. **Rev. Geophys. Space Phys.** 16: 721-732; 1978. 149 refs, 1630-1978.

Review of observations of solar rotation and large-scale velocity fields.

080.021 Howard, R.; Yoshimura, H. Differential rotation and global-scale velocity fields. **IAU Symp.** 71: 19-35; 1976. 56 refs, 1897-1975.

Review of the observational and theoretical background of global-scale velocity fields on the solar surface.

080.022 Kaplan, S. A.; Pikel'ner, S. B.; Tsytovich, V. N. Plasma physics of the solar atmosphere. **Phys. Rep. Phys. Lett. C.** 15C: 1-82; 1974. 414 refs, 1917-75.

Gives a general picture of the phenomena in the upper observable layers of the Sun, based both on the interpretation of observations and on the theoretical considerations of the different plasma motions. The discussion includes problems of heating of solar atmosphere by convection, magnetic structure of the solar atmosphere, solar flares, and radio bursts.

080.023 Kocharov, G. E. Neutrino astrophysics. **Bull. Acad. Sci. USSR, Phys. Ser.** 41(9): 129-152; 1977. Transl. of *Izv. Akad. Nauk SSSR, Ser. Fiz.* 41: 1916-1948; 1977. 53 refs, 1964-77.

Discussion of the current state of research on solar neutrino astrophysics.

080.024 Leibacher, J. W. Velocity fields in the solar atmosphere: theory. **Mem. Soc. Astron. Italiana** 48: 475-497; 1977. 243 refs, 1908-1977.

Includes a cursory discussion of convection, the generation and propagation of waves, and a discussion of models of the "five-minute" oscillation, with an excellent bibliography that includes titles.

080.025 Lin, R. P. Non-relativistic solar electrons. **Space Sci. Rev.** 16: 189-256; 1974. 104 refs, 1949-74.

This review provides a physically meaningful picture through a synthesis of many different types of observations.

080.026 Manno, V.; Page, D. E., eds. **Intercorrelated Satellite Observations Related to**
C **Solar Events.** Dordrecht, Holland: D. Reidel; distr., New York: Springer-Verlag; 1970. 627p. (Astrophys. Space Sci. Libr., 19). 51 papers.

Third ESLAB/ESRIN Symposium. Noordwijk, The Netherlands: 16-19 Sep 1969. The first half of this conference volume is devoted to review lectures on interrelated phenomena occurring on the Sun, in interplanetary space, and in the Earth's magnetosphere. These phenomena include, but are not limited to, flares, solar cosmic rays, solar protons, solar wind, geomagnetic storms, etc. The second half of the book deals with reports of data obtained from satellites during the solar events of February 25, 1969.

080.027 Massey, H., et al., eds. A discussion on solar studies with special reference to
C space observations. **Philos. Trans. R. Soc. London, Ser. A.** 270: 1-195; 1971. 20 papers.

21-22 Apr 1970. Arranged by the British National Committee on Space Research. Topics include the photosphere and low chromosphere, chromosphere and corona, solar activity, and stellar chromospheres and coronae.

080.028 Massey, H.; Sweet, P. A.; Gabriel, A. H., eds. A discussion on the physics of
C the solar atmosphere. **Philos. Trans. R. Soc. London, Ser. A.** 281: 293-513; 1976. 21 papers.

Held 14-15 Jan 1975. Arranged by the British National Committee on Space Research. Arranged in three sections (the quiet Sun, active regions, and flares), the purpose of the discussion was to review recent observations within the framework of a study of the solar atmosphere as a whole.

080.029 Noyes, R. W. Ultraviolet studies of the solar atmosphere. **Annu. Rev. Astron. Astrophys.** 9: 209-236; 1971. 75 refs, 1950-71.

A review of observational methods and results, followed by a discussion of the mean structure of the solar atmosphere in both quiet and active regions. Included is summary data on the temperatures in various parts of the atmosphere.

080.030 Radiation and structure of the solar atmosphere. **IAU Trans. Rep. Astron.** 17A(2): 49-75; 1979. + refs, 1976-79.

An overview of research activity worldwide, with an emphasis on the previous three years, as reported by IAU Commission 12 (Radiation and Structure of the Solar Atmosphere). Additional reports: 14A: 111-124; 1970. no refs. 15A: 129-154; 1973. 412 refs, 1952-73. 16A(2): 78-83; 1976. 551 refs, 1964-75.

080.031 Reynolds, John M.; Snoddy, William C., eds. **Apollo Telescope Mount: A Partial Listing of Scientific Publications and Presentations.** Marshall Space Flight Center, AL: NASA; 1976. 74p. (NASA TM-X-73300; N76-23581). 553 refs, 1968-76.

A bibliography of journal articles, reports, and conference papers written by principal investigators working on the ATM project (the Skylab solar observatory facility). Papers cited, therefore, concentrate on solar data gathered by Skylab between May 28, 1973 and February 8, 1974. Arrangement is by type of item (article, paper, etc.), and there is an author index. There are three supplements: 1) 1977. 51p. (NASA-TM-X-73393; N72-24495). 371 refs, 1968-77. 2) 1978. 46p. (NASA-TM-78183; N78-30582). 368 refs, 1975-78. 3) 1979. 43p. (NASA-TM-78303; N79-27500). 294 refs, 1975-79.

080.032 C Rösch, J., ed. **Pleins feux sur la physique solaire; Contexte coronal des éruptions solaires.** Paris: Éditions du CNRS; 1978. 429p. 35 papers. English with French abstracts.

A conference volume (details below) reviewing current theories and research in solar astronomy. The following general topics are addressed: the internal structure of the Sun; solar oscillations; the flare coronal context; solar wind.

Pleins feux sur la physique solaire. Toulouse, France: 8-10 Mar 1978. Second European Solar Meeting (Deuxième assemblée europèene de physique solaire).

Contexte coronal des éruptions solaires. Toulouse, France: 9-10 Mar 1978. (Colloque international du Centre National de la Recherche Scientifique, no. 282).

080.033 Ross, J. E.; Aller, L. H. The chemical composition of the Sun. **Science** 191: 1223-1229; 1976. 151 refs, 1929-75.

Review of the history of spectroscopic and solar wind studies which permit conclusions about the solar chemical composition and solar nebula. Includes a table of solar elemental abundances.

080.034 Solar activity. **IAU Trans. Rep. Astron.** 17A(2): 11-48; 1979. + refs, 1976-79.

An overview of research activity worldwide, with an emphasis on the previous three years, as reported by IAU Commission 10 (Solar Activity). Topics: solar physics; solar radio astronomy; etc. Additional reports: 14A: 71-110; 1970. 456 refs, 1967-70. 15A: 75-128; 1973. 868 refs, 1961-74. 16A(2): 13-54; 1976. + refs, 1973-76.

080.035 Somov, B. V.; Syrovatskiĭ, S. I. Physical processes in the solar atmosphere associated with flares. **Soviet Phys. Usp.** 19: 813-835; 1976. Transl. of *Usp. Fiz. Nauk* 120: 217-257; 1976. 164 refs, 1955-76.

The relative importance of solar energy channels (accelerated particles, heat fluxes, matter fluxes, and radiation) is reviewed, and the reaction of the solar atmosphere to the corresponding energy fluxes is considered.

080.036 Stein, R. F.; Leibacher, J. Waves in the solar atmosphere. **Annu. Rev. Astron. Astrophys.** 12: 407-435; 1974. 142 refs, 1909-1974.

Includes a description of the oscillatory motions that can occur, a review of observations of these, discussion of how they may be generated, and the role they play in heating the chromosphere and corona.

080.037 Strickland, A. C., gen. ed. **Annals of the IQSY (International Years of the Quiet Sun).** Cambridge, MA: MIT Press; 1968-70. 7v.

In 1957-58, the period of greatest sunspot activity ever recorded, a worldwide venture known as the International Geophysical Year (IGY) collected a very large amount of data on solar events. It became evident that the data could be fully appreciated only if it were compared to information gathered during a period of minimum solar activity (few sunspots). In 1962, the International Year of the Quiet Sun (IQSY) was organized by a special committee of the International Council of Scientific Unions (ICSU) as the project for minimal solar activity study during 1964-65, a predicted period of solar inactivity. The results are published in seven volumes, all of which are cited in full in this bibliography. They are 1) *Geophysical Measurements: Techniques, Observational Schedules, and Treatment of Data* (1968); 2) *Solar and Geophysical Events 1960-65 (Calendar Record)* (1968); 3) *The Proton Flare Project (The July 1966 Event)* (1969); 4) *Solar-Terrestrial Physics: Solar Aspects (Proceedings of the Joint IQSY/COSPAR Symposium, London, 1967, Part I)* (1967); 5) *Solar-Terrestrial Physics: Terrestrial Aspects (Proceedings of the Joint IQSY/COSPAR Symposium, London, 1967, Part II)* (1969); 6) *Survey of IQSY Observations and Bibliography* (1970); 7) *Sources and Availability of IQSY Data* (1970).

080.038 The Sun, a tool for stellar physics. **Mem. Soc. Astron. Italiana** 48: 335-581;
C 1977. 13 papers.

Sixth course of the Advanced School of Astronomy, "E. Majorana" Centre for Scientific Culture. Erice, Italy: 8-21 Aug 1976. Deals with problems of the solar system symbiosis.

080.039 Tandberg-Hanssen, E. Solar activity. **Rev. Geophys. Space Phys.** 11: 469-504; 1973. 188 refs, 1844-1973.

Discussion of the changes in the Sun's surface and atmosphere, divided into the problems of the generation of magnetic fields and of the solar cycle, and the surface manifestations of solar activity.

080.040 Williams, D. J., ed. **Physics of Solar Planetary Environments.** Washington, DC:
C American Geophysical Union; 1976. 2v. 1,038p. 71 papers.

International Symposium on Solar-Terrestrial Physics. Boulder, CO, USA: 7-18 Jun 1976. (Proceedings). An examination of energy, momentum, and mass transfer processes throughout the solar system. Included are some basic review papers with substantial numbers of references touching on the following topics: solar cycle, solar flares and particle emission, solar wind, interplanetary matter, the Earth's magnetosphere, and solar-terrestrial relations.

Bibliography of IQSY publications. *085.003.*

Contexte coronal des éruptions solaires. 080.032.

Coronal Disturbances. 074.023.

Energetic particles from the Sun. *078.003.*

Interactions of energetic nuclear particles in space with the lunar surface. *094.054.*

Introduction to Solar Radio Astronomy and Radio Physics. 141.035.

Physics of the Solar System. 091.052.

Small scale solar magnetic fields: theory. *075.012.*

Solar Physics: A Journal for Solar Research and the Study of Solar Terrestrial Physics. R005.020.

Study of Travelling Interplanetary Phenomena/1977. 106.013.

EARTH

081 Structure, Figure, Gravity, Orbit, Geophysics (General), etc.

081.001 Bell, P. M., ed. U.S. national report to the International Union of Geodesy and Geophysics. **Rev. Geophys. Space Phys.** 13(3): 1-1106; 1975. (Special issue, paged separately from remainder of volume.) 131 articles.

A review of progress in all areas of geophysics and a presentation of an extensive bibliography of the geophysical literature published during the quadrennium (1971-74). Most of the bibliographies cite titles. Of particular interest are the following [all page numbers refer to special issue no. 3 of vol. 13, 1975]:

Barnes, A. Plasma processes in the expansion of the solar wind and in the interplanetary medium. 1049-1053, 1066-1072. 300 refs, 1970-74.

Crooker, N. U. Solar wind-magnetosphere coupling. 955-958, 1021-1025. 271 refs, 1970-75.

Eather, R. H. Advances in magnetospheric physics: Aurora. 925-943, 994-1002. 360 refs, 1962-75.

Gierasch, P. J. Planetary atmospheres. 790-793, 862-871. 344 refs, 1944-75.

Gose, W. A.; Butler, R. F. Magnetism of the Moon and meteorites. 189-193, 226-228. 119 refs, 1961-74. (Univ. Texas Marine Sci. Inst. Contr. 51).

Gosling, J. T. Large-scale inhomogeneities in the solar wind of solar origin. 1053-1058, 1072-1076. 209 refs, 1971-75.

Hirshberg, J. Composition of the solar wind: present and past. 1059-1063, 1077-1081. 204 refs, 1966-74.

Roelof, E. C.; Krimigis, S. M. Low-energy solar cosmic rays: a bibliography. 1092-1094, 1099-1104. 242 refs, 1971-75.

Ward, W. R. Cosmogony of the solar system. 422-424, 427-430. 231 refs, 1968-75.

Warner, J. Mineralogy, petrology, and geochemistry of the lunar samples. 107-113, 163-168. 178 refs, 1969-75.

Wasson, J. T. Meteorite research. 113-116, 168-173. 191 refs, 1967-75.

081.002 C Langel, R. A., convenor. Analysis, processing and interpretation of geophysical data—a symposium. **Phys. Earth Planet. Inter.** 12: 93-289; 1976. 21 papers.

Grenoble, France: 1-2 Sep 1975. IUGG Interdisciplinary Symposium IS9, as part of the Sixteenth General Assembly of the International Union of Geodesy and Geophysics.

081.003 C Mather, R. S.; Angus-Leppan, P. V., eds. **Proceedings of Symposium on Earth's Gravitational Field and Secular Variations in Position.** Sydney: School of Surveying, University of New South Wales; 1974. 726p. 59 papers.

Sydney, Australia: 26-30 Nov 1973. Reports of research in the following areas: astrogeodetic methods; combination methods; geoid determinations; secular variations in gravity; boundary value problem in physical geodesy; very long baseline interferometry; lunar ranging; statistical methods; etc.

081.004 Melchior, Paul. **The Tides of Planet Earth.** Oxford, UK; New York: Pergamon Press; 1978. 609p. 17 chapters. 2,549 refs, 1800-1977.

A comprehensive survey on an advanced level covering the geophysical, astronomical, and geodetical aspects of the subject. In addition to the thorough treatment of Earth tides, there is coverage of tidal effects in astronomy; Earth tides, satellite orbits and space navigation; solid tides on the Moon; etc. The bibliography is claimed to be complete through 1976.

081.005 Proverbio, E., ed. **Atti del Convegno Internazionale ulla 'Rotazione della Terra**
C **e Osservazioni Satelliti Artificiali.'** Bologna, Italy: Graficoop Soc. Tipografica Editoriale; 1974. 120p. (Rend. Semin. Fac. Sci. Cagliari Suppl., 44). 13 papers. Italian/French/English.

International Meeting on "Earth's Rotation by Satellite Observations." Cagliari, Italy: 16-18 Apr 1973. Papers reviewing "astronomical and mathematical aspects of the Earth's rotation, polar motion, continental drift, and ... new techniques and instruments: astronomical, dopperl and laser."

081.006 Runcorn, S. K., ed. **The Application of Modern Physics to the Earth and Plane-**
C **tary Interiors.** London; New York: Wiley-Interscience; 1969. 692p. 49 papers.

NATO Advanced Study Institute. Newcastle-upon-Tyne, UK: 29 Mar—4 Apr 1967. Primarily concerned with Earth geophysics, this book contains half a dozen astronomy-related papers. These include cosmology and the gravitational constant; the oblateness of the Sun; and the Apollo lunar sample analysis program.

081.007 Runcorn, S. K., ed. **International Dictionary of Geophysics.** Oxford, UK; New York: Pergamon Press; 1967. 2v. 1,728p.

More an encyclopedia than a dictionary, this reference book is a collection of brief essays on a wide variety of topics written by experts in the field. Entries are words and phrases, arranged alphabetically, ranging from one paragraph to several pages, the latter being more frequent. These "papers" are generally accompanied by key references to the literature, equations, and some illustrations. For the graduate student and scientist, it also includes two maps of the ocean floor.

Cosmology and Geophysics. 162.034.

082 Atmosphere (Refraction, Scintillation, Extinction, Airglow, Site Testing)

082.001 Divari, Nikolai B., ed. **Atmospheric Optics.** New York: Consultants Bureau; 1970, 1972. 2v. v. 1: 178p. 31 papers.; v. 2: 164p. 38 papers.

Two collections of conference papers describing the effects of the atmosphere on astronomical seeing. Emphasized are the various aspects of daylight, twilight, and atmospheric interference, as well as methods of combatting these problems with respect to different types of observations of various types of celestial objects. Many references are included. Translation of *Atmosfernaia͡ optika.* Moscow: Nauka; 1968, 1970. Translators: Stephen B. Dresner (v. 1) and Frank L. Sinclair (v. 2).

082.002 Dovgalevskaia͡, I. S. **Bibliograficheskiĭ ukazatel' otechestvennykh rabot po opticheskoĭ nestabil'nosti zemnoĭ atmosfery (1945-1975).** [Bibliography of Russian Works on Optical Instability of the Earth's Atmosphere (1945-1975).] Kiev: Naukova Dumka; 1978. 67p. 835 refs, 1945-75.

A list of citations in Russian, by author, under 18 subject categories. No abstracts are included, but there is an author index.

082.003 Hunten, D. M. Airglow—introduction and review. In: McCormac, B. M., ed. **The Radiating Atmosphere.** Dordrecht, Holland: D. Reidel; 1971: 3-16. 95 refs, 1961-71.

A description of recent development and results. Includes a detailed table of 34 known airglow emissions for the Earth, and briefly mentions airglow on other planets.

082.004 Ingham, M. F. The light of the night sky and the interplanetary medium. **Rep. Prog. Phys.** 34: 875-912; 1971. 295 refs, 1826-1972.

Review of integrated starlight, zodiacal light, airglow, and diffuse galactic light, up to August 1971.

082.005 Light of the night sky. **IAU Trans. Rep. Astron.** 17A(1): 145-152; 1979. 82 refs, 1976-78.

An overview of research activity worldwide, with an emphasis on the previous three years, as reported by IAU Commission 21 (Light of the Night Sky). Topics: airglow; zodiacal light; etc. Additional reports: 14A: 193-206; 1970. + refs, 1967-70. 15A: 241-252; 1973. + refs, 1970-73. 16A(1): 131-139; 1976. + refs, 1973-76.

082.006 McCormac, B. M., ed. **Atmospheres of Earth and the Planets.** Dordrecht,
C Holland; Boston: D. Reidel; 1975. 454p. (Astrophys. Space Sci. Libr., 51). 33 papers.

University of Liege Summer Advanced Study Institute. Liege, Belgium: 29 Jul—9 Aug 1974. (Proceedings). A review of theoretical and direct studies, primarily of the Earth's atmosphere, covering physical processes in the upper atmosphere, aurorae, modeling, atmospheric chemistry, and more. There are eight papers devoted to the atmospheres of Venus, Mars, Jupiter, Saturn, and Io.

082.007 McCormac, B. M., ed. **Physics and Chemistry of Upper Atmospheres.** Dord-
C recht, Holland; Boston: D. Reidel; 1973. 389p. (Astrophys. Space Sci. Libr., 35). 34 papers.

Summer Advanced Study Institute. Orléans, France: 31 Jul—11 Aug 1972. Review papers and reports of research on five topics: 1) structure and composition of the atmosphere; 2) physical processes; 3) chemical processes and models; 4) experimental results and interpretations; 5) other planets. The latter includes papers on the atmospheres of Mars, Venus, and Jupiter. Emphasis, however, is on the Earth.

082.008 Roach, F. E.; Gordon, Janet L. **The Light of the Night Sky.** Dordrecht, Holland; Boston: D. Reidel; 1973. 125p. (Geophys. Astrophys. Monogr., 4). 7 chapters w/refs.

A brief overview for students and laypersons examining the various sources of and effects of night-time light. Topics: star counts and starlight; zodiacal light and Gegenschein; night airglow or nightglow; diffuse galactic light; interplanetary and interstellar dust; cosmic light and cosmology.

082.009 Roosen, R. G. The gegenschein. **Rev. Geophys. Space Phys.** 9: 275-304; 1971. 149 refs, 1854-1971.

A comprehensive review of the observational history of the gegenschein with emphasis on its physical parameters: size, shape, brightness, spectrum, position, and parallax.

082.010 Teleki, G., ed. The present state and future of the astronomical refraction investigations. **Publ. Obs. Astron. Beograd** 18: 1-234; 1974. 15 papers.

This publication contains the Report on the Activity of the IAU Commission 8 Study Group on Astronomical Refraction and the papers on the present state and future of astronomical refraction investigations.

082.011 Waters, W. H. **Bibliography: Scientific Papers Resulting from the Canadian Upper Atmosphere Research Program, 1965-68.** Ottawa: National Research Council of Canada, Space Research Facilities Branch; 1969. 21p. 232 refs, 1965-68.

A listing of articles and conference papers arranged by year and by author's name within each year.

082.012 Weinberg, J. L., et al. **A Pictorial Atlas of Low Latitude 5577Å and 6300Å Air-**
A **glow Line Emission.** Albany, NY: Dudley Observatory; 1973. 185p. (Dudley Obs. Rep., 5).

A chronologically arranged set of strip charts illustrating airglow observations which were obtained at Haleakala Observatory on 164 nights between April 1965 and November 1968. Includes a table of data for all observatories, listing page number, date, program, filter, elevation, azimuth, starting and ending times, and remarks.

Atmospheric Emissions. 084.029.

Daylight and Its Spectrum. 022.042.

The Encyclopedia of Atmospheric Sciences and Astrogeology. R009.002.

083 Ionosphere

083.001 Anastassiades, Michael, ed. **Solar Eclipses and the Ionosphere.** New York; Lon-
C don: Plenum Press; 1970. 309p. 19 papers. English/French.

NATO Advanced Study Institute. Lagonissi, Greece: 26 May—4 Jun 1969. Lectures reviewing studies of the ionosphere during solar eclipses, emphasizing the contribution of the solar corona to the ionospheric layers, the chemistry of ionospheric layers, aeronomical problems, etc. Topics: theory; D-region chemistry; solar radiation and layer theory; radio astronomy observations; eclipse studies by rocket and satellite; ionospheric eclipse effects.

083.002 Gordon, Charlotte W.; Canuto, V.; Axford, W. Ian, eds. **Handbook of Astronomy, Astrophysics and Geophysics, Vol. 1: The Earth—The Upper Atmosphere, Ionosphere and Magnetosphere.** New York: Gordon and Breach; 1978. 412p. 6 papers.

Not a reference book, as the title might indicate, this volume contains review papers relying heavily upon mathematics. Topics: upper atmosphere hydrogen; the equatorial electrojet; electron plasma resonances in the topside ionosphere; auroral particle precipitation; polar-cap absorption; the Van Allen Belt.

083.003 Kaiser, T. R., ed. Dynamics, chemistry and thermal processes in the ionosphere
C and thermosphere. **J. Atmos. Terr. Phys.** 36: 1705-2066; 1974. 39 papers.

Kyoto, Japan: Sep 1973. Papers on energy input, minor constituents, and transport processes, with an emphasis on the complex chain of interactions between these processes.

083.004 Kaiser, T. R., ed. Physics of the plasmapause. **J. Atmos. Terr. Phys.** 38:
C 1039-1236; 1976. 28 papers.

Grenoble Symposium. Grenoble, France: Aug/Sep 1975. Papers addressing various aspects of magnetospheric physics. Selected topics: steady state plasmapause positions; shocks, solitrons, and the plasmapause; instability phenomena in detached plasma regions; Pc 1; VLF emissions; satellite observations; and much more.

083.005 National Research Council. Geophysics Study Committee. **The Upper Atmosphere and Magnetosphere.** Washington, DC: National Academy of Sciences; 1977. 168p. (Stud. Geophys.). 10 papers.
C

American Geophysical Union Symposium. San Francisco, CA, USA: Dec 1975. A collection of papers surveying various aspects of the subject, including solar wind interactions; thermosphere; ionosphere; turbulence; ozone; etc. The purpose of the papers was to briefly review topics of current interest in the scientific community and make recommendations for further study. Includes overall summary and committee recommendations.

083.006 Ratcliffe, J. A. **An Introduction to the Ionosphere and Magnetosphere.** London; New York: Cambridge University Press; 1972. 256p. 10 chapters. 176 refs, 1931-72.

An overview of the physical processes taking place in the upper atmosphere. Aimed at graduate students and nonspecialists, the book emphasizes the principles involved, not the mechanisms causing the various phenomena. Topics: the formation and nature of the ionosphere and magnetosphere; the ionospheric layers; geomagnetism, ionospheric currents, and storms; experimental methods; collisions and diffusion; movements of charged particles in magnetic fields; electromagnetic, hydromagnetic, and electro-acoustic waves; using radio waves to explore the ionosphere; experiments in space vehicles.

083.007 Whitten, R. C.; Poppoff, I. G. **Fundamentals of Aeronomy.** New York: John Wiley & Sons; 1971. 446p. 11 chapters w/refs.

A non-elementary, basic text on the science of the upper atmosphere. Aimed at advanced students, the book covers basic fundamental principles; physical, chemical, and fluid aeronomy; optical phenomena; structure of disturbances in, properties of, and electromagnetic waves in the ionosphere. Each chapter contains a substantial number of both general and specific references to the literature.

083.008 Yeh, K. C.; Liu, C. H. Propagation and application of waves in the ionosphere. **Rev. Geophys. Space Phys.** 10: 631-709; 1972. 304 refs, 1902-1972.

Review of the propagation of waves, especially radio waves in the ionosphere. Topics covered include electrodynamics of material media, waves in fluid plasmas in the absence of a steady magnetic field, waves in fluid plasmas with a steady magnetic field, and ionospheric applications.

084 Aurorae, Geomagnetic Field, Magnetosphere

084.001 Akasofu, S.-I., ed. **Dynamics of the Magnetosphere.** Dordrecht, Holland; Boston: D. Reidel; 1980. 658p. (Astrophys. Space Sci. Libr., 78). 32 papers.
C

A. G. U. Chapman Conference "Magnetospheric Substorms and Related Plasma Processes." Los Alamos, NM, USA: 9-13 Oct 1978. (Proceedings). Review papers and reports of recent results in the field of magnetospheric substorms. Aimed at students and specialists, the book deals with both theory and experiment. Topics: interplanetary magnetic field and the magnetosphere; magnetosphere-ionosphere coupling; plasma processes in the magnetosphere; ring current formation; substorm mechanisms; etc.

084.002 Akasofu, Syun-Ichi. **Physics of Magnetospheric Substorms.** Dordrecht, Holland; Boston: D. Reidel; 1977. 599p. (Astrophys. Space Sci. Libr., 47). 9 chapters w/refs.

A comprehensive volume for the student and specialist, with hundreds of references to the literature, on the various components and physical processes of the phenomenon. Topics: open magnetosphere and the auroral oval; auroras and auroral particles; distribution of plasmas in the magnetosphere; responses of the magnetosphere to interplanetary disturbances; magnetospheric substorms: introduction; magnetotail phenomena during magnetospheric substorms; magnetospheric currents during substorms; penetrating convection electric field, plasma ejection, and plasmasphere disturbances; solar-terrestrial relations and magnetospheric substorms.

084.003 Brice, N. M. Magnetosphere. **EOS Trans. American Geophys. Union** 52: IUGG 531-539, 1971. 354 refs, 1967-70.

A report on research progress during 1967-70 covering magnetopause; magnetospheric structure, magnetotail, and plasma sheet; convection and the plasmapause; storms and substorms, ring current, and energetic particles.

084.004 C Carovillano, R. L.; McClay, J. F.; Radoski, H. R., eds. **Physics of the Magnetosphere.** Dordrecht, Holland: D. Reidel; distr., New York: Springer-Verlag; 1968. 686p. (Astrophys. Space Sci. Libr., 10). 24 papers.

Boston, MA, USA: 19-28 Jun 1967. (Proceedings). A combination of tutorial lectures and invited papers, the latter primarily being reports of magnetospheric research including observation. The lectures cover dynamical properties of the magnetosphere; solar wind interaction with the magnetosphere; whistlers and VLF emissions; particle description of the magnetosphere; waves and particles in the magnetosphere.

084.005 Carovillano, R. L.; Siscoe, G. L. Energy and momentum theorems in magnetospheric processes. **Rev. Geophys. Space Phys.** 11: 289-353; 1973. 291 refs, 1892-1971.

Review concerned with the several energy and momentum theories relating to magnetospheric processes that have been developed. The region of primary consideration is the magnetospheric domain that extends between the ionosphere and the interplanetary medium.

084.006 Chapman, S. Auroral physics. **Annu. Rev. Astron. Astrophys.** 8: 61-86; 1970. 147 refs, 1658-1969.

A history and description of the aurora phenomena, along with a discussion of its origin and spectrum.

084.007 Cornwall, J. M.; Schultz, M. Physics of heavy ions in the magnetosphere. In: Lanzerotti, Louis J.; Kennel, Charles F.; Parker, E. N., eds. **Solar System Plasma Physics. Volume III.** Amsterdam: North-Holland; 1979: 165-210. 177 refs, 1963-79.

A review of current understanding and future research on what can be learned about magnetospheric processes from studying heavy ions in space. Topics: auroral magnetospheric coupling and the O^+ problem; heavy ions as probes for magnetospheric processes; multi-ion plasma physics; active plasma-injection experiments; heavy-ion physics at Jupiter; etc.

084.008 C Durney, A. C.; Ogilvie, K. W., eds. Advances in magnetospheric physics with GEOS-1 and ISEE-1 and 2. **Space Sci. Rev.** 22: 321-812; 1978; 23: 3-133; 1979. 38 papers.

Thirteenth ESLAB Symposium. Innsbruck, Austria: 1978. Papers in volume 22 concern the results of magnetopause and magnetosheath observations, while those in volume 23 concern the results of solar wind and bow shock observations.

084.009 Feldstein, Y. I. Auroras and associated phenomena. In: Dyer, E. R., ed. **Solar-Terrestrial Physics/1970.** Dordrecht, Holland: D. Reidel; 1972: pt. III, 152-191. 167 refs, 1960-71.

An examination of the auroral oval and its dependency on DP intensity. Not a general review of aurorae.

084.010 Feldstein, Y. I. Night-time aurora and its relation to the magnetosphere. **Ann. Géophys.** 30: 259-272; 1974. 89 refs, 1962-74.

Review of studies on the global distribution of auroras in order to gain information about the large-scale magnetospheric structure and dynamics under disturbed conditions, and on the auroral particle source locations in the magnetosphere. Topics covered include sudden commencements, auroral ovals, auroral and magnetic substorms, auroras and magnetospheric structure, magnetic flux in the magnetospheric tail, and auroras and the IMF.

084.011 Formisano, V., ed. **The Magnetospheres of the Earth and Jupiter.** Dordrecht,
C Holland; Boston: D. Reidel; 1975. 485p. (Astrophys. Space Sci. Libr., 52). 36 papers.

Neil Brice Memorial Symposium. Frascati, Italy: 28 May—1 Jun 1974. A review of current knowledge, both theoretical and observational. Topics: 1) Earth: bow shock, solar wind interaction, aurora, magnetospheric convection; 2) Jupiter: radio astronomy, x-ray emission, upper atmosphere, plasma, energetic electrons, etc.

084.012 Fredricks, R. W. Wave-particle interaction in the outer magnetosphere: a review. In: Formisano, V., ed. **Magnetospheres of the Earth and Jupiter.** Dordrecht, Holland: D. Reidel; 1975: 113-152. 123 refs, 1953-74.

An overview addressing the following topics: wave-particle interactions in the Earth's bow shock; wave-particle interactions and the magnetopause; VLF and ELF electromagnetic waves in the magnetosphere; electrostatic wave turbulence in the magnetosphere; current-driven instabilities and anomalous resistivity; other wave-particle interactions.

084.013 Gadsden, M. Airglow. **EOS Trans. American Geophys. Union** 52: IUGG 522-526; 1971. 120 refs, 1967-70.

Review of the literature, covering nightglow, twilight, and dayglow.

084.014 Gerard, J. C.; Harang, O. E. Auroral excitation and time variation. **Ann. Geophys.** 30: 273-283; 1974. 108 refs, 1955-74.

Review of recent progress in theories and observations of excitation processes of electron aurora. Topical sections include auroral zones—particle zones, direct processes, indirect excitation processes, and time-varying aurora.

084.015 Hasegawa, A., ed. IAGA symposium on micropulsations theory and new exper-
C imental results. **Space Sci. Rev.** 16: 329-458; 1974. 7 papers.

Held as part of the general scientific assembly of the International Association of Geomagnetism and Aeronomy, September 1973, in Kyoto, Japan. Organized to study the origin of geomagnetic pulsations, space-ground correlations, and the role of pulsations in geophysical phenomena.

084.016 Hasegawa, A. Plasma instabilities in the magnetosphere. **Rev. Geophys. Space Phys.** 9: 703-772; 1971. 141 refs, 1957-71.

Introductory review of theories of plasma instabilities that are relevant to magnetospheric plasmas, and a summary of work relating to actual plasma instabilities in the magnetosphere.

084.017 Hultqvist, B. The aurora. In: Formisano, V., ed. **Magnetospheres of the Earth and Jupiter.** Dordrecht, Holland: D. Reidel; 1975: 77-111. 77 refs, 1939-74.

A report of the more recent developments in the field of auroral physical processes research. Topics: processes determining the geographic distribution of the diffuse aurora; some characteristics of the turbulence that scatters auroral particles into the atmosphere; interaction of the hot plasma with the ionosphere.

084.018 Jones, Alister V. **Aurora.** Dordrecht, Holland; Boston: D. Reidel; 1974. 301p. (Geophys. Astrophys. Monogr., 9). 6 chapters. 792 refs, 1935-74.

An introduction to and review of the research done in the study of the aurora. Written for the nonspecialist and graduate student, the book covers the techniques of observation; occurrence and morphology; optical emissions from aurora; aurora and the ionosphere; mechanisms of precipitation of auroral particles. Includes a quantitative atlas of the auroral spectrum.

084.019 Kennel, Charles F.; Lanzerotti, Louis J.; Parker, E. N., eds. **Solar System Plasma Physics. Volume II. Magnetospheres.** Amsterdam; New York: North-Holland; 1979. 402p. 8 papers.

A collection of review papers on recent research and current knowledge of the field. Topics: Earth's magnetosphere; magnetosphere, ionosphere and atmosphere interactions; Mercury's magnetosphere; the interaction of the solar wind with Mars, Venus, and Mercury; interaction of the solar wind with comets; planetary dynamos; comparative mangetospheric theory.

084.020 Knott, K.; Battrick, B., eds. **The Scientific Satellite Programme during the**
C **International Magnetospheric Study.** Dordrecht, Holland; Boston: D. Reidel; 1976. 463p. (Astrophys. Space Sci. Libr., 57). 30 papers.

Tenth ESLAB Symposium. Vienna, Austria: 10-13 Jun 1975. (Proceedings). Reports of research resulting from the IMS, whose objective was "the achievement of a comprehensive, quantitative understanding of the dynamical processes operating in the Earth's plasma and field environment." Topics: IMS satellites: orbits and scientific potential; quasistatic magnetospheric phenomena and particles; particles and magnetospheric instabilities; substorms and conjugated phenomena; near-Earth experiments related to the IMS satellite programs.

084.021 Kovalevsky, J. V. The interplanetary medium and its interaction with the Earth's magnetosphere. **Space Sci. Rev.** 12: 187-257; 1971. 402 refs, 1908-1971.

Review of the principal results of direct measurements of the plasma and magnetic field by spacecraft close to Earth.

084.022 Lanzerotti, L. J.; Southwood, D. J. Hydromagnetic waves. In: Lanzerotti, Louis J.; Kennel, Charles F.; Parker, E. N., eds. **Solar System Plasma Physics. Volume III.** Amsterdam: North-Holland; 1979: 109-135. 94 refs, 1860-1978.

A review of variations in the Earth's magnetic field intensity known as geomagnetic pulsations (relatively fast, low-amplitude changes). Topics: theoretical and observational background; wave localization and wave modes; the ionosphere and the atmosphere and boundary conditions; etc.

084.023 McCormac, B. M. Aurora. **EOS Trans. American Geophys. Union** 52: IUGG 527-531; 1971. 127 refs, 1940-71.

A brief overview of research and results for 1967-70. Topics covered include morphology and theory, atomic and molecular processes, auroral electric fields, particle precipitation, radio frequency observations, and auroral red arcs.

084.024 McCormac, B. M., ed. **Earth's Magnetospheric Processes.** Dordrecht, Holland:
C D. Reidel; 1972. 417p. (Astrophys. Space Sci. Libr., 32). 39 papers.

NATO Advanced Study Institute and Ninth ESRO Summer School. Cortina, Italy: 30 Aug—10 Sep 1971. A review of theoretical and observational studies of the particles, fields, and physics of the magnetosphere. Topics: magnetospheric structure; electric fields and plasma convection; acceleration and diffusion of particles; magnetospheric substorms.

084.025 McCormac, B. M., ed. **Magnetospheric Particles and Fields.** Dordrecht, Hol-
C land; Boston: D. Reidel; 1976. 331p. (Astrophys. Space Sci. Libr., 58). 29 papers.

A Summer Advanced Study School. Graz, Austria: 4-15 Aug 1975. (Proceedings). With more emphasis on observation than theory, the theme of this volume is "the interface between the geomagnetic field and the solar wind plasma—the outermost regions of the magnetosphere and the various boundary layers and phenomena found therein." Topics: the Earth's bow shock system; plasma in the magnetosphere; aurora and related phenomena, the magnetic fields of Jupiter, and the solar system; etc.

084.026 McCormac, B. M., ed. **Magnetospheric Physics.** Dordrecht, Holland; Boston:
C D. Reidel; 1974. 399p. (Astrophys. Space Sci. Libr., 44). 33 papers.

A Summer Advanced Study Institute. Sheffield, UK: 13-24 Aug 1973. (Proceedings). A series of lectures describing current knowledge and unsolved problems. Topics: structure and characteristics of the magnetosphere; plasma convection and electric fields; wave-particle interactions; substorm phenomena.

084.027 McCormac, B. M., ed. **Particles and Fields in the Magnetosphere.** Dordrecht,
C Holland: D. Reidel; 1970. 453p. (Astrophys. Space Sci. Libr., 17). 39 papers.

Santa Barbara, CA, USA: 4-15 Aug 1969. (Proceedings). Reports of research and observations, and overviews of current knowledge in the following areas: magnetospheric models; bow shock; magnetospheric particles; magnetic and electric fields; wave-particle interactions; radiation belt observations; acceleration and motions of particles.

084.028 McCormac, B. M., ed. **The Radiating Atmosphere.** Dordrecht, Holland: D.
C Reidel; 1971. 455p. (Astrophys. Space Sci. Libr., 24). 38 papers.

Summer Advanced Study Institute. Kingston, ON, Canada: 3-14 Aug 1970. (Proceedings). Reviews of current research and past studies on aurorae and airglow. Topics: atmospheric airglow emissions; atmospheric processes; aurora; auroral interpretations; particle precipitation; radio observations; auroral morphology.

084.029 McCormac, B. M.; Omholt, A., eds. **Atmospheric Emissions.** New York: Van
C Nostrand Reinhold; 1969. 563p. 49 papers.

NATO Advanced Study Institute on Aurora and Airglow. Agricultural College, Norway: 29 Jul—9 Aug 1968. (Proceedings). Edited lectures reviewing the basic physics of the aurora and airglow phenomena, with mention of important observational results. Topics: orientation, observations, particles and their effects, fundamental causes, etc.

084.030 Nishida, A. **Geomagnetic Diagnosis of the Magnetosphere.** New York; Berlin: Springer-Verlag; 1978. 256p. (Phys. Chem. Space, 9). 5 chapters. 338 refs, 1931-78.

An introduction to magnetospheric physics, this advanced text examines both the experimental and the theoretical basis of the diagnostic uses of geomagnetic data. Topics: sudden impulses; interplanetary field effects; magnetic substorms; the DR Field; magnetic pulsations; the magnetotail; effects of solar wind; the structure of the inner magnetosphere; etc.

084.031 Omholt, A. **The Optical Aurora.** New York: Springer-Verlag; 1971. 198p. (Phys. Chem. Space, 4). 9 chapters w/refs.

Aimed at the nonspecialist and advanced student, this book intends to review the studies done on the aurora in the visible light spectrum. Topics: occurrence and cause; the electron aurora; the proton aurora; optical spectrum; physics of the optical emissions; temperature determinations; pulsing aurora; etc.

084.032 Pfotzer, G. Review of important problems in physics of the auroral zone. **ESA Sci. Tech. Rev.** 2: 181-197; 1976. 71 refs, 1881-1976.

Brief discussion of auroral phenomena in terms of magnetospheric topology and dynamics, followed by a review of four problem areas: diffuse auroras and particle injection in the keV range, electric fields, currents, and partition and effects of energy input.

084.033 Roederer, J. G. Earth's magnetosphere: global problems in magnetospheric plasma physics. In: Kennel, Charles F.; Lanzerotti, Louis J.; Parker, E. N., eds. **Solar System Plasma Physics. Volume II.** Amsterdam: North-Holland; 1979: 1-56. 152 refs, 1952-77.

A review of current plasmaphysical problems of a global nature, i.e., involving the magnetosphere as a whole. Past developments are not covered in great detail, and no original data is given: emphasis is on controversial or mutually contradictory results. Topics: effect of the interplanetary magnetic field on the magnetospheric boundary; mechanisms for entry of the solar wind plasma into the magnetosphere; plasma storage, acceleration, and release mechanisms in the magnetospheric tail; etc.

084.034 Russell, C. T.; McPherron, R. L.; Coleman, P. J., Jr. Fluctuating magnetic fields in the magnetosphere 1: ELF and VLF fluctuations. **Space Sci. Rev.** 12: 810-856; 1972. 110 refs, 1962-71.

Review of satellite observations of signals of natural origin.

084.035 Rycroft, M. J.; Lemaire, J., eds. The Earth's magnetopause regions. **J. Atmos.**
C **Terr. Phys.** 40: 227-394; 1978. 16 papers.

Amsterdam, The Netherlands: 7-10 Sep 1976. Deals with boundary layers in magnetospheric physics, considering phenomena occurring in the outermost regions of the magnetosphere.

084.036 Shepherd, G. G. Auroral spectroscopy and excitation. **Ann. Géophys.** 28: 99-107; 1972. 114 refs, 1951-71.

A short review, covering $O(^1S)$ excitation, measurements of its lifetime and its pulsations, emission rate ratios, composition changes, ultraviolet and infrared emissions, latitude variations and auroral temperature measurements.

084.037 Spreiter, J. R.; Alksne, A. Y. Plasma flow around the magnetosphere. **Rev. Géophys.** 7: 11-50; 1969. 82 refs, 1918-69.

Review of the salient features of the flow of solar plasma past the magnetosphere, as revealed by observations in space and by theory.

084.038 Unwin, R. S.; Baggaley, W. J. The radio aurora. **Ann. Géophys.** 28: 111-127; 1972. 93 refs, 1956-71.

A review of theories and results of experimental work during the previous five years compared with the predictions of theory.

084.039 Williams, D. J.; Mead, G. D., eds. International symposium on the physics of
C the magnetosphere. **Rev. Géophys.** 7: 1-459; 1969. 15 papers.

Washington, DC, USA: 3-13 Sep 1968. Cosponsored by COSPAR, IUGG-IAGA, URSI, IUPAP, and the IAU. A summary of the past decade's research, beginning in 1958 with the discovery of the Van Allen radiation belts. Selected topics: solar wind interaction with the geomagnetic field; plasma flow around the magnetosphere; auroras and polar substorms; charged particles in the atmosphere; magnetospheric plasma; etc.

084.040 Willis, D. M. Structure of the magnetopause. **Rev. Geophys. Space Phys.** 9: 953-985; 1971. 97 refs, 1931-71.

Review of the internal structure of the thin boundary layer that separates the distorted geomagnetic field in the magnetosphere from the flow of solar plasma in the magnetosheath.

Flare build-up study. *073.024.*

Handbook of Astronomy, Astrophysics and Geophysics, Vol. 1: The Earth—The Upper Atmosphere, Ionosphere and Magnetosphere. 083.002.

An Introduction to the Ionosphere and Magnetosphere. 083.006.

Magnetospheric plasma waves. *062.020.*

Physics of the Hot Plasma in the Magnetosphere. 062.010.

Shock systems in collisionless space plasmas. *062.007.*

The solar wind and magnetospheric dynamics. *074.028.*

Space Physics. 051.046.

The Upper Atmosphere and Magnetosphere. 083.005.

085 Solar-terrestrial Relations

085.001 Akasofu, Syun-Ichi; Chapman, Sydney. **Solar-Terrestrial Physics.** Oxford, UK: Clarendon Press, Oxford University Press; 1972. 901p. (Int. Ser. Monogr. Phys.). ~3,200 refs, 1889-1971.

An advanced text and reference for specialists and graduate students, concerned with "solar influences on the earth's upper atmosphere and electromagnetic environment partly through solar ultraviolet and x-rays, but particularly through corpuscular radiations." Topics: the Sun and interplanetary space; internal structure and magnetic field of the Earth; the terrestrial atmosphere; photochemistry; dynamics of the upper atmosphere and dynamo action; the formation of the magnetosphere; energetic particles, plasma, and electromagnetic waves in the magnetosphere; solar storms; magnetospheric storms.

085.002 Axford, W. I. A survey of interplanetary and terrestrial phenomena associated with solar flares. In: Manno, V.; Page, D. E., eds. **Intercorrelated Satellite Observations Related to Solar Events.** Dordrecht, Holland: D. Reidel; 1970: 7-22. 102 refs, 1931-70.

A review concentrating on the effects of flares which can be detected from satellites, rockets, and space probes. Specific phenomena discussed include disturbances in the solar wind; x-ray bursts; energetic solar particles; galactic cosmic rays; ionospheric effects; geomagnetic storms.

085.003 Booth, G., comp. Bibliography of IQSY publications. In: Strickland, A. C., ed. **Survey of IQSY Observations and Bibliography.** Cambridge, MA: MIT Press; 1970: 319-549. (Ann. Int. Years Quiet Sun, 6). 5,492 refs, 1957-70.

An extensive list of references to papers resulting from or relevant to the IQSY scientific program during 1964 and 1965. Entries are arranged by author in the following categories: general, aeronomy, geomagnetism, aurora and airglow, ionosphere, solar activity, cosmic radiation, and space research. A list of pertinent conference proceedings is appended.

085.004 Cook, F. E.; McCue, C. G. Solar-terrestrial relations and short-term ionospheric forecasting. **Radio Electron. Eng.** 45: 11-30; 1975. 223 refs, 1859-1973.

Topics covered include forecasting solar activity (including the international arrangements for collecting and disseminating solar/geophysical data), relations between solar activity and geophysical disturbances, and the effects of geophysical disturbances on the ionosphere and ionospheric radio communications.

085.005 de Feiter, L. D.; Schindler, K., eds. International solar-terrestrial physics symposium. **Space Sci. Rev.** 17: 171-614; 1975. 16 papers.
C

São Paulo, Brazil: 17-24 Jun 1974. Topics include the Sun, interplanetary medium, and the magnetosphere. An additional 12 papers on atmospheric physics from this conference appeared in *J. Atmos. Terr. Phys.*, June 1975.

085.006 Dyer, E. R., ed. **Solar-Terrestrial Physics/1970.** Dordrecht, Holland: D. Reidel;
C 1972. 944p. (4v. bound together). (Astrophys. Space Sci. Libr., 29, pt. I-IV).

Leningrad, USSR: 12-19 May 1970. (Proceedings). An extensive compendium of reviews (primarily) covering the broad topic of solar emissions and their effects on the interplanetary medium, the magnetosphere, and the upper atmosphere. Includes many references. Part I: de Jager, C., ed. *The Sun.* 181p. 13 papers. Part II: Dyer, E. R.; Roederer, J. G.; Hundhausen, A. J., eds. *The Interplanetary Medium.* 205p. 8 papers. Part III: Dyer, E. R.; Roederer, J. G., eds. *The Magnetosphere.* 317p. 11 papers. Part IV: Bowhill, S. A., ed. *The Upper Atmosphere.* 211p. 11 papers.

085.007 Kaiser, T. R.; Kendall, P. C., eds. Conference on the solar terrestrial environ-
C ment. **Planet. Space Sci.** 17: 293-584; 1969. 25 papers.

University of Sheffield, UK: 10-12 Jul 1968. Held in honor of Sydney Chapman. Selected topics: a historical survey; radio investigation of the solar plasma; the magnetopause; the magnetosphere; the radio auroral zone; ionospheric current systems; and much more.

085.008 Lincoln, J. V., comp. **Solar and Geophysical Events 1960-65 (Calendar Record).** Cambridge, MA: MIT Press; 1968. 297p. (Ann. Int. Years Quiet Sun, 2).

A day-by-day record of solar activity and effects (see also *080.037*).

085.009 Minnis, C. M., ed. **Geophysical Measurements: Techniques, Observational Schedules, and Treatment of Data.** Cambridge, MA: MIT Press; 1968. 398p. (Ann. Int. Years Quiet Sun, 1). 19 chapters.

A description of the areas of concern during the IQSY project of observing the Sun and various solar phenomena. Topics: methods of observation; solar activity; SIDs; solar radio emission; comets; ionosphere (vertical incidence soundings, absorption measurements, drift measurements); aurora; airglow; cosmic rays; meteorology; etc. (See also *080.037*).

085.010 Mitra, A. P. **Ionospheric Effects of Solar Flares.** Dordrecht, Holland; Boston: D. Reidel; 1974. 294p. (Astrophys. Space Sci. Libr., 46). 11 chapters. 317 refs, 1930-74.

A technical volume describing Sudden Ionospheric Disturbances (SIDs) and other effects of flares on the upper atmosphere. The origin and physics of these effects are covered in detail, along with various observational techniques and a description of the SID phenomenology.

085.011 Ness, N. F., ed. Proceedings of the symposium on solar terrestrial physics. **Space**
C **Sci. Rev.** 23: 135-538; 1979. 16 papers.

Innsbruck, Austria: 29 May—3 Jun 1978. Deals with interactions between the Sun, interplanetary medium, magnetosphere, and atmosphere. Includes overviews of recent results regarding solar wind interaction with Venus and the magnetosphere of Jupiter.

085.012 Page, D. E., ed. **Correlated Interplanetary and Magnetospheric Observations.**
C Dordrecht, Holland; Boston: D. Reidel; 1974. 662p. (Astrophys. Space Sci. Libr., 42). 45 papers.

Seventh ESLAB Symposium. Saulgau, West Germany: 22-25 May 1973. A summary of observations and theory on the subject of Earth-Sun relationships. Satellite data and review papers comprise the major portion of the book. Topics: response of the magnetosphere to changes in the interplanetary medium; the nature of magnetospheric boundaries; magnetospheric plasma flow and electric fields; solar particle propagation; solar particle entry to Earth; the August 1972 solar events; future study topics.

085.013 **Solar-Terrestrial Physics: Solar Aspects (Proceedings of the Joint IQSY/**
C **COSPAR Symposium, London, 1967, Part I).** Cambridge, MA: MIT Press; 1969. 414p. (Ann. Int. Years Quiet Sun, 4). 25 papers.

July 1967. Conference papers discussing the many aspects of the IQSY studies. Topics: activity of the quiet Sun; interplanetary space; cosmic radiation; the radiation belts; aurora and airglow; etc. (see also *080.037*).

085.014 Solar-terrestrial physics symposium. **J. Atmos. Terr. Phys.** 41: 681-916; 1979.
C 14 papers.

Innsbruck, Austria: 29 May—3 Jun 1978. Emphasis on the dynamics and composition of the middle atmosphere, solar activity effects on the atmosphere and on systems, and chemistry and dynamics of the thermosphere and ionosphere.

085.015 **Solar-Terrestrial Physics: Terrestrial Aspects (Proceedings of the Joint IQSY/**
C **COSPAR Symposium, London, 1967, Part II).** Cambridge, MA: MIT Press; 1969. 460p. (Ann. Int. Years Quiet Sun, 5). 23 papers.

July 1967. Topics: meteorology; ionospheric measurements; ionospheric processes; the Earth's atmosphere; geomagnetism (see also *080.037*).

085.016 **Sources and Availability of IQSY Data.** Cambridge, MA: MIT Press; 1970. 345p.

A description of data and a listing of sources and addresses (see also *080.037*).

085.017 Strickland, A. C., ed. **Survey of IQSY Observations and Bibliography.** Cambridge, MA: MIT Press; 1970. 589p. (Ann. Int. Years Quiet Sun, 6). 17 papers.

In addition to the lengthy bibliography of IQSY literature (*085.003*), this volume contains 17 review chapters with references, surveying the data obtained and the conclusions drawn from IQSY observing programs. Topics: meteorology; geomagnetism; aurora; airglow; the ionosphere; solar activity; cosmic rays; and space research. A detailed history of the IQSY program is also included.

085.018 Sýkora, J., ed. **Solar Activity and Solar-Terrestrial Relations.** Bratislava,
C Czechoslovakia: VEDA, Publishing House of the Slovak Academy of Sciences; 1976. 402p. (Contrib. Astron. Obs. Skalnaté Pleso, 6). 52 papers. English/Russian.

Seventh Regional Consultation on Solar Physics. Starý Smokovec, Czechoslovakia: 24-28 Sep 1973. (Proceedings). Papers covering research and theory in the following areas: I) August 1972 activity on the Sun; II) solar flares and prominences; III) scientific results of the Intercosmos project; IV) magnetic fields, sunspots, large structures and periodicity; V) radioastronomy; VI) corona, solar wind, and magnetosphere; VII) theoretical astrophysics and observational results.

085.019 Xanthakis, J., ed. **Solar Activity and Related Interplanetary and Terrestrial**
C **Phenomena.** Berlin: Springer-Verlag; 1973. 195p. 29 papers.

First European Astronomical Meeting. Athens, Greece: 4-9 Sep 1972. (Proceedings, v. 1). Reports of research by European astronomers. Selected topics include interplanetary solar phenomena; solar flares; sunspots; solar radio emissions; solar eclipses; and planetary atmospheres. There are also several short reports from observatories, institutes, and joint research efforts.

Annals of the IQSY (International Years of the Quiet Sun). 080.037.

Illustrated Glossary for Solar and Solar-Terrestrial Physics. 080.004.

Intercorrelated Satellite Observations Related to Solar Events. 080.026.

Solar flare emissions and geophysical disturbances. *073.021.*

Solar Physics: A Journal for Solar Research and the Study of Solar Terrestrial Physics. R005.020.

Solar Radiations and the Earth. 074.034.

PLANETARY SYSTEM

091 Physics of the Planetary System (Dynamics, Figure, Rotation, Interiors, Atmospheres, Magnetic Fields, etc.)

091.001 Adams, J. B., et al. Strategy for scientific exploration of the terrestrial planets. **Rev. Géophys.** 7: 623-661; 1969. 99 refs, 1755-1968.

Presents a framework relating experiments to be performed on the planets and satellites to the major goal of determining the origin and history of the solar system.

091.002 Bauer, Siegfried J. **Physics of Planetary Ionospheres.** New York; Berlin: Springer-Verlag; 1973. 230p. (Phys. Chem. Space, 6). 9 chapters. 280 refs, 1925-73.

An advanced monograph concentrating "on the fundamental physical and chemical processes in an idealized planetary ionosphere as a general abstraction, with actual planetary ionospheres representing special cases." Topics: neutral atmospheres; sources of ionization; thermal structure of planetary ionospheres; chemical processes; model ionospheres; experimental techniques; observed properties of planetary ionospheres (Earth, Mars, Venus, Jupiter); etc.

091.003 **A Bibliography of Planetary Geology Principal Investigators and Their Associates, 1976-1978.** Washington, DC: NASA; 1978. 104p. (NASA TM-79732; N78-28027). 764 refs, 1976-78.

An unannotated list of papers and other publications resulting from NASA-sponsored research. Entries are arranged by author in 12 subject sections, including solar system formation, satellites, planetary interiors, planetary cartography, instrumentation, and more. An author/editor index is included.

091.004 Bibliography of Soviet solar system space research in 1969. **Icarus** 15: 140-146; 1971. 152 refs, 1969.

Primarily journal articles and conference proceedings from Russian literature sources, this bibliography is taken from the report of the Thirteenth Plenary Session of the Committee for Space Research (COSPAR) in Leningrad. Entries are arranged in the following categories: comets, planets, the Moon, meteors and meteorites, interplanetary space, solar wind, and Earth-Sun interactions.

091.005 Bolt, B. A.; Derr, J. S. Free bodily vibrations of the terrestrial planets. **Vistas Astron.** 11: 69-102; 1969. 103 refs, 1829-1967.

A comparison is made of the work in geophysics, astronomy, and atomic physics on free oscillations of spherical models. Observations of and models for the planets and Moon are presented.

091.006 Boyce, J. M. **A Bibliography of Planetary Geology Principal Investigators and Their Associates 1978-1979.** Washington, DC: NASA; 1979. 83p. (NASA TM-80540; N79-31111). 450 refs.

See entry *091.003*.

091.007 Bullen, K. E. The interiors of the planets. **Annu. Rev. Astron. Astrophys.** 7: 177-200; 1969. 120 refs, 1743-1968.

Review from a geophysical point of view. Emphasis is on information about the terrestrial planets, with the Earth used for comparison. Internal mechanical properties (density, pressure, etc.) are given special emphasis.

091.008 Burns, J. A., ed. **Planetary Satellites.** Tucson, AZ: University of Arizona Press;
C 1977. 598p. (IAU Colloq., 28). 27 papers. 907 refs, 1839-1976.

Ithaca, NY, USA: 18-21 Aug 1974. A collection of revised review papers from IAU Colloquium no. 28 along with other contributions, providing a much-needed reference and resource volume. Besides a comprehensive introduction and overview (including basic physical data), there are chapters addressing orbits and dynamical evolution, physical properties, objects, and satellite origin. Additional papers from the meeting: *Icarus* 24(4): 1975, 15 papers; 25(3): 1975, 13 papers; 25(4): 1975, 3 papers; *Celestial Mechanics* 12: 1-110; 1975, 12 papers.

091.009 Busse, F. H. Theory of planetary dynamos. In: Kennel, Charles F.; Lanzerotti, Louis J.; Parker, E. N., eds. **Solar System Plasma Physics. Volume II.** Amsterdam: North-Holland; 1979: 293-317. 84 refs, 1910-78.

A comparative theory of planetary magnetism is explored, with newer information on the planets compared to the fairly well established Earth model. Topics: hydrodynamics of planetary cores; an outline of dynamo theory; planetary dynamos.

091.010 Cassen, P., ed. Solid convection in the terrestrial planets. **Phys. Earth Planet.**
C **Inter.** 19: 107-207; 1979. 6 papers.

Moffett Field, CA, USA: 12-13 Dec 1977. Review papers and contributions on planetary convection and evolution workshop.

091.011 Chamberlain, Joseph W. **Theory of Planetary Atmospheres: An Introduction to Their Physics and Chemistry.** New York: Academic; 1978. 330p. (Int. Geophys. Ser., 22). 7 chapters w/refs.

An overview of atmospheric structure and physical processes for graduate students and scientists. Primarily concerned with the Earth, Venus, Mars, and Jupiter, the book includes references to important papers, books, and review articles in the field. Topics: vertical structure of an atmosphere; hydrodynamics of atmospheres; chemistry and dynamics of Earth's stratosphere; planetary astronomy; ionospheres; airglows and aeronomy; stability of planetary atmospheres.

091.012 Cole, G. H. A. **The Structure of the Planets.** London: Wykeham Publications; 1978. 232p. (Wykeham Sci. Ser.). 14 chapters. 32 refs, 1932-77.

An undergraduate text covering the various aspects of planetary interiors. After introductory material covering general principles of planetary make-up, the book examines all the planets and the Earth's moon. Emphasis is on the Earth and its satellite.

091.013 Conference on solar system plasmas and x-ray astronomy. **Proc. Astron. Soc.**
C **Australia** 1: 123-173; 1968. 28 papers and abstracts.

Adelaide, Australia: 12-14 Aug 1968. Topics include radio waves in astrophysical plasmas, theoretical cosmic rays, x-ray astronomy.

091.014 de Vaucouleurs, G. Photométrie des surfaces planétaires. [Photometry of planetary surfaces.] In: Dollfus, A., ed. **Surfaces and Interiors of Planets and Satellites.** London: Academic; 1970: 225-316. 119 refs, 1865-1969. In French.

A review of theory, technique, and past research. Introductory material covers photometric units and measurement, as well as several photometric parameters. The photometry of each planet, except for Pluto, is presented next, with much numerical data. Phase function theory and detailed photometry are also presented.

091.015 Dolginov, S. S. Planetary magnetism: a survey. **Geomagn. Aeron.** 17: 391-406; 1977. 105 refs, 1932-76.

Includes a review of theories of how a planet's magnetic field is produced; the current data on the magnetic fields of Jupiter, Mars, Mercury, Venus, the Moon, and Earth; discussion of the general properties of planetary magnetic fields; and a comparison of the fields in terms of planetary dynamo mode models.

091.016 Dollfus, A. Diamètres des planètes et satellites. [Diameters of planets and satellites.] In: Dollfus, A., ed. **Surfaces and Interiors of Planets and Satellites.** London: Academic; 1970: 45-139. 38 refs, 1906-1970. In French.

A wealth of data is presented in this paper in the form of tables and diagrams, providing the most up-to-date and best values for planetary diameters. After a brief review of classical methods, the latest techniques (optical, radio, electrical, and space probe) are given. The majority of the paper, however, is concerned with presenting diameter data on each of the planets and many satellites, reviewing past research and important studies.

091.017 Dollfus, A., ed. **Surfaces and Interiors of Planets and Satellites.** London; New York: Academic; 1970. 569p. 10 chapters w/refs. English/French.

A review volume for specialists and nonspecialists, summarizing current knowledge and past studies. Topics: mass determination of the planets and satellites; planetary diameters; radar studies; thermal radio emission; photometry of planetary surfaces and of asteroids; physical properties of Saturn's rings; internal constitution of terrestrial planets; planetary magnetic fields; surface environment and exobiology of Mars.

091.018 Duncombe, Raynor L., ed. **Dynamics of the Solar System.** Dordrecht, Holland;
C Boston: D. Reidel; 1979. 330p. (IAU Symp., 81). 56 papers.

Tokyo, Japan: 23-26 May 1978. Cosponsored by COSPAR and IUTAM. Sections include stability, n- and 3-body problems, variable mass; planetary and lunar theories; ephemerides, equinox, and occultations; satellites and rings; minor planets; and comets.

091.019 Evans, J. V. Radar studies of planetary surfaces. **Annu. Rev. Astron. Astrophys.** 7: 201-248; 1969. 118 refs, 1948-68.

Included in this review are factors affecting radar observation of planets, as well as results of observations of the Moon, Mercury, Venus, and Mars.

091.020 **Évolution des atmosphères planétaires et climatologie de la terre.** [Evolution of
C Planetary Atmospheres and Climatology of the Earth.] Toulouse, France: Centre National d'Etudes Spatiales, Departement des Arraires Universitaires; 1979. 574p. 61 papers. French/English.

Nice, France: 16-20 Oct 1978. Seven of these conference papers are concerned with planetary atmospheres, but the majority of this work is concerned with data and modelling related to long- and short-term climatic changes and their effects on the Earth. The astronomy-related reports cover Viking observations of the Martian atmosphere; evolution and origin of planetary atmospheres; chemical evolution of model atmospheres; etc.

091.021 Fox, K. High-resolution infrared spectroscopy of planetary atmospheres. In: Rao, K. Narahari; Mathews, C. Weldon, eds. **Molecular Spectroscopy: Modern Research.** London: Academic; 1972: 79-114. 226 refs, 1922-71.

A review of recent results, as well as techniques employed in such studies. Emphasis is given to rotation-vibration spectra of molecules in the range 0.7-2.5 μm, or about 4,000-14,000 cm^{-1}. Examples of various molecular species in the atmospheres of planets are presented, with an emphasis on Jupiter, Venus, and Mars. Saturn, Uranus, Neptune, Earth, and Mercury are also considered.

091.022 Gehrels, T., ed. **Planets, Stars, and Nebulae: Studied with Polarimetry.** Tucson,
C AZ: University of Arizona Press; 1974. 1,134p. (IAU Colloq., 23). 74 papers.

Tucson, AZ, USA: 15-17 Nov 1972. A combination of invited review papers and papers from IAU Colloquium no. 23 covering recent developments in polarimetry, including elliptical polarization. Specific areas covered in this comprehensive volume are general theory and techniques, surfaces and molecules, and polarimetry applied to stars and nebulae. Contains hundreds of references.

091.023 Goody, R. Motions of planetary atmospheres. **Annu. Rev. Astron. Astrophys.** 7: 303-352; 1969. 161 refs, 1941-68.

Consideration of the lower atmospheres of Venus, Mars, and Jupiter.

091.024 Gubbins, D. Theories of the geomagnetic and solar dynamics. **Rev. Geophys. Space Phys.** 12: 137-154; 1974. 156 refs, 1919-74.

Review of recent advances in dynamo theory in relation to the problem of the generation of the Earth's and the Sun's magnetic fields. Some relevant observations and estimates of physical quantities are discussed, and the lack of knowledge about the dynamical state of the Earth's core is emphasized.

091.025 Hansen, J. E.; Travis, L. D. Light scattering in planetary atmospheres. **Space Sci. Rev.** 16: 527-610; 1974. 140 refs, 1862-1974.

Review of scattering theory required for analysis of light reflected by planetary atmospheres.

091.026 Hide, R. Motions in planetary atmospheres. **Q. J. R. Meteor. Soc.** 102: 1-23; 1976. 234 refs, 1686-1975.

Discussion of motions and underlying dynamical processes.

091.027 Hide, R. Planetary magnetic fields. In: Dollfus, A., ed. **Surfaces and Interiors of Planets and Satellites.** London: Academic; 1970: 511-534. 106 refs, 1919-68.

A discussion of the Earth and Jupiter, the only two planets known to have magnetic fields in 1969. The emphasis of this article is on the Earth, for which there was substantial data available. Topics: description and analysis of the main geomagnetic field; the Earth's interior; theories of the main geomagnetic field; Jupiter's magnetic field.

091.028 Hill, T. W.; Michel, F. C. Planetary magnetospheres. **Rev. Geophys. Space Phys.** 13(3): 967-974, 1033-1036; 1975. 148 refs, 1964-75.

Review for the past four years, discussing intrinsic slowly and rapidly rotating magnetospheres and induced magnetospheres. Particular emphasis is on knowledge gained through planetary probes.

091.029 Huguenin, R. L.; McCord, T. B. The surfaces of solar-system bodies. **EOS Trans. American Geophys. Union** 52: IUGG 471-483; 1971. Also: **MIT Planet. Astron. Lab. Contr.** 22. 487 refs, 1965-71.

Review for 1966-70 considering results obtained from 1) remote sensing at ultraviolet, optical, infrared, microwave, and radio wavelengths; 2) theoretical and laboratory investigations; and 3) in situ measurement and surface exploration, and laboratory analyses of returned samples.

091.030 Hunten, D. M.; Donahue, T. M. Hydrogen loss from the terrestrial planets. **Annu. Rev. Earth Planet. Sci.** 4: 265-292; 1976. 93 refs, 1925-75.

Review of the processes leading to hydrogen loss from the upper atmosphere, followed by the application of this theory to the Earth, Venus, and Mars.

091.031 Izakov, M. N. Comparison of structure and dynamics of the Earth's, Mars', and Venus' thermospheres. **J. Atmos. Terr. Phys.** 38: 847-862; 1976. 186 refs, 1960-75.

Review of the planets' upper atmospheres based on ground-based and spacecraft observations; reveals their similarities and differences, and constitutes an attempt to mention both the reliable facts and the problems still to be studied.

091.032 Jehle, H. Distribution of the mean motions of planets and satellites and the development of the solar system. **Vistas Astron.** 21: 265-287; 1977. 142 refs, 1755-1975.

Discussion of the distribution of orbital elements on a purely gravitational basis.

091.033 Kaula, William M. **An Introduction to Planetary Physics: The Terrestrial Planets.** New York; London: John Wiley & Sons; 1968. 490p. (Space Sci. Text Ser.). 9 chapters. 586 refs, 1897-1968.

A graduate text concerned with the following subjects: the Earth's interior; mechanical and thermal aspects of a planetary interior; planetary magnetism; dynamics of the Earth-Moon system; dynamics of the solar system; observations of planetary surfaces; geology of the Moon and Mars; meteorites and tektites; constitution and the origin of the terrestrial planets.

091.034 Kennel, C. F., ed. Proceedings of the NASA/JPL workshop on the physics of
C planetary and astrophysical magnetospheres. **Space Sci. Rev.** 24: 369-634; 1979. 7 papers.

Snowmass, CO, USA: 30 Jul – 4 Aug 1978. Topics considered include magnetospheres of Earth and Jupiter, pulsar magnetospheres, x-ray source magnetospheres, and coherent radio emissions in space and astrophysics.

091.035 King, Elbert A. **Space Geology: An Introduction.** New York: John Wiley & Sons; 1976. 349p. 11 chapters w/refs.

Intended for advanced students, this text presents a broad overview of the subject, covering the Moon, meteorites, tektites, craters, terrestrial impact craters, impact metamorphism, Mars, asteroids, comets, comparative planetology, etc. Contains many photographs, references, numerical data, and a glossary.

091.036 Kovalevsky, J. Détermination des masses des planètes et satellites. [Determination of masses of planets and satellites.] In: Dollfus, A., ed. **Surfaces and Interiors of Planets and Satellites.** London: Academic; 1970: 1-44. 77 refs, 1833-1969. In French.

An overview of current knowledge and past studies. In addition to the data on individual planets, the asteroids, and selected satellites, methods used in obtaining the data are reviewed.

091.037 Levin, B. J. Internal constitution of terrestrial planets. In: Dollfus, A., ed. **Surfaces and Interiors of Planets and Satellites.** London: Academic; 1970: 462-510. 129 refs, 1937-70.

A summary of current information and recent research programs, with an emphasis on the construction of model planetary interiors. Concerned with gross composition and different layers of planetary interiors, this article covers the composition of meteorites (and what they tell us about planetary interiors); the nature of the Earth's core; models of the interiors of Mercury, Venus, and Mars; evolution of internal structures; comparative analysis of the composition of the terrestrial planets.

091.038 Levy, E. H. Generation of planetary magnetic fields. **Annu. Rev. Earth Planet. Sci.** 4: 159-185; 1976. 119 refs, 1934-75.

Review of theory and current understanding of the problem, as well as the limits and uncertainties faced by scientists.

091.039 Lewis, J. S. Chemistry of the planets. **Annu. Rev. Phys. Chem.** 24: 339-351; 1973. 55 refs, 1950-73.

A brief review of the chemical composition of the planets and the identifiable trends associated with chemical and physical properties of these bodies.

091.040 Lyttleton, R. A. The structures of the terrestrial planets. **Adv. Astron. Astrophys.** 7: 83-145; 1970. 26 refs, 1930-68.

A review of the physical properties and internal composition of Earth, Venus, Mars, Mercury, and the Moon. Various planetary models are set up and discussed. A comparison of the internal structures of each celestial body concludes the review.

091.041 McElroy, M. B. The composition of planetary atmospheres. **J. Quant. Spectrosc. Radiat. Transfer** 11: 813-825; 1971. 40 refs, 1952-70.

The present status of research on the composition of planetary atmospheres is reviewed, emphasizing those aspects of the subject which involve applications of radiative transfer theory. Much of the paper is concerned with difficulties involved in the interpretation of i.r. spectra.

091.042 Meadows, A. J. The atmospheres of the Earth and the terrestrial planets: their origin and evolution. **Phys. Rep. Phys. Lett. C.** 5C: 197-235; 1972. 142 refs, 1951-72.

A comparison of the atmospheres of the Earth, Venus and Mars, showing that a consistent model of their evolution can be obtained on the basis of degassing from the solid planet.

091.043 Münch, G. Highlights in planetary spectroscopy 1962-75. **Mem. Soc. R. Sci. Liège** 6e ser. 9: 87-100; 1976. 75 refs, 1930-75.

A review of research carried out by spacecraft. Emphasis is on spectroscopic studies in the photomultiplier and infrared spectral ranges.

091.044 National Academy of Sciences. National Research Council. **Planetary Astronomy: An Appraisal for Ground-based Opportunities.** Washington, DC: National Academy of Sciences; 1968. 76p. (NAS Publ., 1688). 8 chapters. no refs.

The report of the Space Science Board's Panel on Planetary Astronomy surveying the present status and future needs of ground-based planetary astronomy. Possible areas of study included dynamics of the planetary system; planetary surfaces; planetary and cometary atmospheres; interiors and magnetic fields; observational techniques and facilities; graduate training in the planetary sciences. Good historical perspective.

091.045 Ness, N. F. The magnetic fields of Mercury, Mars, and Moon. **Annu. Rev. Earth Planet. Sci.** 7: 249-288; 1979. 76 refs, 1961-78.

A review of experimental observations and interpretations of data obtained by spacecraft or instruments which have landed or been orbited around the planets. A history of the various space missions by both the United States and the USSR is presented, along with an analysis and summary of the data gathered by each. The nature and origin of each planet's magnetic field are also discussed.

091.046 Peale, S. J. Orbital resonances in the solar system. **Annu. Rev. Astron. Astrophys.** 14: 215-246; 1976. 53 refs, 1829-1975.

A review of theoretical developments and discussion of properties of a resonance, plus a summary of current ideas and identification of known orbital resonances.

091.047 Peale, S. J. Planetary dynamics. **EOS Trans. American Geophys. Union** 52: IUGG 464-468; 1971. 102 refs, 1967-71.

Literature review for 1967-70, covering range determinations and ephemerides, tidal evolution, resonances, and miscellaneous subjects.

091.048 Pettengill, G. H. Physical properties of the planets and satellites from radar observations. **Annu. Rev. Astron. Astrophys.** 16: 265-292; 1978. 80 refs, 1960-78.

Review of radar astronomy with an emphasis on observations of asteroids, Galilean satellites, and Saturn's rings.

091.049 Physical studies of planets and satellites. **IAU Trans. Rep. Astron.** 17A(1): 103-111; 1979. - refs, 1976-78.

An overview of research activity worldwide, with an emphasis on the previous three years, as reported by IAU Commission 16 (Physical Studies of Planets and Satellites). Additional reports: 14A: 153-168; 1970. - refs, 1967-69. 15A: 191-201; 1973. - refs, 1970-72. 16A(1): 85-100; 1976. + refs, 1973-76.

091.050 Pollack, J. B. Climatic change on the terrestrial planets. **Icarus** 37: 479-553; 1979. 116 refs, 1941-78.

Review of the observational data on climatic change; discussions of the basic factors that influence climate; and examination of the manner in which these factors may have been responsible for some of the known changes.

091.051 Ponnamperuma, Cyril, ed. **Comparative Planetology.** New York; London: Aca-
C demic; 1978. 275p. 16 papers.

Third College Park Colloquium on Chemical Evolution. College Park, MD, USA: 29 Sep—1 Oct 1976. (Proceedings). Reports of research on planetary interiors; crustal evolution; the origin of planetary atmospheres; the origin of life; the origin of continents; etc. Emphasis is on Mars, but the Moon and Earth are also considered.

091.052 Rasool, S. I., ed. **Physics of the Solar System.** Washington, DC: NASA; 1972.
C 523p. (NASA SP-300). 13 papers.

Fourth Summer Institute for Astronomy and Astrophysics. Stony Brook, NY, USA: 17 Jun—15 Jul 1970. Papers based on lectures covering "a broad range of topics in the physics of the Sun, the structure of the planets and their atmospheres, and the origin and evolution of the solar system and of planetary atmospheres." Topics: solar rotation; outer solar atmosphere; interplanetary plasma; radio studies of the planets; lunar rock samples; etc.

091.053 **Reports of Accomplishments of Planetology Programs, 1975-76.** Washington, DC: NASA; 1976. 282p. (NASA TM-X-3364). 105 abstracts.

Brief summaries of NASA-sponsored research in the following areas: solar systems, comets, and asteroids; geophysics; comparative planetary geology; techniques and instrument development; aeolian phenomenon; tectonics, geomorphology, and volcanology; channels; craters; surface processes and features; Mars and Mercury geologic mapping; Venus; etc. Continued by *Reports of Planetary Geology Program* (NASA, 1977-) (see *091.054*).

091.054 **Reports of Planetary Geology Program.** Washington, DC: NASA. 1977- . Annual.

Abstracts (one to three pages) of papers presented at the annual meetings of the Planetary Geology Principal Investigators. These reports of NASA-sponsored research cover the following areas: solar system formation; planetary interiors; asteroids, comets, and moons; cratering; volcanic processes; aeolian processes; mapping programs; instrument development and techniques; planetary cartography; geochemistry; tectonics; etc.

1976-77: Arvidson, R.; Wahmann, R., comps. 1977. 294p. (NASA TM-X-3511). 115 abstracts. St. Louis, MO: 23-26 May 1977.

1977-78: Strom, R.; Boyce, J., comps. 1978. 342p. (NASA TM-79729). 125 abstracts. Tucson, AZ: 30 May—1 Jun 1978.

1978-79: Boyce, J.; Collins, P. S., comps. 1979. 461p. (NASA TM-80339). 169 abstracts. Providence, RI: 6-8 Jun 1979.

1979-80: Wirth, P.; Greeley, R.; D'Alli, R., comps. 1980. 406p. (NASA TM-81776). 145 abstracts. Phoenix, AZ: 14-16 Jan 1980.

091.055 C Runcorn, S. K., ed. High pressure physics and planetary interiors. **Phys. Earth Planet. Inter.** 6: 1-209; 1972. Also: **Lunar Science Inst. Contr.** 89. 33 papers.

Houston, TX, USA: 1-3 Mar 1972. Review papers and reports of research on the measurement of the properties of solids at high pressure and quantum mechanics of compressed solids along with models of planetary interiors. "Of particular interest is the internal structure of the major planets where the equations of state are intermediate in complexity between the gas laws applying to stellar interiors and the solids of high molecular weight and complex crystal structure of the interior of the terrestrial planets."

091.056 C Sagan, Carl; Owen, Tobias C.; Smith, Harlan J., eds. **Planetary Atmospheres.** Dordrecht, Holland; Boston: D. Reidel, 1971. 408p. (IAU Symp., 40). 45 papers, 7 abstracts.

Marfa, TX, USA: 26-31 Oct 1969. The program divided into three parts: Venus (17 papers), Mars (26 papers), and the outer planets (9 papers).

091.057 Salisbury, J. W., ed. AFCRL bibliography for the first quarter of 1969. **Icarus** 11: 118-138; 1969.

An extensive, often lengthy annotated list of articles from the literature on lunar and planetary topics. Published quarterly by the Air Force Cambridge Research Laboratory from 1969 to mid-1974, this substantial effort documents well the research in the field. Selected topics include astrobiology; comets; the Moon; meteors and meteorites; the planets; origin of the solar system; etc. With nearly 2,000 references for the six-year period, this is a major source not to be overlooked. Complete list of references: 11: 118-138; 1969. 12: 258-312; 1970. 12: 457-497; 1970. 13: 114-151; 1970. 13: 305-353; 1970. 14: 126-178; 1971. 14: 431-484; 1971. 16: 581-637; 1972. 17: 234-264; 1972. 17: 265-288; 1972. 19: 247-286; 1973. 19: 287-317; 1973. 19: 429-472; 1973. 19: 576-603; 1973. 20: 72-120; 1973. 20: 356-404; 1973. 21: 369-386; 1974. 22: 476-496; 1974.

091.058 Schubert, G. Subsolidus convection in the mantles of terrestrial planets. **Annu. Rev. Earth Planet. Sci.** 7: 289-342; 1979. 199 refs, 1935-79.

Overview of the role of heat transport by solid state mantle convection in the Earth, Moon, Mercury, Venus, and Mars. Reviews the physical properties of planetary mantles and the various approaches that have been taken in the study of mantle convection.

091.059 Seidelmann, P. K. Planetary theory developments, 1973-76. **Celestial Mech.** 17: 103-112; 1978. 56 refs, 1858-1976.

Emphasis is placed on the efforts to prepare new general theories; on new techniques of numerical integration; and on studies of minor planets.

091.060 Short, Nicholas M. **Planetary Geology.** Englewood Cliffs, NJ: Prentice-Hall; 1975. 361p. 13 chapters. 462 refs, 1883-1973.

A text for undergraduate science students majoring in geology and earth sciences. With an emphasis on the Moon, selected topics include meteorites, the solar system, origin of the planets, the lunar surface, impact cratering, lunar igneous processes; etc.

091.061 Siscoe, G. L. Towards a comparative theory of magnetospheres. In: Kennel, Charles F.; Lanzerotti, Louis J.; Parker, E. N., eds. **Solar System Plasma**

Physics. Volume II. Amsterdam: North-Holland; 1979: 319-402. 194 refs, 1955-77.

A review of current information on the various planetary magnetospheres aimed at developing a theory relating all such phenomena, their origin, and nature. Topics: an overview of current knowledge; independent variables affecting planetary magnetospheres; dependent magnetospheric variables; long-term aspects.

091.062 Smith, E. J.; Wolfe, J. H. Fields and plasmas in the outer solar system. **Space Sci. Rev.** 23: 217-252; 1979. 69 refs, 1955-78.

A review based on Pioneer 10 and 11 investigations.

091.063 **The Solar System.** San Francisco: W. H. Freeman; 1975. 145p. 12 articles w/refs.

A collection of articles for the layperson and scientist, providing an overview of our planetary system gained from manned and unmanned spacecraft, as opposed to Earth-based study. Excerpted from the September 1975 issue of *Scientific American*, the papers include an overview of 18 years of space study of the solar system; cosmogony; the Sun; Mercury; Venus; Earth; Mars; Jupiter; the outer planets; asteroids; moons; and solar wind. Includes many close-up photos from spacecraft.

091.064 Solomon, S. C. Formation, history and energetics of cores in the terrestrial planets. **Phys. Earth Planet. Inter.** 19: 168-182; 1979. 116 refs, 1937-78.

A synthesis of the known or inferable answers to questions about the characteristics and history of central metallic cores in the terrestrial planets.

091.065 Sonnerup, U. Ö. Magnetic field reconnection. In: Lanzerotti, Louis J.; Kennel, Charles F.; Parker, E. N., eds. **Solar System Plasma Physics. Volume III.** Amsterdam: North-Holland; 1979: 45-108. 144 refs, 1958-78.

"A concise qualitative summary of the present state of reconnection theory and observations, with special reference to the Earth's magnetosphere."

091.066 Stevenson, D. Planetary magnetism. **Icarus** 22: 403-415; 1974. 62 refs, 1883-1974.

Discussion of origin and maintenance of planetary magnetic fields.

091.067 Strobel, D. F. Aeronomy of the major planets: photochemistry of ammonia and hydrocarbons. **Rev. Geophys. Space Phys.** 13: 372-382; 1975. 90 refs, 1937-74.

Detailed discussion of primary dissociation paths, chemical kinetics, aeronomical models, the formation of complex hydrocarbons, and the role these constituents and their photolysis products play in the energy balance of planetary atmospheres.

091.068 Symposium on planetary atmospheres and surfaces. **Icarus** 17: 289-524, 540-542;
C 1972. 13 papers.

Madrid, Spain: 10-13 May 1972. Papers on Mariner 9 and Mars 2 and 3 spaceprobes.

091.069 Symposium on planetary atmospheres and surfaces. **Radio Sci.** 5: 121-533; 1970.
C 37 papers, 7 abstracts.

Woods Hole, MA, USA: 11-15 Aug 1969. Cosponsored by URSI, IAU, and COSPAR. Four sections on the Moon, and others on atmospheres of Venus and Mars; surfaces of Mercury, Venus, and Mars; and radio observations of Jupiter and the giant planets.

091.070 A symposium on planetary science in celebration of the quincentenary of
C Nicolaus Copernicus 1473-1543. **Proc. R. Soc. London, Ser. A.** 336: 1-114; 1974. 6 papers.

London, UK: 25 Jan 1973. Review of the developments taking place in the knowledge of the solar system. There are individual papers on the Moon, Mars and Venus, Jupiter and Saturn, and planetary dynamics.

091.071 Teyfel', V. G., ed. **Physical Characteristics of the Giant Planets.** Washington, DC: NASA; 1972. 216p. (NASA TT-F-717). 723 refs, 1904-1970.

A handbook of data on the atmospheres, radio emissions, internal structures, and more of Jupiter, Saturn, Uranus, and Neptune. Primarily tables and graphs with some explanatory text, the book contains an extensive bibliography primarily covering 1950-1970. Other topics include planetary size, mass, optical characteristics, chemical composition, unusual features (e.g., Saturn's rings), etc. Translation of *Fizicheskie kharakteristiki planetgigantov* (Alma-Ata, USSR: Nauka; 1971).

091.072 Thomsen, L. Theoretical foundations of equations of state for the terrestrial planets. **Annu. Rev. Earth Planet. Sci.** 5: 491-513; 1977. 75 refs, 1907-1977.

Examination of theoretical synthesis with regard to the constitution and structure of the terrestrial planets.

091.073 Whipple, Fred L. **Earth, Moon, and Planets.** 3rd ed. Cambridge, MA: Harvard University Press; 1968. 297p. (Harvard Books Astron.). 14 chapters. 40 refs, 1951-66.

A descriptive volume on the solar system for students and laypersons, combining observation (ground-based and spacecraft), theory, and history. Topics: gravity; planetary discoveries; lunar influence on the Earth; the Moon; the giant planets; the terrestrial planets; Mars; origin and evolution of the solar system; etc.

091.074 Wood, J. A. Planetary interiors. **EOS Trans. American Geophys. Union** 52: IUGG 468-471; 1971. 67 refs, 1966-70.

A brief review of the literature, concentrating on studies of the Moon's interior.

091.075 Woszcyk, A.; Iwaniszewska, C., eds. **Exploration of the Planetary System.**
C Dordrecht, Holland; Boston: D. Reidel; 1974. 564p. (IAU Symp., 65). 46 papers, 5 abstracts.

Torun, Poland: 5-8 Sep 1973. Topics include origin and general physics of the planetary system, terrestrial planets, outer planets and their satellites, and future explorations of the solar system.

091.076 Zharkov, V. N.; Trubitsyn, V. P. **Physics of Planetary Interiors.** Tucson, AZ: Pachart Publishing; 1978. 388p. (Astron. Astrophys. Ser., 6). 4 chapters. 501 refs, 1879-1977.

A review of current knowledge and past research, drawing upon both Russian and Western sources. This highly technical book deals with 1) physics of the Earth (useful for drawing parallels to and understanding the outer planets); 2) high-pressure equations of state; 3) the theory of figures (rotating planets in hydrostatic equilibrium); 4) interior structure of the planets (emphasizing modeling). Translated, edited, and additional material by W. B. Hubbard.

Analytical Chemistry in Space. 022.077.

Atmospheres of Earth and the Planets. 082.006.

Bibliography. *094.026.*

Chemical Petrology: With applications to the terrestrial planets and meteorites. 022.064.

Dynamical astronomy of the solar system. *042.004.*

The Encyclopedia of Atmospheric Sciences and Astrogeology. R009.002.

Geochemical Exploration of the Moon and Planets. 051.001.

ISAS Lunar and Planetary Symposium. 094.019.

ISAS Symposium on Lunar and Planetary Science. 094.020.

Index to the Proceedings of the Lunar and Planetary Science Conferences: Houston, Texas, 1970-1978. 094.036.

The interaction of the solar wind with Mars, Venus and Mercury. *074.027.*

Light Scattering in Planetary Atmospheres. 022.073.

Lunar & Planetary Information Bulletin. 094.032.

Lunar and Planetary Science Conference(s). 094.033; 094.034.

Mass spectroscopy in solar system exploration. *031.517.*

Methods of calculating infrared transfer—a review. *022.035.*

The Moon and the Planets. Index to Volumes 1-20. R005.016.

Origins of the Moon and satellites. *094.006.*

Physics and Chemistry of Upper Atmospheres. 082.007.

Physics of Solar Planetary Environments. 080.040.

Radar studies of the planets. *031.508.*

Radio Emission of the Sun and Planets. 141.080.

Radiophysical Investigations of the Planets: A Bibliography: 1960-1973. 141.050.

Solar System Plasma Physics. Volume II. Magnetospheres. 084.019.

Solar System Plasma Physics. Volume III. Solar System Plasma Processes. 062.013.

Solar-Wind Interaction with the Planets Mercury, Venus, and Mars. 074.022.

The Soviet-American Conference on Cosmochemistry of the Moon and Planets. 022.069.

Symposium on Jupiter and the outer planets. *099.023.*

U.S. national report to the International Union of Geodesy and Geophysics. *081.001.*

Use for Space Systems for Planetary Geology and Geophysics. 051.009.

092 Mercury

092.001 Comparisons of Mercury and the Moon. **Phys. Earth Planet. Inter.** 15: 113-312;
C 1977. Also: **Lunar Science Inst. Contr.** 274. 16 papers.

Houston, TX, USA: 15-17 Nov 1976. Reports of recent research aided and conducted by spacecraft and astronauts. Selected topics: planetary magnetism and interiors; convection in Mercury; cratering; global seismic effects; optical observations interpretation; etc.

092.002 Davies, M. E., et al. **Atlas of Mercury.** Washington, DC: NASA; 1978. 128p.
A (The 1: 5,000,000 Map Ser.; NASA SP-423). 39 refs, 1896-1977.

A photographic atlas resulting from the flight of Mariner 10 in 1974 and 1975, this book includes 20 pages of text describing the mission and its experiments, providing historical perspective and presenting physical data about the planet. Also covered are discussions of past surface mapping, topographic features and surface history, and a description of the photographs.

092.003 Gault, D. E., et al. Mercury. **Annu. Rev. Astron. Astrophys.** 15: 97-126; 1977. 127 refs, 1950-77.

Review of results of Mariner flybys, theoretical studies, and ground-based observations on Mercury's mass and size, atmospheric composition and density, charge-particle environment, infrared thermal radiation, and the existence of a planetary magnetic field.

092.004 Guest, J. E.; O'Donnell, W. P. Surface history of Mercury: a review. **Vistas Astron.** 20: 273-300; 1977. 29 refs, 1970-76.

This review relates mainly to research by the Mariner 10 television team.

092.005 Morrison, David, ed. International colloquium on Mercury. **Icarus** 28: 429-609;
C 1976. (IAU Colloq., 34). 19 papers.

Pasadena, CA, USA: 10-16 Jun 1975. Reports of research projects presented at an IAU meeting following the successful encounters with Mercury by Mariner 10. Topics: origin; rotation; magnetic field; evolution; volcanism; atmosphere; etc. Proposed IAU nomenclature for the planet is outlined.

092.006 Ness, N. F. The magnetosphere of Mercury. In: Kennel, Charles F.; Lanzerotti, Louis J.; Parker, E. N., eds. **Solar System Plasma Physics. Volume II.** Amsterdam: North-Holland; 1979: 183-206. 29 refs, 1970-78.

A summary of Mariner 10 data and its interpretation. Topics: Mercury I encounter; Mercury III encounter; magnetosphere models; magnetic tail and polar cap; etc.

093 Venus

093.001 The atmosphere of Venus. **J. Atmos. Sci.** 25: 533-671; 1968. Also in: **Contrib.**
C **Kitt Peak Natl. Obs.**, no. 377, 1968. 27 papers.

Second Arizona Conference on Planetary Atmospheres. Tucson, AZ, USA: 11-13 Mar 1968. (Proceedings). Technical papers devoted primarily to presenting data obtained by Mariner 5 and Venera 4 in 1967. Topics: mission reviews; chemical composition of the Venusian atmosphere; radio astronomical measurements; the upper atmosphere; spectrum of Venus; optical polarization of Venus; etc.

093.002 Bury, J. S. The planet Venus. **J. Brit. Interplanet. Soc.** 32: 123-155; 1979. 44 refs, 1928-78.

Elementary review paper.

093.003 Hansen, James E., ed. **The Atmosphere of Venus.** Washington, DC: NASA;
C 1975. 198p. (NASA SP-382). no refs.

New York, NY, USA: 15-17 Oct 1974. (Proceedings). Primarily abstracts and discussions of recent findings and observational results; no formal papers are included. Topics: clouds; cloud motions; dynamics and atmospheric structure; aeronomy; and evolution of the atmosphere.

093.004 Hunten, D. M.; McGill, G. E.; Nagy, A. F. Current knowledge of Venus. **Space Sci. Rev.** 20: 265-282; 1977. 103 refs, 1963-77.

An introduction to a special issue on Venus, summarizing current knowledge with an emphasis on recent progress and on contributions expected from the Pioneer Venus missions.

093.005 Izakov, M. N. Structure and dynamics of the upper atmospheres of Venus and Mars. **Soviet Phys. Usp.** 19: 503-529; 1976. Transl. of *Usp. Fiz. Nauk* 119: 295-342; 1976. 253 refs, 1931-76.

Summary and critical analysis of the existing experimental data and the theoretical models constructed to describe them.

093.006 Jastrow, R.; Rasool, S. I., ed. **The Venus Atmosphere.** New York: Gordon and
C Breach; 1969. 604p. 33 papers.

Second Arizona Conference on Planetary Atmospheres. Tucson, AZ, USA: 11-13 Mar 1968. Primarily reports of data obtained by Venera 4 and Mariner 5 on their missions in October 1967.

093.007 Knollenberg, R. G., et al. The clouds of Venus. **Space Sci. Rev.** 20: 329-354; 1977. 65 refs, 1926-77.

The current state of knowledge is reviewed, with an overview of Pioneer cloud experiments and the expected scientific results.

093.008 Marov, M. Y. Results of Venus missions. **Annu. Rev. Astron. Astrophys.** 16: 141-169; 1978. 120 refs, 1960-77.

Review of the data on the atmosphere as well as properties of the surface and planetary environment obtained from the Russian Venera series and the Mariner flybys.

093.009 Marov, M. Y. Venus: a perspective at the beginning of planetary exploration. **Icarus** 16: 415-461; 1972. 170 refs, 1924-71.

Physical characteristics of Venus, based on Venera 4, 5, 6, and 7, and Mariner 5 data are reviewed. Brief discussion of some evidence on the origin and evolution of Venus as a planet and prospective problems for further study.

093.010 Moroz, V. I. The atmosphere of Venus. **Soviet Phys. Usp.** 14: 317-340; 1971. Transl. of *Usp. Fiz. Nauk* 104: 255-296; 1971. 203 refs, 1928-71.

Review of current information with emphasis on results from Venera 4, 5, 6, and 7, and Mariner 5.

093.011 Young, L. G. Infrared spectra of Venus. **IAU Symp.** 65: 77-160; 1974. 325 refs, 1760-1973.

A historical account of observations of Venus and their interpretation. Emphasis is on how infrared spectroscopy assists in studying the motions and composition of Venus' cloud cover.

The atmospheres of Mars and Venus. *097.008.*

The ionospheres of Mars and Venus. *097.026.*

Mars and Venus. *097.005.*

Planetary Atmospheres. 091.056.

Planetary exploration: accomplishments and goals. *053.002.*

Results of recent investigations of Mars and Venus. *053.004.*

094 Moon (Dynamics, General Aspects)

094.001 Allenby, R. J. Lunar orbital science. **Space Sci. Rev.** 11: 5-53; 1970. 126 refs, 1960-69.

Describes the scientific results of the lunar orbital flights by the United States and the USSR.

094.002 Alter, Dinsmore, ed. **Lunar Atlas.** New York: Dover Publications; 1968. 343p.
A 154 plates.

A reprint with a 33% reduction of plate size from the 1964 edition issued by North American Aviation, Inc. One of the last Earth-based atlases, this volume concentrates on the side of the Moon seen from Earth. Each plate includes a detailed page of descriptive material, pointing out landmarks and formations of interest. Also included are background material, a glossary, a table of lunar formations, and appendices of lunar physical data and data on the plates.

094.003 **Apollo 11 Lunar Science Conference.** Washington, DC: American Association
C for the Advancement of Science; 1970. 358p. 144 papers.

Shorter versions of most of the papers published in the proceedings of the same conference (see *094.515*). This volume was issued to provide fast distribution of information to the scientific community. Also published as a special Moon issue of *Science* 167: 447-784; 1970.

094.004 Bowker, David E.; Hughes, J. Kenrick. **Lunar Orbiter Photographic Atlas of**
A **the Moon.** Washington, DC: NASA; 1971. 41p. (NASA SP-206). 675 plates. 33 refs, 1957-70.

A massive compilation of photographs, covering the entire lunar surface, with introductory matter describing techniques, lunar features, and arrangement of the atlas.

094.005 Burnett, D. S. Lunar science: the Apollo legacy. **Rev. Geophys. Space Phys.** 13: 13-34; 1975. 169 refs, 1964-75.

A general review of lunar science, with an emphasis on fundamental problems relating to the composition, structure, and history of the Moon, along with a discussion of unanticipated results obtained from Apollo lunar science.

094.006 C Contopoulos, G., ed. Origins of the Moon and satellites. **Highlights Astron.** 3: 467-489; 1974. 5 papers.

Joint discussion at the Fifteenth General Assembly of the IAU. Topics: growth of the Earth-Moon system; formation of small particulate matter; gravitational collapse and the formation of the solar nebula.

094.007 El-Baz, F. The Moon after Apollo. **Icarus** 25: 495-537; 1975. 130 refs, 1880-1975.

Summary of the state of the art of lunar science, with emphasis on important results of the six Apollo lunar surface exploration missions.

094.008 Evans, J. V.; Hagfors, T. Radar studies of the Moon. **Adv. Astron. Astrophys.** 8: 29-105; 1971. 105 refs, 1946-70.

Review of major contributions of radar to lunar studies. Topics include surface statistical properties, mapping, and lunar motions.

094.009 Fielder, G., ed. **Geology and Physics of the Moon: A Study of Some Fundamental Problems.** London; New York: Elsevier Publishing; 1971. 159p. 11 chapters w/refs.

A collection of review papers aimed at the scientist interested in lunar studies. Topics: recent lunar exploration; lava flows and origin of small craters; sinuous rilles; geology of the farside crater Tsiolkovsky; photometric, polarimetric, and thermal studies; origins of craters; etc.

094.010 Freeberg, Jacquelyn H. **Bibliography of the Lunar Surface: Final Report.** Washington, DC: Geological Survey, U.S. Department of Interior; 1970. 344p. >4,500 refs, ~1800-1970.

References cover surface features and materials, and telescope observations in this compilation, which includes 21 citations to older lunar bibliographies. Includes subject and locality (features and areas) indexes. Topics: landforms, geomorphology, lunar geology, etc.

094.011 Fuller, M. Lunar magnetism. **Rev. Geophys. Space Phys.** 12: 23-70; 1974. 128 refs, 1949-74.

Covers the identification and determination of the magnetic phases and the characteristics of the ferro-magnetic phases that carry remanence. Magnetic viscosity is also considered.

094.012 Gavrilov, I. V.; Dovgalevskai͡a, I. S. **Selenodezicheskie issledovanii͡a v SSSR; bibliograficheskii ukazatel' 1928-1975 gg.** [Selenodetic Investigations in the USSR: A Bibliography, 1928-1975.] Kiev: Naukova Dumka; 1978. 38p. 184 refs. In Russian.

A brief review of selenodetic research followed by a bibliography of Soviet works. Each entry is in Russian, followed by an English translation of author and title. An author index is included.

094.013 A Gavrilov, I. V.; Kislyuk, V. S.; Duma, A. S. **Consolidated System of Selenodetic Coordinates of 4900 Points of the Lunar Surface (Near Side of the Moon).** Kiev: Naukova Dumka; 1977. 25 refs, 1911-73. In Russian.

Rectangular coordinates are given in a common selenodetic system. Coordinates of the 2,390 basic points have been determined from primary data published in 11 selenodetic catalogs. Those of the 2,510 additional points were determined from data in two additional catalogs and from hypsometric characteristics of the lunar surface obtained by analysis of absolute heights of the basic points.

094.014 Guest, J. E.; Greeley, R. **Geology on the Moon.** London: Wykeham Publishing; distr., New York: Crane, Russak & Co.; 1977. 235p. (Wykeham Sci. Ser.). 13 chapters. 131 refs, 1893-1977.

An overview of the geologic processes at work on the Moon, gained from an analysis of lunar samples and various photographs taken by astronauts and satellites. Topics: circular basins, maria, impact cratering mechanics, large craters, small craters, erosion, regolith, stratigraphy, etc. Emphasis is on geology rather than a chemical analysis of samples.

094.015 Gutschewski, Gary L.; Kinsler, Danny C.; Whitaker, Ewen. **Atlas and Gazetteer**
A **of the Near Side of the Moon.** Washington, DC: NASA; 1971. 538p. (NASA SP-241). 10 refs, 1935-69.

An atlas of 404 photographs with tabular compilation of named lunar features. Based on Lunar Orbiter IV photographs, it includes a short history of lunar nomenclature and a guide to the use of the book. There are also five indexes to assist the reader in locating specific lunar topographic features.

094.016 Hagfors, T. Microwave studies of thermal emission from the Moon. **Adv. Astron. Astrophys.** 8: 1-28; 1971. 72 refs, 1930-68.

Included are observational methods and results. A summary lists lunar physical characteristics based on radio observations.

094.017 Hagfors, T. Remote probing of the Moon by infrared and microwave emissions and by radar. **Radio Sci.** 5: 189-227; 1970. 119 refs, 1930-69.

Review of the results of three remote sensing techniques.

094.018 IAU-COSPAR Julius Schmidt symposium on 100 years of lunar mapping.
C **Moon Planets** 20: 103-210; 1979. 8 papers.

Lagonissi, Greece: 25-27 May 1978. Includes a good, brief overview of the NASA space program by J. E. Naugle.

094.019 **ISAS Lunar and Planetary Symposium.** 11th. Tokyo: Institute of Space and
C Aeronautical Science, University of Tokyo; 1978. 281p. English/Japanese.

Tokyo, Japan: 4-7 Jul 1978. (Proceedings). See *094.020* for details.

094.020 **ISAS Symposium on Lunar and Planetary Science.** 10th. Tokyo: Institute of
C Space and Aeronautical Science, University of Tokyo; 1977. 238p. English/ Japanese.

Tokyo, Japan: 11-13 Jul 1977. (Proceedings). A collection of brief papers reporting on research in Japan. The first volume in which English-language papers appeared.

094.021 Jaffe, L. D. Recent observations of the Moon by spacecraft. **Space Sci. Rev.** 9: 491-616; 1969. 1,027 refs, 1960-69.

Review of lunar observations made by spacecraft subsequent to Luna 9; particular attention is paid to the Surveyor landers. The bibliography contains sections on Surveyor, Lunar Orbiter, Explorer 35, Ranger, and Soviet spacecrafts.

094.022 Kaula, W. M. The gravity and shape of the Moon. **EOS Trans. American Geophys. Union** 56: 309-316; 1975. 100 refs, 1968-75.

Discussion of the current knowledge of the gravity and geometry of the Moon and their implications as to lunar structure. Emphasis is on work since 1970.

094.023 Kaula, W. M.; Harris, A. W. Dynamics of lunar origin and orbital evolution. **Rev. Geophys. Space Phys.** 13: 363-371; 1975. 80 refs, 1879-1975.

Discussion based on Apollo data.

094.024 Kokurin, Y. L. Current status and future prospects of lunar laser-ranging studies (review). **Soviet J. Quantum Electron.** 6: 645-657; 1976. Transl. of *Kvantovaya Elektron., Moskva.* 3: 1189-1210; 1976. 81 refs, 1960-75.

Review of lunar laser ranging investigations published from 1962 to 1975.

094.025 Kopal, Z.; Moutsoulas, M.; Salisbury, J. W.; Waranius, F. B. Bibliography. **Moon** 1: 278-292; 1970.

A selected, regular listing of journal articles with abstracts on the various aspects of lunar studies. The 30 bibliographies in the series contain about 4,000 total entries. Arrangement in each list is topical under the following headings: lunar motion; shape and gravitational field; internal structure; thermal and stress history; chemical composition; exosphere; lunar coordinates and mapping; morphology of lunar surface; origin and stratigraphy of lunar formations; physical structure of lunar surface; photometry; thermal emission of the lunar surface; electromagnetic properties of the Moon; exploration by spacecraft. A complete listing of bibliographies follows: 1: 278-292; 1970. 1: 403-425; 1970. 1: 512-517; 1970. 2: 104-125; 1970. 2: 211-260; 1970. 2: 365-401; 1971. 2: 474-490; 1971. 3: 90-156; 1971. 3: 239-262; 1971. 3: 352-362; 1971. 3: 472-491; 1972. 4: 250-268; 1972. 4: 507-535; 1972. 5: 233-250; 1972. 5: 457-481; 1972. 6: 212-228; 1973. 6: 414-512; 1973. 9: 457-519; 1974. 10: 207-303; 1974. 11: 445-479; 1974. 12: 113-124; 1975. 12: 479-510; 1975. 13: 495-519; 1975. 15: 183-201; 1976. 15: 485-514; 1977. 16: 351-386; 1977. 16: 465-492; 1977. 17: 179-201; 1977. 17: 309-339; 1977. 17: 425-470; 1977.

094.026 Kopal, Z.; Moutsoulas, M.; Waranius, F. B. Bibliography. **Moon Planets** 18: 251-262; 1978.

Continues the bibliography begun in *The Moon*, the journal's previous title. New topics covered, in addition to the previously cited subjects, are lunar petrology, mineralogy, crystallography, general reviews, planets (general), individual planets, asteroids, comets, meteorites, etc. A total of about 1,000 references are contained in the six references cited below: 18: 251-262; 1978. 20: 321-368; 1979. 20: 468-506; 1979. 21: 199-254; 1979. 21: 361-387; 1979. 21: 463-526; 1979.

094.027 Kopal, Zdeněk. **The Moon.** Dordrecht, Holland: D. Reidel; 1969. 525p. 24 chapters. 1,169 refs, 1610-1969.

An introductory, pre-Apollo text for advanced students and scientists covering four major areas: motion of the Moon and dynamics of the Earth-Moon system; internal constitution of the lunar globe; topography of the Moon; radiation of the Moon. Includes a historical overview of lunar mapping with illustrations of early maps (seventeenth century and later). This book was first published as *An Introduction to the Study of the Moon* (1966).

094.028 Kopal, Zdeněk. **The Moon in the Post-Apollo Era.** Dordrecht, Holland; Boston: D. Reidel; 1974. 223p. (Geophys. Astrophys. Monogr., 7). 10 chapters w/refs.

A summary of lunar exploration from 1957 to 1972, with particular emphasis on the Apollo program. A coherent review of the astronomy, physics, chemistry, geology, and mineralogy of the Moon, the book condenses an immense amount of data into a compact, well-written text, understandable by both layperson and scientist. Topics: manned exploration; physical features and figures; shape and gravitational field; internal structure; morphology of lunar formations; surface structure and chemical composition; exosphere; origin and evolution; stratigraphy and chronology of the lunar surface.

094.029 Kopal, Zdeněk; Carder, Robert W. **Mapping of the Moon: Past and Present.** Dordrecht, Holland; Boston: D. Reidel; 1974. 237p. (Astrophys. Space Sci. Libr., 50). 11 chapters w/refs.

Of interest to both astronomers and laypersons, this book attempts to provide an overview of the efforts to map the topography of the lunar surface, from pre-telescopic astronomy to the present. Topics: history of lunar mapping: 1600-1960; lunar rotation and librations; selenographic coordinates; shape of the Moon; relative elevations; U.S. and Soviet efforts; National Geographic Lunar Mapping. Includes detailed maps and photographs.

094.030 Kopal, Zdeněk; Strangway, David, eds. **Lunar Geophysics.** Dordrecht, Holland; Boston: D. Reidel; 1972. 639p. (Lunar Science Inst. Contr., 86). 49 papers.
C

Houston, TX, USA: 18-21 Oct 1971. Reviews of recent research combining three areas: 1) physical properties of lunar samples; 2) analysis of geophysical experiments conducted on the Moon; 3) lunar models and history. Topics: seismology; magnetism; electrical properties; thermal properties; figure of the Moon and moments of inertia; surface properties. This collection was first printed in two issues of *The Moon* 4: 3-249; 1972, and 5: 3-160; 1972.

094.031 Lindsay, John F. **Lunar Stratigraphy and Sedimentology.** Amsterdam; New York: Elsevier; 1976. 302p. (Dev. Sol. Syst. Space Sci., 3). 6 chapters w/refs.

An overview based on Apollo lunar samples data of the information about the sedimentary rocks forming the lunar crust, with comparison to the terrestrial sedimentary environment. Topics: the Moon as a planet; energy at the lunar surface; hypervelocity impact; stratigraphy and chronology of the Moon's crust; lithology and depositional history of major lunar material units; the lunar soil.

094.032 **Lunar & Planetary Information Bulletin.** Houston, TX: Lunar and Planetary Institute (NASA). no. 15- , 1978- . Four/year. Continues *Lunar Science Information Bulletin* (nos. 1-14, 1974-1978) and continues its numbering.

Primarily a newsletter of information on NASA activities such as Viking, Voyager, and the Space Shuttle, this publication also contains a lengthy bibliography in each issue which highlights recent important publications in the field. Entries are arranged by author under several subject categories including, but not limited to, the Moon, the planets, asteroids, comets, meteorites, etc. Each issue also contains news about NASA-sponsored and other conferences on solar system and lunar studies, as well as information on programs and activities of the Lunar and Planetary Institute.

094.033 Lunar and Planetary Institute, comp. **Lunar and Planetary Science Conference.** 9th. New York; Oxford, UK: Pergamon Press; 1978. 3v. 3,973p. (Geochim. Cosmochim. Acta Suppl. 10). 211 papers.
C

Houston, TX, USA: 13-17 Mar 1978. (Proceedings). A compendium of papers in three volumes: 1) Petrogenetic Studies: the Moon and Meteorites; 2) Lunar and Planetary Studies; 3) The Moon and the Inner Solar System. Topics: origin and evolution of the Moon; mare basalts; non-mare rocks; breccias; meteorite studies; regolith studies; impact phenomena; optical, x-ray, and gamma-ray remote sensing; magnetic and electrical properties; morphology and processes; volcanoes; structure and tectonics; seismology; craters; etc. *Mare Crisium: The View from Luna 10* (*094.526*) is an unofficial companion fourth volume to this set.

094.034 Lunar and Planetary Institute, comp. **Lunar and Planetary Science Conference.** 10th. New York; Oxford, UK: Pergamon Press; 1979. 3v. 3,077p. (Geochim. Cosmochim. Acta Suppl. 11). 177 papers.
C

Houston, TX, USA: 19-23 Mar 1979. (Proceedings). Reports of recent research in three volumes: 1) Meteorites and Lunar Rocks; 2) Early Solar System and Lunar Regolith; 3) Planetary Interiors and Surfaces. Topics: basalt studies; highland rock studies; meteorite studies; lunar regolith studies; compositional remote sensing; origin and evolution of the solar system; geophysical investigations of the planets; impact processes; volcanic processes; other surface features and processes.

094.035 Massey, H., et al., eds. The Moon—a new appraisal from space missions and laboratory analyses. **Philos. Trans. R. Soc. London, Ser. A.** 285: 1-606; 1977. 66 papers.
C

A Royal Society discussion, held June 9-12, 1975, arranged by the British National Committee on Space Research. The meeting was organized to present the findings of

European and Commonwealth scientists who had participated in the analysis of Apollo and Luna samples.

094.036 Masterson, Amanda R., comp. **Index to the Proceedings of the Lunar and Planetary Science Conferences: Houston, Texas, 1970-1978.** New York: Pergamon Press; 1979. 261p. (Geochim. Cosmochim. Acta. Index to Geochimica et Cosmochimica Acta Supplements 1-8 & 10).

A cumulative index to the nine sets (27 volumes; 1,932 papers; 30,198p.) of proceedings which constitute one of the most comprehensive, massive collections of information on one topic (lunar and planetary science) ever published in astronomy. There are four indexes (keyword, mission, sample, and author) and complete tables of contents for all nine proceedings. All authors are indexed.

094.037 Masursky, Harold; Colton, G. W.; El-Baz, Farouk. **Apollo over the Moon: A View from Orbit.** Washington, DC: NASA; 1978. 255p. (NASA SP-362). 7 chapters. 61 refs, 1962-76.

A collection of some of the best lunar photographs taken during the Apollo 15, 16, and 17 missions. These orbital photos are arranged by lunar feature or area, and are accompanied by an interesting, interpretive text. Topics: mission overview; camera systems; regional views; the Terrae; the Maria; craters; rimae; unusual features.

094.038 The Moon. **IAU Trans. Rep. Astron.** 17A(1): 115-122; 1979. 85 refs, 1966-79.

An overview of research activity worldwide, with an emphasis on the previous three years, as reported by IAU Commission 17 (The Moon). Topics: lunar research; figure and motion of the Moon; etc.

094.039 C **The Moon—A New Appraisal from Space Missions and Laboratory Analyses.** London: The Royal Society; 1977. 606p. Also in: **Philos. Trans. R. Soc. London, Ser. A.** 285: 1-606; 1977. 66 papers.

A Royal Society Discussion. London, UK: 9-12 Jun 1975. Reviews of current understanding of and reports of recent research on the Moon, its origin, composition, evolution, internal processes, interaction with its environment; etc.

094.040 **The Moon, Volume 1, A DDC Bibliography, Jan 1953—Dec 1968.** Alexandria, VA: Defense Documentation Center; 1969. 243p. (AD-694500; DDC-TAS-69-60-1-vol-1; N70-17203). 452 refs.

A computer-produced listing from DDC files, in reverse chronological order. References include authors, title, descriptors, abstract, and AD number. Indexes include author, subject, corporate author/monitoring agency. Volume 2 was classified.

094.041 C Moutsoulas, M., ed. Colloquium on lunar dynamics and observational coordinate systems. **Moon** 8: 433-556; 1973. (IAU Colloq., 24). 14 papers. Revised abstracts appeared in *Lunar Science Inst. Contr.*, 135.

Houston, TX, USA: 15-17 Jan 1973. Reports of research and reviews of theory. Topics: analytical lunar theory; selenographic control; lunar rotation; lunar physical libration and lunar ranging; etc.

094.042 Murthy, V. R.; Banerjee, S. K. Lunar evolution: how well do we know it now? **Moon** 7: 149-171; 1973. 101 refs, 1924-72.

Examination of the known astronomical, geochemical, and geophysical data in order to argue for an accretional layering in the Moon.

094.043 Mutch, Thomas A. **Geology of the Moon: A Stratigraphic View.** rev. ed. Princeton, NJ: Princeton University Press; 1972. 391p. 12 chapters. 421 refs, 1610-1972.

Aimed at upper-level undergraduates, beginning graduate students, and geologists, this work attempts to synthesize some of the early and quickly growing data on the structure and composition of the Moon. Profusely illustrated with black-and-white photos, it surveys recent studies and provides historical background. The primary change in this edition over the 1970 version is the revision and expansion of chapter 12, which reports on the Apollo missions. Covered here are early rock and soil data with many surface photographs. Topics: lunar shape and motion; remote sensing techniques; craters; Imbrium basin; crater stratigraphy; volcanic stratigraphy; highland stratigraphy; age of lunar materials; etc.

094.044 Origin and evolution of the lunar regolith. **Moon** 13: 1-359; 1975. Also in:
C **Lunar Science Inst. Contr.** 207. 21 papers.

Houston, TX, USA: 13-15 Nov 1974. Topics include origin, structure, and transport of regolith; deposition and turnover; present and future studies of lunar cores; and physical and chemical evolution of the regolith.

094.045 Rukl, A. **Maps of Lunar Hemispheres.** Dordrecht, Holland; Boston: D. Reidel;
A 1972. 24p. (Astrophys. Space Sci. Libr., 33). 6 maps.

Six detailed maps produced from lunar-orbiting satellites, accompanied by an index and some introductory material. The maps provide views of the Eastern and Western hemispheres, the near and far sides, and the Northern and Southern hemispheres. The introduction covers coordinates of places on the lunar surface; construction of the maps; and lunar nomenclature.

094.046 Runcorn, S. K.; O'Reilly, W.; Srnka, L. J., eds. Recent activity in the Moon.
C **Phys. Earth Planet. Inter.** 14: 186-332; 1977. 12 papers.

Houston, TX, USA: 16 Mar 1976. Special symposium at the Seventh Lunar Science Conference. Topics included lunar transient phenomena, moonquakes, lunar gas emissions, physical processes related to recent activity on the Moon, etc.

094.047 Runcorn, S. K.; Urey, H. C., eds. **The Moon.** Dordrecht, Holland: D. Reidel;
C 1972. 471p. (IAU Symp., 47). 37 papers, 3 abstracts.

Newcastle-upon-Tyne, UK: 22-26 Mar 1971. Topics include lunar mechanics, surface, Apollo missions, petrological studies, tectonics, physical properties of lunar samples, interior, evolution of the Moon's orbit, and origin and evolution of the Moon.

094.048 Schubert, G.; Lichtenstein, B. R. Observations of Moon-plasma interactions by orbital and surface experiments. **Rev. Geophys. Space Phys.** 12: 592-626; 1974. 145 refs, 1939-74.

Concentration on the lunar interaction with the solar wind, based on lunar explorations from the launch of Luna 2 in 1959 through the Apollo series.

094.049 Taylor, Stuart Ross. **Lunar Science: A Post-Apollo View.** New York: Pergamon Press; 1975. 372p. 7 chapters. 36 refs, 1969-74.

A synthesis and interpretation, especially with reference to the evolution and origin of the Moon, of the great mass of data resulting from the manned lunar missions. Intended for the scientist and educated layperson, the book covers geology, surface, maria, highlands, interiors, and origin and evolution. Besides the brief bibliography of major sources, the individual chapters contain several hundred references. The chemistry and mineralogy of the Moon are summarized well in this book, which includes, but does not emphasize, tabular and numerical data.

094.050 Toksöz, M. N. Geophysical data and the interior of the Moon. **Annu. Rev. Earth Planet. Sci.** 2: 151-177; 1974. 98 refs, 1966-73.

An overview of current information with an emphasis on physical properties. Also presented are detailed models of the lunar interior based on seismic data, and a discussion of the thermal state and evolution of the Moon.

094.051 Toksöz, M. N., et al. Structure of the Moon. **Rev. Geophys. Space Phys.** 12: 539-567; 1974. 118 rcfs, 1959-74.

Review of seismic data from the Apollo seismic network, followed by discussion of the problem of lunar structure.

094.052 Toksöz, M. N.; Johnston, D. H. The evolution of the Moon. **Icarus** 21: 389-414; 1974. 112 refs, 1951-73.

The thermal evolution of the Moon, as it can be defined by the available data and theoretical calculations, is discussed.

094.053 Toksöz, M. N.; Solomon, S. C. Thermal history and evolution of the Moon. **Moon** 7: 251-278; 1973. 112 refs, 1951-72.

Investigation of the thermal evolution and present-day temperature models for the Moon as currently constrained by the vast amount of data from the lunar missions.

094.054 Walker, R. M. Interactions of energetic nuclear particles in space with the lunar surface. **Annu. Rev. Earth Planet. Sci.** 3: 99-128; 1975. 115 refs, 1961-74.

Review of the properties and effects of solar wind particles, solar flare particles, and galactic cosmic rays, followed by a summary of information about this radiation gained from Apollo missions. Lunar surface dynamics are also considered.

094.055 Winter, D. F. Infrared emission from the surface of the Moon. **Adv. Astron. Astrophys.** 9: 203-243; 1972. 80 refs, 1929-71.

A review of lunar temperature and the thermal properties of the lunar surface.

AFCRL bibliography for the first quarter of 1969. *091.057.*

Comparisons of Mercury and the Moon. *092.001.*

Conference on interactions of the interplanetary plasma with the modern and ancient Moon. *106.003.*

Moon and Planets II. 053.001.

The Moon and Planets. Index to Volumes 1-20. R005.016.

New and Full Moons 1001 B.C. to A.D. 1651. 047.003.

Phases of the Moon 1801-2010. 047.005.

094.5 Moon (Local Properties)

094.501 Ahrens, L. H., ed. Chemistry of the Moon. **Phys. Chem. Earth** 10: 145-214; 1977. 2 papers.

This special issue contains two (only) review articles: "Potassium-argon Chronology of the Moon" (G. Turner, pp. 145-195), and "The Irradiation History of the Lunar Soil" (G. Crozaz, pp. 197-214).

094.502 Anders, E.; Albee, A. L., eds. Luna 20: a study of samples from the lunar high-
C lands returned by the unmanned Luna 20 spacecraft. **Geochim. Cosmochim. Acta** 37: 719-1109; 1973. 36 papers.

A compendium of papers reporting on scientific analyses of the Luna 20 samples, in particular a core soil sample from the landing site. Topics include mineralogy, petrology, chemical composition, thermal history, and so on.

094.503 Cherkasov, I. I.; Shvarev, V. V. **Lunar Soil Science: Physicomechanical Properties of Lunar Soils.** Jerusalem: Israel Program for Scientific Translation; 1975. 170p. 5 chapters w/refs.

An overview of current knowledge "of the genesis, structure and properties of lunar soils based on astronomical and radiophysical studies." Coverage includes information gathered by spacecraft photography, astronauts, and unmanned lunar landers from the United States and the Soviet Union. Topics: loose volcanic deposits and terrestrial analogs of lunar soils; effects of lunar environment on properties of lunar rocks and soil; composition and properties of first returned samples; etc. Translation by N. Kaner of *Nachala gruntovedeniya Luny. Fiziko-mekhanicheskie svoǐstva lunnykh gruntov* (Moscow: Nauka; 1971).

094.504 Dyal, P.; Parkin, C. W.; Daily, W. D. Magnetism and the interior of the Moon. **Rev. Geophys. Space Phys.** 12: 568-591; 1974. 120 refs, 1934-74.

Review of the lunar magnetometer data analysis from the 11 magnetometers sent to the Moon 1961-72, with emphasis on the lunar interior.

094.505 El-Baz, F. Surface geology of the Moon. **Annu. Rev. Astron. Astrophys.** 12: 135-165; 1974. 90 refs, 1846-1975.

Summarizes current knowledge of the surface geology of the Moon, with an emphasis on recent results of the Apollo missions.

094.506 Frondel, Judith W. **Lunar Mineralogy.** New York: John Wiley & Sons; 1975. 323p. 11 chapters. 82 refs, 1951-74.

Intended as an aid to future mineralogical study and as a summary of data through March 1974, this technical work describes the various lunar samples from a chemical and geological point of view. Arranged by type of material, the book covers native elements; sulfides, phosphides, and carbides; simple and multiple oxides; carbonates and phosphates; the SiO_2 minerals; silicates; etc. Includes many black-and-white photographs and dozens of references in the text, in addition to the bibliographic references cited above.

094.507 Fruchter, J. S.; Arnold, J. R. Chemistry of the Moon. **Annu. Rev. Phys. Chem.** 23: 485-508; 1972. 76 refs, 1956-72.

A review dealing primarily with the chemical and isotropic composition of various lunar materials.

094.508 Heiken, G. Petrology of lunar soils. **Rev. Geophys. Space Phys.** 13: 567-587; 1975. 136 refs, 1954-74.

Discussion is restricted to the finer-grained fractions ($<$ 1 cm) and their physical and petrographic characteristics as known one and one-half years after Apollo 17.

094.509 Hinners, N. W. The new Moon: a view. **Rev. Geophys. Space Phys.** 9: 447-522; 1971. 207 refs, 1955-71.

Review of Apollo 11, 12, and 14 and Luna 16 data.

094.510 Horz, F.; Brownlee, D. E.; Fechtig, H. Lunar microcraters: implications for the micrometeoroid complex. **Planet. Space Sci.** 23: 151-172; 1975. 98 refs, 1961-74.

The contributions of lunar microcrater studies to the understanding of the overall micrometeoroid environment are summarized and compared to satellite data. This report attempts to summarize the most significant contributions to date on shape, density, chemistry, mass frequency, and flux of micrometeoroids.

094.511 Howard, K. A.; Wilhelms, D. E.; Scott, D. H. Lunar basin formation and highland stratigraphy. **Rev. Geophys. Space Phys.** 12: 309-327; 1974. 150 refs, 1893-1974.

Review of landforms and possible origins of basin-related features, the gross stratigraphy of the highlands, and the role of basins in shaping the highlands.

094.512 Kiesl, W.; Malissa, H., Jr., eds. **Analyse extraterrestrischen Materials.** [Analyses C of Extraterrestrial Materials.] Wien; New York: Springer-Verlag; 1974. 326p. 25 papers. German/English.

Vienna, Austria: 2-3 Oct 1973. A discussion of analyses of moon rocks and meteorites. Primarily German, with English and German abstracts.

094.513 King, Elbert A., Jr.; Heymann, Dieter; Criswell, David R., eds. **Lunar Science C Conference.** 3rd. Cambridge, MA: MIT Press; 1972. 3v. 3,263p. (Geochim. Cosmochim. Acta, Suppl. 3). 228 papers.

Houston, TX, USA: 10-13 Jun 1972. (Proceedings). Reports of research on lunar samples from Apollo 11, 12, 14, and 15, as well as Luna 16. Topics: general geology; surface and orbital observations; igneous type rocks; pyroxenes; plagioclases; breccias; glasses; fines samples; etc. Volume titles: 1) Mineralogy and Petrology; 2) Chemical and Isotope Analyses; Organic Chemistry; 3) Physical Properties.

094.514 Langevin, Y.; Arnold, J. R. The evolution of the lunar regolith. **Annu. Rev. Earth Planet. Sci.** 5: 449-489; 1977. 130 refs, 1964-76.

A review of knowledge about the broken-up surface layer of the Moon, based on studies of soil samples collected by astronauts and unmanned Soviet spacecraft. Included are discussions of impacting bodies such as meteorites, the cratering process, the properties of the regolith, and regolith evolution models.

094.515 Levinson, A. A., ed. **Apollo 11 Lunar Science Conference.** New York; Oxford, C UK: Pergamon Press; 1970. 3v. 2,492p. (Geochim. Cosmochim. Acta, Suppl. 11). 180 papers.

Houston, TX, USA: 5-8 Jan 1970. (Proceedings). Scientific papers concerned with the study of the first Moon rocks and soil brought back to Earth by astronauts. There are three volumes: 1) Mineralogy and Petrology; 2) Chemical and Isotope Analyses (including organic chemistry); 3) Physical Properties.

094.516 Levinson, A. A., ed. **Lunar Science Conference.** 2nd. Cambridge, MA: MIT C Press; 1971. 3v. 2,818p. (Geochim. Cosmochim. Acta, Suppl. 2). 222 papers.

Houston, TX, USA: 11-14 Jan 1971. (Proceedings). Scientific reports of research on the Apollo 11 and 12 lunar samples, the Luna 16 samples, and the Surveyor III probe, parts of which were later returned to Earth by the Apollo 12 astronauts. The papers are in three volumes: 1) Mineralogy and Petrology; 2) Chemical and Isotope Analysis; Organic Chemistry; and 3) Physical Properties; Surveyor III.

094.517 Levinson, Alfred A.; Taylor, S. Ross. **Moon Rocks and Minerals.** New York: Pergamon Press; 1971. 222p. 9 chapters. no refs.

A review of the early scientific results of the Apollo 11 and 12 lunar samples, with an emphasis on Apollo 11. Topics: rocks, soils, minerals, chemistry of the samples, bioscience and organic matter, petrology, age, physical properties, and origin of the Moon. Includes some color photos and a lengthy glossary.

094.518 Lucas, John W., ed. **Thermal Characteristics of the Moon.** Cambridge, MA: MIT Press; 1972. 340p. (Progress Astronaut. Aeronaut., 28). 11 papers w/refs.

A summary of results obtained by early manned and unmanned lunar exploration. The 11 survey articles cover Earth-based measurements; in situ measurements; thermal characteristics of lunar-type materials; geophysical interpretation.

094.519 **Lunar Science Conference.** 4th. New York; Oxford, UK: Pergamon Press;
C 1973. 3v. 3,290p. (Geochim. Cosmochim. Acta, Suppl. 4). 228 papers.

Houston, TX, USA: 5-8 Mar 1973. (Proceedings). Reports of research on the lunar rock and soil samples from the missions of Apollo 11, 12, 14, 15, 16, and 17, and Lunar 16 and 20, as well as results of experiments set up or conducted on the Moon by astronauts. Volume titles: 1) Mineralogy and Petrology; 2) Chemical and Isotope Analyses; Organic Chemistry; 3) Physical Properties.

094.520 **Lunar Science Conference.** 5th. Elmsford, NY: Pergamon Press; 1974. 3v.
C 3,134p. (Geochim. Cosmochim. Acta, Suppl. 5). 207 papers.

Houston, TX, USA: 18-22 Mar 1974. (Proceedings). Research results on the study of lunar rocks and soil arranged in three volumes: 1) Mineralogy and Petrology; 2) Chemical and Isotope Analyses; Organic Chemistry; 3) Physical Properties. Topics: formation and evolution of lunar basins; lunar breccias; lunar basalts; lunar regolith; lunar mineralogy; chemical composition of lunar rocks and soils; geochemical evolution of the Moon and its rocks; lunar chronology; noble gases; soils and surface properties; fission tracks and microcraters; magnetism and paleomagnetism; orbital mapping; etc.

094.521 Lunar Science Institute, comp. **Lunar Science Conference.** 6th. New York; Ox-
C ford, UK: Pergamon Press; 1975. 3v. 3,637p. (Geochim. Cosmochim. Acta, Suppl. 6). 219 papers.

Houston, TX, USA: 17-21 Mar 1975. (Proceedings). Research reports on recent studies of lunar rocks, minerals, and soil in three volumes: 1) Mineralogical and Petrological Studies; 2) Chemical and Isotopic Studies; 3) Physical Studies. Topics: lunar samples; anorthositic rocks; soils and breccias; experimental petrologic studies; lunar evolution; chemical composition of lunar material; lunar chronology; volatile meteoritic elements; agglutinates and regolith processes; selenography; craters; remote sensing; selenophysics; lunar surface fields and particles; magnetic properties of lunar samples; solar system regoliths and cosmic rays; etc.

094.522 Lunar Science Institute, comp. **Lunar Science Conference.** 7th. New York;
C Oxford, UK: Pergamon Press; 1976. 3v. 3,651p. (Geochim. Cosmochim. Acta, Suppl. 7). 208 papers.

Houston, TX, USA: 15-19 Mar 1976. (Proceedings). Papers reporting on recent studies of lunar rocks and soil, and of meteorites, Mercury, and Mars in three volumes: 1) Regolith Studies; 2) Petrogenetic Studies of Mare and Highland Rocks; 3) The Moon and Other Bodies. Topics: lunar core studies; surface soil studies; surface process studies; mare basalts; breccia and highland samples; remote sensing and photogeology; physical properties; comparative planetology.

094.523 Lunar Science Institute, comp. **Lunar Science Conference.** 8th. New York;
C Oxford, UK: Pergamon Press; 1977. 3v. 3,965p. (Geochim. Cosmochim. Acta, Suppl. 8). 228 papers.

Houston, TX, USA: 14-18 Mar 1977. (Proceedings). Reports of research on recent solar system studies. Topics: asteroids, meteorites, and the early solar system; origin and evolution of the Moon; planetary geophysics; lunar magnetism; remote observations of lunar and planetary surfaces; physical properties of lunar materials; mare basalts; highland rocks; materials of KREEP composition; Venus; regional lunar studies; surface processes; etc. Volume titles: 1) The Moon and the Inner Solar System; 2) Petrogenetic Studies of Mare and Highland Rocks; 3) Planetary and Lunar Surfaces.

094.524 Malina, Frank J., ed. **Research in Physics and Chemistry.** Oxford, UK: Perga-
C mon Press; 1969. 145p. 9 papers.

Third Lunar International Laboratory Symposium. Belgrade, Yugoslavia: 28 Sep 1967. A series of papers discussing various potential physical and chemical experiments that could be carried out on the Moon. Includes a summary paper describing a survey of the Moon's chemical and physical properties, the creation of living and laboratory space on the Moon, and the initiation of research on the Moon.

094.525 Mason, Brian; Melson, William G. **The Lunar Rocks.** New York: John Wiley & Sons; 1970. 179p. 8 chapters. 147 refs, 1919-70.

One of the earliest books of its kind, this volume attempts to summarize the data obtained from the first analysis of lunar materials obtained by Apollo. Intended for scientists, students, and laypersons, the book covers mineralogy, petrology, and geochemistry, as well as a summary of the Apollo 11 and 12 missions and pre-Apollo background. A review of the elements and chemical compounds found in the samples, the book also compares the rocks with terrestrial rocks, meteorites, and tektites.

094.526 Merrill, R. B.; Papike, J. J., eds. **Mare Crisium: The View from Luna 24.** New
C York; Oxford, UK: Pergamon Press; 1978. 709p. (Geochim. Cosmochim. Acta, Suppl. 9). 45 papers.

Conference on Luna 24. Houston, TX, USA: 1-3 Dec 1977. (Proceedings). Reports of studies made on lunar soil samples returned to Earth by Luna 24, the Soviet automated spacecraft. The papers, written by American scientists who studied a part of the 160-cm core sample, cover four major areas: regional studies; regolith studies; petrology; geochemistry. Since this meeting was held in conjunction with the Ninth Lunar and Planetary Science Institute, this volume is considered (unofficially) part of that set (*094.033*).

094.527 Oberbeck, V. R. The role of ballistic erosion and sedimentation in lunar stratigraphy. **Rev. Geophys. Space Phys.** 13: 337-362; 1975. 96 refs, 1962-75.

Review of the origin and applications of base surge and ballistic concepts to the study of lunar stratigraphy.

094.528 Öpik, E. J. Cratering and the Moon's surface. **Adv. Astron. Astrophys.** 8: 107-337; 1971. 164 refs, 1879-1968.

A complete mechanical and statistical analysis of the lunar surface. Topics include origin of craters, origin of the Moon, lunar soil, erosion, and so on. Includes 53 tables of data.

094.529 Öpik, E. J. The Moon's surface. **Annu. Rev. Astron. Astrophys.** 7: 473-526; 1969. 60 refs, 1936-69.

Review of the author's work on the origin and evolution of the lunar surface. Included is a discussion of the theory of impact craters, used to help predict the structure of the Moon's surface.

094.530 Papike, J. J., et al. Mare basalts: crystal chemistry, mineralogy, and petrology. **Rev. Geophys. Space Phys.** 14: 475-540; 1976. 218 refs, 1954-76.

Synthesization of the major element chemistry, petrology, mineral chemistry, and crystal chemistry of mare basalts.

094.531 Pillinger, C. T. Solar-wind exposure effects in the lunar soil. **Rep. Prog. Phys.** 42: 897-961; 1979. 337 refs, 1955-78.

Discussion of solar wind exposure effects revealed by the study of lunar samples returned to Earth by the Apollo program.

094.532 Pillinger, C. T.; Eglinton, G. The chemistry of carbon in the lunar regolith. **Philos. Trans. R. Soc. London, Ser. A.** 285: 369-377; 1977. 50 refs, 1963-76.

Discussion concluding that carbon in the lunar regolith appears to derive mainly from extralunar sources, predominantly the solar wind.

094.533 A Rode, O. D., et al. **Atlas of photomicrographs of the surface structures of lunar regolith particles.** Dordrecht, Holland; Boston: D. Reidel; 1979. 76p. 164 plates. 48 refs, 1970-79. Russian/English.

"A first attempt at a systematic survey of morphological observations based on the results of the Luna 16 and Luna 20 sample investigations." The book is primarily a collection of 164 black-and-white photographs of microscopic views of the lunar soil and rock samples, with accompanying descriptive paragraphs in Russian and English. The introductory material discusses general characteristics of the lunar regolith particles, methods of investigations, and details of the morphological characteristics of specific subclasses of the samples.

094.534 Schmidt, P. Das wichtigste internationale Schrifttum über Mondbeben, Mondseismologie und Mondseismik 1960-1972. [The most important international literature about Moonquakes and Moon seismology 1960-1972.] **Sterne** 49: 247-252; 1973. 131 refs, 1959-72.

A bibliography emphasizing recent investigations.

094.535 Smith, J. V.; Steele, I. M. Lunar mineralogy: a heavenly detective story. Part II. **American Mineral.** 61: 1059-1116; 1976. 362 refs, 1915-76.

Concentration on crystal-chemical, petrologic, and geochemical aspects of lunar mineralogy.

094.536 Sohel, M. S. Bibliography on the measurement of electrical parameters of layered lunar/earth surfaces. **IEEE Trans. Aerosp. Electron. Syst.** AES-9: 320-323; 1973. 112 refs, 1949-71.

A listing of journal articles and technical reports dealing with measurement by electromagnetic sounding techniques. No abstracts.

094.537 Summary of Apollo 11 Lunar Science Conference. **Science** 167: 449-784; 1970. 143 papers.

Compilation of the first systematic studies of lunar samples returned by Apollo 11.

Apollo 11 Lunar Science Conference. 094.003.

Bibliography. *094.025.*

Bibliography of the Lunar Surface: Final Report. 094.010.

Index to the Proceedings of the Lunar and Planetary Science Conferences: Houston, Texas, 1970-1978. 094.036.

Lunar Stratigraphy and Sedimentology. 094.031.

Origin and evolution of the lunar regolith. *094.044.*

Planetary interiors. *091.074.*

095 Lunar Eclipses

095.001 Link, F. **Eclipse Phenomena in Astronomy.** Berlin; New York: Springer-Verlag; 1969. 271p. 7 chapters w/refs.

A review of the theoretical and experimental aspects of eclipses with some historical background. An advanced text, and the only one available on this topic, the work is divided into the following chapters: lunar eclipses; eclipses of artificial Earth satellites; twilight phenomena; occultation and eclipses by other planets; transits of planets over the Sun; eclipse phenomena in radio astronomy; Einstein's deflection of light. No mention of solar eclipses.

095.002 Link, F. Lunar eclipses. **Adv. Astron. Astrophys.** 9: 67-148; 1972. 179 refs, 1644-1968.

Review covering history, characteristics, observations, data, and allied phenomena. Concludes with a summary of important scientific notions arrived at or confirmed by the observation of lunar eclipses.

095.003 Meeus, Jean; Mucke, Hermann. **Canon of Lunar Eclipses -2002 to +2526.** Vienna: Astronomisches Buro; 1979. 244p. English/German.

Covering all eclipses (10,936) from July 2002 B.C. to 2526 A.D., these computer-produced tables include the following data for each event: lunation number; date and instant of maximum phase; semiduration in minutes of the partial and total phases; greatest magnitude of the eclipse; location (latitude and longitude) of the center of the Moon at maximum; Saros series. Also included are brief comments on accuracy of data and a statistical summary. German title: *Canon der Mondfinsternisse -2002 bis +2526.*

095.004 Shorthill, R. W.; Saari, J. M. Infrared observation on the eclipsed Moon. **Adv. Astron. Astrophys.** 9: 149-201; 1972. 14 refs, 1963-70.

A detailed discussion of infrared observations of a 1967 lunar eclipse. Includes several charts of the Moon showing locations of thermal anomalies.

096 Lunar and Planetary Occultations

096.001 C Deeming, T. J., ed. Photo-electric observations of stellar occultations. **Highlights Astron.** 2: 585-722; 1971. 19 papers.

Joint discussion at the Fourteenth General Assembly of the IAU.

096.002 Elliot, J. L. Stellar occultation studies of the solar system. **Annu. Rev. Astron. Astrophys.** 17: 445-475; 1979. 134 refs, 1890-1979.

Review covering the principles, observational procedures, and results relating to occultations of stars by solar system bodies other than the Moon. Topics include physical processes involved, prediction and observational techniques, radii, atmospheres, rings, stellar diameters, and future programs.

096.003 Morrison, L. V. An analysis of lunar occultations in the years 1943-1974 for corrections to the constants in Brown's theory, the right ascension system of the FK4, and Watts' lunar-profile datum. **Mon. Not. R. Astron. Soc.** 187: 41-82; 1979. 50 refs, 1882-1978.

The times of 62,000 occultations of stars by the Moon observed during 1943-74 are analyzed to determine corrections to the lunar ephemeris $j = 2$ (based on Brown's theory), the right ascension system of the FK4, and the datum of the lunar-profile in Watts' charts.

Eclipse Phenomena in Astronomy. 095.001.

097 Mars, Mars Satellites

097.001 Barth, C. A. The atmosphere of Mars. **Annu. Rev. Earth Planet. Sci.** 2: 333-367; 1974. 94 refs, 1963-74.

Overview of current knowledge based on data gathered by Mariner spacecraft and ground-based observations. Topics include composition, structure and changes in the lower atmosphere, the upper atmosphere and ionosphere, photochemistry, and more.

097.002 Batson, R. M.; Bridges, P. M.; Inge, J. L. **Atlas of Mars.** Washington, DC:
A NASA; 1979. 146p. (The 1:5,000,000 Map Ser.; NASA SP-438). 100 refs, 1930-78.

A photographic atlas compiled primarily from Mariner 9 data, with minor input from the Viking orbiters. The 72 plates are arranged according to 30 regions or quadrangles. In addition to an introduction and the maps, there are five useful appendices: nomenclature, planimetric mapping procedures, contour mapping, Mars mapping team, and bibliography of Mars cartography.

097.003 Burns, J. A. Dynamical characteristics of Phobos and Deimos. **Rev. Geophys. Space Phys.** 10: 463-483; 1972. 75 refs, 1878-1971.

Orbital properties of the two small Martian satellites are discussed, as well as those dynamical constants of Mars that can be determined from the satellite orbits.

097.004 Chapman, C. R.; Jones, K. L. Cratering and obliteration history of Mars. **Annu. Rev. Earth Planet. Sci.** 5: 515-540; 1977. 69 refs, 1962-77.

Based primarily on pre-Viking information, this overview examines the formation of Martian craters and how they are subsequently affected by erosion and deposition. The article concludes with a look at the geomorphological evolution of Mars.

097.005 Goody, R. M. Mars and Venus. **Proc. R. Soc. London, Ser. A.** 336: 35-61; 1974. Also in: **Vistas Astron.** 19: 197-214; 1975. 54 refs, 1965-73.

Discussion of interiors; crustal dynamics; water, carbon dioxide, and the early history of an atmosphere; life on the inner planets; comparative meteorology; photochemical processes; and planetary environments based on Earth-based measurements and the first decade of space exploration.

097.006 Gornitz, Vivien, ed. **Geology of the Planet Mars.** Stroudsburg, PA: Dowden, Hutchinson & Ross; 1979. 414p. (Benchmark Pap. Geol., 48). 36 papers. 295 refs, 1609-1978.

A collection of key source papers from the literature covering a broad range of topics including, but not limited to, geomorphology; degradation processes; geochemistry; geophysics; volcanism and tectonics; cratering. Includes an excellent introduction covering historical perspective and summarizing space exploration through the Viking missions, and introductory comments for each of the 14 papers.

097.007 Hartmann, William K.; Raper, Odell. **The New Mars: The Discoveries of Mariner 9.** Washington, DC: NASA; 1974. 179p. (NASA SP-337). 13 chapters. 82 refs, 1971-73.

Aimed at laypersons and nonspecialists, this work highlights the Mariner 9 space mission by presenting many spectacular photographs along with a descriptive text. Topics: Mars before Mariner; topographical and geological features; an overview of the mission; volcanic regions; polar caps; atmosphere; Phobos and Deimos.

097.008 Ingersoll, A. P.; Leovy, C. B. The atmospheres of Mars and Venus. **Annu. Rev. Astron. Astrophys.** 9: 147-182; 1971. 203 refs, 1951-71.

Comparison of the compositions, thermal structures, and dynamics of the lower and upper atmospheres of Mars and Venus, and a consideration of some of the problems of their evolution.

097.009 International Colloquium on Mars. **Icarus** 22(1): May 1974; 22(3): Jul 1974.
C Pasadena, CA, USA: 28 Nov–1 Dec 1973. These issues of *Icarus* contain several contributed papers on cratering, dust storms, crater morphology, Earth-Mars comparisons, climatology, etc.

097.010 **International Colloquium on Mars.** 2nd. Washington, DC: NASA; 1979. 90p.
C (NASA CP-2072). 110 abstracts.

Pasadena, CA, USA: 15-18 Jan 1979. Abstracts of NASA-supported research papers. Topics: geology; meteorology; atmospheres; surface features; internal structure; spacecraft studies; Phobos and Deimos; mapping; etc.

097.011 Johnston, D. H.; Toksöz, M. N. Internal structure and properties of Mars. **Icarus** 32: 73-84; 1977. 61 refs, 1937-77.

Thermal models are briefly presented, followed by density and seismic velocity models. The final section summarizes the main features of the Martian structure.

097.012 Leovy, C. B. Martian meteorology. **Annu. Rev. Astron. Astrophys.** 17: 387-413; 1979. 120 refs, 1735-1979.

Presentation of the structure and dynamical processes, with an emphasis on data gathered by Viking orbiters and landers. Includes comparisons among the Martian atmosphere, the atmosphere of the Earth, and other rotating differentially heated fluid systems.

097.013 Magnolia, L. R., comp. **The Planet Mars: A Selected Bibliography.** Redondo Beach, CA: TRW Systems Group; distr., Springfield, VA: National Technical Information Service; 1973. 21p. (Spec. Lit. Surv., 61). (99900-7717-TU-00; AD-759291). 231 refs, 1965-73.

Comprised primarily of citations from the journal literature, this list covers spacecraft exploration and results. Arranged by author; no abstracts.

097.014 Masursky, H. An overview of geological results from Mariner 9. **J. Geophys. Res.** 78: 4009-4030; 1973. 48 refs, 1924-73.

Provides a summary of Mariner 9 mission operations insofar as they affected acquisition of geologic data.

097.015 Moroz, V. I. The atmosphere of Mars. **Space Sci. Rev.** 19: 763-843; 1976. 242 refs, 1941-76.

Review, up to the middle of 1975, of the composition of the Martian atmosphere.

097.016 Mutch, T. A.; Saunders, R. S. The geologic development of Mars: a review. **Space Sci. Rev.** 19: 3-57; 1976. 95 refs, 1895-1976.

An overview, relying heavily on data obtained by Mariner missions.

097.017 Mutch, Thomas A., et al. **The Geology of Mars.** Princeton, NJ: Princeton University Press; 1976. 400p. 9 chapters. 586 refs, 1839-1975.

A compilation of data from the Mariner 6, 7, and 9 missions, presented in the context of two goals: the compilation of a complete topographic/geologic atlas, and an examination of Mars in the context of interplanetary comparisons. The result is a basic handbook or text for specialists and nonspecialists alike, with dozens of photographs showing all major landforms: craters, channels, volcanos, and faults. Other topics: Mars missions; physiographic provinces; geophysics and structure; effects of wind and water; a summary of Martian geologic history. Includes 24 pages of crater maps.

097.018 **The Planet Mars, Volume I, Report Bibliography, Jan 1953—Dec 1968.** Alexandria, VA: Defense Documentation Center; 1969. 116p. 85 refs, 1953-69.

A computer-produced list arranged by AD number of references from the DDC collection. Entries include complete bibliographic data, descriptors, and an abstract. Title, subject, personal and corporate author indexes are included.

097.019 Pollack, J. B.; Sagan, C. Studies of the surface of Mars (very early in the era of spacecraft reconnaissance). **Radio Sci.** 5: 443-464; 1970. 61 refs, 1957-70.

Rapid review, in eight parts, of the surface environment of Mars with some emphasis on current enigmas; interpretation of some recent radar and radio observations of Mars; and a discussion of current data and debates on elevation differences on Mars.

097.020 Sagan, C. The surface environment and possible biology of Mars. In: Dollfus, A., ed. **Surfaces and Interiors of Planets and Satellites.** London: Academic; 1970: 535-556. 76 refs, 1892-1968.

A progress report on recent Martian studies, including a discussion of the prospects of life. Topics: atmosphere and radiation field; temperature and polar caps; surface composition; etc.

097.021 Schorn, R. A. The spectroscopic search for water on Mars: a history. **IAU Symp.** 40: 223-236; 1971. 57 refs, 1873-1970.

Historical review from mid-1800s to the verifiable observations in the 1960s.

097.022 **Scientific Results of the Viking Project.** Washington, DC: American Geophysical Union; 1977[?]. 722p. 53 papers.

An overview of the Viking Mars mission, the resultant data and its interpretation in the first year after the successful landing. Topics: orbiter imaging; atmospheric water detection; infrared thermal mapping; radio science; entry science; lander imaging; physical properties; seismology; magnetic properties; meteorology; inorganic chemistry; molecular analysis. The volume is a reprint of the special issue of *The Journal of Geophysical Research* (82: 3959-4680; 1977) devoted to Viking results.

097.023 Seidelmann, P. K., ed. The satellites of Mars. **Vistas Astron.** 22: 119-220; 1978.
C 6 papers.

Washington, DC, USA: 11 Aug 1977. The purpose of the meeting was to present a thorough review of our knowledge of all aspects of the moons of Mars.

097.024 Tombaugh, C. W. Geology of Mars. **Adv. Space Sci. Technol.** 10: 45-74c; 1970. 118 refs, 1932-69.

Discussion of what was known of the Martian geology at the time, taking into account the seasonal behavior of the polar caps and other seasonal phenomena.

097.025 Wells, R. A. **Geophysics of Mars.** Amsterdam; New York: Elsevier; 1979. 678p. (Dev. Sol. Syst. Space Sci., 4). 8 chapters w/refs.

This advanced text integrates data from both ground-based and spacecraft observation, centering on "two principal themes: (1) light scattering from small mineral dust grains in atmospheric suspensions; and (2) the origin and geophysical structure of the surface and interior based on the geopotential field of the planet." Topics: physical interaction of the atmosphere with the surface; blue haze and seasonal wave of surface darkening; surface features; topography; gravity field and Mars' interior; Viking; etc.

097.026 Whitten, R. C.; Colin, L. The ionospheres of Mars and Venus. **Rev. Geophys. Space Phys.** 12: 155-192; 1974. 173 refs, 1965-74.

Discussion of the measurements and interpretations of the ionospheres and atmospheres made in the Mariner and Venera series. Specifically, it summarizes the characteristics and use of UV radiometry, magnetometers, and ion probes for determining solar wind properties near planets.

Planetary Atmospheres. 091.056.

Planetary exploration: accomplishments and goals. *053.002.*

Results of recent investigations of Mars and Venus. *053.004.*

Solar system planets (Mars). *053.003.*

Structure and dynamics of the upper atmospheres of Venus and Mars. *093.005.*

Viking Lander Imaging Investigation. 053.008.

098 Minor Planets

098.001 Bender, D., et al. The Tucson revised index of asteroid data. **Icarus** 33: 630-631; 1978. no refs.
Attention is called to the availability of the TRIAD file, a compilation of all reliable physical parameters for minor planets.

098.002 Chapman, C. R. The asteroids: nature, interrelations, origin, and evolution. In: Gehrels, Tom, ed. **Asteroids.** Tucson, AZ: University of Arizona Press; 1979: 25-60. 55 refs, 1805-1979.
An introductory review of the physical and chemical nature of the asteroids, along with their physical and orbital distribution. Specific subjects include asteroid classes, population characteristics, and origin theories.

098.003 Chapman, C. R.; Williams, J. G.; Hartmann, W. K. The asteroids. **Annu. Rev. Astron. Astrophys.** 16: 33-75; 1978. 160 refs, 1889-1978.
Primarily a review of the recent advances in understanding asteroid dynamics.

098.004 Chebotarev, G. A.; Shor, V. A. The structure of the asteroid belt. **Fundam. Cosmic Phys.** 3: 87-138; 1978. 76 refs, 1862-1976.
Study of distribution of orbits and masses of asteroids in order to construct models of the behavior of the asteroid system.

098.005 C Cristescu, C.; Klepczynski, W. J.; Milet, B., eds. **Asteroids, Comets, Meteoric Matter.** Bucuresti: Editura Academiei Republicii Socialiste Romania; distr., Flushing, NY: Scholium International; 1974. 333p. (IAU Colloq., 22). 44 papers. English/French.
Nice, France: 4-6 Apr 1972. Reports of recent research which may be useful in planning close-up spacecraft studies of asteroids, comets, etc. Major topics include the motion and origin of the celestial bodies in question as well as photographic astrometry.

098.006 Dohnanyi, J. S. Interplanetary objects in review: statistics of their masses and dynamics. **Icarus** 17: 1-48; 1972. 211 refs, 1923-71.
Review of spatial distribution of masses and evolution of interplanetary bodies, such as comets, asteroids, meteoroids, etc.

098.007 Gehrels, T. Asteroids and comets. **EOS Trans. American Geophys. Union** 52: IUGG 453-459; 1971. 146 refs, 1929-70.
Literature review for 1967-70 covering asteroids: astrometry, masses, statistics, photometry, the Trojans, shape, spectra, polarimetry, radar measurements, Icarus, origin; comets: origin, orbital elements, motion, observations, spectroscopy, infrared measurements, particles, models, plasma flow, etc.

098.008 Gehrels, T. The asteroids: history, surveys, techniques, and future work. In: Gehrels, Tom, ed. **Asteroids.** Tucson, AZ: University of Arizona Press; 1979: 3-24. 33 refs, 1596-1979.

A brief introduction to the field, including "a chronology and historical sketch of asteroid studies ... for the period 1766-1978."

098.009 Gehrels, T. Photometry of asteroids. In: Dollfus, A., ed. **Surfaces and Interiors of Planets and Satellites.** London: Academic; 1970: 317-375. 71 refs, 1897-1970.

A review of current techniques and results. Topics: photographic and photoelectric photometry; reflectivities; light curves; colors; phase and aspect variators; future work. Included is a table of magnitudes (mean opposition and absolute) for the 1,735 numbered minor planets known in 1967.

098.010 Gehrels, T., ed. **Physical Studies of Minor Planets.** Washington, DC: NASA;
C 1971. 687p. (NASA SP-267). (IAU Colloq., 12). 69 papers.

Tucson, AZ, USA: 8-10 Mar 1971. Reports of observational research projects, discussions of the origin of asteroids, and reviews of the various aspects of an asteroid space mission. Physical data on asteroids is presented, and their interrelations with comets, meteors, and meteorites are explored.

098.011 Gehrels, Tom, ed.; Matthews, Mildred S. **Asteroids.** Tucson, AZ: University
C of Arizona Press; 1979. 1,181p. 45 papers.

Asteroids and Planets X (meeting). Tucson, AZ, USA: 6-10 Mar 1979. A collection of invited papers and reviews, revised and organized into a major sourcebook for specialists and nonspecialists alike, especially graduate students. There are hundreds of references to the literature in this reference work, which is divided into seven parts: introduction; exploration; interrelation (with comets, planetary satellites, and other asteroids); configuration; composition; evolution; and tabulation. An additional 25 contributed papers from this meeting appeared in a special asteroid issue of *Icarus* (Dec 1979) (see *098.017*).

098.012 Morrison, D. Asteroid sizes and albedo. **Icarus** 31: 185-220; 1977. 52 refs, 1902-1977.

A review of the radiometric method of determining diameters of asteroids, and a synthesis of radiometric and polarimetric measurements of the diameters and geometric albedos of a total of 187 asteroids are presented.

098.013 Morrison, David; Wells, William C., eds. **Asteroids: An Exploration Assess-**
C **ment.** Washington, DC: NASA; 1978. 300p. (NASA CP-2053). 19 papers.

Workshop. Chicago, IL, USA: 19-21 Jun 1978. A review and assessment of the present state of asteroid science, as well as a consideration of future studies. Topics: origin and relation to meteorites; physical characteristics; mineralogy; nearby asteroids; Earth-based observations; space missions; remote sensing.

098.014 Pilcher, Frederick; Meeus, Jean. **Tables of Minor Planets.** Jacksonville, IL: F.
A Pilcher, Illinois College; 1973. 104p. 13 refs, 1935-73.

Lists over 1,800 asteroids and associated numerical data including orbital elements and magnitudes. Also included are discoverers, place and date of discovery, and families of minor planets.

098.015 Samoĭlovoĭ-I͡Akhontovoĭ, N. S. Malye planety. [Minor Planets]. **Byull. Inst. Teor. Astron. Akad. Nauk SSSR.** Irregular. In Russian.

This approximately annual article reports on observational and ephemeris work on asteroids, including reports of newly discovered objects. A brief literature review and bibliography for the year in question is included, along with tables of appropriate data. A summary is printed in English. Recent references include: (1967): 12: 641-648; 1971. 55

refs. (1971): 13: 469-476; 1974. 66 refs. (1972): 14: 1-7; 1975. 47 refs. (1973 & 74): 14: 197-205; 1976. 59 refs. (1975): 14: 329-333; 1978. 30 refs. (1976): 14: 393-397; 1979. 18 refs.

098.016 van Houten, C. J., comp. Index of minor planets: 1921-1969. **Bull. Astron. Inst. Netherlands.** 20: 406-413; 1969.
A cumulative index with references arranged numerically by asteroid number.

098.017 Zellner, Benjamin, ed. Asteroids and Planets X. **Icarus** 40: 319-530; 1979. 25
C papers.
Tucson, AZ, USA: 6-10 Mar 1979. A special issue containing contributed conference papers. Topics: orbital dynamics, physical observations, and parent bodies of meteorites.

Comets, Asteroids, Meteorites: Interrelations, Evolution and Origins. 102.003.

Physical study of comets, minor planets and meteorites. *102.018.*

099 Jupiter, Jupiter Satellites

099.001 The atmosphere of the Jovian planets. **J. Atmos. Sci.** 26: 795-1001; 1969. Also
C in: **Contrib. Kitt Peak Natl. Obs.** 435. 31 papers.
Third Arizona Conference on Planetary Atmospheres. Tucson, AZ, USA: 30 Apr–2 May 1969. Conference papers devoted to surveys and analyses of the major planets, primarily Jupiter and Saturn. Articles are arranged under the following topics: general reviews, composition, structure, theoretical and laboratory spectra, and dynamics.

099.002 Berge, G. L.; Gulkis, S. Earth-based radio observations of Jupiter: millimeter to meter wavelengths. In: Gehrels, T., ed. **Jupiter.** Tucson, AZ: University of Arizona Press; 1976: 621-692. 195 refs, 1955-76.
A review of the study of continuous radio emission in the 1mm to 5m range. Topics: history; observational data; separation into thermal and nonthermal components; interpretation of both components; comparison of Earth-based observations with Pioneer 10 and 11 data. Also included is a detailed table of 124 different measurements of Jovian radio emission at various wavelengths and temperatures.

099.003 Carr, T. D.; Gulkis, S. The magnetosphere of Jupiter. **Annu. Rev. Astron. Astrophys.** 7: 577-618; 1969. 153 refs, 1926-68.
Review of information gained from radio observations about the Jovian magnetic field and magnetosphere. Also covered are results of theoretical investigations of interactions between the magnetic field and plasma in the Jovian magnetosphere.

099.004 Gehrels, T., ed. **Jupiter: Studies of the Interior, Atmosphere, Magnetosphere**
C **and Satellites.** Tucson, AZ: University of Arizona Press; 1976. 1,254p. (IAU Colloq., 30). 44 papers.
Tucson, AZ, USA: 18-23 May 1975. A compendium of review papers covering the planet's origin and interior, atmosphere and ionosphere, magnetosphere and radiation belts, and satellites. The result is a comprehensive sourcebook summarizing Jovian knowledge obtained from Earth-bound and spacecraft observations. Hundreds of references are included, as are a short glossary and a summary of physical data. Topics: photometry and polarimetry; infrared spectrum; origin; stratosphere; meteorology; particles and fields; non-optical studies. Thirty-six additional contributed papers from this conference can be found in three sources: *Icarus* 27: 335-459; 1976; *Icarus* 29: 165-328; 1976; *J. Geophys. Res.*, 81: 3373-3422; 1976.

099.005 Hubbard, W. B. Interior of Jupiter and Saturn. **Annu. Rev. Earth Planet. Sci.** 1: 85-106; 1973. 67 refs, 1935-72.

Review combining theoretical work and ground-based observational data. Various interior models are constructed and discussed.

099.006 Hunt, G. E. The atmospheres of the outer planets. **Adv. Phys.** 25: 455-487; 1976. 136 refs, 1944-76.

Review of current understanding of the atmospheres of Jupiter, Saturn, Uranus, and Neptune. Concerned with results of recent observational studies of the dynamical atmospheres of the outer planets and two of their satellites.

099.007 Hunt, G. E. The structure, composition and motions of Jupiter's atmosphere. **Vistas Astron.** 19: 329-340; 1976. 67 refs, 1958-75.

A discussion of current knowledge of the Jovian atmosphere concentrating on the meteorological features exhibited.

099.008 Ingersoll, A. P. The atmosphere of Jupiter. **Space Sci. Rev.** 18: 603-639; 1976. Also in: **California Inst. Technol. Div. Geol. Planet. Sci. Contr.** 2652. 116 refs, 1951-75.

Review of current information on the neutral atmosphere, covering both composition and thermal structure, and markings and dynamics.

099.009 Johnson, T. V. The Galilean satellites of Jupiter: four worlds. **Annu. Rev. Earth Planet. Sci.** 6: 93-125; 1978. 85 refs, 1610-1977.

Review of the Jovian moons, based on ground-based observations and on data gathered by Pioneer 10 and 11. Includes discussion of physical properties, atmospheres, surface characteristics, evolution, and interior structure.

099.010 Kennel, C. F.; Coroniti, F. V. Jupiter's magnetosphere. **Annu. Rev. Astron. Astrophys.** 15: 389-436; 1977. 105 refs, 1955-77.

Concentrates on the hydromagnetic structure of Jupiter's middle and outer magnetosphere, with emphasis on theoretical questions, plus a review of Pioneer 10 and 11 experimental results.

099.011 Kennel, C. F.; Coroniti, F. V. Jupiter's magnetosphere and radiation belts. In: Kennel, Charles F.; Lanzerotti, Louis J.; Parker, E. N., eds. **Solar System Plasma Physics. Volume II.** Amsterdam: North-Holland; 1979: 105-181. 200 refs, 1955-78.

A review of recent theory and observation, with an emphasis on Pioneer spacecraft data. Models of Jupiter's outer magnetosphere are presented in some detail.

099.012 Kuiper, G. P. Interpretation of the Jupiter Red Spot, I. **Commun. Lunar Planet. Lab.** 9: 249-313; 1973. 110 refs, 1923-73.

A review summarizing the LPL studies aimed at understanding Jupiter's atmosphere and clouds. Included is a description of the LPL Jupiter research programs, along with the lengthy discussion of the Red Spot, its nature, origin, and characteristics.

099.013 Maxworthy, T. A review of Jovian atmospheric dynamics. **Planet. Space Sci.** 21: 623-641; 1973. 69 refs, 1951-73.

A short review of fluid motions in Jupiter's atmosphere.

099.014 Mendis, D. A.; Axford, W. I. Satellites and magnetospheres of the outer planets. **Annu. Rev. Earth Planet. Sci.** 2: 419-474; 1974. 131 refs, 1927-73.

The known physical properties of the moons of Jupiter, Saturn, and Neptune are reviewed, along with a consideration of the effects of the planets' magnetic fields on their satellites. Variations in brightness, atmospheres, and other topics are covered as well.

099.015 Morrison, D.; Burns, J. A. The Jovian satellites. In: Gehrels, T., ed. **Jupiter.** Tucson, AZ: University of Arizona Press; 1976: 991-1034. 146 refs, 1914-76.

With an emphasis on the Galilean moons, this paper discusses the physical and dynamical properties of Jupiter's satellites. After an overview of the topic, each satellite is described separately, then compared with the others. Both ground-based and spacecraft data are presented in this overview of surface properties, orbits, internal composition, and evolution.

099.016 Newburn, R. L., Jr.; Gulkis, S. A survey of the outer planets Jupiter, Saturn, Uranus, Neptune, Pluto and their satellites. **Space Sci. Rev.** 14: 179-271; 1973. 389 refs, 1928-72.

A review of current knowledge (through May 1, 1972), including best available numerical values for physical parameters.

099.017 Owen, T. The outer solar system: perspectives for exobiology. **Origins of Life** 5: 41-55; 1974. 53 refs, 1944-74.

Summary of current knowledge about the composition and structure of outer planet atmospheres with special emphasis on Jupiter, Saturn, and Titan.

099.018 Pierce, D. A. Observations of Jupiter's satellites, 1662-1972. **Publ. Astron. Soc. Pacific** 86: 998-1000; 1974. no refs.

A brief description of the results of a project which collected nearly 26,000 observations of the known Jovian satellites. Statistics are presented on the types of observations, and a 400-entry bibliography is cited: Jet Propulsion Laboratory Technical Memorandum 900.672 (1974).

099.019 Prinn, R. G.; Owen, T. Chemistry and spectroscopy of the Jovian atmosphere. In: Gehrels, T., ed. **Jupiter.** Tucson, AZ: University of Arizona Press; 1976: 319-371. 153 refs, 1863-1976.

A thorough review covering atmospheric structure, thermochemical equilibrium models, disequilibrating processes, composition from remote spectroscopy, and spectroscopic abundances. The latter chapter summarizes the history of the discovery and study of the various elements found in the Jovian atmosphere.

099.020 Sagan, C. The solar system beyond Mars: an exobiological survey. **Space Sci. Rev.** 11: 827-866; 1971. 120 refs, 1934-71.

This review concentrates on Jupiter, with some lesser attention to Saturn, Uranus, and Neptune.

099.021 Schulz, M. Jupiter's radiation belts. **Space Sci. Rev.** 23: 277-318; 1979. 196 refs, 1958-78.

A review emphasizing data from Pioneers 10 and 11.

099.022 Smith, E. J.; Gulkis, S. The magnetic field of Jupiter: a comparison of radio astronomy and spacecraft observations. **Annu. Rev. Earth Planet. Sci.** 7: 385-415; 1979. 57 refs, 1936-77.

Review restricted to a discussion of the decimetric emission and to the spacecraft measurements of the inner magnetosphere.

099.023 Symposium on Jupiter and the outer planets. **Icarus** 10: 353-427; 1969. 9 papers.
C Dallas, TX, USA: 29-30 Dec 1968. Sessions included atmospheric composition, atmospheric structure, atmospheric dynamics, and problems of planetary origin and subsequent atmospheric evolution.

099.024 Teyfel', V. G. **The Atmosphere of the Planet Jupiter.** Washington, DC: NASA; 1970. 168p. (NASA TT F-617). 5 chapters. 215 refs, 1874-1968.

An overview of current knowledge and results of optical and radiometric observations, and the interpretation of the results. Topics: chemical composition; temperature conditions; structure of the atmosphere; the cloud cover; circulation and atmospheric processes in the atmosphere. Translation of *Atmosfera Planety Iupiter*, Moscow: Nauka; 1969.

099.025 Van Allen, J. A. On the magnetospheres of Jupiter, Saturn, and Uranus. **Highlights Astron.** 4(1): 195-224; 1977. 58 refs, 1947-76.

Brief overview of current knowledge, with a discussion of planned missions that will greatly expand knowledge in the field.

099.026 Wallace, L.; Hunten, D. M. The Jovian spectrum in the region 0.4-1.1 μm: the C/H ratio. **Rev. Geophys. Space Phys.** 16: 289-319; 1978. 158 refs, 1909-1978.

Review of the interpretations of CH_4 and H_2 absorptions in the Jovian spectrum. The historical overview is followed by laboratory and theoretical considerations and a review of the various determinations of CH_4 and H_2 amounts.

The Magnetospheres of the Earth and Jupiter. 084.011.

The radioastronomy of Jupiter. *141.065.*

Symposium on very long baseline interferometry. *033.008.*

100 Saturn, Saturn Satellites

100.001 Bobrov, M. S. Physical properties of Saturn's rings. In: Dollfus, A., ed. **Surfaces and Interiors of Planets and Satellites.** London: Academic; 1970: 376-461. 122 refs, 1802-1968.

A review of current information and past studies, including methodology. Topics: structural details; astrophysical and radioastronomical data; typical ring particle features; physical thickness; dynamics of the rings; etc.

100.002 Cook, A. F.; Franklin, F. A.; Palluconi, F. D. Saturn's rings—a survey. **Icarus** 18: 317-337; 1973. 63 refs, 1885-1973.

Discussion of the dimensions of major ring features and of the disk of the planet and possible physical models of the ring system.

100.003 Gross, S. H. The atmosphere of Titan. **Rev. Geophys. Space Phys.** 12: 435-446; 1974. 56 refs, 1925-74.

Present knowledge, speculations, and problems concerning Titan's atmosphere are reviewed. Physical characteristics, details of measurements, theoretical work, and possibility of future efforts are given.

100.004 Hunten, Donald M., ed. **The Atmosphere of Titan.** Washington, DC: NASA;
C 1974. 177p. (NASA SP-340). 17 papers. 101 refs, 1908-1973.

Workshop. Moffett Field, CA, USA: 25-27 Jul 1973. Papers from a NASA-sponsored meeting to discuss information useful in planning a possible space mission to study the only planetary satellite with an atmosphere. Topics: photometric, polarimetric, and spectrophotometric studies; the interior of Titan; organic chemistry in the atmosphere; atomic hydrogen distribution; optical, infrared, and ultraviolet observations.

100.005 Hunten, Donald M.; Morrison, David, eds. **The Saturn System.** Washington,
C DC: NASA; 1978. 420p. (NASA CP-2068). 25 papers.

Workshop. Reston, VA, USA: 9-11 Feb 1978. A review of current knowledge and possible topics for scientific investigation to be carried out by a space mission to Saturn in the early 1980s. Topics: origin and evolution of Saturn and its moons; the ring system; thermal structure; atmospheres of Saturn and Titan; aeronomy; magnetosphere; etc.

100.006 Palluconi, F. D.; Pettengill, G. H., eds. **The Rings of Saturn.** Washington, DC:
C NASA; 1974. 222p. (NASA SP-343). 17 papers.

Saturn's Rings Workshop. Pasadena, CA, USA: 31 Jul—1 Aug 1973. (Proceedings). Reports of research, review of properties, and transcripts of discussions. Topics: physical characteristics; ring particles; space missions for detailed study; infrared brightness temperature; radio and microwave observations; recent spacecraft findings; etc.

100.007 Pierce, D. A. Observations of Saturn's satellites 1789-1972. **Publ. Astron. Soc. Pacific** 87: 785-787; 1975. 1 ref, 1974.

A brief description of a project developed for the purpose of collecting more than 22,000 observations of the known satellites of Saturn. Statistics are presented on the types of observations (photographic and micrometer), and a 200-entry bibliography is referenced: Jet Propulsion Laboratory Technical Memorandum 900-968 (1975).

100.008 Pollack, J. B. The rings of Saturn. **Space Sci. Rev.** 18: 3-93; 1975. 94 refs, 1847-1974.

Review of observations of the rings of Saturn at visual, infrared, and radio wavelengths, followed by a consideration of their origin and evolution.

100.009 Teifel, V. G. The atmosphere of Saturn. **IAU Symp.** 65: 415-440; 1974. 56 refs, 1926-74.

The results of recent studies of the optical properties, temperature, chemical composition, and probable structure of Saturn's atmosphere are reviewed.

The atmosphere of the Jovian planets. *099.001.*

The atmospheres of the outer planets. *099.006.*

Interior of Jupiter and Saturn. *099.005.*

On the magnetospheres of Jupiter, Saturn, and Uranus. *099.025.*

The outer solar system: perspectives for exobiology. *099.017.*

Satellites and magnetospheres of the outer planets. *099.014.*

A survey of the outer planets Jupiter, Saturn, Uranus, Neptune, Pluto and their satellites. *099.016.*

101 Uranus, Neptune, Pluto, Transplutonian Planets

[No entries; cross-references only.]

The atmospheres of the outer planets. *099.006.*

On the magnetospheres of Jupiter, Saturn, and Uranus. *099.025.*

Satellites and magnetospheres of the outer planets. *099.014.*

A survey of the outer planets Jupiter, Saturn, Uranus, Neptune, Pluto and their satellites. *099.016.*

102 Comets (Origin, Structure, Atmospheres, Dynamics)

102.001 C Axford, W. I.; Fechtig, H.; Rahe, J., eds. **Cometary Missions.** Bamberg, West Germany: [s.n.]; 1979. 223p. (Kleine Veröff. Remeis Sternwarte, 12, no. 132). 30 papers and abstracts.

Workshop. Bamberg, West Germany: 20-22 Feb 1979. (Proceedings). Papers focusing on the instrumentation of a comet probe, dealing with its feasibility and prospects for research. Topics: the proposed ESA/NASA mission to fly by Halley's Comet and to rendezvous with Comet Tempel-2; observation of cometary plasma; magnetic fields; cometary atmospheres; cometary dust; mass spectrometry; etc.

102.002 C Chebotarev, G. A.; Kazimirchak-Polonskaya, E. I.; Marsden, B. G., eds. **The Motion, Evolution of Orbits, and Origin of Comets.** Dordrecht, Holland: D. Reidel; New York: Springer-Verlag; 1972. 521p. (IAU Symp., 45). 62 papers, 23 abstracts.

Leningrad, USSR: 4-11 Aug 1970. Sections include observations and ephemerides, general methods of orbit theory, motions of the short-period comets, physical processes in comets, origin and evolution of comets, and relationships with meteors and minor planets.

102.003 C Delsemme, A. H., ed. **Comets, Asteroids, Meteorites: Interrelations, Evolution and Origins.** Toledo, OH: University of Toledo Press; 1977. 587p. (IAU Colloq., 39). 74 papers.

Conference: Relationships between Comets, Minor Planets and Meteorites. Lyon, France: 17-20 Aug 1976. Contributed conference papers in the following areas: physical nature of comets; orbital evolution of comets; meteors and meteoroids; physical nature of asteroids; orbital evolution and fragmentation of asteroids; primitive meteorites; differentiated meteorites; origin of comets; the primitive solar nebula.

102.004 Delsemme, A. H. Gas and dust in comets. In: Cameron, A. G. W., ed. **Cosmochemistry.** Dordrecht, Holland: D. Reidel; 1973: 89-101. Also in: **Space Sci. Rev.** 15: 89-101; 1973. 98 refs, 1836-1973.

A review of "the origin, the nature and the fate of the material, either gaseous or dusty, that appears during the transient cometary phenomena." Topics: the tail, the coma, gas emitted in the coma, molecules in comets, dust grains, etc.

102.005 Dobrovolskii, O. V. Comet astronomy in the U.S.S.R. **Astron. Vestn.** 6: 137-152; 1972. 192 refs, 1784-1972. In Russian.

Review of research, including motion, physical theory, photometric methods, and experimental modelling.

102.006 C Donn, B., et al., eds. **The Study of Comets.** Washington, DC: NASA; 1976. 1,083p. 2v. (IAU Colloq., 25). (NASA SP-393). 65 papers.

Greenbelt, MD, USA: 28 Oct–1 Nov 1974. Review papers, reports of recent and current projects, and discussions of all aspects of cometary research. Selected topics include photometry, spectra, data collections, specific observations, tails, orbits, the nucleus, formation, etc.

102.007 Herzberg, G. Cometary spectra and related topics. **Mem. Soc. R. Sci. Liège.** 6e ser. 9: 115-132; 1976. 64 refs, 1935-75.

Summary of current knowledge, emphasizing the role of laboratory analysis.

102.008 Kaymakov, E. A. Probable parent molecules in cometary nuclei. **Probl. Kos. Fiz.** 9: 141-155; 1974. 63 refs, 1873-1972. In Russian.

Experimental data is reviewed to demonstrate that cometary ice may contain some organic and inorganic compounds, in particular vast amounts of amino acids which can provide the spectral components observed as well as the optically active dust component.

102.009 Keller, H. U. The interpretations of ultraviolet observations of comets. **Space Sci. Rev.** 18: 641-684; 1976. 104 refs, 1931-75.

A summary of results and their implications for physical properties and composition of comets.

102.010 Kuiper, G. P.; Roemer, E., eds. **Comets: Scientific Data and Missions.** Tucson,
C AZ: Lunar and Planetary Laboratory, University of Arizona; 1972. 222p. 24 papers.

Tucson Comet Conference. Tucson, AZ, USA: 8-9 Apr 1970. (Proceedings). A review of current knowledge along with suggestions for spacecraft study. Topics: cometary nuclei; nature and origin of cometary heads; tails (type I & II); spectra; orbits; photometry; spectroscopy; etc.

102.011 Marsden, B. G. Comet catalogues. **Int. Comet Q.** 1: 13-16; 1979. 25 refs,
A 1687-1979.

A brief history and complete list of catalogs containing cometary orbital data.

102.012 Marsden, B. G. Cometary motions. **Celestial Mech.** 9: 303-314; 1974. 69 refs, 1543-1973.

A history of the study of cometary motions is presented, from pre-Copernican to the present, with emphasis on the determination of orbits, the calculation of special perturbations, and the analysis of nongravitational effects.

102.013 Marsden, B. G. Comets. **Annu. Rev. Astron. Astrophys.** 12: 1-21; 1974. 167 refs, 1893-1974.

Review of current information known about comets, including statistics of oscillating orbits, perturbations and true distribution of orbits, nongravitational effects on cometary motions, photometric observations, spectroscopic and polarimetric results, and comet structure.

102.014 Marsden, Brian G. **Catalogue of Cometary Orbits.** 3rd ed. Cambridge, MA:
A Central Bureau for Astronomical Telegrams (IAU), Smithsonian Astrophysical Observatory; 1979. 18 refs, 1894-1973.

Contains descriptive data and orbital elements for 1,027 cometary apparitions observed through the end of 1978, including 658 individual comets, of which 113 are periodic. The descriptive data includes comet designation, perihelion time, perihelion distance (in A.U.), eccentricity, revolution period (years), argument of perihelion, longitude of ascending node, inclination, osculation date, number of observations, time span required to calculate the orbit. Also included are names of recoverers and rediscoverers, 15 pages of bibliographic references (more than 1,200), lists of periodic comets and the number of visits, nongravitational parameters, and 23 pages of tables of orbital elements. The previous editions of this very useful work, also by Marsden, are 1972 (70p.) and 1975 (83p.). Available from the Central Bureau for Astronomical Telegrams, Smithsonian Astrophysical Observatory, 60 Garden St., Cambridge, MA 02138.

102.015 Mendis, D. A.; Ip, W.-H. The ionospheres and plasma tails of comets. **Space Sci. Rev.** 20: 145-190; 1977. 135 refs, 1910-76.

The present understanding of cometary ionospheres and plasma tails is critically evaluated. Following a brief introduction of the significance of the study, the observational statistics and spectroscopic observations are summarized.

102.016 Mendis, D. A.; Ip, W.-H. The neutral atmospheres of comets. **Astrophys. Space Sci.** 39: 335-385; 1976. 115 refs, 1910-75.

A review of present knowledge of the neutral gas atmospheres of comets, including relevant photometric and spectroscopic observations, followed by a tentative picture of the nucleus that seems to be indicated.

102.017 Neugebauer, M., et al., eds. **Space Missions to Comets.** Washington, DC:
C NASA; 1979. 226p. (NASA CP-2089). 7 papers.

Greenbelt, MD, USA: Oct 1977. Possible scientific results of a spacecraft visit to a comet are presented. Topics: the scientific need for a space mission; comets and space plasmas; the connection of comets with carbonaceous meteorites; interstellar molecules and the origin of life; Comet Halley; scientific returns of comet missions. Includes a review paper on Halley's Comet and other important comets in history.

102.018 Physical study of comets, minor planets and meteorites. **IAU Trans. Rep. Astron.** 17A(1): 73-102; 1979. + refs, 1976-78.

An overview of research activity worldwide, with an emphasis on the previous three years, as reported by IAU Commission 15 (Physical Study of Comets, Minor Planets and Meteorites). Additional reports: *14A: 149-152; 1970. 121 refs, 1966-70. *15A: 179-190; 1973. 47 refs, 1963-72. 16A(1): 61-84; 1976. + refs, 1972-75. (*Commission title was "Physical Studies of Comets.")

102.019 Rahe, Jürgen; Donn, Bertram; Wurm, Karl. **Atlas of Cometary Forms, Vol. 1:**
A **Structures Near the Nucleus.** Washington, DC: NASA; 1969. 128p. (NASA SP-198).

A combination of drawings from nineteenth- and twentieth-century observations and photographs from recent years, this volume concentrates on three of the brightest comets of this century: Morehouse 1908 III, Halley 1910 II, and Humason 1962 VIII. Other comets are covered as well, in an attempt to illustrate various types of comets and their characteristics. Includes physical data, as well as information on each drawing and photograph.

102.020 Tomanov, V. P. Catalogue of periodic comets. **Astron. Vestn.** 13: 94-98; 1979.
A 2 refs, 1958-68. In Russian.

The catalog contains 118 cometary orbits with semimajor axis $a < 30$ A.U. Orbital elements, Tisserand's criterion, and absolute magnitudes are given, as well as orbit parameters, area constants, heliocentric distances of orbit nodes, and the velocity.

102.021 van Houten, C. J., comp. Index of comets: 1921-1969. **Bull. Astron. Inst. Netherlands** 20: 414; 1969.

A cumulative index arranged alphabetically by comet name.

102.022 Vanýsek, V. The structure and formation of comets. In: Elvius, A., ed. **From Plasma to Planet.** Stockholm: Almqvist & Wiksell; 1972: 233-259. 58 refs, 1910-71.

A review of present knowledge of the behavior of neutral molecules and dust particles in cometary bodies.

102.023 Whipple, F. L. Comets. In: McDonnell, J. A. M., ed. **Cosmic Dust.** Chichester, UK: John Wiley & Sons; 1978: 1-73. 159 refs, 1823-1976.

A review of current knowledge based on observational evidence and theory leading to a presently accepted cometary model. Topics: cometary motion; luminosity, bursts, and splitting; radiation processes; physical processes in the cometary head; dust tails and grains; plasma tails; the nucleus; origin; history; etc.

102.024 Whipple, F. L.; Huebner, W. F. Physical processes in comets. **Annu. Rev. Astron. Astrophys.** 14: 143-172; 1976. 201 refs, 1877-1976.
Discussion of physical processes and structure of comets, relating them to cometary observations.

Asteroids and comets. *098.007.*

Asteroids, Comets, Meteoric Matter. 098.005.

The interaction of the solar wind with comets. *074.003.*

Interplanetary objects in review: statistics of their masses and dynamics. *098.006.*

103 Comets (Individual Objects)

103.001 Gary, Gilmer Allen, ed. **Comet Kohoutek.** Washington, DC: NASA; 1975.
C 272p. (NASA SP-355). 42 papers.
Workshop. Huntsville, AL, USA: 13-14 Jun 1974. A collection of very brief reports of research conducted on and observations made of the decade's most celebrated comet. Topics: tail form, structure, and evolution; H_2O-related observations; molecules and atoms in the coma and tail; and photometry and radiometry. Data presented was obtained from Skylab, aircraft, rockets, and ground-based stations.

103.002 Maran, S. P., ed. Comet Kohoutek. **Icarus** 23: 489-631; 1972. 22 papers.
C A special issue devoted to studies of Comet Kohoutek. Reports of recent observations, both ground-based and from space, including data obtained from Skylab. Papers include optical, infrared, and spectroscopic studies.

104 Meteors, Meteor Streams

104.001 Halliday, I.; McIntosh, B. A., eds. **Solid Particles in the Solar System.** Dord-
C recht, Holland; Boston: D. Reidel; 1980. 441p. (IAU Symp., 90). 23 papers, 59 abstracts.
Ottawa, ON, Canada: 27-30 Aug 1979. Sections include zodiacal light; meteors and meteorites; the interplanetary dust complex—sources, evolution, and dynamics; physical properties; and particles and planets.

104.002 Hemenway, C. L.; Millman, P. M.; Cook, C. F., eds. **Evolutionary and Physi-**
C **cal Properties of Meteoroids.** Washington, DC: NASA; 1973. 377p. (IAU Colloq., 13). (NASA SP-319). 40 papers.
Albany, NY, USA: 14-17 Jun 1971. Reports of research including photography of fireballs, laboratory analysis, meteor spectra, meteor streams, ablation in meteors, cosmic dust, origin, etc. Includes many citations to the literature.

104.003 Hughes, D. W. Meteors. In: McDonnell, J. A. M., ed. **Cosmic Dust.** Chichester, UK: John Wiley & Sons; 1978: 123-185. 138 refs, 1866-1976.

A review concerned with the motions and observations of meteors in the Earth's atmosphere. Topics: meteor streams; meteors and micrometeoroids; meteor observation techniques; results of different observations; meteor size distribution; meteor influx; orbital parameters; meteoroid densities; the meteor in the Earth's atmosphere.

104.004 Meteors and interplanetary dust. **IAU Trans. Rep. Astron.** 17A(1): 153-170; 1979. 87 refs, 1966-79.

An overview of research activity worldwide, with an emphasis on the previous three years, as reported by IAU Commission 22 (Meteors and Interplanetary Dust). Additional reports: *14A: 207-223; 1970. 15 refs, 1967-70. *15A: 253-273; 1973. 53 refs, 1967-73. 16A(1): 141-158; 1976. + refs, 1973-76. (*Commission title was "Meteors and Meteorites.")

Asteroids, Comets, Meteoric Matter. 098.005.

105 Meteorites, Meteorite Craters

105.001 Anders, E. Meteorites and the early solar system. **Annu. Rev. Astron. Astrophys.** 9: 1-34; 1971. 162 refs, 1930-71.

Review of primitive meteorites, their formation, composition, and chemistry.

105.002 Ashworth, D. G. Lunar and planetary impact erosion. In: McDonnell, J. A. M., ed. **Cosmic Dust.** Chichester, UK: John Wiley & Sons; 1978: 427-526. 307 refs, 1610-1977.

An extensive review of the frequency, processes, and results of impact cratering throughout the solar system. The sizes and relative numbers of impact craters are analyzed in detail to provide a picture of the flux of particulate matter in the solar system. Although the Earth, Mars, and Mercury are considered, the major emphasis is on the Moon and the particulate and larger matter striking its surface.

105.003 Axon, H. J. The metallurgy of meteorites. **Prog. Mater. Sci.** 13: 183-228; 1967. 137 refs, 1877-1967.

A review of the origin, formation, composition, and analysis of meteorites. Topics: chondritic stones; iron meteorites; the equilibrium diagram in relation to meteorites; quantitative aspects of meteorite structures; metamorphism of iron meteorites.

105.004 Barnes, Virgil E.; Barnes, Mildred A., eds. **Tektites.** Stroudsburg, PA: Dowden, Hutchinson and Ross; 1973. 445p. (Benchmark Pap. Geol.). 46 papers.

A collection of selected important papers in the field, covering 1934-72 with comments on each by the editors, providing historical perspective and a synthesis of subject matter. Topics: geology, petrology, and mineralogy; physical and chemical data; ages of tektites; sculpture of tektites; origin; bibliography prior to 1959.

105.005 Bogard, D. D. Noble gases in meteorites. **EOS Trans. American Geophys. Union** 52: IUGG 429-435, 1971. 122 refs, 1960-70.

A review of how the five stable noble gases have provided information on past physical events in the solar system and how they are used in studies of elemental abundances, high- and low-energy nuclear reactions in the solar system, ages of meteorites and the elements, etc.

105.006 Buchwald, Vagn F. **Handbook of Iron Meteorites.** Berkeley, CA: University of California Press; 1975. 3v. (boxed). 1,426p. 2,090 refs, 1716-1974.

Subtitled "their history, distribution, composition, and structure," this set represents the most comprehensive study ever made. Volume 1 is a complete survey of the various aspects of meteorites, covering their origin, fall, craters, statistics, history, shapes and surface characteristics, classification, chemical composition, mineralogy, structure, and ages. Eight appendices include lists of well-examined iron meteorites, a lengthy bibliography, and numerous tables of data. The second and third volumes list alphabetically, and describe in great detail, 532 different iron meteorites. Besides the lengthy descriptions, there are more than 2,000 photographs and numerous tables, sketches, and diagrams. Descriptions cover find location, history, physical characteristics, chemical analysis, etc. An outstanding compendium: *the* source for this type of information.

105.007 Burnett, D. S. Formation times of meteorites and lunar samples. **EOS Trans. American Geophys. Union** 52: IUGG 435-440; 1971. 91 refs, 1960-70.

Review of recent work on crystallization ages based on parent-daughter isotropic relationships resulting from the decay of naturally occurring radioactive nuclei. The ages of chondrites, achondrites, iron meteorites, and lunar samples are covered.

105.008 Cooper, H. F., Jr. A summary of explosion cratering phenomena relevant to meteor impact events. In: Roddy, D. J.; Pepin, R. O.; Merrill, R. B., eds. **Impact and Explosion Cratering.** New York: Pergamon Press; 1977: 11-44. 68 refs, 1961-77.

A review of "theoretical and experimental studies of cratering and related phenomena from buried explosions." Phenomena considered include crater volume and shape, ejecta, airblast, and ground shock. Includes a discussion of the mechanics of explosive cratering.

105.009 Dodd, R. T. Chondrites. **EOS Trans. American Geophys. Union** 52: IUGG 447-453; 1971. 195 refs, 1963-70.

Topics include classification, metamorphism, ages and thermal history, chondrite-achondrite relationships, origin, and organic compounds.

105.010 Dohnanyi, J. S. Micrometeoroids. **EOS Trans. American Geophys. Union** 52: IUGG 459-464; 1971. 188 refs, 1946-71.

A review for 1967-71 of ground-based observations, flux measurement by satellites, particle collection experiments, and theory.

105.011 Herndon, J. M.; Rowe, M. W. Magnetism in meteorites. **Meteoritics** 9: 289-305; 1974. 40 refs, 1864-1974.

A somewhat simplified overview of the subject. Includes a glossary of magnetism terminology, a discussion of techniques of study, a review of results, and a brief critical analysis.

105.012 King, E. A. The origin of tektites: a brief review. **American Sci.** 65: 212-218; 1977. 82 refs, 1844-1976.

Review of the steps that led to the conclusion that tektites are produced from terrestrial rocks melted by hypervelocity impacts of large, extraterrestrial objects.

105.013 LaPaz, Lincoln. **Topics in Meteoritics. Hunting Meteorites: Their Recovery, Use, and Abuse from Paleolithic to Present.** Albuquerque, NM: University of New Mexico Press; 1969. 184p. (Univ. New Mexico Publ. Meteoritics, 6). 12 papers.

A collection of papers on a variety of subjects of interest to specialists. Topics: probabilistic population criteria; paleometeoritics; historical meteorites; the Evans Meteorite; frequency of falls; etc.

105.014 Larimer, J. W. Meteorites: relics from the early solar system. In: Dermott, S. F., ed. **The Origin of the Solar System.** Chichester, UK: John Wiley & Sons; 1978: 347-393. 133 refs, 1952-78.

A detailed review of the chemical composition and formation of meteorites, and a discussion of how this information aids in the formation of solar system origin theories. Following an introduction to meteorites, there are three sections to this paper: 1) chondrites and the chemistry of the solar nebula; 2) carbonaceous and enstatite chondrites—the extremes; 3) differentiated meteorites—implications for the evolution of planetary matter.

105.015 Leonardi, Piero. **Volcanoes and Impact Craters on the Moon and Mars.** Amsterdam; New York: Elsevier; 1976. 463p. 18 chapters. ~1,500 refs, 1787-1974.

Overflowing with color and black-and-white photographs of the Moon and Mars, this volume provides a thorough look at craters and volcanoes, their origin and characteristics. Topics: history; lunar soil; lunar tectonics; craters; present volcanic activity; terrestrial meteorite craters; etc. Intended for the scientist and educated layperson, this book emphasizes the Moon; only one chapter addresses Mars.

105.016 Lewis, C. F.; Moore, C. B. Chemical analysis of thirty-eight iron meteorites. **Meteoritics** 6: 195-205; 1971. 51 refs, 1824-1971.

A continuation of chemical analyses carried out by the authors in 1969.

105.017 McCall, G. J. H. **Meteorite Craters.** Stroudsburg, PA: Dowden, Hutchinson & Ross; 1977. 364p. (Benchmark Pap. Geol., 36). 29 papers.

A collection of key papers addressing three different classes of craters: 1) proved meteorite impact-fragmentation craters; 2) craters showing shock effects, but no material association of meteorites; 3) craters showing the characteristic physiographic form and structure of meteorite craters, but no shock metamorphism, meteorite material, or boilde association. Topics: falls, formation, material in and near craters, etc. Many specific examples of (large) craters are presented in detail.

105.018 McCall, G. J. H. **Meteorites and Their Origins.** New York: John Wiley & Sons; 1973. 352p. 23 chapters. 192 refs, 1792-1971.

A not-too-technical summary text on meteoritics for students, scientists, and certain laypersons, providing a broad overview of the subject. Topics: history; place in the solar system; meteorite flight, falls, and impacts; morphology; classification; mineralogy; origin; review of the various types of meteorites: iron, stony, chondrules, tektites; etc.

105.019 Mapper, D.; Smales, A. A. Recovered extraterrestrial materials. In: Wainerdi, Richard E., ed. **Analytical Chemistry in Space.** Oxford, UK: Pergamon Press; 1970; 209-275. 472 refs, 1884-1968.

A thorough review of the chemical analysis techniques used in studying the composition of meteorites, cosmic dust, tektites, and lunar material brought back to Earth. Topics: collection of extraterrestrial material; preparation of samples for analysis; activation analysis; mass spectrometry; emission spectroscopy; flame photometry; flourescence spectroscopy; etc.

105.020 Mason, Brian, ed. **Handbook of Elemental Abundances in Meteorites.** New York: Gordon and Breach; 1971. 555p. (Ser. Extraterr. Chem., 1).

The result of a systematic study of the composition of meteorites started in 1923, this volume presents data on approximately 70 elements. Most chapters concentrate on one element, describing the research done and detailing the results. The volume contains a wealth of useful numerical data and bibliographic references.

105.021 O'Keefe, John A. **Tektites and Their Origin.** Amsterdam; New York: Elsevier Publishing; 1976. 254p. (Dev. Petrol., 4). 10 chapters. 652 refs, 1844-1975.

An advanced monograph providing a detailed review of the physical nature of tektites and a presentation of evidence (pro and con) on the question of terrestrial or extra-terrestrial origin. Topics: distribution; shapes; internal structure; physical properties; chemical composition; isotopes, fission tracks, and cosmic ray tracks; origin question; conclusions.

105.022 C Roddy, D. J.; Pepin, R. O.; Merrill, R. B., eds. **Impact and Explosion Cratering: Planetary and Terrestrial Implications.** New York: Pergamon Press; 1977. 1,301p. 64 papers.

Symposium on Planetary Cratering Mechanics. Flagstaff, AZ, USA: 13-17 Sep 1976. (Proceedings). A collection of contributed papers dealing with current research in the following areas: cratering phenomenology (experimental studies, planetary cratering, etc.); material properties and shock effects; theoretical cratering mechanics; ejecta; scaling.

105.023 Sears, D. W. **The Nature and Origin of Meteorites.** Bristol, UK: Adam Hilger; New York: Oxford University Press; 1978. 186p. (Monogr. Astron. Subj., 5). 6 chapters. 392 refs, 1772-1978.

A brief introduction and overview for the scientist, advanced student, and advanced layperson. Topics: historical introduction; meteorite falls and associated phenomena; classification, mineralogy, and petrology; compositional properties; physical properties and processes; history and origin.

105.024 Shaw, D. M.; Harmon, R. S. Factor analysis of elemental abundances in chondritic and achondritic meteorites. **Meteoritics** 10: 253-282; 1975. 72 refs, 1959-75.

A description of the use of factor analysis in meteorite geochemistry, along with a compilation of data on element abundances in 55 stony meteorites.

105.025 **Soviet Literature on Meteorite Craters.** Washington, DC: NASA; 1976. 5p. (NASA TT F-17147; N76-32088). 52 refs, 1969-75.

A translation of "Sovetskai͡a literatura po meteoritnym krateram," an unpublished report of the Academy of Sciences USSR, Moscow, 1976. A list of references with no abstracts.

105.026 Vdovykin, G. P. The Canyon Diablo meteorite. **Space Sci. Rev.** 14: 758-831; 1973. 280 refs, 1891-1971.

Review of the many studies of this particular iron meteorite, which is unique in a number of scientifically important ways.

105.027 Wasson, J. Differentiated meteorites. **EOS Trans. American Geophys. Union** 52: IUGG 441-447; 1971. 233 refs, 1967-71.

A brief overview of meteorites other than chondrites and the research carried out from 1967-70. Covered are achondrites, stony irons, and iron meteorites. Each group is broken down into subclasses, and appropriate studies are then cited.

105.028 Wasson, J. T. Formation of ordinary chondrites. **Rev. Geophys. Space Phys.** 10: 711-759; 1972. 113 refs, 1916-72.

Review of the chemical and mineralogical properties of the ordinary chondrites and the fractionation events that led to their formation.

105.029 Wasson, John T. **Meteorites: Classification and Properties.** New York: Springer-Verlag; 1974. 316p. (Miner. Rocks, 10). 18 chapters. 804 refs, 1745-1974.

Intended for scientists and advanced students, this volume intends to be both an introduction to meteorite science and a handbook on meteorite classification. Including a wealth of tabular numeric data and an extensive list of references, the book provides a good summary for the new reader and a handy reference for the experienced. Topics:

classification; bulk composition; mineralogy; petrology; trace elements; orbits; morphology; magnetic properties; etc. The appendices include a glossary and four lists of classified meteorites with an additional 184 references.

105.030 Wetherill, G. W. Solar system sources of meteorites and large meteoroids. **Annu. Rev. Earth Planet. Sci.** 2: 303-331; 1974. 103 refs, 1823-1973.

A review which rejects the once-common hypothesis that all meteorites originated from a common source or parent body. The article discusses the planets, comets, and asteroids as possible origins of the objects, and looks at the orbits of the meteoroids.

Analyses of Extraterrestrial Materials. 094.512.

Asteroids and Planets X. *098.017.*

Chemical Petrology: With Applications to the Terrestrial Planets and Meteorites. 022.064.

Comets, Asteroids, Meteorites: Interrelations, Evolution and Origins. 102.003.

Interplanetary objects in review: statistics of their masses and dynamics. *098.006.*

Physical study of comets, minor planets and meteorites. *102.018.*

Solid Particles in the Solar System. 104.001.

106 Interplanetary Matter, Interplanetary Magnetic Field, Zodiacal Light

106.001 Axford, W. I. Observations of the interplanetary plasma. **Space Sci. Rev.** 8: 331-365; 1968. 265 refs, 1923-68.

Observations bearing on the nature and properties of the interplanetary plasma are reviewed. A fairly complete set of references up to September 1967 is given for the cases of comet tail, radar, radio source scattering and scintillation, and space-probe measurements.

106.002 Burch, J. L. Observations of interactions between interplanetary and geomagnetic fields. **Rev. Geophys. Space Phys.** 12: 363-378; 1974. 131 refs, 1959-74.

Summary of present knowledge regarding the role of the solar wind, or interplanetary, magnetic, and electric fields in producing large-scale magnetospheric reconfigurations.

106.003 Conference on interactions of the interplanetary plasma with the modern and
C ancient Moon. **Moon** 14: 3-207; 1975. 17 papers.

Sponsored by the Lunar Science Institute and held at the Lake Geneva Campus of George Williams College, WI, USA: 30 Sep—4 Oct 1974. Selected topics: bow shock protons in the lunar environment; lunar remnant magnetic field mapping; magnochemistry of the Apollo landing sites; formation of the lunar atmosphere; etc.

106.004 Elsasser, H.; Fechtig, H., eds. **Interplanetary Dust and Zodiacal Light.** Berlin:
C Springer-Verlag; 1976. 496p. (IAU Colloq., 31). (Lecture Notes Phys., 48). 89 papers.

Heidelberg, Germany: 10-13 Jun 1975. Review papers and research reports covering observations of zodiacal light (from space and ground-based); zodiacal light models; spacecraft measurement of interplanetary dust; lunar microcraters; cometary dust; meteors; dynamics and evolution of interplanetary dust.

106.005 Hollweg, J. V. Hydromagnetic waves in interplanetary space. **Publ. Astron. Soc. Pacific** 86: 561-594; 1974. 114 refs, 1956-74.

Review of recent observational and theoretical work seeking to explain the origin of waves and their observed nonlinear properties and possible roles the waves may play in modifying the thermal and dynamic properties of the solar wind itself.

106.006 Holzer, T. E. Neutral hydrogen in interplanetary space. **Rev. Geophys. Space Phys.** 15: 467-490; 1977. 146 refs, 1958-78.

A critical review of theory and observations relevant to the interplanetary H problem.

106.007 Jokipii, J. R. Turbulence and scintillations in the interplanetary plasma. **Annu. Rev. Astron. Astrophys.** 11: 1-28; 1973. 93 refs, 1952-73.

Consideration of direct and indirect observations of broadband fluctuations observed in the solar wind, and placement of these observations in some kind of overall perspective.

106.008 Leinert, C. Zodiacal light—a measure of the interplanetary environment. **Space Sci. Rev.** 18: 281-339; 1975. 229 refs, 1730-1975.

Review of research, with an emphasis on recent zodiacal light measurements, including space observations, and on methods and results of interpretation.

106.009 Piddington, J. H. **Cosmic Electrodynamics.** New York: John Wiley & Sons; 1969. 305p. (Intersci. Monogr. Texts Phys. Astron., 23). 13 chapters w/refs.

A survey of the nature and effects of interplanetary and intergalactic magnetic fields, aimed at the nonspecialist. Topics: cosmic plasmas; electrodynamic effects of universal occurrence; solar activity; the Earth's magnetosphere; geomagnetic disturbances; galactic forms and activities; etc.

106.010 Rao, U. R. Solar modulation of galactic cosmic radiation. **Space Sci. Rev.** 12: 719-809; 1972. 242 refs, 1949-71.

An integrated view of the solar modulation process that causes time variation of cosmic ray particles.

106.011 Roosen, R. G. An annotated bibliography on the gegenschein. **Icarus** 13: 523-539; 1970. 166 refs, 1954-70.

A comprehensive, annotated bibliography including both Western and Russian sources. Citations include author, title, language, source, year, volume, pages, and annotation. Brightness and position data have been compiled from the papers in the article and are presented in both tabular and graphic format.

106.012 Schatten, K. H. Large-scale properties of the interplanetary magnetic field. **Rev. Geophys. Space Phys.** 9: 773-812; 1971. 47 refs, 1951-70.

Review of Parker's early theoretical work through more recent work on the influence of the Sun's polar field on the interplanetary field.

106.013 Shea, M. A.; Smart, D. F.; Wu, S. T., eds. **Study of Travelling Interplanetary**
C **Phenomena/1977.** Dordrecht, Holland; Boston: D. Reidel; 1977. 439p. (Astrophys. Space Sci. Libr., 71). 20 papers.

L. D. de Feiter Memorial Symposium. Tel Aviv, Israel: 7-10 Jun 1977. (Proceedings). Review papers on a variety of subjects in the field of the interplanetary medium: solar physics; solar radio astronomy; interplanetary scintillations; solar wind; comets; solar cosmic rays. Contributed papers from this meeting appeared in AFGL-TR-77-0309; Special Reports, 209.

106.014 Smith, E. J.; Sonett, C. P. Extraterrestrial magnetic fields: achievements and opportunities. **IEEE Trans. Geosci. Electron.** GE-14: 154-171; 1976. 151 refs, 1908-1976.

The major scientific achievements associated with the measurement of magnetic fields in space over the past 15 years are reviewed. Aspects of space technology relevant to magnetic-field observations are discussed.

106.015 Weinberg, J. L.; Sparrow, J. G. Zodiacal light as an indicator of interplanetary dust. In: McDonnell, J. A. M., ed. **Cosmic Dust.** Chichester, UK: John Wiley & Sons; 1978: 75-122. 190 refs, 1730-1976.

An overview with a critical evaluation of existing observations and an examination of the data from those observations. Topics: observational parameters and observing geometry; methods of observation; errors and absolute calibration; results; information content on the nature and distribution of the dust; future problems and requirements.

The Dusty Universe. 131.019.

The interaction of the solar wind with the interstellar medium. *074.001.*

Light of the night sky. *082.005; 082.008.*

The light of the night sky and the interplanetary medium. *082.004.*

Meteors and interplanetary dust. *104.004.*

Solar and Interplanetary Dynamics. 074.007.

Solid Particles in the Solar System. 104.001.

107 Cosmogony

107.001 Alfvén, H.; Arrhenius, G. Structure and evolutionary history of the solar system. I. **Astrophys. Space Sci.** 8: 338-421; 1970. 77 refs, 1893-1970. II. **Astrophys. Space Sci.** 9: 3-33; 1970. 27 refs, 1951-70. III. **Astrophys. Space Sci.** 21: 117-176; 1973. 100 refs, 1905-1973. IV. **Astrophys. Space Sci.** 29: 63-159; 1974. 167 refs, 1939-74.

A series of papers aimed at clarification of the evolutionary history of the solar system. Parts I and II deal with mechanical processes, III and IV with plasma processes and the hydromagnetic aspects.

107.002 Alfvén, Hannes; Arrhenius, Gustaf. **Evolution of the Solar System.** Washington, DC: NASA; 1976. 599p. (NASA SP-345). 27 chapters. 442 refs, 1883-1975.

An advanced text emphasizing the likely chemical and physical processes involved and the interrelation between the two areas. The volume is divided into five parts: present state and basic laws; the accretion of celestial bodies; plasma and condensation; physical and chemical structure of the solar system; special problems. A detailed table of physical and orbital data for each planet and its satellites is included.

107.003 Alfvén, Hannes; Arrhenius, Gustaf. **Structure and Evolutionary History of the Solar System.** Dordrecht, Holland; Boston: D. Reidel; 1975. 276p. (Geophys. Astrophys. Monogr., 5). 4 chapters w/refs.

Based on papers originally published in *Astrophysics and Space Science*, this volume examines existing theories of cosmogony in light of both physical and chemical considerations. Topics: general principles and observational facts (orbits, small bodies, resonance structure, tides, etc.); accretion of celestial bodies (formation, spin, accretion of planets and satellites); the plasma phase (plasma physics and hetegony, models); chemical differentiation (chemical composition, meteorites, mass distribution and critical velocity, group structure).

107.004 Arrhenius, G. Chemical aspects of the formation of the solar system. In: Dermott, S. F., ed. **The Origin of the Solar System.** Chichester, UK: John Wiley & Sons; 1978: 521-581. 145 refs, 1935-78.

A survey of "some of the ways in which modern knowledge of processes in space can be used as a basis for reconstruction of the formation of the solar system." Among the many subjects covered in this detailed and thorough paper are chemical differentiation aspects, nucleation and growth of interstellar grains, effects of temperature, and a number of chemical and physical processes.

107.005 Barshay, S. S.; Lewis, J. S. Chemistry of primitive solar material. **Annu. Rev. Astron. Astrophys.** 14: 81-94; 1976. 48 refs, 1956-76.

Review of the chemical processes that occurred in the cooler, outer regions of the primitive solar nebula at the time of intimate chemical contact between preplanetary condensate and nebular gas.

107.006 Clayton, R. N. Isotopic anomalies in the early solar system. **Annu. Rev. Nucl. Sci.** 28: 501-522; 1978. 80 refs, 1954-78.

Review of measurements of isotopic abundances of primitive solar system materials, primarily meteorites, in order to show that anomalies of isotopic abundances indicate that the solar nebula was not completely homogenized prior to the formation of the planetary system.

107.007 Conference on protostars and planets. **Moon Planets** 20: 3-101; 1979. 4 papers.
C Planetary Science Institute, University of Arizona, Tucson, AZ, USA: 3-7 Jan 1978. Topics: survey of problems in planetary cosmogony; pre-stellar interstellar dust; grain motions in the solar nebula; tidal theory for the origin of the solar nebula.

107.008 Dermott, S. F., ed. **The Origin of the Solar System.** Chichester, UK; New York:
C John Wiley & Sons; 1978. 668p. 29 papers.

Newcastle upon Tyne, UK: 29 Mar—9 Apr 1976. Review papers examining basic theories of cosmogony as well as the formation and evolution of individual planets. Various observational evidence is also reviewed in support of the various origin theories such as the solar nebula and the capture theory.

107.009 Elvius, A., ed. **From Plasma to Planet.** Stockholm: Almqvist & Wiksell; 1972.
C 389p. 18 papers.

Twenty-first Nobel Symposium. Saltsjöbaden, Sweden: 6-10 Sep 1971. (Proceedings). "A discussion of a number of physical and chemical processes which are likely to have been essential to the formation and evolution of the Solar System." Specific areas covered include the formation of meteorites, comets, and asteroids; cosmic grains; atomic and molecular reactions; chemical studies of the Moon and meteorites; plasma dynamics. Also considered is the role of space exploration in advancing our understanding of the origin of the solar system.

107.010 Gehrels, T., ed. **Protostars and Planets.** Tucson, AZ: University of Arizona
C Press; 1978. 756p. (IAU Colloq., 52). 39 papers.

Tucson, AZ, USA: 3-7 Jan 1978. A sourcebook in cosmogony consisting of review papers from the IAU colloquium. It is divided into seven parts: 1) introduction and overviews; 2) grains and chemistry; 3) clouds and fragmentation; 4) associations and isotopes; 5) protoplanets and planetesimals; 6) cores and stellar winds; 7) glossary, acknowledgments, and index. Contributed papers from the meeting appeared in two issues of *The Moon and Planets* 19: 109-315; 1978 (22 papers), and 20: 3-101; 1979 (4 papers).

107.011 Grossman, L.; Larimer, J. W. Early chemical history of the solar system. **Rev. Geophys. Space Phys.** 12: 71-101; 1974. 193 refs, 1934-74.

Review of recent literature on chemical fractionations during the condensation of the solar system, and on their consequences to the establishment of chemical differences between the different classes of chondrites and between the planets.

107.012 Kirsten, T. Time and the solar system. In: Dermott, S. F., ed. **The Origin of the Solar System.** Chichester, UK: John Wiley & Sons; 1978: 267-346. 235 refs, 1955-77.

"An attempt to demonstrate how radiometric dating has contributed to our present understanding of the origin and evolution of the solar system." Following a summary of the principles and problems in the measurement of elpased time using radiometric dating, the author covers ages of the elements; planetary ages and rock ages (meteorite ages, lunar chronology, and geochronology); and the duration and sequence of solar system formation.

107.013 Kuiper, G. P. On the origin of the solar system. I. **Celestial Mech.** 9: 321-348; 1974. 73 refs, 1895-1973.

A broad review covering protostars, star formation, post-formation dynamics, and evolution.

107.014 Middlehurst, B. M., ed. The origin of the Earth and planets. **Highlights Astron.**
C 2: 191-243; 1971. 3 papers.

Joint discussion during the Fourteenth General Assembly of the IAU. Includes three invited discourses and an open discussion on the origin of the solar nebula; internal constitution and thermal histories of the terrestrial planets; and the internal constitution of the giant planets.

107.015 The Moon and planets. **Phys. Earth Planet. Inter.** 4: 153-287; 1971. 20 papers.
C Eighth NATO Advanced Study Institute held by the School of Physics, the University of Newcastle upon Tyne, UK: 9-16 Apr 1970. Sections include the figure of the Moon, the Moon's surface, Jupiter, meteorites and tektites, and Mars.

107.016 Podosek, F. A. Isotopic structures in solar system materials. **Annu. Rev. Astron. Astrophys.** 16: 293-334; 1978. 67 refs, 1929-78.

Concentrates on the bearing of isotopic studies for understanding the formation and earliest evolution of the solar system and its planets.

107.017 Reeves, H., ed. **Symposium sur l'origine du système solairè.** [Symposium on the
C Origin of the Solar System.] Paris: Edition du Centre National de la Recherche Scientifique; 1972. 383p. 44 papers. In English.

Nice, France: 3-7 Apr 1972. Papers reviewing the various cosmogonic models and reports of results which support, disprove, or do not affect selected model theories. There is an overview chapter on concepts and theories of the origin of the solar system, as well as individual papers on the primitive solar nebula theory, among others on comets, asteroids, chondrites, and individual planets.

107.018 Ringwood, A. E. **Origin of the Earth and Moon.** New York: Springer-Verlag; 1979. 295p. 12 chapters. 626 refs, 1755-1979.

An advanced text reviewing the problem of cosmogony from a geochemical point of view. Divided into three parts: composition and constitution of the Earth; origin of the

Earth; and the Moon and planets, this work presents a detailed examination of the chemical processes that were, or were thought to be, involved in the formation of the Earth and Moon and other terrestrial planets.

107.019 Tai Wen-Sai; Chen Dao-Han. Critical review of theories on the origin of the solar system. **Chinese Astron.** 1: 165-182; 1977. Transl. of *Acta Astron. Sinica.* 17: 93-105; 1976. 85 refs, 1745-1976.

Forty theories on the origin of the solar system are critically reviewed, especially with regard to two fundamental problems: the source of planetary material and the mode of planetary formation.

107.020 Williams, I. P. **The Origin of the Planets.** London: Adam Hilger; 1975. 108p. (Monogr. Astron. Subj.). 5 chapters. 157 refs, 1644-1974.

A review with in-depth interpretation of the various cosmogonic theories. Aimed at students, laypersons, and scientists, the book provides an overview of the planetary system before presenting tidal and related theories, the accretion theories, and processes for the formation of the solar system.

107.021 Williams, I. P. Planetary formation. **IAU Symp.** 65: 3-12; 1974. 47 refs, 1912-73.

Existing theories of planetary formation based on a continuous solar nebula or on discrete objects are briefly reviewed.

107.022 Williams, I. P. A survey of current problems in planetary cosmogony. **Moon Planets** 20: 3-13; 1979. 56 refs, 1916-78.

Current theories for planetary formation are briefly reviewed, with the intention of providing a general survey of the whole problem rather than a discussion of details.

Meteorites: relics from the early solar system. *105.014.*

STARS

111 Parallaxes, Proper Motions, Radial Velocities, Space Motions, Distances

111.001 Abt, Helmut A.; Biggs, Eleanor S. **Bibliography of Stellar Radial Velocities.** Tucson, AZ: Kitt Peak National Observatory; 1972. 502p.

This compilation contains over 44,000 references to the literature which contain descriptive information and numerical data on about 25,000 stars and their measured radial velocities. Arranged in ascending right ascension order, the book contains star name or designation, HR and HD numbers, R.A. and Dec., visual magnitude, spectral class, average velocity (km/sec), bibliographic reference, and notes.

111.002 Breakiron, L. A.; Upgren, A. R. A catalog of parallax stars with MK spectral
A classifications. **Astrophys. J. Suppl. Ser.** 41: 709-741; 1979. 28 refs, 1963-79.

Spectral classes and photometry are presented for 3,264 of the stars listed in the *General Catalogue of Trigonometric Stellar Parallaxes* and its supplement. Data are used to derive absolute magnitudes.

111.003 Eggen, O. J. Catalogs of proper-motion stars. I. Stars brighter than visual mag-
A nitude 15 and with annual proper motion of 1" or more. **Astrophys. J. Suppl. Ser.** 39: 89-101; 1979. 5 refs, 1949-75.

Approximately 400 stars are listed with photometric and astrographic data.

111.004 Giclas, H. L.; Burnham, R., Jr.; Thomas, N. G. Lowell proper motion survey.
A Southern hemisphere catalog 1978. **Lowell Obs. Bull.** 8: 89-144; 1978. 74 refs, 1894-1973.

Summary of the Lowell Proper Motion Survey for the Southern Hemisphere as completed to mid-1978. The catalog gives the position, motion, magnitude, and color of 2,758 stars from the Lowell program.

111.005 Gliese, W.; Jahreiss, H. Nearby star data published 1969-1978. **Astron.**
A **Astrophys. Suppl. Ser.** 38: 423-448; 1979. 83 refs, 1943-78.

From trigonometric parallaxes and spectroscopic and photometric data published since 1969, four lists are compiled: 1) stars with parallaxes $\pi \geq 0''.045$ not contained in the catalog of nearby stars; 2) suspected nearby stars; 3) improved trigonometric parallaxes for 377 GL stars; and 4) additional faint companions to 9 GL stars.

111.006 Harrington, R. S., et al. Fifth catalog of trigonometric parallaxes of faint stars.
A **Publ. United States Naval Obs.** 24(4): 1-33; 1978. 22 refs, 1947-78.

A compilation of relative parallaxes and UBV photometry for 95 stars in 85 systems. Continuation of the first through the fourth catalogs, *Publ. United States Naval Obs.* 20: pts. 3 & 6, and 24: pts. 1 & 3.

111.007 Hube, D. P. The radial velocities of 335 late B-type stars. **Mem. R. Astron.**
A **Soc.** 72: 233-280; 1970. 23 refs, 1953-67.

Individual plate velocities are given along with MK spectral types where available for 335 stars of spectral type B8 or B9 from the *Catalogue of Bright Stars.*

111.008 Luyten, W. J. **LHS Catalogue.** 2nd ed. Minneapolis, MN: University of
A Minnesota; 1979. 100p.

A revised edition of the 1976 catalog of stars with proper motions exceeding $0''.5$ annually.

111.009 Luyten, W. J. **NLTT Catalogue.** Minneapolis, MN: University of Minnesota;
A 1979-80.

"New Luyten Two-Tenths" catalogs. volume I: +90° to +30°; volume II: +30° to 0°; volume III: 0° to -30°; volume IV: -30° to -90°. Material is divided into zones of 10 degrees declination, comparable to the SAO catalog. Includes over 50,000 entries of stars with proper motions exceeding 0.″2 annually.

111.010 Luyten, W. J., ed. **Proper Motions.** Minneapolis, MN: University of Minnesota
C Press; 1970. 200p. (IAU Colloq., 7). 26 papers.

Minneapolis, MN, USA: 21-23 Apr 1970. Reports of research and methods of observation, including a description of a computerized proper motion measuring machine. Topics include proper motion surveys and catalogs.

111.011 Luyten, W. J.; Albers, H. **LHS Atlas.** Minneapolis, MN: University of Minne-
A sota; 1979. 154p.

Finding charts for the five-tenths catalog (*111.008*).

111.012 Mikami, T. Compiled data of C- and M-type stars in solar neighborhood. **Ann.**
A **Tokyo Astron. Obs.** 17: 1-49; 1978. 49 refs, 1922-78.

The physical and kinematical data for 321 carbon stars and 1,490 M-type stars are compiled from various catalogs and papers.

111.013 Radial velocities. **IAU Trans. Rep. Astron.** 17A(2): 165-171; 1979. 18 refs, 1977-80.

An overview of research activity worldwide, with an emphasis on the previous three years, as reported by IAU Commission 30 (Radial Velocities). Additional reports: 14A: 335-342; 1970. 87 refs, 1967-69. 15A: 407-414; 1973. 69 refs, 1956-72. 16A(2): 157-160; 1976. + refs, 1973-76.

111.014 van de Kamp, P. The nearby stars. **Annu. Rev. Astron. Astrophys.** 9: 103-126; 1971. 82 refs, 1910-71.

A review of stars located within 5.2 parsecs, with an emphasis on astrometric aspects. The techniques and methods of long-focus photographic astrometry are discussed and used to derive various properties of this group of objects. A table includes parallax, orbital motions, perturbation, luminosities, and masses for 45 of these stars.

111.015 Vasilevskis, S. Stellar proper motions with reference to galaxies. **Vistas Astron.** 15: 145-160; 1973. 73 refs, 1718-1971.

Review of the study of fundamental proper motions that are affected by errors in precession versus proper motions with reference to galaxies that are free from this effect and should therefore serve better for research in stellar astronomy.

111.016 Woolley, R., et al. Catalogue of stars within twenty-five parsecs of the Sun.
A **R. Obs. Ann.** 5: 1-227; 1970. 175 refs, 1847-1969.

Continuation and extension of Gleise's *Catalogue of Nearby Stars*. The main catalog consists of 1,744 systems within 24pc of the Sun, of which 1,566 have trigonometric parallaxes and 177 have spectroscopic parallaxes.

A catalogue of radial velocities in the Large Magellanic Cloud. *159.001*.

Fundamental systems of positions and proper motions. *041.003*.

112 Circumstellar Matter (Shells, Dust, Masers, Stellar Winds, etc.)

112.001. Cassinelli, J. P. Stellar winds. **Annu. Rev. Astron. Astrophys.** 17: 275-308; 1979. 115 refs, 1956-79.

Review concerned with theories explaining the continual expansion of the outer atmospheric layers of luminous early- and late-type stars. The primary emphasis is on theory and modeling, with a substantial portion devoted to mass loss from evolved late-type stars and stellar winds of early-type stars.

112.002 Coyne, G. V. Polarization in Be stars. **IAU Symp.** 70: 233-260; 1976. 73 refs, 1934-76.

Review of polarization produced in the extended circumstellar disks about Be stars. Emphasis is on recently discovered polarization effects in the emission lines and on a discussion of models.

112.003 Greenberg, J. M.; van de Hulst, H. C., eds. **Interstellar Dust and Related Top-**
C **ics.** Dordrecht, Holland; Boston: D. Reidel; 1973. 584p. (IAU Symp., 52). 76 papers, 3 abstracts.

Albany, NY, USA: 29 May – 2 Jun 1972. Sections include extinction, diffuse features, reflection nebulae and diffuse galactic light, interstellar polarization, distribution of dust and gas, physical processes, molecules, dust and H II regions, and circumstellar dust.

Be and Shell Stars. 064.025.

Models for the circumstellar envelopes of Be stars. *064.018.*

113 Photometric Properties

113.001 Brunet, J. P. UBV photometry for supergiants of the Large Magellanic Cloud. **Astron. Astrophys.** 43: 345-358; 1975. 71 refs, 1953-75.

UBV photometric results from the study of 603 stars in the direction of the Large Magellanic Cloud.

113.002 Canterna, R. Broad-band photometry of G and K stars: the C, M, T_1, T_2 photometric system. **Astron. J.** 81: 228-244; 1976. 70 refs, 1952-76.

Presentation of an alternative broad-band system for accurate photometric abundances of G and K stars.

113.003 Deutschman, W. A.; Davis, R. J.; Schild, R. E. The galactic distribution of
A interstellar absorption as determined from the Celescope catalog of ultraviolet stellar observations and a new catalog of UBV, H-beta photoelectric observations. **Astrophys. J. Suppl. Ser.** 30: 97-225; 1976. 98 refs, 1859-1974.

New UBV data for 2,846 stars and H-beta photometry for 2,099 stars.

113.004 Dubyago, I. A. Catalogue of stellar magnitudes and colour indices of 1906 stars
A in the neighborhood of NGC 6866. **Izv. Astron. Ehngel'gardt. Obs.** 43: 67-87; 1978. 6 refs, 1961-78. In Russian.

113.005 Eggen, O. J. Intermediate-band photometry of late-type stars. IV. The catalog.
A **Astrophys. J. Suppl. Ser.** 37: 251-263; 1978. no refs.

The paper presents a photometric catalog based on some 4,000 (u, v, b, y), 5,000 (R, I), and 1,600 (UBV) observations of some 1,000 stars.

113.006 Epps, E. A. UBV photoelectric observations. I. Stars within 25 parsecs of the Sun. II. Stars in quasar and galaxy fields. III. Stars in Kapteyn selected areas. IV. Miscellaneous stars. **R. Obs. Bull.** 20: 127-145; 1972. 51 refs, 1918-72.

UBV photometry is presented for 115 nearby stars, for bright secondary standards in 10 quasar and 10 galaxy fields, and for stars in 11 Kapteyn selected areas.

113.007 Gronbech, B.; Olsen, E. H. Four-colour uvby photometry for bright O to G0
A type stars south of declination +10°. **Astron. Astrophys. Suppl. Ser.** 25: 213-270; 1976. 32 refs, 1963-76.

Four-color photometry for nearly all stars fainter than $V = 4\overset{m}{.}5$ earlier than G 1 and south of +10° in the *Catalogue of Bright Stars.* Provides information on 2,771 stars.

113.008 Gronbech, B.; Olsen, E. H. Photoelectric Hβ photometry for bright O to G0
A type stars south of declination +10°. **Astron. Astrophys. Suppl. Ser.** 27: 443-462; 1977. 27 refs, 1963-76.

Photoelectric Hβ photometry for nearly all stars fainter than $V = 4\overset{m}{.}5$ earlier than G1 and south of +10° in the *Catalogue of Bright Stars.* β indices are given for 2,742 stars.

113.009 Hauck, B.; Westerlund, B. E., eds. **Problems of Calibration of Absolute Magni-**
C **tudes and Temperatures of Stars.** Dordrecht, Holland; Boston: D. Reidel; 1973. 304p. (IAU Symp., 54). 24 papers, 11 abstracts.

Geneva, Switzerland: 12-15 Sep 1972. Topics include absolute magnitudes from trigonometric and statistical parallaxes and galactic clusters and associations; calibration of spectroscopic parallaxes; absolute magnitude determinations from hydrogen-line photometry; ground-based and extraterrestrial observations of stellar flux; the use of model atmospheres for temperature, stellar temperature scale, and bolometric corrections; and choice of standard stars.

113.010 Jamar, C., et al. **Ultraviolet Bright-Star Spectrophotometric Catalogue. A**
A **Compilation of Absolute Spectrophotometric Data Obtained with the Sky Survey Telescope (S2/68) on the European Astronomical Satellite TD-1.** Paris: European Space Agency; 1976. 489p. (ESA SR-27). 16 refs, 1918-75.

The catalog contains data for 1,356 stars based on 5,330 individual observations. A statistical summary is presented in two tables; one listing the number of stars of each spectral type in various magnitude ranges, and the other listing the number of stars of each spectral type in various luminosity classes. The major portion of the catalog is divided into primary data providing HD number, name, position, spectral type, and so on for each star, and ultraviolet data giving the ultraviolet fluxes and spectrum.

113.011 Knude, J. Intrinsic uvbyβ indices, distances and color excesses of 644 B, A and
A F stars in 63 selected areas. **Astron. Astrophys. Suppl. Ser.** 33: 347-366; 1978. 37 refs, 1938-77.

113.012 Knude, J. K. Photoelectric uvby and Hβ photometry of 750 A and F stars in
A 63 selected areas with $|b| < 30°$. **Astron. Astrophys. Suppl. Ser.** 30: 297-305; 1977. 13 refs, 1929-77.

Catalog stars are spectral type in the range A3-F5, brighter than 9.5 m_{pg}, and close to the galactic plane.

113.013 Landolt, A. U. UBV photoelectric sequences in the celestial equatorial selected
A areas 92-115. **Astron. J.** 78: 959-981; 1973. Also in: **Louisiana State Univ. Obs. Contr.** 87. 52 refs, 1926-72.

UBV photoelectric observations of 642 stars in selected areas 92-115. Nearly all of the stars are within ±1° of the celestial equator.

113.014 Neckel, H. Photoelectric catalogue of 1030 BD M-type stars located along the
A galactic equator. **Astron. Astrophys. Suppl. Ser.** 18: 169-234; 1974. 14 refs, 1951-73.

Presentations of magnitudes and colors (v, B-V) of 1,031 early M-type stars which are located in galactic longitude from 6° to 235° and in latitude between +6° and -6°.

113.015 Nicolet, B. Catalogue of homogeneous data in the UBV photoelectric photo-
A metric system. **Astron. Astrophys. Suppl. Ser.** 34: 1-49; 1978. 11 refs, 1966-78.

Includes the V magnitude and the B-V and U-B color indices for 53,845 stars measured from 1953 through 1975.

113.016 Philip, A. G. D.; Cullen, M. F.; White, R. E. UBV color-magnitude diagrams of
A galactic globular clusters. **Dudley Obs. Rep.** 11: 1-26, 1-186; 1976. 113 refs, 1955-75.

UBV observations of over 37,000 stars were compiled from the literature to yield 165 color-magnitude and color-color diagrams. Included are a list of bibliographical references to the diagrams presented for each cluster and a table of data for 115 clusters summarizing much of the current data concerning globular clusters.

113.017 Philip, A. G. D.; Hayes, D. S., eds. **Multicolor Photometry and the Theoreti-**
C **cal HR Diagram.** Albany, NY: Dudley Observatory; 1975. 523p. (Dudley Obs. Rep., 9). 48 papers.

Albany, NY, USA: 24-27 Oct 1974. (Proceedings). Reports of research on the measurement of stellar radiation using a variety of photometric systems, and the application of the data to different forms of the H-R diagram. Included throughout are results of spectroscopy of many different stars. A list of photometric references for 666 standard stars can be found on pages 503-512.

113.018 Philip, A. G. D.; Perry, C. L. Photometric bibliographies (bibliography of the Strömgren four-color and H Beta systems, 1950-1976). **Vistas Astron.** 22: 279-306; 1978. 775 refs, 1950-76.

Essentially, a culling out of the subject area from *Astronomy and Astrophysics Abstracts*. Papers are separated into the *AAA* subject headings and listed in order of abstract number.

113.019 Rufener, F. Catalogue of the photometric parameters for stars measured in the
A Geneva Observatory System. **Publ. Obs. Genève, Ser. B.** Fasc. 2; 1977. 4+76p. 53 refs, 1906-1976.

The most frequently used photometric parameters and color indices are computed for 4,670 stars. Concise bibliographic information as well as the V magnitude are given.

113.020 Stellar photometry. **IAU Trans. Rep. Astron.** 17A(2): 83-96; 1979. 232 refs, 1977-79.

An overview of research activity worldwide, with an emphasis on the previous three years, as reported by IAU Commission 25 (Stellar Photometry). Topics: narrow-band and intermediate-band photometry; stellar polarimetry; etc. Additional reports: 14A: 231-247; 1970. 119 refs, 1960-70. 15A: 285-295; 1973. - refs, 1970-72. 16A(2): 95-107; 1976. + refs, 1973-76.

113.021 Thompson, G. I., et al. **Catalogue of Stellar Ultraviolet Fluxes. A Compila-**
A **tion of Absolute Stellar Fluxes Measured by the Sky Survey Telescope (S2/68) Aboard the ESRO Satellite TD-1.** London: The Science Research Council; 1978. 449p. 12 refs, 1966-77.

Contains the absolute fluxes in four passbands for 31,215 stars. Stars were selected subject to the constraint that the signal to noise rates should be at least 10.0 in any of the four passbands (centered at 2740Å, 2365Å, 1965Å, and 1565Å). For each star, the catalog provides HD or other identifying number, visual magnitude, spectral type, and absolute flux in each of the passbands.

113.022 Voroshilov, V. I.; Kolesnik, L. N. Catalogue of B, V magnitudes and spectral
A classes of 720 stars centered at $\alpha 1950 = 2^h 17^m.9$, $\delta 1950 = +58°59'$. **Astrometr. Astrofiz.** 33: 21-29; 1977. 9 refs, 1956-76. In Russian.

113.023 Warren, W. H., Jr.; Hesser, J. E. A photometric study of the Orion OB1 asso-
A ciation. I. Observational data. **Astrophys. J. Suppl. Ser.** 34: 115-206; 1977. **(Publ. Goethe Link Obs.** 187). 147 refs, 1935-76.

An extensive catalog of observational data for stars in the region of the young stellar association Orion OB 1. Includes new photoelectric observations on the uvbyβ and UBV systems and a compilation of previous photoelectric and spectroscopic data for all 526 stars in the program.

Catalog of Far-Ultraviolet Objective-Prism Spectrophotometry. 114.017.

Problems of calibration of multicolor photometric systems. *031.525.*

Spectral Classification and Multicolour Photometry. 114.010.

114 Spectra, Temperatures, Chemical Composition, etc.

114.001 Bappu, M. K. V.; Sahade, J., eds. **Wolf-Rayet and High-Temperature Stars.**
C Dordrecht, Holland: D. Reidel; 1973. 263p. (IAU Symp., 49). 13 papers.

Buenos Aires, Argentina: 9-14 Aug 1971. Topics include overviews of observations, classification, distribution, spectra, relationships to other kinds of stars, evolution, spectra theory, and problems and conclusions on the nature and physical structure of Wolf-Rayet stars.

114.002 Berger, J.; Fringant, A.-M. A search for faint blue stars in high galactic lati-
A tudes. I. Nine PSS fields near the North Galactic Pole. **Astron. Astrophys. Suppl. Ser.** 28: 123-152; 1977. 60 refs, 1930-75.

A catalog of 4,431 stars and 84 compact objects found in nine PSS fields scattered around the North Galactic Pole is presented, with the 1,950 positions and estimated magnitudes and color classes.

114.003 Bertaud, C.; Floquet, M. Nouveau catalogue des étoiles a à sprectre particulier
A (Ap) et à raies métalliques (Am). [New catalogue of A stars with peculiar spectra (Ap) and with metallic lines (Am).] **Astron. Astrophys. Suppl. Ser.** 16: 71-153; 1974. 668 refs, 1955-71. In French.

Lists 1,238 stars, of which 1,049 are seen to be well classified and 189 stars of more doubtful classification.

114.004 Boesgaard, A. M. Stellar abundances of lithium, beryllium, and boron. **Publ. Astron. Soc. Pacific** 88: 353-356; 1976. 99 refs, 1940-77.

Concerned with the stellar abundances of Li, Be, and B that are relevant to both the origin of the light elements and to the details of stellar structure.

114.005 Buscombe, W., comp. **MK Spectral Classification. Third General Catalogue.**
A Evanston, IL: Northwestern University; 1977. 263p. no refs.

A computer printout of a file of MK spectral types and UBV photoelectric photometry maintained at the Dearborn Observatory. Arranged in right ascension order, for each star is provided MK spectral type, luminosity class, peculiarity, reference number, v magnitude, B-V color index, U-B color index, reference number for photometry, and notes. Others in the series are the 1974 edition compiled by P. M. Kennedy and W. Buscombe, and the *Fourth General Catalogue*, compiled by W. Buscombe, both published by Northwestern University.

114.006 Code, A. D.; Meade, M. R. Ultraviolet photometry from the Orbiting
A Astronomical Observatory. XXXII. An atlas of ultraviolet stellar spectra. **Astrophys. J. Suppl. Ser.** 39: 195-289; 1979. 25 refs, 1932-79.

Ultraviolet stellar fluxes are presented for 164 bright stars in the spectral region from 1200 to 3600 Å in a graphical and tabular form.

114.007 Cowley, Charles R. **The Theory of Stellar Spectra.** New York; London: Gordon and Breach; 1970. 260p. 6 chapters. 182 refs, 1913-69.

An advanced text for graduate students and scientists, dealing with two broad subjects: the extent to which theory is supported by observation, and a discussion of the modern theories of line broadening. Topics: theory of the line-absorption coefficient; spectral lines in stellar atmospheres; statistical mechanics; quantitative chemical analysis of a stellar atmosphere; theory of line broadening; quantum-mechanical treatment of pressure broadening.

114.008 Cruz-Gonzalez, C., et al. A catalogue of galactic O stars and the ionization of
A the low density interstellar medium by runaway stars. **Rev. Mexicana Astron. Astrofis.** 1: 211-259; 1974. 85 refs, 1943-74.

A catalog of 664 O stars, including visual magnitudes, B-V magnitudes, spectral type, distance, radial velocity, radial component of the peculiar velocity, possible multiplicity, whether inside or outside the faintest H II region detectable on the Palomar Sky Survey, and identification of the H II region where the star is projected.

114.009 Curchod, A.; Hauck, B. Second catalogue of Am stars with known spectral
A types. **Astron. Astrophys. Suppl. Ser.** 38: 449-461; 1979. 70 refs, 1948-79.

Catalog of 1,334 stars with known spectral types with respect to K-line, hydrogen lines, and metallic lines.

114.010 Fehrenbach, Charles Max; Westerlund, Bengt E., eds. **Spectral Classification**
C **and Multicolour Photometry.** Dordrecht, Holland: D. Reidel; 1973. 314p. (IAU Symp., 50). 45 papers, 7 abstracts.

Villa Carlos Paz, Argentina: 18-24 Oct 1971. Sections included classification of slit spectra, classification of objective-prism spectra, photometric classification, and catalogs and documentation.

114.011 Goy, G. Un nouveau catalogue général d'étoiles de type O. [A new general O
A type stars catalogue.] **Astron. Astrophys. Suppl. Ser.** 12: 277-311; 1973. 74 refs, 1927-73. In French.

An up-to-date edition of Hiltner's 1956 catalog. Contains fundamental data (position, identification, spectral type, polarization, H II region membership) for each star.

114.012 Greenstein, J. L.; Sargent, A. I. The nature of faint blue stars in the halo. II. **Astrophys. J. Suppl. Ser.** 28: 157-209; 1974. 125 refs, 1932-74.

Spectra and colors of 189 hot (FB) stars selected colorimetrically and mostly within 30° of the galactic poles are analyzed quantitatively for surface gravity and effective temperatures.

114.013 Hack, M., ed. Stellar abundances and stellar rotation. **Highlights Astron.** 5:
C 809-837; 1980. 5 papers.

Joint commission meeting during the XVIIth General Assembly of the IAU.

114.014 Hack, Margherita; Struve, Otto. **Stellar Spectroscopy: Normal Stars.** Trieste, Italy: Osservatorio Astronomico di Trieste; 1969. 203p. 4 chapters. 235 refs, 1914-68.

A survey for graduate students, providing "the main results obtained in stellar physics through the study of stellar spectra." Topics: stellar atmospheres and methods for the study of stellar spectra; spectral classification; stellar rotation; quantitative analysis of stellar spectra; the HR diagram; chemical composition and models of stellar atmospheres.

114.015 Hack, Margherita; Struve, Otto. **Stellar Spectroscopy: Peculiar Stars.** Trieste, Italy: Osservatorio Astronomico di Trieste; 1970. 317p. 3 chapters. 329 refs, 1913-70.

A review of the characteristics, structure, and spectral classification of certain non-main sequence stars of population I. In particular, the following topics are addressed: emission lines in spectra of hot stars and related problems; novae and other explosive variables; magnetic, metallic-line, and related stars.

114.016 Hauck, B.; Keenan, P. C., eds. **Abundance Effects in Classification.** Dordrecht,
C Holland; Boston: D. Reidel; 1976. 264p. (IAU Symp., 72). 28 papers, 14 abstracts.

Lausanne-Dorigny, Switzerland: 8-11 Jul 1975. Topics include influence of abundances upon stellar atmosphere calculations; derivation of abundances through photometric and spectroscopic methods; abundance effects in spectral classification; abundances in stellar populations; and a catalog of [Fe/H] determinations.

114.017 Henize, K. G., et al. **Catalog of Far-Ultraviolet Objective-Prism Spectro-**
A **photometry: Skylab Experiment S-019, Ultraviolet Stellar Astronomy.** Washington, DC: NASA; 1979. 536p. (NASA RP., 1031). 167 refs, 1958-78.

Ultraviolet stellar spectra at wavelengths from 1300 to 5000 Å were photographed during the three manned Skylab missions using a 15cm aperture objective-prism telescope. Approximately 1,000 spectra representing 500 stars were measured; spectrophotometric results are tabulated from these. Most of the stars are of spectral class B, with a number of O and A stars and a sampling of WG, WN, F, and G types.

114.018 Herbig, G. H., ed. **Spectroscopic Astrophysics: An Assessment of the Contributions of Otto Struve.** Berkeley, CA: University of California Press; 1970. 462p. 20 papers.

A collection of 10 scientific papers by Mr. Struve, reprinted to commemorate his many contributions to the field of spectroscopy. Each paper is accompanied by a commentary on the present status of those same areas, by a colleague who is active in astrophysical spectroscopy. Selected subjects include spectral classification; hydrogen lines in normal spectra; peculiar spectra; interstellar materials; spectroscopic binaries; etc.

114.019 Houk, N. **Michigan Catalogue of Two-dimensional Spectral Types for the HD**
A **Stars. Volume 2. Declinations -53°0 to -40°0.** Ann Arbor: Department of Astronomy, University of Michigan; 1978. 395p. 17 refs, 1918-74.

The second volume of a program of systematic reclassification of the Henry Draper stars on the MK system. Of the 30,400 HD stars in the catalog, 29,555 were classifiable. Previously published types are listed for the 3% not classifiable.

114.020 Houk, N.; Hartoog, M. R.; Cowley, A. P. O stars and supergiants south of
A declination -53.0°. **Astron. J.** 81: 116-121; 1976. 4 refs, 1964-75.

Lists are provided of 750 O stars and 281 supergiants from volume 1 of the *University of Michigan Catalogue of Two-dimensional Spectral Types for the HD Stars.*

114.021 Houk, Nancy; Cowley, Anne P. **University of Michigan Catalogue of Two-**
A **dimensional Spectral Types for the HD Stars. Volume 1. Declinations -90° to -53°0.** Ann Arbor: Department of Astronomy, University of Michigan; 1975. 452p. 27 refs, 1918-75.

Part of a program of systematic reclassification of the HD stars on the MK system. Contains all HD stars south of and including δ_{1900} = -53°0. Of the 36,382 HD stars in the catalog, 34,886 were classifiable. Accompanied by: Houk, N.; Irvine, N. J.; Rosenbush, D. *An Atlas of Objective-Prism Spectra* (Ann Arbor: University of Michigan; 1974), which consists of 82 spectra compiled to accompany the Michigan catalog.

114.022 Houziaux, L., ed. The impact of ultraviolet observations on spectral classifica-
C tion. **Highlights Astron.** 4(2): 277-369; 1977. 14 papers.

Joint discussion no. 7 of the Sixteenth General Assembly of the IAU. Selected topics: spectral classification of early-type stars; spectral classification with objective-prism spectra from Skylab; extreme ultraviolet observations of white dwarfs; etc.

114.023 Houziaux, L.; Butler, H. E., eds. **Ultraviolet Stellar Spectra and Related**
C **Ground-Based Observations.** Dordrecht, Holland: D. Reidel; New York: Springer-Verlag; 1970. 361p. (IAU Symp., 36). 40 papers, 12 abstracts.

Lunteren, The Netherlands: 24-27 Jun 1969. The main purpose of this meeting was to bring together the space astronomers working in the far-ultraviolet, and ground-based astronomers studying related problems. Sections include stellar fluxes, stellar line spectra, and interstellar absorption and emission.

114.024 Jarzebowski, T. Binary systems with an x-ray component. **Postepy Astron.** 23: 33-53; 1975. 93 refs, 1967-74. In Polish.

A review of all hitherto known x-ray binaries (10).

114.025 Jaschek, M.; Jaschek, C. The CNO stars. **Astron. Astrophys.** 36: 401-408; 1974. 49 refs, 1912-74.

Summary of the knowledge of a group of hot stars exhibiting anomalies in the behavior of the elements CNO.

114.026 Jordan, S. D.; Avrett, E. H., eds. **Stellar Chromospheres.** Washington, DC:
C NASA; 1973. 318p. (IAU Colloq., 19). (NASA SP-317). 7 papers.

Greenbelt, MD, USA: 21-24 Feb 1972. Review papers and discussions on the theoretical and observational questions related to the outer layers of stars. There are four parts to the book: 1) spectroscopic diagnostics of chromospheres and the chromospheric energy balance; 2) observational evidence for stellar chromospheres; 3) mechanical heating and its effect on the chromospheric energy balance; 4) variation of chromospheric properties with stellar mass and age.

114.027 Keenan, P. C.; McNeil, R. C. **An Atlas of Spectra of the Cooler Stars, Types**
A **G, K, M, S, and C.** Columbus: Ohio State University Press; 1976. 26p. 32 plates. 29 refs, 1940-77.

Designed to help those actually working in spectral classification to make use of the greater understanding of stellar spectra achieved in the more than 30 years that have elapsed since the appearance of the Yerkes atlas.

114.028 Khozov, G. V. Infrared stars, the review of observational data. **Astrophysics** 12: 468-485; 1976. Transl. of *Astrofizika.* 12: 705-732; 1976. 122 refs, 1937-76.

The basic data of photometric, spectral, and polarization observations of cold stars are considered. Observations performed in optical, infrared, and radio ranges during 1965-75 are included.

114.029 McCarthy, M. F.; Philip, A. G. D.; Coyne, G. V., eds. **Spectral Classification of**
C **the Future.** Vatican City: Vatican Observatory; 1979. 575p. (IAU Colloq., 47). (Ric. Astron., 9, special N°). 45 papers.

Vatican City: 11-15 Jul 1978. Reviews and reports of past research with an emphasis on future developments and results. Topics: the MK classification (criteria and applications); correlation of spectroscopic and photometric data; space spectroscopy; detectors; automatic classification; etc.

114.030 Macau-Hercot, D., et al. **Supplement to the Ultraviolet Bright-Star Spectro-**
A **photometric Catalogue.** A compilation of absolute spectrophotometric data obtained with the Sky Survey Telescope (S2/68) on the European Astronomical Satellite TD-1. Paris: European Space Agency; 1978. 163p. (ESA SR-28). 18 refs, 1918-75.

Supplement to ESA SR-27, 1976. Contains data obtained in 1973 and 1974. Spectra listed are those of stars not seen during the first observation periods (for the main catalog). Contains data for 435 stars.

114.031 McCook, G. P.; Sion, E. M. A catalogue of spectroscopically identified white
A dwarfs. **Villanova Univ. Obs. Contrib.** 2: 1-50; 1977. 82 refs, 1952-77.

Includes white dwarfs with published spectroscopic identification. The aim is to collect the available data on white dwarfs into a general, comprehensive catalog.

114.032 Merrill, K. M.; Ridgway, S. T. Infrared spectroscopy of stars. **Annu. Rev. Astron. Astrophys.** 17: 9-41; 1979. 242 refs, 1963-79.

Current overview covering the application of IR techniques in stellar classification, studies of stellar photospheres, elemental and isotopic abundances, and the nature of remnant and ejected material in the near circumstellar region.

114.033 Mihalas, D. Progress towards an interpretation of stellar spectra. **Astron. J.** 79: 1111-1121; 1974. 94 refs, 1913-74.

Warner lecture for 1974 reviewing recent work on the interpretation of stellar spectra, mainly concerning stars of the earliest spectral types.

114.034 Mihalas, D.; Athay, R. G. The effects of departures from LTE in stellar spectra. **Annu. Rev. Astron. Astrophys.** 11: 187-218; 1973. 156 refs, 1931-73.

Review of recent calculations of stellar spectra by simultaneous solution of transfer equations and equations of statistical equilibrium.

114.035 Morel, M., et al. A catalogue of [Fe/H] determinations. **IAU Symp.** 72: 223-
A 259; 1976. 242 refs, 1948-75.

Catalog resulting from the compilation of published values of iron/hydrogen abundances for 515 stars.

114.036 Morgan, W. W.; Abt, H. A.; Tapscott, J. W. **Revised MK Spectral Atlas for**
A **Stars Earlier than the Sun.** Williams Bay, WI: Yerkes Observatory, University of Chicago; Tucson, AZ: Kitt Peak National Observatory; 1978. 14p. 32 plates. 7 refs, 1931-78.

An improved version of the 1943 *Atlas of Stellar Spectra*, by Morgan, Keenan, and Kellman, with a spectral range extended to the neighborhood of lambda 3500.

114.037 Morgan, W. W.; Keenan, P. C. Spectral classification. **Annu. Rev. Astron. Astrophys.** 11: 29-50; 1973. 49 refs, 1943-73.

Concerned with a reexamination of the validity of the MK system. Gives a revised frame of reference for the early-type stars, and a revision of the MK system for giants and supergiants of classes G, K, and M.

114.038 Pagel, B. E. J. Chemical composition of old stars. **Vistas Astron.** 12: 313-333; 1970. 119 refs, 1914-67.

Review of the theories of element formation derived from studies of abundances in the oldest stars.

114.039 Pagel, B. E. J. Observational evidence for atmospheric chemical composition peculiarities relevant to stellar evolution. **Highlights Astron.** 4(2): 119-135; 1977. 104 refs, 1957-76.

Abundance peculiarities in successive stages of stellar evolution are reviewed.

114.040 Rublev, S. V. The Wolf-Rayet stars. **IAU Symp.** 67: 259-274; 1975. 90 refs, 1929-74.

Discussion of the spectral classification, the absolute magnitude, the position in stellar systems, the physical properties, and the evolution of WR stars.

114.041 Seitter, Waltraut Carola. **Atlas for Objective Prism Spectra.** Bonn: Ferd.
A Dummlers Verlag; Part 1: 1970. 56p. 65 plates. 11 refs, 1939-68. Part 2: 1975. 24p. 61 plates. 8 refs, 1953-75.

A compilation and discussion of objective prism spectra of standard stars taken with three different resolutions. Part 1 presents photographs of spectra with a reciprocal linear dispersion of 240 Å/mm at Hγ; part 2 is a collection of side-by-side spectral photographs taken at 645 and 1280 Å/mm at Hγ. The purpose of these sets is "to link the classification of objective prism spectra to the MK system thus extending spectral standardization to generally smaller dispersions" and "to approach the classification accuracy of the MK system as closely as possible within the instrumental limits." Also known as the *Bonner Spectral Atlas I* and *II* or *Bonner Spektral-Atlas* (German).

114.042 Spectral classifications and multi-band colour indices. **IAU Trans. Rep. Astron.** 17A(2): 235-246; 1979. + refs, 1976-79.

An overview of research activity worldwide, with an emphasis on the previous three years, as reported by IAU Commission 45 (Spectral Classifications and Multi-Band Colour Indices). Additional reports: 14A: 547-558; 1970. no refs. 15A: 697-716; 1973. + refs, 1970-73. 16A(2): 229-240; 1976. + refs, 1973-76.

114.043 Spinrad, H. Abundances in stellar populations. **IAU Symp.** 72: 183-204; 1976. 90 refs, 1922-75.

Stellar abundances are reviewed, with emphasis on large-scale objects that may yield clues to galactic structure and evolution.

114.044 Spinrad, H.; Wing, R. F. Infrared spectra of stars. **Annu. Rev. Astron. Astrophys.** 7: 249-302; 1969. 173 refs, 1924-69.

A review considering atomic lines, molecular spectra, and selected areas of astrophysical interest. Coverage includes the one-micron region, the two-micron region, and the far infrared.

114.045 Stellar spectra. **IAU Trans. Rep. Astron.** 17A(2): 151-163; 1979. + refs, 1976-79.

An overview of research activity worldwide, with an emphasis on the previous three years, as reported by IAU Commission 29 (Stellar Spectra). Additional reports: 14A: 319-333; 1970. + refs, 1967-70. 15A: 387-405; 1973. + refs, 1970-73. 16A(2): 147-156; 1976. + refs, 1976-79.

114.046 Stephenson, C. B. A general catalogue of S stars. **Publ. Warner Swasey Obs.**
A 2: 21-53; 1976. 61 refs, 1935-75.

The catalog contains 741 S stars having known positions of at least roughly the precision of the Henry Draper Catalog. Provides right ascension, declination, magnitude, spectrum, designations from various finding lists, and galactic longitude and latitude for each star.

114.047 Les transitions interdites dans les spectres des astres. [Forbidden transitions in
C stellar spectra.] **Mém. Soc. R. Sci. Liège.** 5e ser. 17: 9-410; 1969. 42 papers, 3 abstracts.

Fifteenth Colloque international d'astrophysique. Liège, Belgium: 24-26 Jun 1968. Reports of recent research activity.

114.048 Underhill, A. B. The interpretation of early-type spectra. **Vistas Astron.** 13: 169-206; 1972. 60 refs, 1918-69.

A review of the spectra of O- and B-stars in terms of physical conditions in the stellar atmosphere and the abundances of the elements.

114.049 Wallerstein, G. The physical properties of carbon stars. **Annu. Rev. Astron. Astrophys.** 11: 115-134; 1973. 87 refs, 1928-73.

Review of the properties of carbon stars, concentrating on developments of the previous 16 years.

114.050 Wallerstein, G.; Conti, P. S. Lithium and Beryllium in stars. **Annu. Rev. Astron. Astrophys.** 7: 99-120; 1969. 88 refs, 1940-69.

A review of research and observations of the two elements in the Sun and other stars. Includes an overview of the processes responsible for their creation in stars.

114.051 Williams, P. M. Stellar compositions from narrow-band photometry—I. Iron abundances in 180 G and K giants. **Mon. Not. R. Astron. Soc.** 153; 171-193; 1971. 78 refs, 1909-1970.

Using narrow-band indices observed with the Cambridge spectrophotometer, red-infrared colors, and independent luminosity estimates, the iron abundances of 180 G and K type giant stars have been determined.

114.052 Wyckoff, S.; Clegg, R. E. S. Molecular spectra of pure S stars. **Mon. Not. R.
A Astron. Soc.** 184: 127-143; 1978. 55 refs, 1932-78.

Includes an atlas of the spectra of pure S stars near minimum light, and a list of the positions of more than 200 molecular band heads.

114.053 Yamashita, Y. The C- classification of the spectra of carbon stars. **Ann. Tokyo Astron. Obs.** 13: 169-217; 1972. 91 refs, 1927-72.

Spectral data on 180 carbon stars is presented, along with a discussion of the various spectral lines and their meaning for certain types of carbon stars. Data include HD and BD numbers, coordinates, visual magnitudes, spectral type, line intensities, etc.

114.054 Yamashita, Yasumasa; Nariai, Kyoji; Norimoto, Yuji. **An Atlas of Representa-
A tive Stellar Spectra.** New York: Halsted Press; 1978. 129p. 47 refs, 1918-76.

The aim of this atlas is to give the basis for the classification of stellar spectra, which has been carried out and will be done at Okayma. The first part consists of 45 plates showing the spectra of MK standard stars, and the second part is 19 plates showing the spectra of peculiar stars of various kinds.

Atmospheres of very late-type stars. *064.029.*

Bonner Spectral Atlas. 114.041.

Catalogue of B, V magnitudes and spectral classes of 720 stars. *113.022.*

Catalogue of Stellar Ultraviolet Fluxes. 113.021.

Compact H II regions and OB star formation. *132.010.*

Problems of Calibration of Absolute Magnitudes and Temperatures of Stars. 113.009.

Spectral Line Broadening by Plasmas. 062.008.

Spectrum Formation in Stars with Steady-State Extended Atmospheres. 064.013.

Ultraviolet Bright-Star Spectrophotometric Catalogue. 113.010.

115 Luminosities, Masses, Diameters, HR and Other Diagrams

115.001 A — Alcaino, G. **Atlas of Galactic Globular Clusters with Colour Magnitude Diagrams.** Santiago: Ediciones Nueva Universidad, Universidad Catolica de Chile; 1973. 108p. 59 refs, 1935-74.

The catalog presents general data for 131 galactic globular clusters and the photometric results for 42 galactic globular clusters for which color-magnitude diagrams on the UBV system are available.

115.002 Henry Norris Russell: bibliography. In: Philip, A. G. D.; DeVorkin, D. H., eds. **In Memory of Henry Norris Russell.** Albany, NY: Dudley Observatory; 1977: 159-170. (Dudley Obs. Rep., 13). 265 refs, 1898-1957.

A complete listing in chronological order of Russell's work. Title, source, volume, first page, and co-authors are given. The volume is the result of sessions I and II of the IAU Symposium no. 80, *The HR Diagram* (*115.004*).

115.003 Larsson-Leander, G. The Hertzsprung-Russell Diagram and stellar ages. **IAU Colloq.** 17: II-1—II-30; 1972[?]. 126 refs, 1925-72.

A review primarily dealing with observational aspects of the use of H-R diagrams to determine stellar ages. The author examines the historical aspects and present-day problems in using the H-R diagram, citing particular stars and stellar characteristics as examples of the difficulty in using this tool to determine stellar ages.

115.004 C — Philip, A. G. D.; Hayes, D. S., eds. **The HR Diagram: The 100th Anniversary of Henry Norris Russell.** Dordrecht, Holland; Boston: D. Reidel; 1978. 474p. (IAU Symp., 80). 64 papers, 17 abstracts.

Washington, DC, USA: 2-5 Nov 1977. Topics include fundamentals; the solar neighborhood; subluminous stars; clusters; horizontal branch; galaxies; and theoretical papers. Sessions I and II of this meeting were published as *In Memory of Henry Norris Russell* in *Dud. Obs. Rep.* no. 13, consisting of 15 papers.

115.005 Scalo, J. M. A composite Hertzsprung-Russell diagram for the peculiar red giants. **Astrophys. J.** 206: 474-489; 1976. 130 refs, 1941-75.

A composite H-R diagram for the peculiar red giants of the disk population is constructed using the available data for stars of types R, N, S, SC, MS, and Ba. Their positions are compared with theoretical evolutionary tracks.

Compiled data of C- and M-type stars in solar neighborhood. *111.012.*

Multicolor Photometry and the Theoretical HR Diagram. 113.017.

116 Magnetic Fields, Polarization, Figure, Rotation, Radio Radiation

116.001 Ferris, G. A. J.; Goldsworthy, F. A., eds. Magnetic problems in astronomy.
C **Q. J. R. Astron. Soc.** 12: 347-446; 1971. 8 papers.
Short review papers presented at a symposium honoring T. G. Cowling on his retirement. Topics include: magnetic fields, magnetohydrodynamic stability problems, time fluctuations of the general magnetic field of the Sun, magnetohydrodynamics of rotating fluids, cosmic rays and interstellar matter, magnetic star theory, comets, and the dynamo problem.

116.002 Fricke, K. J.; Kippenhahn, R. Evolution of rotating stars. **Annu. Rev. Astron. Astrophys.** 10: 45-72; 1972. 155 refs, 1919-71.
Review of general transport mechanisms, instabilities and induced transport mechanisms, and the implications of these processes.

116.003 Fujita, Y., ed. Stellar infrared spectroscopy. **Highlights Astron.** 3: 233-363; 1974. 8 papers.
Joint discussion at the Fifteenth General Assembly of the IAU. Topics: high resolution interferometry of cool stars; high resolution spectra of late-type stars; narrow band photometry; Fourier transform spectrophotometry; etc.

116.004 Heiles, C. The interstellar magnetic field. **Annu. Rev. Astron. Astrophys.** 14: 1-22; 1976. 120 refs, 1951-76.
Summarizes effects of polarization of radiation relevant to the interstellar field. The effects reviewed include Faraday rotation, synchrotron polarization, starlight polarization, and Zeeman splitting.

116.005 Henize, K. G. Observations of southern emission-line stars. **Astrophys. J. Suppl. Ser.** 30: 491-550; 1976. 92 refs, 1921-75.
Catalog of 1,929 stars showing emission at H alpha. Survey covers the southern sky south of -25° to a red limiting magnitude of about 11.0.

116.006 Slettebak, A., ed. **Stellar Rotation.** Dordrecht, Holland: D. Reidel; 1970. 355p.
C (IAU Colloq., 4). 39 papers.
Columbus, OH, USA: 8-11 Sep 1969. (Proceedings). Papers describing the physics, kinematics, and other aspects of stellar rotation. Four major areas are covered: 1) the effects of rotation on stellar interiors and evolution; 2) effects on stellar atmospheres; 3) rotation in binaries, clusters, and special objects; statistics of stellar rotation; 4) rotation of the Sun.

116.007 Strittmatter, P. A. Stellar rotation. **Annu. Rev. Astron. Astrophys.** 7: 665-684; 1969. 91 refs, 1923-68.
Review of the problems involved in the theoretical study of rotating stars.

116.008 Tassoul, Jean-Louis. **Theory of Rotating Stars.** Princeton, NJ: Princeton University Press; 1978. 506p. (Princeton Ser. Astrophys.). 16 chapters w/refs.
An advanced, theoretical text which includes a historical overview, observational evidence, and hundreds of references to the literature. The main thrust, however, is theoretical, with the author emphasizing those topics that may be treated by Newtonian mechanics. Other topics include rotating white dwarfs; stellar magnetism and rotation; and rotation in close binaries.

116.009 Uesugi, A.; Fukuda, I. A catalog of rotational velocities of the stars. **Mem. Fac.**
A **Sci. Kyoto Univ.** 33: 205-250; 1970. Also in: **Kyoto Univ. Inst. Astrophys. Obs. Contr.** 189. 59 refs, 1927-68.

A compilation of 3,951 stars from a variety of sources, using the Slettebak system of observation as the standard for calibration. Each entry includes HD number, spectral type, rotational velocity, and reference source.

116.010 Walter, H. G. Positions of radio stars. **Astron. Astrophys. Suppl. Ser.** 30: 381-386; 1977. 75 refs, 1963-77.

First epoch positions of identified and probable radio stars existent to date are compiled, accompanied by supplemental data on stellar magnitude and radio flux. Tables include a total of 66 true or probable radio stars.

116.011 Weiss, W. W.; Jenkner, H.; Wood, H. J., eds. **Physics of Ap-Stars.** Vienna:
C Universitäts-sternwarte Wien mit Figl-Observatorium f. Astrophysik; 1977. 754p. (IAU Colloq., 32). 64 papers.

Vienna, Austria: 8-11 Sep 1975. Review papers and reports of research on the various aspects of magnetic stars. Topics: stellar structure, peculiar atmospheres, spectroscopic investigations, photometry, observational aspects of magnetic fields, and Am stars.

116.012 Wendker, H. J. A catalogue of radio stars. **Abh. Hamburger Sternw.** 10: 3-41;
A 1978. 222 refs, 1963-78.

A catalog of 607 stars was compiled to provide convenient access to the data on radio stars. The list is complete through December 1977.

Polarization in Be stars. *112.002.*

Stellar abundances and stellar rotation. *114.013.*

117 Close Binaries (Observations, Theory)

117.001 Batten, A. H. Discussion of observations of the flow of matter within binary systems. **IAU Symp.** 51: 1-21; 1973. 90 refs, 1924-73.

Observational determinations of density, dimensions, temperatures, and velocities of circumstellar features are surveyed and discussed, with a view to establishing limiting values that could be useful in any theoretical treatment of circumstellar structure.

117.002 Binnendijk, L. The orbital elements of W Ursae Majoris systems. **Vistas Astron.** 12: 217-256; 1970. 71 refs, 1919-67.

General methods of orbital determination are outlined, with special emphasis on very close binary systems.

117.003 Cester, B., et al. A catalogue of modern light curve synthesis photometric solu-
A tions of close binary systems. **Mem. Soc. Astron. Italiana.** 50: 551-800; 1979. 38 refs, 1912-79.

Includes modern light curve synthesis photometric solutions of 166 close binary systems.

117.004 Close binary stars. **IAU Trans. Rep. Astron.** 17A(2): 211-234; 1979. + refs, 1976-79.

An overview of research activity worldwide, with an emphasis on the previous three years, as reported by IAU Commission 42 (Close Binary Stars). Additional reports: 14A: 491-513; 1970. + refs, 1967-70. 15A: 647-667; 1973. + refs, 1970-73. 16A(2): 207-227; 1976. + refs, 1973-76.

117.005 Eggleton, Peter; Mitton, Simon; Whelan, John, eds. **Structure and Evolution of**
C **Close Binary Systems.** Dordrecht, Holland; Boston: D. Reidel; 1976. 414p. (IAU Symp., 73). 36 papers, 10 abstracts.

Cambridge, UK: 28 Jul–1 Aug 1975. Papers discuss the observational and theoretical aspects of close binary stars.

117.006 Gyldenkerne, K.; West, R. M., eds. **Mass Loss and Evolution in Close Binaries.**
C Copenhagen: Copenhagen University Publications Fund; 1970. 238p. (IAU Colloq., 6). 3 papers.

Elsinore, Denmark: 15-19 Sep 1969. Included are three review papers and four discussion transcripts. Papers: 1) Knowledge of masses and radii of eclipsing binaries and their accuracy (D. M. Popper); 2) Observational "facts" of binary mass loss (R. H. Koch); 3) Close binaries: theoretical computations (B. Paczyński). Includes a binary star index.

117.007 Kopal, Z. Evolution in close binary systems. **Publ. Astron. Soc. Pacific** 83: 521-538; 1971. 60 refs, 1922-71.

Comparison of the consequences of current theories of stellar evolution with known observational aspects of close binary systems.

117.008 Kopal, Z. The Roche model and its applications to close binary systems. **Adv. Astron. Astrophys.** 9: 1-65; 1972. 43 refs, 1849-1971.

A rigorous, comprehensive outline of the theory based on research done on close binary stars. Topics: Roche equipotentials; Roche coordinates; stability of the Roche Model.

117.009 Kopal, Zdeněk. **Dynamics of Close Binary Systems.** Dordrecht, Holland; Boston: D. Reidel; 1978. 510p. (Astrophys. Space Sci. Libr., 68). 8 chapters. 583 refs, 1671-1976.

A thorough and rigorous examination of current knowledge of the theory of dynamical behavior exhibited by close binary stars, and an interpretation of observational data to predict evolutionary trends for these systems. Topics: figures of equilibrium; dynamical tides; generalized rotation; the Roche model; stability of components; origin and evolution of binary systems; etc.

117.010 Larsson-Leander, G., ed. Close binaries and stellar activity. **Highlights Astron.**
C 5: 839-865; 1980. 8 papers.

Joint commission meeting at the Seventeenth General Assembly of the IAU.

117.011 Martynov, D. Y. Close binary stars and their significance for the theory of stellar evolution. **Soviet Phys. Usp.** 15: 786-803; 1973. Transl. of *Usp. Fiz. Nauk*. 108: 701-732; 1972. 170 refs, 1897-1971.

Brief review article.

117.012 Paczynski, B. Evolutionary processes in close binary systems. **Annu. Rev. Astron. Astrophys.** 9: 183-208; 1971. 158 refs, 1941-71.

Presentation of current ideas about the evolution of close binaries, with an emphasis on model computations with mass exchange.

117.013 Plavec, M. J.; Popper, D. M.; Ulrich, R. K., eds. **Close Binary Stars: Observa-**
C **tions and Interpretation.** Dordrecht, Holland; Boston: D. Reidel; 1980. 598p. (IAU Symp., 88). 59 papers, 42 abstracts.

Toronto, ON, Canada: 7-10 Aug 1979. Topics include general reviews of binaries; mass transfer and loss; massive binary systems, Algol and Algols, x-ray binaries, contact binaries, RS Canum Venaticorum stars, cataclysmic variables and polars; plus symbiotics, supergiants, planetaries, and Population II systems.

117.014 Sahade, J. Symbiotic objects. **Mém. Soc. R. Sci. Liège.** 6e ser. 9: 303-318; 1976. 82 refs, 1945-75.

Reviews the work performed on these objects and presents certain aspects relating to individual stars.

117.015 Sahade, Jorge; Wood, Frank Bradshaw. **Interacting Binary Stars.** Oxford, UK: Pergamon Press; 1978. 186p. 781 refs, 1824-1977.

An overview of double star systems whose components are so close "that the evolutionary history of each of the two components at some stage begins to depart appreciably from the evolution of single stars." A comprehensive volume with a complete bibliography, the book covers history, zero-velocity surfaces, photometric analysis, W Ursae Majoris systems, changes in period, evolution, etc.

117.016 Thomas, H.-C. Consequences of mass transfer in close binary systems. **Annu. Rev. Astron. Astrophys.** 15: 127-151; 1977. 164 refs, 1941-77.

Discussion of the theory of stellar structure as applied to binary evolution.

117.017 Żytkow, Anna N., ed. **Nonstationary Evolution of Close Binaries.** Warsaw:
C PWN—Polish Scientific Publishers; 1978. 169p. 14 papers.

Second Symposium of the Problem Commission "Physics and Evolution of Stars." Warsaw, Poland: 20-25 Jun 1977. A review of theory, including stellar models, and observational results. Topics: special problems, x-ray binaries, disks and accretion, eclipsing binaries, novae.

118 Visual Binaries, Multiple Stars, Astrometric Binaries

118.001 Batten, Alan H. **Binary and Multiple Systems of Stars.** Oxford, UK; New York: Pergamon; 1973. 275p. (Int. Ser. Monogr. Nat. Philos., 51). 10 chapters. 539 refs, 1785-1971.

A broad overview, with an emphasis on double stars, for advanced students and nonspecialists, providing basic subject matter. Topics: frequency of binary systems; multiple star systems; periods of binary stars; stellar mass and radius determination; apsidal motion; circumstellar matter in binary systems; evolution and origin of binary systems.

118.002 Heintz, W. D., ed. La cohesion entre les procedes d'observation des etoiles
C doubles visuelles. [Coordination of observing techniques of visual double stars.] **Astrophys. Space Sci.** 11: 1-188; 1971. (IAU Colloq., 5). 23 papers. English/French.

Nice, France: 8-10 Sep 1969. Short review papers and reports of research covering the measurement of visual doubles, astrometry of double stars, spectroscopic observations, photometry, and errors in observation.

118.003 Heintz, W. D., ed. Orbital and physical parameters of double stars. **J. R.**
C **Astron. Soc. Canada** 67: 49-87; 1973. (IAU Colloq., 18). 7 papers.

Swarthmore, PA, USA: 12-15 Apr 1972. Includes seven introductory papers which are summarized or presented in full, followed by summaries of discussion sessions. Topics: orbital parameters; dynamical parallaxes; luminosity-color relation for visual binaries; multiple star systems; photometry and spectroscopy of binaries; etc.

118.004 Heintz, Wulff D. **Double Stars.** Dordrecht, Holland; Boston: D. Reidel; 1978. 174p. (Geophys. Astrophys. Monogr., 15). 56 chapters. 350 refs, 1879-1976.

Suitable for scientists and advanced students, this book provides a brief overview of the many aspects of double and multiple star knowledge and research, covering the literature through late 1976. Topics: orbits; observations; masses; radial velocities; light curves; origin; mass loss; x-ray binaries; etc. (A revised translation of *Dopplesterne*, 1971).

118.005 Huang, S.-S. Extrasolar planetary systems. **Icarus** 18: 339-376; 1973. 218 refs, 1912-73.

Review of work dealing with the occurrence of planetary systems in the universe.

118.006 Huang, S.-S. Occurrence of planetary systems in the universe as a problem in stellar astronomy. **Vistas Astron.** 11: 217-263; 1969. 198 refs, 1923-67.

Review of the interrelationship between double stars, rotating stars, and planetary systems. Special emphasis is placed on the frequency of the occurrence of planetary systems in the universe and how their existence can be inferred.

118.007 Martin, A. R. The detection of extrasolar planetary systems. I. Methods of detection. **J. Brit. Interplanet. Soc.** 27: 643-659; 1974. 41 refs, 1944-73.

Review of various methods which have been or might be used to detect extrasolar planets.

118.008 Martin, A. R. The detection of extrasolar planetary systems. II. Discussion of astrometric results. **J. Brit. Interplanet. Soc.** 27: 881-906; 1974. 48 refs, 1938-74.

Presentation of observational data from previous attempts to detect planets.

118.009 Micrometer measures of double stars. **U.S. Naval Obs. Publ. 2nd Ser.**

A A series of measurements made with several telescopes at the U.S. Naval Observatory. For each double star, the following data are included: ADS number, discoverer's number, right ascension, declination, date of observation, measured position angle, measured separation, position of the eyes, seeing, hour angle, magnifying power, estimated magnitude difference, telescope used. Mean values and notes are also included. The most recent sets of measurements are listed below. Item no. 2 includes citations to sets of measurements before 1966.

1. Walker, R. L., Jr. 18(4); 1966. 37p. 256 stars. 1,013 measurements.
2. Worley, C. E. 18(6); 1967. 133p. 1,164 stars. 4,586 measurements.
3. Walker, R. L., Jr. 22(1); 1969. 55p. 463 stars. 1,965 measurements.
4. Worley, C. E. 22(2); 1971. 136p. 1,343 stars. 5,581 measurements.
5. Walker, R. L., Jr. 22(5); 1972. 66p. 618 stars. 2,326 measurements.
6. Behall, A. L. 24(2); 1976. 39p. 267 stars. 1,514 measurements.
7. Worley, C. E. 24(6); 1978. 186p. 1,980 stars. 7,359 measurements.

118.010 Salukvadze, G. N. A new list of Trapezium-type multiple systems. **Byull.**
A **Abastuman. Astrofiz. Obs.** 49: 39-68; 1978. 15 refs, 1929-74. In Russian.

List of 412 Trapezium systems according to spectral class, based on the Index Catalogue of Visual Double Stars.

118.011 van de Kamp, P. Unseen astrometric companions of stars. **Annu. Rev. Astron. Astrophys.** 13: 295-333; 1975. 64 refs, 1891-1975.

Summary of the available information on unseen companions of stars. A list is given of well-established as well as provisional and uncertain results.

Black holes and neutron stars: evolution of binary systems. *066.508.*

Celestial binary x-ray sources. *142.002.*

Observational parameters and dynamical evolution of multiple stars. *042.006.*

X-ray Binaries. 142.050.

119 Eclipsing Binaries

119.001 C Analytical procedures for eclipsing binary light curves. **Astrophys. Space Sci.** 21: 7-12; 1973. Also in: **Villanova Univ. Obs. Contr.**, no. 1, 1975. (IAU Colloq., 16).

Philadelphia, PA, USA: 8-11 Sep 1971. Only two papers were published from this conference, one in each source above. Respectively, they are "Correlations among Parameters of the Spherical Models for Eclipsing Binaries," by S. Sobieski and J. White; and "Analyses of TX UMa and MR Cyg on the Extended Russell Model," by E. F. Guinan.

119.002 Kopal, Zdeněk. **Language of the Stars: A Discourse on the Theory of the Light Changes of Eclipsing Variables.** Dordrecht, Holland; Boston: D. Reidel; 1979. 280p. (Astrophys. Space Sci. Libr., 77). 7 chapters. 163 refs, 1783-1979.

An advanced text providing "an introduction to the interpretation of the observed light changes of eclipsing binary stars and their analysis for the elements of the respective systems." Topics: spherical stars; close eclipsing systems; special functions of the theory of light curves; analysis of light changes in the time domain and in the frequency domain (spherical and distorted stars); error analysis.

119.003 McCluskey, G. E.; Wood, F. B. Wolf-Rayet-, AO Cassiopeia-, and U Geminorum- stars. **Vistas Astron.** 12: 257-270; 1970. 80 refs, 1918-67.

A summary of observational data and chief theoretical interpretations of three types of eclipsing binaries.

119.004 Szafraniec, R. Henry Norris Russell's contribution to the study of eclipsing variables. **Vistas Astron.** 12: 7-20; 1970. 44 refs, 1899-1966.

A survey of Russell's work, emphasizing theory of orbits; the problem of structure of stellar interiors; variations of periods; and other effects.

119.005 Tsesevich, V. P., ed. **Eclipsing Variable Stars.** Jerusalem: Israel Program for Scientific Translations; distr., New York: John Wiley & Sons; 1973. 310p. (IPST Libr.). 9 chapters w/refs.

A review of current knowledge with emphasis on methods of observation and study. Topics: photometric phases of eclipses; eclipses of spheroidal stars moving in circular orbits; limb darkening of spherical stars; element determination using computers; eclipsing systems with deformed components; unique systems; etc. Translation by R. Hardin of *Zatmennye peremennye zvezdy* (Moscow: Nauka; 1971).

120 Spectroscopic Binaries

120.001 C Batten, A. H., ed. **Extended Atmospheres and Circumstellar Matter in Spectroscopic Binary Systems.** Dordrecht, Holland; Boston: D. Reidel; 1973. 291p. (IAU Symp., 51). 6 papers, 10 discussion sessions.

Struve Memorial Symposium. Parksville, BC, Canada: 6-12 Sep 1972. Topics include observations of the flow of matter within binary systems; gaseous motion around stars; envelopes; stellar spectra related to extended atmospheres; theory of extended and expanding atmospheres; and evolutionary aspects of circumstellar matter in binary systems.

120.002 A Batten, A. H.; Fletcher, J. M.; Mann, P. J. Seventh catalogue of the orbital elements of spectroscopic binary systems. **Publ. Dominion Astrophys. Obs.** 15:

121-295; 1978. Also in: **NRC** no. 16671. 26 refs, 1934-77, plus refs to each of the 978 binaries.

The catalog contains orbital elements for 978 spectroscopic binary systems and critical notes on most of them.

121 Early-stage Stars (T Tauri Stars, Herbig-Haro Objects, etc.)

121.001 C Gebbie, Katharine B.; Thomas, Richard N., eds. **Wolf-Rayet Stars.** Washington, DC: U.S. Department of Commerce, National Bureau of Standards; 1968. 277p. (NBS SP, 307). 4 papers.

Boulder, CO, USA: 10-14 Jun 1968. (Proceedings). Reviews of problem areas, recent research, and transcripts of discussion. Topics: distribution, physical properties and evolutionary status of Wolf-Rayet stars; the detailed features of their spectra; the interpretation of these features and the models on which they are based; and finally a survey of the material and ideas arising out of the symposium itself.

121.002 Klimek, Z.; Kreiner, J. M. The variability of period of Beta Lyrae. **Acta Astron.** 23: 331-365; 1973. 171 refs, 1844-1972.

A comprehensive catalog of the observed minima of Beta Lyrae from the time of discovery up to 1971.

121.003 Wright, K. O. The Zeta Aurigae stars. **Vistas Astron.** 12: 147-182; 1970. 107 refs, 1918-69.

A review of orbits, dimensions, spectra, and outer atmospheres, along with other physical characteristics.

Variable Stars and Stellar Evolution. 065.035.

122 Intrinsic Variables (Pulsating Variables, Spectrum Variables, etc.)

122.001 Baglin, A., et al. Delta Scuti stars. **Astron. Astrophys.** 23: 221-240; 1973. 138 refs, 1926-72.

Observational and theoretical data are collected and analyzed. Properties of pulsation are discussed, and relationships between δ Scuti stars and other types of pulsators and non-variable stars are established.

122.002 C Bateson, F. M.; Smak, J.; Urch, I. H., eds. **Changing Trends in Variable Star Research.** Hamilton, New Zealand: University of Waikato; 1979. 527p. (IAU Colloq., 46). 58 papers.

Hamilton, New Zealand: 27 Nov–1 Dec 1978. Reports of recent research on cataclysmic variables, red variables, symbiotic and flare stars, Cepheids, and a variety of other variables.

122.003 Breger, M. Delta Scuti and related stars. **Publ. Astron. Soc. Pacific** 91: 5-26; 1979. 179 refs, 1935-79.

An extensive review of the current status of knowledge of the stars in the lower instability strip, with an emphasis on current problems. Includes an interesting discussion of the controversial naming of groups of variable stars.

122.004 Cox, J. P. Pulsating stars. **Rep. Prog. Phys.** 37: 563-698; 19074. 622 refs, 1912-74.

A review with the main emphasis on purely radial pulsations of spherically symmetric, non-rotating, non-magnetic stars. Reviews relevant stellar time scales, summarizes main observational data, and presents elements of the basic theory of pulsating stars.

122.005 Cox, J. P. Some recent developments in stellar pulsation theory. **Bull. Astron. Soc. India** 7: 4-11; 1979. 121 refs, 1939-79.

Reviews in the areas of interaction between stellar pulsation and convection; the instability mechanisms for β Cephei stars; Cepheid masses and beat Cepheids; nonlinear Cepheid pulsation calculations; and oscillations of various types.

122.006 Detre, L., ed. **Non-Periodic Phenomena in Variable Stars.** Budapest: Academic;
C Dordrecht, Holland: D. Reidel; 1969. 490p. 67 papers.

Fourth Colloquium on Variable Stars (IAU). Budapest, Hungary: 5-9 Sep 1968. Papers review the theory and research related to irregular changes in variables. Among the subjects covered are statistical and physical interpretations; observational techniques and analysis; intrinsic irregular variables; irregular activity in periodic variables; non-periodic phenomena in binary systems; etc.

122.007 Feast, M. W. The R Coronae Borealis type variables. **IAU Symp.** 67: 129-141; 1975. 77 refs, 1933-74.

A general review of observational data and theory of RCB stars.

122.008 Fernie, J. D., ed. **Variable Stars in Globular Clusters and in Related Systems.**
C Dordrecht, Holland; Boston: D. Reidel; 1973. 234p. (IAU Colloq., 21). 33 papers.

Toronto, ON, Canada: 29-31 Aug 1972. Reports of research and review papers on both theoretical and observational aspects. The volume is divided into four parts: 1) general problems of variables in Population II systems; 2) RR Lyrae variables in Population II systems; 3) slow variables in Population II systems; 4) theoretical considerations of Population II variables.

122.009 Fischel, David; Sparks, Warren M., eds. **Cepheid Modeling.** Washington, DC:
C NASA; 1975. 332p. (NASA SP-383). 10 papers.

Conference and workshop. Greenbelt, MD, USA: 29 July 1974. (Proceedings). A review and comparison of modeling techniques, as well as a report on ultraviolet observations by satellite of classical Cepheids. The second half of the volume is a transcription of the workshop discussions that followed the formal papers.

122.010 Fitch, W. S., ed. The absolute magnitudes of the RR Lyrae stars. **Highlights Astron.** 2: 769-793; 1971. 5 papers.

Short papers presented at a joint meeting of commissions 24, 27, 30, 33, and 37 at the Fourteenth General Assembly of the IAU.

122.011 Fitch, W. S., ed. **Multiple Periodic Variable Stars.** Dordrecht, Holland: D.
C Reidel; 1976. 348p. (Astrophys. Space Sci. Libr., 60). (IAU Colloq., 29). 15 papers.

Budapest, Hungary: 1-5 Sep 1975. Review papers and reports of research on variables which exhibit more than one light curve. Discussed are β Canis Majoris stars; magnetic and Ap variables; Cepheids; RR Lyrae stars; δ Scuti stars; close binaries; etc. Most papers contain a large number of references. Contributed papers published in a separate volume: Fitch, W. S., ed. *Multiple Periodic Variable Stars*. Budapest: Akadémiai Kiadó; 1976. 356p.

122.012 Gershberg, R. E. Flare activity of UV Ceti stars. **Astrophysics** 13: 310-331; 1977. Transl. of *Astrofizika*. 13: 553-589; 1977. 165 refs, 1939-77.

Review of different aspects of the flare activity of red dwarf stars. Three sections: flares, activity outside flares, and interpretation of flare observations and a physical model of flare activity.

122.013 Gershberg, R. E. Some results of the cooperative photometric observations of the UV Cet-type flare stars in the years 1967-71. **Astrophys. Space Sci.** 19: 75-92; 1972. 132 refs, 1967-72.

A list of cooperative photometric observations of the UV Cet-type flare stars that have been organized during the years 1967-71 by the Working Group on Flare Stars of IAU Commission 27.

122.014 Glasby, John S. **The Nebular Variables.** Oxford, UK; New York: Pergamon Press; 1974. 208p. (Int. Ser. Monogr. Nat. Philos., 69). 25 chapters w/refs.

An in-depth examination of those irregular variables associated with bright or dark nebulae: RW Aurigae variables, T Orionis variables, T Tauri variables, and peculiar nebular variables. A large number of bibliographic references are included in this review volume.

122.015 Kippenhahn, R.; Rahe, J.; Strohmeier, W., eds. **The Interaction of Variable Stars with Their Environment.** Bamberg, Germany: Schadel; 1977. 649p. (Veröff. Remeis-Sternwarte Bamberg, Astron. Inst. Univ. Erlangen-Nürnberg; Bd. XI, 121). 65 papers.

Reports of research on stars which receive mass from or eject mass into their environment. Topics: young objects such as T Tauri stars and flare stars; cataclysmic variables such as certain novae; evolved stars such as OH/IR stars and long-period variables.

122.016 Kukarkin, B. V., ed. **Pulsating Stars.** Jerusalem: Israel Program for Scientific Translations; distr., New York: John Wiley & Sons; 1975. 320p. (IPST Astrophys. Libr.). 10 chapters w/refs.

An overview of current knowledge and recent research on intrinsic variable stars. The properties and processes of eight types of pulsating stars are presented in great detail: classical Cepheids; RR Lyrae stars; RV Tauri stars; Delta Scuti stars; dwarf Cepheids; Beta Cephei stars; long-period variables of the Mira Ceti type; and semiregular and irregular variables. Translation by R. Hardin of *Pul'siruiushchie Zvezdy* (Moscow: Nauka; 1970).

122.017 Kunkel, W. E. Solar neighborhood flare stars—a review. **IAU Symp.** 67: 15-46; 1975. 123 refs, 1905-1974.

A review of the astronomical aspects of flare activity, such as where and under what circumstances flare activity is found in the solar neighborhood.

122.018 Lesh, J. R.; Aizenman, M. L. The observational status of the β Cephei stars. **Annu. Rev. Astron. Astrophys.** 16: 215-240; 1978. 140 refs, 1950-79.

Summary of the observational data for β Cephei group stars; discussion of general physical properties; and discussion of results from recently developed observational methods. Emphasis is on work performed between 1970 and 1977.

122.019 Lesh, J. R.; Aizenman, M. L. The statistics of the β Cephei stars. **Astron. Astrophys.** 26: 1-9; 1973. 49 refs, 1953-73.

A statistical study of the Beta Cephei variables drawn from a homogeneous and complete sample of early B stars.

122.020 Orlov, M. Y. Review of observational results on the R Coronae Borealis stars. **Perem. Zvezdy** 19: 501-513; 1975. 47 refs, 1923-73. In Russian.

Basic observational data on the R CrB stars are summarized, with emphasis on the spectral behavior of these stars.

122.021 Pettersen, B. R. Catalogue of flare star data. **Inst. Theor. Astrophys., Blindern-**
A **Oslo, Rep.** 46: 1-22; 1976. 101 refs, 1947-75.

List of 68 flare stars in the solar neighborhood. For each star, there is position, magnitude, spectral type, nine-color photometrical data, kinematical data, and spectroscopic data.

122.022 Robinson, E. L. The structure of cataclysmic variables. **Annu. Rev. Astron. Astrophys.** 14: 119-142; 1976. 137 refs, 1934-76.

Discussion of observations that bear most directly on the structure of cataclysmic variables at minimum light. Emphasis is on developments after 1971.

122.023 Rodono, M. Flare active binary systems. EQ Peg and its brighter companion BD+ 19 5116 A. **Astron. Astrophys.** 66: 175-185; 1978. 98 refs, 1939-76.

An appendix of binary or multiple systems located within 25pc of the Sun and having at least one member of UV Cet- or BY Dra-type. Includes about 60% of the known flare stars in the solar neighborhood.

122.024 Strohmeier, W., ed. **New Directions and New Frontiers in Variable Star**
C **Research.** Bamberg, West Germany: [?]; 1971. 326p. (IAU Colloq., 15). (Kleine Veröff. Remeis Sternwarte Bamberg, 9, no. 100). 44 papers.

Fifth Colloquium on Variable Stars. Bamberg, West Germany: 31 Aug—3 Sep 1971. Reports on recent research and future areas of study. Topics: I) New phenomena in very cool, or very young, or very peculiar variables (polarization, microwave emission, other processes); II) Flare stars; III) Rapid and ultrarapid variables (including x-ray sources); IV) Duplicity and its consequences among the intrinsic variable stars (including eruptive variables and contact binaries); V) Variable star observations from outside the Earth's atmosphere.

122.025 Strohmeier, W. (Meadows, A. J., ed.). **Variable Stars.** Oxford, UK; New York: Pergamon Press; 1972. 279p. (Int. Ser. Monogr. Nat. Philos., 50). 9 chapters. 744 refs, 1899-1971.

A broad overview of the various classes of variable stars and the sources of their brightness fluctuation, along with a review of recent studies. Aimed at students, educated laypersons, and nonspecialists, the book covers variability in the form of lower-energy outbursts; variability in young stars; variability due to pulsation; semiregular and irregular variability; variability with extensive convection; variability due to geometrical and physical factors; variability and magnetism; variability of an entire galaxy. The emphasis is on intrinsic variables.

122.026 Tsesevich, V. P. **RR Lyrae Stars.** Jerusalem: Israel Program for Scientific Translations; 1969. 357p. 5 chapters. 378 refs, 1902-1963.

Containing a wealth of data, this volume summarizes half a million magnitude observations of RR Lyrae stars carried out primarily in the Soviet Union. Topics: physical properties; stable and non-stable oscillations; regular variation of periods; irregular variation of periods; extreme and strange stars. Data on many individual variables and subgroups of stars are presented. Translation by Z. Lerman of *Zvezdy tipa RR Liry* (Kiev: Naukova Dumka; 1966).

122.027 Variable stars. **IAU Trans. Rep. Astron.** 17A(2): 103-149; 1979. + refs, 1976-79.

An overview of research activity worldwide, with an emphasis on the previous three years, as reported by IAU Commission 27 (Variable Stars). Topics: flare stars; Cepheids; novae; supernovae; RR Lyrae variables; binary x-ray stars; etc. Additional reports: 14A: 259-299; 1970. 52 refs, 1968-70. 15A: 313-356; 1973. + refs, 1970-73. 16A(2): 115-146; 1976. + refs, 1973-76.

122.028 Warner, B. Observations and interpretation of cataclysmic variables. **IAU Colloq.** 46: 1-23; 1979. 120 refs, 1954-78.

A review of recent observations relating to the structure and mechanisms of cataclysmic variables, such as novae and dwarf novae. Rapid pulsations and x-ray observations are addressed, too.

Variable Stars and Stellar Evolution. 065.035.

123 Variable Stars (Surveys, Lists of Observations, Charts, etc.)

123.001 Cacciari, C.; Renzini, A. A graphical catalogue of RR Lyrae variables. **Astron.**
A **Astrophys. Suppl. Ser.** 25: 303-363; 1976. 32 refs, 1935-75.

The catalog gives the period-amplitude diagram, the period frequency distribution, and the period cumulative diagram for the RR Lyrae variables in the galactic globular clusters, in the solar neighborhood, in some samples situated near the galactic center, and in some extragalactic systems.

123.002 Clube, S. V. M.; Evans, D. S.; Jones, D. H. P. Observations of southern RR
A Lyrae stars. **Mem. R. Astron. Soc.** 72: 101-184; 1969. 32 refs, 1938-68.

Contains photometric and spectroscopic data for 62 RR Lyrae variables brighter than B = 14 at minimum.

123.003 Haro, G. An observational approach to stellar evolution. I. Flare stars and related objects. **Bol. Inst. Tonantzintla** 2: 3-54; 1976. 114 refs, 1935-75.

Review of 20 years of flare star observations in order to develop theories about their evolution. Concentration is on the stellar aggregates in Orion and NGC 2264 and flare stars in the Pleiades region.

123.004 Heck, A.; Lakaye, J. M. A bibliographical catalogue of field RR Lyrae stars (magnetic tape). **Astron. Astrophys. Suppl. Ser.** 30: 397-398; 1977. 8 refs, 1958-77.

A brief description of the catalog, which is available on magnetic tape and microfiche from the Strasbourg Stellar Data Center. The tape includes 6,607 bibliographical references (1972-76) for 5,855 field RR Lyrae stars.

123.005 Kukarkin, B. V., et al. **Obshchii katalog peremennykh zvezd.** [General Catalog
A of Variable Stars.] 3rd ed. Moscow: Akademiia Nauk SSSR; 1969-71. 3v. 5,216 refs, 1843-1968. In Russian.

The most authoritative and extensive compilation ever constructed on the subject, this source contains data on 20,437 variables. The result of years of careful observations, the work's entries include star name or designation, 1900.0 coordinates, galactic coordinates, bibliographic information, type of variability, maximum and minimum magnitudes, period, spectrum, and much more. The following supplements have been published: 1 (1971), 2 (1974), 3 (1976); special supplements: 1 (1974) and 2 (1976).

123.006 Payne-Gaposchkin, C. The development of our knowledge of variable stars. **Annu. Rev. Astron. Astrophys.** 16: 1-13; 1978. no refs.

Survey of the changes that the subject of variable stars has undergone from 1918 to the present.

123.007 Van Houten, C. J., comp. Index of variable stars: 1921-1969. **Bull. Astron. Inst. Netherlands** 20: 395-405; 1969.

A cumulative index arranged by constellation designation (e.g., SY Aur or DI CrA).

General Catalog of Variable Stars. 123.005.

Stellar Instability and Evolution. 065.022.

124 Novae

124.001 Arhipova, V. P.; Mustel, E. R. Novae. **IAU Symp.** 67: 305-333; 1975. 105 refs, 1920-74.

A review, with emphasis on interpretation of observational data, structure of the principal envelopes, infrared emission, and the problems of magnetic fields in novae.

124.002 de Jager, C., ed. The outer layers of novae and supernovae. **Highlights Astron.**
C 3: 499-574; 1974. 7 papers.

Joint discussion at the Fourteenth General Assembly of the IAU. Topics: chemical composition of novae envelopes; spectrophotometry of supernovae; supernova remnants; x-ray emission from supernova remnants; etc.

124.003 Friedjung, M., ed. **Novae and Related Stars.** Dordrecht, Holland; Boston: D.
C Reidel; 1977. 228p. (Astrophys. Space Sci. Libr., 65). 11 papers.

Paris, France: 7-9 Sep 1976. (Proceedings). A combination of review papers and brief reports of research on novae. Topics: novae, dwarf novae, and similar objects at minimum light; observations of novae and related objects during outburst; nebular stage of novae; transient x-ray sources; observations of Nova Cygni 1975; theories of the causes of outbursts.

124.004 Gallagher, J. S.; Starrfield, S. Theory and observations of classical novae. **Annu. Rev. Astron. Astrophys.** 16: 171-214; 1978. 260 refs, 1920-78.

Review concentrating on the physical nature and eruptive behavior of the classical nova outburst.

124.005 Mallama, A. D.; Trimble, V. L. Novae versus dwarf novae: energy sources and systematics. **Q. J. R. Astron. Soc.** 19: 430-441; 1978. 61 refs, 1934-78.

Review of observational and theoretical work on classical novae and dwarf novae that bears on the question of why binary systems with very similar component stars and orbits should have different outburst behavior.

124.006 Payne-Gaposchkin, C. Past and future novae. In: Friedjung, M., ed. **Novae and Related Stars.** Dordrecht, Holland: D. Reidel; 1977: 3-32. 173 refs, 1921-76.

A survey of the various classes of novae and their physical characteristics. Novae in general are also discussed: distribution, spectra, variation in brightness, binary systems. There are several useful tables in the book, including a list of new novae to be added to existing catalogs.

124.007 Robinson, E. L. Pre-eruption light curves of novae. **Astron. J.** 80: 515-524; 1975. 107 refs, 1912-74.

A list of 33 novae on which one or more measurements of their pre-eruption magnitudes have been made. The catalog was compiled from all of the published pre-eruption photometry of novae.

124.008 Warner, B. Observations of dwarf novae. **IAU Symp.** 73: 85-140; 1976. 263 refs, 1922-75.

Review of spectroscopic and photometric observations relevant to the structure of classical novae, recurrent novae, and dwarf novae. Details are given of optical, infrared, satellite ultraviolet, x-ray, and radio observations.

125 Supernovae, Supernova Remnants

125.001 Apparao, K. M. V. The Crab Nebula. **Astrophys. Space Sci.** 25: 3-116; 1973. 400 refs, 1818-1973.

Extensive review of the present knowledge of the nebula and the pulsar.

125.002 Brancazio, Peter J.; Cameron, A. G. W., eds. **Supernovae and Their Remnants.**
C New York: Gordon and Breach; 1969. 240p. 13 papers.

Conference on Supernovae. New York, NY, USA: 2-3 Nov 1967. (Proceedings). Reports of observational research, including surveys and theoretical studies. Selected topics: general properties; hydrodynamics; exploding star models; neutron stars; x-ray emission; the Crab Nebula.

125.003 Chevalier, R. A. The interaction of supernovae with the interstellar medium. **Annu. Rev. Astron. Astrophys.** 15: 175-196; 1977. 123 refs, 1934-77.

This review concentrates on the theoretical understanding of the interaction of the energy release in the supernova explosion with the interstellar medium. Observations of individual supernova remnants are not comprehensively reviewed.

125.004 Clark, D. H.; Caswell, J. L. A study of galactic supernova remnants, based on Molonglo-Parkes observational data. **Mon. Not. R. Astron. Soc.** 174: 267-305; 1976. 130 refs, 1958-75.

General properties of SNRs are reviewed from a statistical analysis of observational data. The first large-scale study of SNRs using a near-homogeneous data set. Includes a catalog of 97 supernova remnants south of $+18°$ and of 23 north of $+18°$.

125.005 Clark, David H.; Stephenson, F. Richard. **The Historical Supernovae.** Oxford, UK; New York: Pergamon Press; 1977. 233p. (Pergamon Int. Libr.). 12 chapters. 211 refs, 1723-1976.

An interdisciplinary study aimed at astronomy and history of science students, sinologists, and nonspecialists. Using historical records, including those of the Far East, the authors illustrate the effects of supernovae throughout the years on science (astronomy, in particular) and culture. A good blend of historical and scientific data, comparing observations from different parts of the globe.

125.006 Colgate, S. A. Supernovae. In: Gursky, Herbert; Ruffini, Remo, eds. **Neutron Stars, Black Holes and Binary X-Ray Sources.** Dordrecht, Holland: D. Reidel; 1975: 13-27. 34 refs, 1920-74.

A review covering physical characteristics, origin, the explosion process, neutron star formation, binding energies, neutrinos, etc.

125.007 Cosmovici, C. B.; D'Anna, E.; Borghesi, A., eds. **Supernovae and Supernova**
C **Remnants.** Dordrecht, Holland; Boston: D. Reidel; 1974. 387p. (Astrophys. Space Sci. Libr., 45). 46 papers.

International Conference on Supernovae. Lecce, Italy: 7-11 May 1973. (Proceedings). Reports of research and reviews of various aspects of the supernova phenomenon. Included are 1) results and techniques of supernova surveys; 2) photometric studies of supernovae; 3) spectra of supernovae and their interpretation; 4) statistics of supernovae; 5) supernova remnants; 6) theories on supernovae and supernova remnants. Also includes a list of 378 supernovae discovered since 1885 (pp. 41-47).

125.008 Crab Nebula symposium. **Publ. Astron. Soc. Pacific** 82: 375-564; 1970. 10
C papers.

Flagstaff, AZ, USA: 18-21 Jun 1969. Topics: optical studies; radio and x-ray observations; the pulsar; photometric optical observations of the central star; etc.

125.009 Danziger, I. J.; Renzini, A., eds. Supernovae and supernova remnants. **Mem.**
C **Soc. Astron. Italiana** 49: 299-628; 1978. 35 papers.

First Workshop of the Advanced School of Astronomy, "E. Majorana" Centre for Scientific Culture, Erice, Italy: 16-27 May 1978. Topics range from model precursor stars, nucleosynthesis, pulsars, cosmic rays, supernova types and statistics, to theories and observations of supernova remnants of different evolutionary ages.

125.010 Davies, Rodney Deane; Smith, Francis Graham, eds. **The Crab Nebula.**
C Dordrecht, Holland: D. Reidel; 1971. 470p. (IAU Symp., 46). 58 papers, 11 abstracts.

Jodrell Bank, UK: 5-7 Aug 1970. Sessions included observations of the Crab Nebula and the Crab Pulsar, observations of other pulsars, relation of the Crab Nebula to other supernova remnants, physics of the Crab Nebula and of the neutron star, energy considerations, and radiation mechanisms.

125.011 Flin, P., et al. Catalogue of supernovae. **Acta Cosmologica** 8: 5-296; 1979.
A Contains information on supernovae, discovered and published through the end of 1976. The article is divided into five parts: 1) list of supernovae; 2) list of suspected and false supernovae; 3) cluster membership of parent galaxies; 4) bibliography; and 5) checklists. The bibliography section is most extensive, and lists articles published about each supernovae, from ancient manuscripts through 1976.

125.012 Gorenstein, P.; Tucker, W. H. Supernova remnants. In: Giacconi, Riccardo; Gursky, Herbert, eds. **X-ray Astronomy.** Dordrecht, Holland: D. Reidel; 1974: 267-297. 64 refs, 1964-74.

A review of the observed x-ray features of supernova remnants. The Crab Nebula, the Cygnus Loop, and other objects are examined in terms of x-ray emissions.

125.013 Maza, J.; van den Bergh, S. Statistics of extragalactic supernovae. **Astrophys. J.** 204: 519-529; 1976. 95 refs, 1896-1974.

Reclassification on the DDO system of all galaxies in which supernovae are known to have occurred, followed by comparisons and discussion of supernova frequency and galaxy type and luminosity.

125.014 Milne, D. K. A new catalogue of galactic SNRs corrected for distance from
A the galactic plane. **Australian J. Phys.** 32: 83-92; 1979. 29 refs, 1950-79.

A new catalog of 125 galactic supernova remnants at 1 GHz is presented.

125.015 Mitton, Simon. **The Crab Nebula.** New York: Scribner; 1979, c1978; London: Faber and Faber; 1979. 194p. 11 chapters. 32 refs, 1921-78.

A non-mathematical overview for students, laypersons, and nonspecialists. Topics: historical background; discovery; structure; energy flow; the pulsar; magnetic fields; observations; supernovae remnants; future study.

125.016 Oke, J. B.; Searle, L. The spectra of supernovae. **Annu. Rev. Astron. Astrophys.** 12: 315-329; 1974. 80 refs, 1926-73.

Concentration on the literature concerned with the description, classification, and interpretation of the spectra of supernovae.

125.017 Schramm, David N., ed. **Supernovae.** Dordrecht, Holland; Boston: D. Reidel;
C 1977. 192p. (Astrophys. Space Sci. Libr., 66). 16 papers.

Special IAU Session on Supernovae. Grenoble, France: 1 Sep 1976. (Proceedings). A review of theory and observation of all aspects of the phenomenon. Papers address supernova remnants, evolution, statistics, nucleosynthesis, explosions of supernovae; etc.

125.018 Tammann, G. A. On the frequency of supernovae as a function of the integral properties of intermediate and late type spiral galaxies. **Astron. Astrophys.** 8: 458-475; 1970. 97 refs, 1898-1970.

Consideration of the frequency of supernovae as a function of the integral properties of some favorable types of galaxies, e.g., Sb and Sc spirals.

125.019 Tammann, G. A. Statistics of supernovae remnants. In: Cosmovici, C. B.; D'Anna, E.; Borghesi, A., eds. **Supernovae and Supernova Remnants.** Dordrecht, Holland: D. Reidel; 1974: 155-185. 97 refs, 1942-74.

"An attempt to determine within a clearly defined sample volume the SN frequency as a function of two principal parameters: the type and the luminosity (or mass) of the parent galaxy...." The population, overall rate, and distribution of SNe are discussed, too, as is the supernova rate in the Milky Way. The sample and selection effects are discussed at some length.

125.020 van den Bergh, S.; Marscher, A. P.; Terzian, Y. An optical atlas of galactic
A supernova remnants. **Astrophys. J. Suppl. Ser.** 26: 19-36, plates 1-24; 1973. 128 refs, 1937-73.

Presents photographs of 23 of the 24 known optical supernova remnants. A tentative classification scheme for these objects is proposed.

125.021 Woltjer, L. Supernova remnants. **Annu. Rev. Astron. Astrophys.** 10: 129-158; 1972. 139 refs, 1942-72.

Review of 43 specific SNRs, giving distance information, kinematical and other data on filamentary nebulae, references to the highest resolution radio maps, and information on x-rays and radio polarization.

The outer layers of novae and supernovae. *124.002.*

Papers presented at the joint Australia-U.S.S.R.-U.S.A. symposium on pulsars and supernova remnants. *141.513.*

Special IAU Session on Supernovae. *125.017.*

X-rays from supernovae remnants. *142.015.*

126 Low-luminosity Stars, Subdwarfs, White Dwarfs, Degenerate Stars

126.001 Anderson, P. H. Subdwarfs. **J. R. Astron. Soc. Canada.** 65: 119-128; 1971. 104 refs, 1926-70.

This review, prepared for a master's thesis, cites recent articles on spectra and photometry, kinematics, nucleosynthesis, models, atmospheres, and composition.

126.002 Angel, J. R. P. Magnetic white dwarfs. **Annu. Rev. Astron. Astrophys.** 16: 487-519; 1978. 145 refs, 1931-78.

List of 16 currently known magnetic white dwarfs, with summary of the observations and literature for each. Includes discussion of origin and evolution of magnetic fields within white dwarfs; review of polarization and spectral changes caused by magnetic fields at their surface; and an overview of the effects of magnetic fields on accretion in white dwarf binary systems.

126.003 Glasby, John S. **The Dwarf Novae.** New York: American Elsevier; 1970. 293p. 15 chapters w/refs.

An in-depth study of one small class of variable stars which are short-period binary systems consisting of a hot white dwarf and a much cooler and slightly more massive late-type main sequence companion. Examples are U Geminorum variables and Z Camelopardalis variables. Topics: discovery, light variations, spectra, masses, relationship to other variables, evolution, etc.

126.004 Greenstein, J. L. Subluminous stars: post-novae, white dwarfs and hot subdwarfs. **Mém. Soc. R. Sci. Liège.** 6e ser. 9: 247-267; 1976. 59 refs, 1954-75.

A review of the spectra and evolution of subluminous stars.

126.005 Kumar, S. S., ed. **Low-Luminosity Stars.** New York: Gordon and Breach; 1969.
C 542p. 46 papers.

Charlottesville, VA, USA: 28-30 Mar 1968. (Proceedings). A discussion of the M dwarfs and other similar stars below the main sequence, the most numerous stars in our galaxy. Topics: subluminous stars; energy output; pulsars; evolution; composition; populations; model atmospheres; physical processes; etc.

126.006 Lockwood, G. Wesley; Dyck, H. Melvin, eds. **Proceedings of the Conference**
C **on Late-type Stars.** Tucson, AZ: Kitt Peak National Observatory; 1971. 247p. (Contrib. Kitt Peak Natl. Obs., 554). 23 papers.

Tucson, AZ, USA: 27-28 Oct 1970. (Proceedings). Papers reporting on recent research, including observational results, with an emphasis on spectroscopic and photometric studies. Many papers later appeared in the journal literature.

126.007 Luyten, W. J. **The Stars of Low Luminosity.** Minneapolis, MN: University of
A Minnesota; 1977. 75p.

A compilation of data for 2,907 stars. A revised edition of a 1970 listing.

126.008 Luyten, W. J. **White Dwarfs II.** Minneapolis, MN: University of Minnesota;
A 1977. 103p.

A continuation of a catalog first published in 1970. Data is given for 3,513 stars.

126.009 Ostriker, J. P. Recent developments in the theory of degenerate dwarfs. **Annu. Rev. Astron. Astrophys.** 9: 353-366; 1971. 94 refs, 1931-71.

A summary addressing equilibrium and dynamics, oscillations and stability, and evolution and thermodynamics.

126.010 Sanduleak, N. Catalog of probable dwarf stars of type M3 and later in the direc-
A tion of the north galactic pole. **Astron. J.** 81: 350-363; 1976. 49 refs, 1939-76.

A catalog, including finding charts, is presented which contains 273 faint stars of spectral type M3 and later, located in an area 190 square degrees near the north galactic pole. A comprehensive review of the controversy concerning the luminosity function and space motions of very low mass stars is given.

126.011 Shipman, H. L. Masses and radii of white-dwarf stars. III. Results for 110 hydrogen-rich and 28 helium-rich stars. **Astrophys. J.** 228: 240-256; 1979. 83 refs, 1961-79.

Provides masses and radii for a large sample of white-dwarf stars, with radii listed in tables and shown in a figure.

126.012 Shulov, O. S. Magnetic fields in white dwarfs. **Astrophysics** 11: 108-124; 1975. Transl. of *Astrofizika.* 11: 163-192; 1975. 98 refs, 1947-75.

Review of quadratic Zeeman effects, circular polarization in spectral lines and continuum, and the interpretation of the effects. A table summarizes the observations of circular polarization in white dwarfs as of the end of 1973.

126.013 Van Horn, H. M.; Weidemann, V., eds. **White Dwarfs and Variable Degenerate**
C **Stars.** Rochester, NY: University of Rochester; 1979. 534p. (IAU Colloq., 53). 78 papers, 34 abstracts.

Rochester, NY, USA: 30 Jul–2 Aug 1979. A series of papers describing current research on single white dwarfs and white dwarfs in cataclysmic binary systems. Topics: evolution of white dwarfs; EUV objects and hot white dwarfs; cool white dwarfs; distribution of white dwarfs; magnetic white dwarfs and AM Her stars; oscillating white dwarfs; white dwarfs in close binary systems. The book's abstracts are from a workshop (following the conference) with the title "Novae, Dwarf Novae, and Other Cataclysmic Variables" (Rochester, NY: Aug 3, 1979).

126.014 Weidemann, V. White dwarfs: composition, mass budget and galactic evolution. In: Baschek, B.; Kegel, W. H.; Traving, G., eds. **Problems in Stellar Atmospheres and Envelopes.** New York: Springer-Verlag; 1975: 173-203. 145 refs, 1938-75.

A review of current research and results. Topics: atmospheres; composition of interiors and envelopes; interpretation of atmospheric composition differences; mass and evolution.

A catalogue of spectroscopically identified white dwarfs. *114.031.*

White Dwarfs. 065.023.

INTERSTELLAR MATTER, NEBULAE

131 Interstellar Matter, Star Formation

131.001 Aannestad, P. A.; Purcell, E. M. Interstellar grains. **Annu. Rev. Astron. Astrophys.** 11: 309-362; 1973. 287 refs, 1949-73.

Emphasis is on observations bearing on the nature and distribution of interstellar dust.

131.002 Andrew, B. H., ed. **Interstellar Molecules.** Dordrecht, Holland; Boston: D.
C Reidel; 1980. 704p. (IAU Symp., 87). 56 papers, 84 abstracts.

Mont Tremblant, PQ, Canada: 6-10 Aug 1979. Topics include characteristics, infrared measurements, formation, physical and dynamical conditions in interstellar clouds, galactic distributions, masers, and more. Included are an index of astronomical objects and a molecule index.

131.003 Andriesse, C. D. Radiating cosmic dust. **Vistas Astron.** 21: 107-190; 1977. 28 refs, 1954-76.

Review of radiating interstellar dust that originated from a seminar on the role of solid particles in the infrared emission from the interstellar medium.

131.004 Balian, R.; Encrenaz, P.; Lequeux, J., eds. **Atomic and Molecular Physics and**
C **the Interstellar Matter.** Amsterdam: North-Holland; distr., New York: American Elsevier; 1975. 632p. 11 papers.

Twenty-sixth Session of the Les Houches Summer School in Theoretical Physics. Grenoble, France: 1 Jul—23 Aug 1974. (Proceedings). Lectures providing a detailed description of the basic physical and chemical processes occurring in the interstellar gas and dust, giving thorough and current accounts of the present status of our observational and theoretical knowledge. Topics: atomic physics; molecular spectroscopy and collisional excitation; physics of fully ionized regions; etc.

131.005 Bandermann, L. W. Interplanetary dust. In: Ogelman, H.; Wayland, J. R., eds. **Lectures in High Energy Physics.** Washington, DC: NASA; 1969: 137-165. 148 refs, 1893-1969.

A brief review of the evidence for, the dynamics of, and the origin of interplanetary dust.

131.006 Beaudet, R. A.; Poynter, R. L. Microwave spectra of molecules of astrophysical interest. XII. Hydroxyl radical. **J. Phys. Chem. Ref. Data.** 7: 311-362; 1978. 447 refs, 1937-76.

Available data on the microwave spectrum of the hydroxyl radical are critically reviewed for information applicable to radio astronomy.

131.007 Brown, R. D. Interstellar molecules, galactochemistry and the origin of life. **Interdisciplinary Sci. Rev.** 2: 124-139; 1977. 149 refs, 1834-1975.

Elementary review of the state of the art of galactochemistry.

131.008 Brownlee, D. E. Microparticle studies by sampling techniques. In: McDonnell, J. A. M., ed. **Cosmic Dust.** Chichester, UK: John Wiley & Sons; 1978: 295-336. 101 refs, 1950-77.

A review of collection methods and study techniques of cosmic dust particles, and a presentation of data from selected experiments. Topics: space collections; atmospheric collection; surface collections; future experiments; classification of particulate matter; analytical techniques and contamination problems.

131.009 Bussoletti, E.; Borghesi, A. Cold clouds and interstellar grains. **Mem. Soc. Astron. Italiana** 47: 125-158; 1976. 97 refs, 1927-75.

Cold clouds distribution in galactic space is presented, and chemical composition and physical properties of the clouds are discussed.

131.010 Bystrova, N. V.; Rakhimov, I. A. Pulkovskii obzor neba v radiolinii
A mezzvezdnogo neitral'noga vodoroda. [Pulkovo sky survey in the interstellar neutral hydrogen radio line.] I. Neitral'nyi vodorod v okrestnostrh zvezdnyh associacii λ Oriona i Edinorog I. [Neutral hydrogen in the neighborhood of stellar associations λ Orionis and Monoceros I.] 11 refs, 1956-74. II. Neitral'nyi vodorod v raione petli II. [Neutral hydrogen in the region of Loop II.] no refs. III. Nabliudeniia v zone $-29° \leq \delta \leq 40°$. Atlas krivyh prohozdeniia $T_A(\alpha)'$ V, δ. [Observations for the zone $-29° \leq \delta \leq +40°$. Atlas of drift scans $T_A(\alpha)'$ V, δ.] 8 refs, 1960-76. Parts I and II: **Astrofiz. Issled. Izv. Spets. Astrofiz. Obs.** 7: 70-95; 1975. Part III: monograph—Leningrad: Spetsial'noi Astrofizicheskoi Observatoriia; 1977. 4p. + 58p. of atlas.

131.011 Cahn, J. H. Interstellar extinction: a calibration by planetary nebulae. **Astron. J.** 81: 407-418; 1976. 86 refs, 1942-75.

From currently available data, a catalog of temperature-dependent absolute Hβ extinctions and distances of planetary nebulae has been prepared. These data have been used to transform FitzGerald's three-dimensional color excess distribution to one in absolute absorption at the frequency of Hβ (λ4861).

131.012 Carruthers, G. R. Visible and ultraviolet observations of the interstellar medium. In: Pinkau, K., ed. **The Interstellar Medium.** Dordrecht, Holland: D. Reidel; 1974: 29-68. 50 refs, 1951-73.

The author considers both absorption and emission spectroscopy as well as the role of interstellar dust in these observations. Besides observations and results, the paper considers the physical processes involved in the production of emission and absorption line spectra.

131.013 Dalgarno, A. The interstellar molecules CH and CH^+. In Burke, P. G.; Moiseiwitsch, B. L., eds. **Atomic Processes and Applications.** Amsterdam: North-Holland; 1976: 109-132. 94 refs, 1937-75.

A review of the evidence for and abundances of CH and CH^+ in the galaxy, along with a detailed description of the formation and destruction mechanisms for the molecules.

131.014 Dalgarno, A.; Black, J. H. Molecule formation in the interstellar gas. **Rep. Prog. Phys.** 39: 573-612; 1976. 265 refs, 1926-76.

Review of the chemical mechanisms that lead to the formation and destruction of molecules in the interstellar gas.

131.015 De, B. R.; Arrhenius, G., eds. Proceedings of a workshop on thermodynamics
C and kinetics of dust formation in the space medium. **Astrophys. Space Sci.** 65: 1-240, 259-395; 1979. Also in: **Lunar Planet. Inst. Contr.** 360. 27 papers.

Houston, TX, USA: 6-8 Sep 1978. Conceived as a means of interaction between experimentalists dealing with the past record in meteorites and observers of present-day processes in space in order to outline the boundary conditions for physical parameters at condensation, and to give observational and theoretical support for interpretation of past and present-day phenomena. The volume of abstracts appeared as *Lunar Planet. Inst. Contr.* 330.

131.016 de Jong, Teive; Maeder, A., eds. **Star Formation.** Dordrecht, Holland; Boston:
C D. Reidel; 1977. 296p. (IAU Symp., 75). 8 papers.

Geneva, Switzerland: 6-10 Sep 1976. Topics include molecular clouds, kinematics and dynamics of dense clouds, infrared observations, radio observations, early stages of formation, theoretical processes, and collapse dynamics and models.

131.017 Dilworth, C. The interstellar medium in x- and gamma-ray astronomy. In: Pinkau, K., ed. **The Interstellar Medium.** Dordrecht, Holland: D. Reidel; 1974: 69-89. 36 refs, 1955-73.

A review paper characterizing the wavelength regions into which x- and gamma-rays fall, and discussing the general types of investigations being carried out and recent research results.

131.018 Field, G. B. Heating and ionization of the interstellar medium: star formation. In: Balian, R.; Encrenaz, P.; Lequeux, J., eds. **Atomic and Molecular Physics and the Interstellar Matter.** Amsterdam: North-Holland; 1975: 467-531. 128 refs, 1915-75.

A review of the physical processes occurring in the interstellar medium which lead to star formation. Topics: gas temperature; ionization; heating and cooling mechanisms; equilibrium; steady-state molecules; self-gravitating clouds; etc.

131.019 Field, G. B.; Cameron, A. G. W., eds. **The Dusty Universe.** New York: Neale
C Watson; 1975. 323p. 13 papers.

Cambridge, MA, USA: 17-19 Oct 1973. A survey of the properties of interstellar and interplanetary dust, and an attempt to explore possible cosmogonic relationships between the two phenomena. Topics: dust in stellar atmospheres; composition of interstellar dust; cometary debris; chondritic meteorites; etc. The symposium was held to honor Fred L. Whipple on his retirement as director of the Smithsonian Astrophysical Observatory.

131.020 Gordon, M. A.; Snyder, Lewis E., eds. **Molecules in the Galactic Environment.**
C New York: Wiley Interscience, John Wiley & Sons; 1973. 475p. 35 papers.

Charlottesville, VA, USA: 4-7 Nov 1971. A review, excluding observational results, of the study of molecular astronomy, covering the following topics: the interstellar medium; observations of molecules; techniques of molecular spectroscopy; physics of molecular extinction; chemistry relevant to the astronomical environment; possible biological relevance.

131.021 Gorenstein, P. Interstellar medium. In: Giacconi, Riccardo; Gursky, Herbert, eds. **X-ray Astronomy.** Dordrecht, Holland: D. Reidel; 1974: 299-319. 20 refs, 1965-74.

A review of the effects, mostly detrimental, of the interstellar medium on x-ray observations.

131.022 Greenberg, J. M. Interstellar dust. In: McDonnell, J. A. M., ed. **Cosmic Dust.** Chichester, UK: John Wiley & Sons; 1978: 187-294. 229 refs, 1934-77.

A review which considers possible origins and composition of interstellar grains. Observational data and theory are combined to infer a grain model that is probably most correct at present. Topics: basic observations; the interstellar medium; basic scattering relationships; the basic grain model; physical problems; photochemistry of dust grains; evolution of dust; infrared emission by galactic dust.

131.023 Heiles, C. Physical conditions and chemical constitution of dark clouds. **Annu. Rev. Astron. Astrophys.** 9: 293-322; 1971. 144 refs, 1927-71.

Consideration of the densest of the relatively large dust clouds, which contain enough molecules to be observable in the radio region of the spectrum. Sections include the chemical composition of the gas, physical conditions, and formation of the molecules.

131.024 Huffman, D. R. Interstellar grains. The interaction of light with a small-particle system. **Adv. Phys.** 26: 129-230; 1977. 177 refs, 1884-1977.

Survey of the observational information about interstellar solids and small particle optical properties as they relate to interstellar grains. Special emphasis is given to a discussion of the 39 unidentified interstellar bands in the visible spectral region.

131.025 Interstellar dust. **Astron. Nachr.** 293: 1-77; 1971. (IAU Colloq., 3). 16 papers.
C Jena, Germany: Aug 1969. Short reports of research on dust grains, interstellar bands, and dust formation.

131.026 Interstellar matter and planetary nebulae. **IAU Trans. Rep. Astron.** 17A(3): 73-114; 1979. + refs, 1975-78.

An overview of research activity worldwide, with an emphasis on the previous three years, as reported by IAU Commission 34 (Interstellar Matter and Planetary Nebulae). Topics: supernovae remnants; H II regions; neutral hydrogen; interstellar absorption; interstellar molecules; interstellar grains; etc. Additional reports: 14A: 387-404; 1970. 15 refs, 1967-69. 15A: 467-506; 1973. + refs, 1969-73. 16A(3): 73-115; 1976. + refs, 1972-75.

131.027 Kaplan, S. A.; Pikel'ner, S. B. **The Interstellar Medium.** Cambridge, MA: Harvard University Press; 1970. 465p. 25 chapters. 694 refs, 1927-67.

A comprehensive, advanced monograph covering the five major aspects of the subject: interstellar hydrogen; the physical state of the interstellar gas; interstellar dust; interstellar magnetic fields and nonthermal radio emission; interstellar gas dynamics and the evolution of the interstellar medium. Physical processes are emphasized, while data and observational techniques are not. Translation of *Mezhzvezdnaya sreda* (Moscow: State Publishing House of Physics-Mathematics Literature; 1963).

131.028 Kegel, W. H. Natural masers: maser emission from cosmic objects. **Appl. Phys.** 9: 1-10; 1976. 59 refs, 1964-75.

Brief review of some properties of cosmic masers, in particular OH, H_2O, and SiO sources. Emphasis is given to problems of radiative transfer in the maser line as well as in other lines of the masing molecule.

131.029 Kleinmann, S. G.; Dickinson, D. F.; Sargent, D. G. Stellar H_2O masers.
A **Astron. J.** 83: 1206-1213; 1978. 50 refs, 1970-78.

Includes an updated catalog of the 82 known H_2O/IR stars.

131.030 Larson, R. B. Processes in collapsing interstellar clouds. **Annu. Rev. Astron. Astrophys.** 11: 219-238; 1973. 57 refs, 1954-73.

Review of recent progress in the development of theoretical models for collapsing clouds, along with observations thereof.

131.031 Lequeux, J. Kinematics and dynamics of dense clouds. **IAU Symp.** 75: 69-94; 1977. 106 refs, 1969-77.

Review of the observational evidences for collapse in dense clouds and for the factors which can play against collapse (turbulence, rotation, magnetic field).

131.032 Lequeux, J. Large-scale distribution of interstellar matter in the galaxy. In: Pinkau, K., ed. **The Interstellar Medium.** Dordrecht, Holland: D. Reidel; 1974: 191-218. 106 refs, 1965-73.

A discussion of current knowledge, along with a consideration of the inherent difficulties in obtaining accurate distance information, data needed to correctly describe the distribution of interstellar matter.

131.033 Lerche, I. Wave phenomena in the interstellar plasma. **Adv. Plasma Phys.** 2: 47-138; 1969. 178 refs, 1935-68.

Discusses some of the basic theoretical dynamical properties of small amplitude perturbations in the interstellar medium and some of the consequences which follow from the existence of such waves.

131.034 Litvak, M. M. Coherent molecular radiation. **Annu. Rev. Astron. Astrophys.** 12: 97-112; 1974. 75 refs, 1954-74.

Discussion of the nature of cosmic maser sources. Topics include non-thermal populations, pumping models, apparent maser size, signal bandwidth, and polarization.

131.035 Lynds, Beverly T., ed. **Dark Nebulae, Globules, and Protostars.** Tucson: University of Arizona Press; 1971. 150p. 13 papers.
C

Tucson, AZ, USA: 26-27 Mar 1970. A collection of papers on galactic astronomy dedicated to Bart Bok and his work. Topics: star counts; dark clouds; globules; polarimetry; 21-cm studies of galactic dust and gas; interstellar molecules; galactic infrared sources; etc.

131.036 McCray, R.; Snow, T. P., Jr. The violent interstellar medium. **Annu. Rev. Astron. Astrophys.** 17: 213-240; 1979. 174 refs, 1890-1979.

An overview of the observational evidence for high-velocity and high-temperature gas as a major part of the interstellar medium. Accompanied by a detailed theoretical interpretation of the physical processes in the non-quiet ISM, the paper also addresses the difficulties in constructing a satisfactory global model of the interstellar medium.

131.037 McDonnell, J. A. M., ed. **Cosmic Dust.** Chichester, UK; New York: John Wiley & Sons; 1978. 693p. 9 papers.

A collection of lengthy review papers summarizing and bringing up to date our understanding of the many aspects of the phenomena. Aimed at advanced students, the book could also serve as a reference work for researchers in the field, since it covers the many facets of the detection, study, and interpretation of the cosmic dust family. Topics: comets; meteors; zodiacal light; interstellar dust; microparticle studies; lunar and planetary impact erosion; particle dynamics; laboratory simulation.

131.038 McDonnell, J. A. M. Microparticle studies by space instrumentation. In: McDonnell, J. A. M., ed. **Cosmic Dust.** Chichester, UK: John Wiley & Sons; 1978: 337-426. 144 refs, 1876-1977.

An overview of the detection and study of cosmic dust performed by artificial satellites, along with studies of lunar samples retrieved by astronauts. A wide variety of data from a number of specifically cited spacecraft missions is presented. Topics: detection techniques in space; measurements in space; characteristics of the cosmic dust complex measured by spacecraft; sources and sinks of microparticles in the solar system.

131.039 McNally, D., ed. Interstellar molecules. **Highlights Astron.** 2: 333-462; 1971. 15 papers.
C

Joint discussion at the Fourteenth General Assembly of the IAU. Ten of the papers review the systematics and the occurrence of the molecules identified thus far, and five are on the competing ideas on the origins of these molecules. In addition, there is a panel discussion led by C. H. Townes on excitation mechanisms.

131.040 Mann, A. P. C.; Williams, D. A. A list of interstellar molecules. **Nature** 283: 721-725; 1980. 163 refs, 1940-79.

A table of 90 interstellar molecules as of September 1979. Includes chemical formula and name, references, sources, spectral line, spectrum, column density, conditions, and comments.

131.041 Martin, P. G. The nature of dust grains. In: Van Woerden, H., ed. **Topics in Interstellar Matter.** Dordrecht, Holland: D. Reidel; 1977: 149-154. 72 refs, 1973-76.

A very brief review of recent investigations. Topics: dust to gas ratio; distance determinations; physical properties; spectra; scattering theory; etc.

131.042 Martin, Peter G. **Cosmic Dust: Its Impact on Astronomy.** Oxford, UK: Clarendon Press, Oxford University Press; 1978. 266p. 12 chapters w/refs.

An advanced text providing a summary of current research and recent results. Subjects addressed include a historical overview and description of the phenomena; a description of its three major features (effects on transmission of radiation in the interstellar medium, role in scattering, and its thermal emission); frequency and location of dust particles; composition; etc.

131.043 Mathis, J. S., ed. Helium in the universe. **Highlights Astron.** 2: 245-331; 1971. 8
C papers.

Joint discussion during the Fourteenth General Assembly of the IAU. The first session was devoted to abundance of helium in various objects, and the second session to what these imply about stellar, galactic, or universal evolution.

131.044 Mezger, P. G. Interstellar matter. In: Setti, G., ed. **Structure and Evolution of Galaxies.** Dordrecht, Holland: D. Reidel; 1975: 143-177. 60 refs, 1955-74.

A short summary of observational results pertaining to interstellar matter and radiation fields. Also covered are selected areas in which recent progress has been made: abundances of light elements, nature of dust grains, clouds and intercloud gas. H II regions and the efficiency of star formation in the galaxy conclude the paper.

131.045 Morton, D. C. The interpretation of space observations of stars and interstellar matter. **Highlights Astron.** 2: 466-475; 1971. 39 refs, 1959-71.

Review of space observations of the interstellar medium and stars, and what can be expected to be learned from instruments planned for the near future.

131.046 Ney, E. P. Star dust. **Science** 195: 541-546; 1977. 58 refs, 1934-76.

A brief, elementary discussion of interstellar dust and some of its possible origins.

131.047 Pinkau, K., ed. **The Interstellar Medium.** Dordrecht, Holland; Boston: D.
C Reidel; 1974. 298p. (NATO Adv. Stud. Inst. Ser., Ser. C., Math. Phys. Sci., 6). 13 papers.

NATO Advanced Study Institute. Schliersee, Germany: 2-13 Apr 1973. A review volume including descriptions of observations in the visible, radio, ultraviolet, x-ray, and gamma-ray portions of the spectrum. Also addressed are H II regions, molecules in dense interstellar clouds, interstellar dust, distribution, star formation, etc.

131.048 Rickett, B. J. Interstellar scattering and scintillation of radio waves. **Annu. Rev. Astron. Astrophys.** 15: 479-504; 1977. 78 refs, 1956-77.

Review of interstellar scattering and scintillation in the light of modern theory for extended spatially homogeneous, power law inhomogeneities.

131.049 Robinson, B. J. Molecular astronomy. **Proc. Astron. Soc. Australia** 3: 12-19; 1976. 67 refs, 1960-76.

Short review of radio observations of molecules, including tables of observed molecules.

131.050 Rosenberg, J., ed. International conference on laboratory astrophysics.
C **Physica** 41: 1-223; 1969. 25 papers.

Lunteren, The Netherlands: 2-7 Sep 1968. Discussions of molecular spectroscopy, physical properties of interstellar matter, and recent developments in astronomical spectroscopic instrumentation.

131.051 Rydbeck, O. E. H., et al. Radio observations of interstellar CH. I. **Astrophys. J. Suppl. Ser.** 31: 333-415; 1976. 316 refs, 1928-75.
The paper contains the results of more than eight months of almost continuous, high-sensitivity observations of the three hyperfine transitions in the $^2\Pi_{1/2}$, $J=1/2$ state of CH.

131.052 Salpeter, E. E. Formation and destruction of dust grains. **Annu. Rev. Astron. Astrophys.** 15: 267-293; 1977. 215 refs, 1946-77.
Review of interstellar dust grains, concentrating on very recent work describing element abundances, chemical physics of the processes, optical properties of grains, circumstellar dust, and the interstellar medium.

131.053 Salpeter, E. E. Planetary nebulae, supernovae remnants, and the interstellar medium. **Astrophys. J.** 206: 673-678; 1976. 57 refs, 1921-76.
Formation and destruction of interstellar dust grains are reviewed. Text of the Henry Norris Russell Lecture, Gainesville, FL, December 11, 1974.

131.054 Savage, B. D.; Mathis, J. S. Observed properties of interstellar dust. **Annu. Rev. Astron. Astrophys.** 17: 73-111; 1979. 216 refs, 1941-79.
Review covering the following topics: interstellar extinction, distribution of dust and dust-to-gas ratio, light scattering by grains, interstellar polarization, heavy element depletion, thermal emission from grains, diffuse interstellar features, and composition of grains.

131.055 Solomon, P. M.; Edmunds, M. G., eds. **Giant Molecular Clouds in the Galaxy.**
C Oxford, UK; New York: Pergamon Press; 1980. 344p. 32 papers.
Third Gregynog Astrophysics Workshop. Cardiff, Wales, UK: Aug 1977. Reviews and reports of research on Giant Molecular Clouds (GMCs), including theoretical and observational work. Topics: radio and millimeter observations; infrared properties; the galactic center; kinematics; maser sources in molecular clouds; H II regions and molecular clouds; star formation; interstellar chemistry; the search for planetary systems.

131.056 Somerville, W. B. Interstellar radio spectrum lines. **Rep. Prog. Phys.** 40: 483-565; 1977. 505 refs, 1932-77.
Covers all aspects of the microwave spectrum line seen from interstellar atoms and molecules. Discussion is concerned with general principles rather than with detailed results for individual sources.

131.057 Spitzer, L., Jr.; Jenkins, E. B. Ultraviolet studies of the interstellar gas. **Annu. Rev. Astron. Astrophys.** 13: 133-164; 1975. 163 refs, 1946-75.
Review of ultraviolet line observations by the Copernicus satellite. Topics include atomic hydrogen, H_2 molecules and other interstellar molecules, abundances of atoms in H I regions, properties of the ionized gas, ionization, and thermal equilibrium.

131.058 Spitzer, Lyman, Jr. **Diffuse Matter in Space.** New York: John Wiley & Sons; 1968. 262p. (Intersci. Tracts Phys. Astron.). 6 chapters. 224 refs, 1904-1968.
A short graduate text primarily concerned with the physics of interstellar dust and gas. Topics: observations of gas and dust grains; interactions among interstellar particles; dynamics of interstellar gas; the role of the interstellar medium in star formation; etc.

131.059 Spitzer, Lyman, Jr. **Physical Processes in the Interstellar Medium.** New York; Chichester, UK: John Wiley & Sons; 1978. 318p. 13 chapters. 369 refs, 1935-77.

A graduate text providing an overview of interstellar dust and gas, emphasizing physical processes and deemphasizing observational data. Topics: overview; elastic collisions and kinetic equilibrium; radiative processes; excitation; ionization and dissociation; kinetic temperature; optical properties of grains; polarization and grain alignment; physical properties of grains; dynamical principles; overall equilibrium; explosive motions; gravitational motion.

131.060 Strom, S. E.; Strom, K. M.; Grasdalen, G. L. Young stellar objects and dark interstellar clouds. **Annu. Rev. Astron. Astrophys.** 13: 187-216; 1975. 87 refs, 1951-75.

Discussion of current observational knowledge, primarily optical and infrared, of pre-main-sequence evolution and the interaction of PMS stars with their dark cloud environment.

131.061 Symposium on solid state astrophysics. **Astrophys. Space Sci.** 34: 3-229;
C 1975. 22 papers.

University College, Cardiff, Wales, UK: 9-12 Jul 1974. Selected topics: interstellar absorption and extinction; diffuse absorption features; dust grains; optical properties of particulates; neutron star cores; etc.

131.062 ter Haar, D.; Pelling, M. A. Interstellar hydroxyl and water masers and formaldehyde masers and dasars. **Rep. Prog. Phys.** 37: 481-561; 1974. 307 refs, 1930-73.

A review concerned with the anomalous emission and absorption by hydroxyl, water, and formaldehyde molecules.

131.063 Thaddeus, P. Molecular clouds. **IAU Symp.** 75: 37-54; 1977. 75 refs, 1964-77.

Discussion of how molecular observations provide data on the physical state of the dense interstellar gas, and the molecular observations of H II regions, stellar associations, and dark nebulae.

131.064 Thomas, G. E. The interstellar wind and its influence on the interplanetary environment. **Annu. Rev. Earth Planet. Sci.** 6: 173-204; 1978. 99 refs, 1962-78.

An examination of the interaction of the solar wind and the local interstellar gas, with an emphasis on those properties of the gas that have been measured, as opposed to those that are mainly of theoretical interest.

131.065 Turner, B. E. General physical characteristics of the interstellar molecular gas. **IAU Symp.** 84: 257-270; 1979. 73 refs, 1953-78.

Short review containing sections on morphology of the interstellar medium, physical properties of the components, heating and cooling of clouds, energetics and evolution, and molecular clouds and the large-scale characteristics of the galaxy.

131.066 Turner, B. E. Interstellar molecules. In: Verschuur, Gerrit L.; Kellermann, Kenneth I., eds. **Galactic and Extragalactic Radio Astronomy.** New York: Springer-Verlag; 1974: 199-255. 124 refs, 1943-74.

A review of the 24 known (in 1974) interstellar molecule species, 21 of which have been observed at radio wavelengths. Topics: observational data; distribution; physical conditions; microwave molecular observations; missing molecules; formation and destruction of interstellar molecules; etc.

131.067 Turner, B. E. Interstellar molecules—a review of recent developments. **J. R. Astron. Soc. Canada** 68: 55-88; 1974. 62 refs, 1956-74.

Presented at the Symposium on Chemical Evolution in the Universe, Kingston, ON, Canada, June 6, 1973.

131.068 Turner, B. E. A survey of OH near the galactic plane. **Astron. Astrophys. Suppl. Ser.** 37: 1-332; 1979. 55 refs, 1946-78.

Survey of all four 18-cm lines of OH near the galactic plane. A total of 1,766 positions are included.

131.069 van den Bergh, S. A preliminary classification scheme for interstellar absorbing clouds. **Vistas Astron.** 13: 265-277; 1972. 21 refs, 1919-67.

Intended to be a starting point for further study, this article presents a scheme primarily based on the appearance of the edges of dark clouds. Included is a table of 424 dark nebulae and their classifications. Distribution of the clouds is also considered.

131.070 Van Woerden, Hugo. **Topics in Interstellar Matter.** Dordrecht, Holland;
C Boston: D. Reidel; 1977. 295p. (Astrophys. Space Sci. Libr., 70). 25 papers.

Meeting of IAU Commission 34 (Interstellar Matter) at the Sixteenth General Assembly of the IAU. Grenoble, France: Aug 1976. Invited review papers reporting on the many aspects of the subject. Five major areas are covered: the hot interstellar gas phase; interaction of stars and interstellar medium; interstellar molecules and dust; large-scale distribution of interstellar matter in the galaxy; interstellar matter in the galaxy; interstellar matter in external galaxies. Selected topics: soft x-ray background; compact H II regions; planetary nebulae; dust grains; gamma-ray astrophysics; galactic warps; etc.

131.071 Vanýsek, V. Reflection nebulae and the nature of interstellar grains. **Vistas Astron.** 11: 189-216; 1969. 51 refs, 1922-66.

Review of recent studies on reflection nebulae and interstellar dust and gas.

131.072 Verschuur, G. L. High-velocity neutral hydrogen. **Annu. Rev. Astron. Astrophys.** 13: 257-293; 1975. 63 refs, 1958-75.

Review of observations of neutral hydrogen emission and discussion of models for the general high-velocity hydrogen phenomenon.

131.073 von Hoerner, S. A review of star formation. In: Wilson, T. L.; Downes, D., eds. **H II Regions and Related Topics.** Berlin: Springer-Verlag; 1975: 53-78. 74 refs, 1948-75.

A history of star formation in the galaxy, covering rate of formation, necessary conditions, observations, and future problems.

131.074 Watson, W. D. Gas phase reactions in astrophysics. **Annu. Rev. Astron. Astrophys.** 16: 585-615; 1978. 135 refs, 1951-78.

Discussion of charge transfer between atoms, and reactions related to the abundance of interstellar molecules.

131.075 Watson, W. D. Interstellar molecule reactions. **Rev. Mod. Phys.** 48: 513-552; 1976. 166 refs, 1945-76.

A lengthy introduction to the interstellar medium is provided for the non-astronomer, followed by a review of the status of information on the basic surface and gas phase processes.

131.076 Watson, W. D. Physical processes for the formation and destruction of interstellar molecules. In: Balian, R.; Encrenaz, P.; Lequeux, J., eds. **Atomic and Molecular Physics and the Interstellar Medium.** Amsterdam: North-Holland; 1975: 177-324. 149 refs, 1948-74.

A discussion of the basic physical processes for normal conditions in the interstellar medium: gas temperature $\leq 100°$ K; grain temperature $\leq 20°$ K; particle density $\leq 10^9$ cm^{-3}. These processes are then compared with observed molecular abundances.

131.077 Werner, M. W.; Becklin, E. E.; Neugebauer, G. Infrared studies of star formation. **Science** 197: 723-732; 1977. 82 refs, 1959-77.

Review of infrared observations at wavelengths from a few micrometers to one millimeter which pertain to the problem of star formation. The infrared data include both observations of large clouds of dust and gas within which groups of stars may be forming, and detailed studies of individual objects within these clouds.

131.078 Wickramasinghe, N. C.; Kahn, F. D.; Mezger, P. G. **Interstellar Matter.**
C Sauverny, Switzerland: Geneva Observatory; 1972. 437p. 3 papers.

Second Advanced Course, Swiss Society of Astronomy and Astrophysics. Saas-Fee, Switzerland: 20-25 Mar 1972. Includes the text of three lectures aimed at graduate students: 1) Interstellar matter: an observer's view (P. G. Mezger, pp. 1-205); 2) Interstellar dust (N. C. Wickramasinghe, pp. 207-340); 3) Interstellar gas dynamics (F. D. Kahn, pp. 341-437).

131.079 Wickramasinghe, N. C.; Morgan, D. J., eds. **Solid State Astrophysics.** Dord-
C recht, Holland; Boston: D. Reidel; 1976. 314p. (Astrophys. Space Sci. Libr., 55). 28 papers.

Cardiff, Wales, UK: 9-12 Jul 1974. (Proceedings). Papers reviewing the application of solid state physics to the problem of interstellar dust and grains, primarily, and to neutron stars. Selected topics on the former include interstellar extinction, physical characteristics and formation of dust grains, optical properties of particulates, polarization properties of grains; etc.

131.080 Winnewisser, G. Molecules in astrophysics. In: Diercksen, G. H. F.; Sutcliffe,
C B. T.; Veillard, A., eds. **Computational Techniques in Quantum Chemistry and Molecular Physics.** Dordrecht, Holland; Boston: D. Reidel; 1975: 529-568. 53 refs, 1958-75.

NATO Advanced Study Institute. Ramsau bei Berchtesgaden, Germany: 4-21 Sep 1974. A variety of topics are addressed in this review: microwave spectral data; interstellar molecular spectroscopy; interstellar and molecular clouds; etc.

131.081 Woodward, P. R. Theoretical models of star formation. **Annu. Rev. Astron. Astrophys.** 16: 555-584; 1978. 145 refs, 1939-78.

Covers the star formation process in a more or less chronological sequence, and discusses how a state of gravitational instability can be approached from a more diffuse gaseous state. Possible chain reactions of star formation are considered, along with a discussion of the collapse of an unstable cloud under gravity.

131.082 Wynn-Williams, C. G. Infrared observations of star formation regions. **IAU Symp.** 75: 105-118; 1977. 63 refs, 1972-76.

Introduction to the different infrared wavelength ranges and the various kinds of infrared objects seen in regions of star formation, followed by a review of recent progress in infrared observations and a detailed discussion of four varied examples of star formation regions.

131.083 Zuckerman, B. Interstellar molecules. **Nature** 268: 491-495; 1977. 59 refs, 1964-77.

Review of some of the more interesting recent developments in the field, both astronomical and interdisciplinary. Includes a table of known interstellar molecules as of January 1977.

131.084 Zuckerman, B.; Palmer, P. Radio radiation from interstellar molecules. **Annu. Rev. Astron. Astrophys.** 12: 279-313; 1974. 145 refs, 1952-74.

Concerned with the astrophysics of interstellar radio molecules. Topics covered include history, equation of transfer, estimates of molecular projected densities, optically dark nebulae for infrared sources and H II regions, the galactic center and galactic structure, isotopic abundances and isotopic background radiation.

131.085 Zwicky, F. Compact and dispersed cosmic matter, Part II. **Adv. Astron. Astrophys.** 7: 227-283; 1970. 86 refs, 1933-69.
A survey of knowledge about the distribution of matter beyond our galaxy, with an emphasis on the characteristics of compact galaxies. Part I appeared in *Adv. Astron. Astrophys.* 5: 267-343; 1967.

Bibliography on Molecular Lines in Galactic Objects. 022.026.

Celestial Masers. 063.003.

H II Regions and Related Topics. 132.018.

The interaction of supernovae with the interstellar medium. *125.003.*

Interstellar Dust and Related Topics. 112.003.

Interstellar Gas Dynamics. 062.009.

Proceedings of the Second European Regional Meeting in Astronomy. *R011.007.*

Some characteristics of interstellar gas in the galaxy. *155.025.*

Symposium on very long baseline interferometry. *033.008.*

Zodiacal light as an indicator of interplanetary dust. *106.015.*

132 H I, H II Regions

132.001 Brown, R. L.; Lockman, F. J.; Knapp, G. R. Radio recombination lines. **Annu. Rev. Astron. Astrophys.** 16: 445-485; 1978. 153 refs, 1920-78.
This review deals primarily with the uses of recombination line observations, specifically those from H II regions and from cool, partially ionized clouds (C II regions).

132.002 Churchwell, E. Considerations on the physical parameters, evolution and ionization structure of H II regions. **Mem. Soc. Astron. Italiana** 45: 259-275; 1974. 102 refs, 1956-74.
Discusses aspects of H II regions that may have an influence on the conditions in the interstellar medium.

132.003 Churchwell, E. Recent radio observations of galactic H II regions. **IAU Symp.** 60: 195-218; 1974. 109 refs, 1955-74.
Structure and physics of H II regions are reviewed.

132.004 Courtes, G. H II regions in galaxies of the Local Group. In: Van Woerden, H., ed. **Topics in Interstellar Matter.** Dordrecht, Holland: D. Reidel; 1977: 209-242. 123 refs, 1925-76.
A review of recent investigations and results. Topics: M31; M33; large-scale galactic H II problems; the irregular galaxies (Magellanic Clouds, NGC 6822, IC 1613).

132.005 Dalgarno, A.; McCray, R. A. Heating and ionization of H I regions. **Annu. Rev. Astron. Astrophys.** 10: 375-426; 1972. 187 refs, 1948-72.
Review directed toward heating and cooling of normal H I regions, although there is a brief section on cooling of high-temperature H II regions and dust clouds.

132.006 Field, G. B. Theoretical description of the interstellar medium. **IAU Symp.** 39: 51-76; 1970. 101 refs, 1950-69.

Consideration of H I and H II clouds and shock waves.

132.007 Goudis, C. A classification of the available astrophysical data of particular H II regions II: Orion Nebula: physical parameters. **Astrophys. Space Sci.** 36: 79-104; 1975. 92 refs, 1942-75.

An attempt to classify all of the data that reveals the physical parameters of the Orion Nebula.

132.008 Goudis, C. A classification of the available astrophysical data of particular H II regions VIII. W3 complex: mapping and physical parameters of various sources. **Astrophys. Space Sci.** 61: 417-476; 1979. 177 refs, 1958-78.

Recapitulation of studies involving the W3 Complex.

132.009 Habing, H. J. H II regions. In: Pinkau, K., ed. **The Interstellar Medium.** Dordrecht, Holland: D. Reidel; 1974: 91-125. 70 refs, 1939-73.

An overview of recent developments covering the following areas: history; group properties; spectrum; neutral matter and dust; compact and ultracompact H II regions.

132.010 Habing, H. J.; Israel, F. P. Compact H II regions and OB star formation. **Annu. Rev. Astron. Astrophys.** 17: 345-385; 1979. 289 refs, 1963-79.

A review of compact objects (H II regions, IR sources, and interstellar masers) and their relation to OB associations. Optical, radio, and IR observations are presented and discussed. Tables include a classification of H II regions and a list of 102 star-forming regions in the galaxy.

132.011 Hodge, P. W. H II regions in galaxies. **Publ. Astron. Soc. Pacific** 86: 845-860; 1974. 100 refs, 1952-74.

Review of the current status of the study of H II regions in other galaxies, especially of their spatial distribution and relation to galactic structure. An extensive table compiles all available data on the H alpha surveys of galaxies.

132.012 Isobe, S. Dust in H II regions. In: Van Woerden, H., ed. **Topics in Interstellar Matter.** Dordrecht, Holland: D. Reidel; 1977: 61-79. 85 refs, 1934-76.

A summary of the observational evidence for dust grains in H II regions, and a discussion of globules and molecular clouds as the source of the dust. Optical, radio, and infrared observations are reviewed.

132.013 Mathews, W. G.; O'Dell, C. R. Evolution of diffuse nebulae. **Annu. Rev. Astron. Astrophys.** 7: 67-98; 1969. 118 refs, 1939-69.

A discussion of the various forms of extended emission regions and their relation to problems of astrophysics, with an emphasis on H I and H II regions.

132.014 Moorwood, A. F. M., ed. **H II Regions and the Galactic Centre.** Noordwijk,
C The Netherlands: ESRO Scientific and Technical Information Branch; 1974. 272p. (ESRO SP-105). 46 papers.

Eighth ESLAB Symposium. Frascati, Italy: 4-7 Jun 1974. (Proceedings). A series of short papers focusing on infrared, optical, and radio observations. Topics: galactic and extragalactic H II regions; dust in H II regions; molecular sources and star formation; the galactic center.

132.015 Pagel, B. E. J., et al. A survey of chemical compositions of H II regions in the Magellanic Clouds. **Mon. Not. R. Astron. Soc.** 184: 569-592, microfiche MN 184/1; 1978. 104 refs, 1930-78.

Measurements of relative line intensities are analyzed to derive electron temperatures and densities and abundances of nitrogen, oxygen, neon, sulphur, chlorine, and argon.

132.016 Peimbert, M. Chemical composition of extragalactic gaseous nebulae. **Annu. Rev. Astron. Astrophys.** 13: 113-131; 1975. 89 refs, 1942-75.

Review with a concentration on normal H II regions, nuclear H II regions of normal galaxies, and planetary nebulae.

132.017 Smith, L. F.; Biermann, P.; Mezger, P. G. Star formation rates in the galaxy. **Astron. Astrophys.** 66: 65-76; 1978. 68 refs, 1955-78.

Data relevant to giant H II regions in the galaxy are collected and analyzed.

132.018 Wilson, T. L.; Downes, D., eds. **H II Regions and Related Topics.** Berlin:
C Springer-Verlag; 1975. 488p. (Lecture Notes Phys., 42). 47 papers.

Mittelberg, Kleinwalsertal, Austria: 13-17 Jan 1975. (Proceedings). A compilation of papers addressing recent radio, infrared, and optical observations of H II regions and associated molecular clouds and maser sources, as well as the interpretation of the resultant data. Topics: molecular clouds and dust clouds; formation of stars and H II regions; IR-sources, masers and compact H II regions; evolved H II regions and OB-stars; distribution of H II regions; etc.

132.019 Wynn-Williams, C. G.; Becklin, E. E. Infrared emission from H II regions. **Publ. Astron. Soc. Pacific** 86: 5-25; 1974. 106 refs, 1961-74.

Covers recent observations from 1.5μ to 1.5mm. Includes a table of 52 H II regions (other than the Orion Nebula) that have been detected or studied at infrared wavelengths.

The Nançay survey of absorption by galactic neutral hydrogen. *141.012.*

Pulkovo sky survey in the interstellar neutral hydrogen radio line. *131.010.*

Radio Recombination Lines. 141.059.

133 Infrared Sources

133.001 Neugebauer, G.; Leighton, R. B. **Two-micron Sky Survey: a preliminary cata-**
A **log.** Washington, DC: NASA; 1969. various paging. (NASA SP-3047).

Data compiled from an infrared sky survey, listing all objects that had a flux density at 2.2μ exceeding $4x10^{-25}$ $W/m^2/Hz$. Covering -33° to +81° declination, the survey was interested in infrared-emitting objects and their apparent intensities, colors, variability, and spatial distributions. Data for more than 5,000 stars are presented in computer printout format.

133.002 Rieke, G. H.; Lebofsky, M. J. Infrared emission of extragalactic sources. **Annu. Rev. Astron. Astrophys.** 17: 477-511; 1979. 159 refs, 1964-79.

An overview of the various emitters of infrared radiation outside the Milky Way, with an interpretation of recent observational results. Objects covered include elliptical galaxies, spiral and irregular galaxies, QSOs, and Seyfert galaxies.

133.003 Rowan-Robinson, M., ed. **Far Infrared Astronomy.** Oxford, UK; New York:
C Pergamon Press; 1976. 335p. 37 papers. Supplement to *Vistas in Astronomy.*

Windsor, UK: 9-11 Jul 1975. (Proceedings). Primarily reports of recent research including quantitative data, this volume outlines the techniques and difficulties in carrying out observations in this relatively new field. Topics: instrumentation and atmospheric constraints; solar and Jovian atmospheres; cosmic microwave background; line astronomy; continuum emission; theoretical models.

133.004 Stein, W. A. Recent revelations of infrared astronomy. **Publ. Astron. Soc. Pacific** 87: 5-16; 1975. 87 refs, 1961-75.

Brief review of a few of the recent developments in infrared astronomy, including consideration of protostars, solid particles in the galaxy, radiation mechanisms of extragalactic sources, and peculiar sources.

Infrared and Submillimeter Astronomy. 061.035.

Infrared Astronomy. 061.016; 061.019; 061.064.

Infrared astronomy instrumentation. *032.009.*

Infrared emission from H II regions. *132.019.*

Infrared: The New Astronomy. 061.001.

134 Emission Nebulae, Reflection Nebulae

134.001 C Maran, Stephen P.; Brandt, John C.; Stecher, Theodore P., eds. **The Gum Nebula and Related Problems.** Washington, DC: NASA; 1973. 236p. (NASA SP-332). 25 papers.

Greenbelt, MD, USA: 18 May 1971. Reports of research on the Gum Nebula, the Vela X supernova remnant, the hot stars gamma Velorum and zeta Puppis, the B-associations in Vela-Puppis, and the pulsar PSR 0833-45, all objects in the area of the sky extensively studied by Colin Gum (1924-60), to whom the conference was dedicated.

134.002 Osterbrock, D. E. Line and continuum problems in gaseous nebulae. **J. Quant. Spectrosc. Radiat. Transfer** 11: 623-639; 1971. 44 refs, 1939-70.

Discussion of nebular-radiative transfer problems, particularly those of the ultraviolet continuum and resonance lines, which fix the physical conditions throughout the nebula.

134.003 Osterbrock, Donald E. **Astrophysics of Gaseous Nebulae.** San Francisco: W. H. Freeman; 1974. 251p. (Ser. Books Astron. Astrophys.). 9 chapters w/refs.

An advanced text for researchers and graduate students emphasizing theory instead of observation, although the latter is included for illustrative purposes. Topics: introduction; photoionization equilibrium; thermal equilibrium; calculation of emitted spectrum; comparison of theory with observation; internal dynamics of gaseous nebulae; interstellar dust; H II regions in the galactic context; planetary nebulae.

134.004 Seaton, M. J. Temperatures of gaseous nebulae—a decade of depression. **Q. J. R. Astron. Soc.** 15: 370-391; 1974. 88 refs, 1927-74.

Review of recent observational and theoretical work that reflects on the inaccuracy of previously accepted low electron temperatures in nebulae.

Evolution of diffuse nebulae. *132.013.*

The Nebular Variables. 122.014.

Reflection nebulae and the nature of interstellar grains. *131.071.*

135 Planetary Nebulae

135.001 Acker, A. A new synthetic distance scale for planetary nebulae. **Astron. Astrophys. Suppl. Ser.** 33: 367-381; 1978. 81 refs, 1950-77.

A new distance scale is provided for planetary nebulae by combining 13 lists published between 1950 and 1976. New distances were calculated for 330 planetary nebulae.

135.002 Acker, A.; Marcout, J. Bibliographical index of planetary nebulae for the period 1965-1976. **Astron. Astrophys. Suppl. Ser.** 30: 217-221; 1977.

A description of, and supplementary information to, a literature survey published on magnetic tape and microcard by the Centre de Données Stellaires de Strasbourg. Included are a table of 148 new planetary nebulae, a cross-identification list of the various names used in the literature to designate the objects, a cross-identification list of central stars, and a list of 53 sources of planetary nebulae information.

135.003 Aller, L. H. Central stars of planetary nebulae. **Mém. Soc. R. Sci. Liège** 6e ser. 9: 271-299; 1976. 135 refs, 1930-76.

Classification of the spectra is reviewed, and a compilation is given of spectral types, magnitudes, and colors.

135.004 Cohen, M.; Barlow, M. J. An infrared photometric survey of planetary nebulae. **Astrophys. J.** 193: 401-418; 1974. 72 refs, 1950-74.

Infrared photometry between 2.2 and 22 μ of 113 planetary nebulae is presented, covering a wide variety of types. Offers evidence which supports an interpretation of this long-wavelength radiation as due to thermal reradiation by dust grains.

135.005 Ford, H. C.; Jacoby, G. H. Planetary nebulae in Local-Group galaxies. VIII. A
A catalog of planetary nebulae in the Andromeda galaxy. **Astrophys. J. Suppl. Ser.** 38: 351-356, plates 19-28; 1978. 4 refs, 1963-78.

Finding charts and equatorial coordinates are presented for 311 planetary nebulae and for nearby reference stars in M31.

135.006 Gurzadyan, G. A. **Planetary Nebulae.** rev. ed. New York: Gordon and Breach; 1969. 314p. 10 chapters. 233 refs, 1916-65.

A comprehensive survey of the available observational material and earlier theoretical work. Much of the text reports on the author's own work in the field. Topics: basic observational data; origin of emission lines; distances and dimensions; continuous spectrum; radiative equilibrium; stability of the forms of gaseous envelopes; magnetic fields; origin of planetary nebulae. Translation of *Planetarnye tumannosti* (1962), by D. G. Hummer. The 1969 edition includes substantial new material by the author.

135.007 Higgs, L. A., comp. **Catalog of Radio Observations of Planetary Nebulae and**
A **Related Optical Data.** Ottawa: Astrophysics Branch, Radio and Electrical Engineering Division, National Research Council of Canada; 1971. 454p. (NRC 12129). 141 refs, 1881-1970.

A compilation of data on planetary nebulae found in the literature through the end of 1970, the majority of this book is devoted to numerical information resulting from radio observations of total flux density. Other useful data includes lists of planetary nebulae, their physical characteristics, optical data, references, etc.

135.008 Kohoutek, L. New and misclassified planetary nebulae. **IAU Symp.** 76: 47-62; 1978. 87 refs, 1955-77.

Table 1, providing a summary of 226 new objects classified as planetary nebulae since the *Catalogue of Planetary Nebulae* (1967), gives designation, names, coordinates and references to the discovery. Table 2 lists misclassified planetary nebulae.

135.009 Miller, J. S. Planetary nebulae. **Annu. Rev. Astron. Astrophys.** 12: 331-358; 1974. 145 refs, 1926-74.

Review of selected areas of investigation as they appeared in mid-1973. Topics include optical spectra and physical processes involved, infrared and radio studies, and problems of ionization and excitation structure of the nebular gas.

135.010 Perinotto, M. Planetary nebulae. **Mem. Soc. Astron. Italiana** 47: 177-209; 1976. 81 refs, 1926-76. In Italian.

Summary of the present knowledge of planetary nebulae, focusing on the interpretation of the optical, radio, and infrared spectra.

135.011 Salpeter, E. E. Central stars of planetary nebulae. **Annu. Rev. Astron. Astrophys.** 9: 127-146; 1971. 91 refs, 1931-71.

Discussion of the theoretical ideas on the evolution of the central stars of planetary nebulae. Questions considered are the nature and history of the star before ejection of the nebula, the causes for ejection, and the reason for the apparent rapid evolution of the central stellar remnant after the explosion.

135.012 Terzian, Y., ed. **Planetary Nebulae: Observations and Theory.** Dordrecht,
C Holland; Boston: D. Reidel; 1978. 376p. (IAU Symp., 76). 29 papers, 60 abstracts.

Ithaca, NY, USA: 6-10 Jun 1977. Topics include distribution, observations, physical processes, central stars, chemical abundances, evolution and morphology, origin, and influence of the galaxy.

135.013 Torres-Peimbert, S.; Peimbert, M. Photometric photometry and physical conditions of planetary nebulae. **Rev. Mexicana Astron. Astrofis.** 2: 181-207; 1977. 81 refs, 1956-77.

Photoelectric spectrophotometry of emission lines in the 3400-7400 Å range is presented for 33 planetary nebulae, covering a galactocentric radial range of about 10 kpc.

135.014 Weinberger, R. A list of possible, probable, and true planetary nebulae detected
A since 1966. **Astron. Astrophys. Suppl. Ser.** 30: 335-341; 1977. 59 refs, 1943-77.

Listed are 335 objects designated since the closing of the Perek and Kohoutek catalogue (1967), with name, designations, equatorial coordinates, galactic coordinates, apparent dimensions, and indications of observations in the optical, infrared, and radio ranges.

The ESO/Uppsala survey of the ESO(B) atlas of the southern sky. *158.020.*

Interstellar extinction: a calibration by planetary nebulae. *131.011.*

Interstellar matter and planetary nebulae. *131.026.*

The Menzel Symposium on Solar Physics, Atomic Spectra, and Gaseous Nebulae. 080.013.

The Revised New General Catalogue of Nonstellar Astronomical Objects. 158.042.

RADIO SOURCES, X-RAY SOURCES, COSMIC RADIATION

141 Radio Sources, Quasars

141.001 Abell, G. O.; Peebles, P. J. E., eds. **Objects of High Redshift.** Dordrecht,
C Holland; Boston: D. Reidel; 1980. 340p. (IAU Symp., 92). 30 papers, 18 abstracts.

Los Angeles, CA, USA: 28-31 Aug 1979. Topics include galaxy counts and colors, clusters and superclusters, quasars, Seyfert galaxies, and the x-ray background.

141.002 Barbieri, C.; Capaccioli, M.; Zambon, M. Catalogue of quasi-stellar objects
A (Edition of June 28, 1975). **Mem. Soc. Astron. Italiana** 46: 461-499; 1975. 326 refs, 1958-75.

The catalog provides name, photometry, redshift L, equatorial coordinates, and references for 508 QSOs. Covers the literature prior to December 31, 1974.

141.003 Boksenberg, A. The physics of QSO absorption line regions. **Phys. Scr.** 17: 205-214; 1978. 89 refs, 1966-77.

Physical properties of the absorption systems observed in high redshift QSOs are described in some detail.

141.004 Bolton, J. G.; Wright, A. E.; Savage, A. The Parkes 2700 MHz survey. (Four-
A teenth part): catalogue for declinations -4° to -15°, right ascensions 10^h to 15^h. **Australian J. Phys., Astrophys. Suppl.** 46: 1-17; 1979. 26 refs, 1966-78.

Catalog of 278 radio sources from a 2700 MHz survey of 0.247 sr.

141.005 Braude, S. Ya., et al. Decametric survey of discrete sources in the northern sky.
A I. The UTR-2 radio telescope. **Astrophys. Space Sci.** 54: 3-36; 1978. 38 refs, 1962-74.

Survey of the northern sky in the declination range -10° and +80°. Other parts of the survey include:

II. Source catalogue in the range of declinations +10° to +20°. *Astrophys. Space Sci.* 54: 37-128; 1978. 12 refs, 1965-78.

III. Low frequency absolute flux scale of discrete radio sources. *Astrophys. Space Sci.* 54: 129-143; 1978. 20 refs, 1963-78.

IV. Spectra of 266 discrete sources in the range 10 to 1400 MHz. *Astrophys. Space Sci.* 54: 145-179; 1978. 17 refs, 1967-78.

141.006 Burbidge, G.; Crowne, A. H. An optical catalog of radio galaxies. **Astrophys.
A J. Suppl. Ser.** 40: 583-655; 1979. 414 refs, 1956-79.

A catalog containing basic optical information on all radio galaxies ($L_{radio} \gtrsim 10^{41}$ ergs s^{-1}) which have been identified and for which redshifts have been measured. The catalog contains 495 galaxies.

141.007 Burbidge, G. R. Problem of the redshifts. **Nature Phys. Sci.** 246: 17-25; 1973. 99 refs, 1951-73.

The observational evidence concerning the nature of the redshifts in quasistellar objects and some peculiar galaxies is reviewed.

141.008 Burbidge, G. R.; Crowne, A. H.; Smith, H. E. An optical catalog of quasistellar
A objects. **Astrophys. J. Suppl. Ser.** 33: 113-188; 1977. 413 refs, 1958-77.

A catalog containing the basic optical information on over 600 QSOs which have been certainly identified as of August 1976.

141.009 Clarke, J. N.; Little, A. G.; Mills, B. Y. The Molonglo radio source catalogue
A 4: the Magellanic Cloud region. **Australian J. Phys., Astrophys. Suppl.** 40: 1-71; 1976. 20 refs, 1956-74.

A total of 1,349 sources are listed to a limiting flux density of 0.2 Jy. The catalog is largely complete above 0.25 Jy.

141.010 Cohen, M. H. High-resolution observations of radio sources. **Annu. Rev. Astron. Astrophys.** 7: 619-664; 1969. 182 refs, 1946-68.

Review of techniques and instrumentation for very-long-baseline interferometry, lunar occultations, and interplanetary scintillations.

141.011 Craine, Eric R. **A Handbook of Quasistellar and BL Lacertae Objects.** Tucson, AZ: Pachart Publishing; 1977. 283p. (Ref. Works Astron.). 528 refs, 1916-76.

A reference volume including all QSOs brighter than $m_V = 17$ and all known BL Lac objects. Arranged in order of ascending right ascension, each object takes up one or two pages of text. Included is a finding chart, a radio spectrum graph, various spectral data, wavelength of emission, and various catalog designations. Other information is occasionally included: optical polarization and variability; light curves; and photometric sequences. Appendix 1 is a comprehensive list of 89 radio source catalogs; appendix 2 is a cross-reference of source names.

141.012 Crovisier, J.; Kazès, I.; Aubry, D. The Nançay survey of absorption by galactic
A neutral hydrogen. I. Absorption towards extragalactic sources. **Astron. Astrophys. Suppl. Ser.** 32: 205-282; 1978. 35 refs, 1958-78.

Catalog of relevant data on 819 sources, as well as emission and absorption profiles towards 386 of the sources.

141.013 Davies, I. M.; Little, A. G.; Mills, B. Y. The Molonglo radio source catalogue
A I. **Australian J. Phys., Astrophys. Suppl.** 28: 1-59; 1973. 14 refs, 1963-72.

Catalog of 1,545 radio sources observed at 408 MHz between -19°.3 and -22°.4.

141.014 De Veny, J. B.; Osborn, W. H.; Janes, K. A catalogue of quasars. **Publ. Astron.**
A **Soc. Pacific** 83: 611-625; 1971. 185 refs, 1957-71.

Catalog of 202 objects with important optical data, and a bibliography for all quasars with redshifts published before June 1971.

141.015 De Young, D. S. Extended extragalactic radio sources. **Annu. Rev. Astron. Astrophys.** 14: 447-474; 1978. 97 refs, 1959-76.

Emphasis is on summarizing and comparing the work done to date on understanding the origins, evolution, and relevant physical processes pertaining to extended extragalactic radio sources, in light of the existing observational information.

141.016 Douglas, J. N., et al. First results from the Texas interferometer: positions of
A 605 discrete sources. **Astron. J.** 78: 1-17; 1973. 18 refs, 1961-72.

Positions of 605 discrete radio sources with average rms uncertainties of 1".9 arc in right ascension and 1".3 arc in declination. The first overall program of optical and radio studies of large samples of identified radio sources.

141.017 Dupree, A. K.; Goldberg, L. Radio-frequency recombination lines. **Annu. Rev. Astron. Astrophys.** 8: 231-264; 1970. 84 refs, 1938-70.

Review of spectral lines in the radio-frequency domain emitted as a result of transitions between atomic levels with high principal quantum numbers. Concentration is on recent developments connected with the theoretical interpretation of observations and on those observations that seem most relevant to the evaluation of theories.

141.018 Elliot, J. L.; Shapiro, S. L. On the variability of the compact nonthermal sources. **Astrophys. J. Lett.** 192: L3-L6; 1974. 54 refs, 1921-74.

Results of a literature search for reported variability in a group of BL Lac objects, quasars, and Seyfert galaxies.

141.019 Ellis, R.; Phillips, S. A catalogue of absorption lines in QSO spectra. **Mon.**
A **Not. R. Astron. Soc.** 183: 271-274, microfiche MN 183/1; 1978. 9 refs, 1971-77.

The catalog contains line lists for 108 objects.

141.020 Engels, D. Catalogue of late-type stars with OH, H_2O or SiO maser emission.
A **Astron. Astrophys. Suppl. Ser.** 36: 337-345; 1979. 114 refs, 1966-79.

Catalog of more than 300 objects showing maser line radio emission from OH, H_2O, and/or SiO molecules.

141.021 Erickson, W. C.; Matthews, T. A.; Viner, M. R. 26.3 MHz radio source survey.
A III. Correlation with extragalactic x-ray sources. **Astrophys. J.** 222: 761-778; 1978. 150 refs, 1931-77.

The low-frequency radio characteristics of radio sources associated with 22 well-identified extragalactic x-ray sources are discussed. Table 1 contains all known x-ray sources within the declination range of the Clark Lake survey except for obvious galactic objects.

141.022 Fricke, K. J. Ausgedehnte und kompakte Radioquellen. [Extended and compact radio sources.] **Mitt. Astron. Ges.** 42: 59-74; 1977. 60 refs, 1953-77. In German.

Observational methods, the North Atlantic Antenna Network, general physical considerations, specific models, discriminatory observations, and model calculations are discussed.

141.023 Ghigo, F. D. Optical identification of 664 Ohio sources using accurate radio and
A optical positions measured by the Texas interferometers. **Astrophys. J. Suppl. Ser.** 35: 359-393; 1977. 78 refs, 1961-77.

Results of optical identification work are reported for 664 radio sources selected from the Ohio 1415 MHz survey.

141.024 Grandi, S. A.; Tifft, W. G. The degree of optical variability of quasistellar objects. **Publ. Astron. Soc. Pacific** 86: 873-884; 1974. 97 refs, 1959-73.

Information on optical variability of QSOs is collected and analyzed to determine the effect of variability on the data that have been used in various statistical arguments.

141.025 Haynes, R. F.; Caswell, J. L.; Simons, L. W. J. A southern hemisphere survey
A of the galactic plane at 5 GHz. **Australian J. Phys., Astrophys. Suppl.** 45: 1-87; 1978. 18 refs, 1959-75.

Presented here are 75 maps showing the 5 GHz emission in the range $\ell = 190° \rightarrow 360° \rightarrow 40°$ for $-2° < b < 2°$.

141.026 Hey, J. S. **The Evolution of Radio Astronomy.** New York: Science History Publications, Neale Watson Academic Publications; 1973. 214p. 8 chapters. 309 refs, 1932-70.

A history for laypersons, students, and scientists, covering early discoveries and research programs, as well as later achievements and future prospects. This nontechnical book was authored by a scientist having first-hand knowledge of the events that shaped radio astronomy. Topics: pioneers; radio telescopes and observatories; the solar system;

galactic radio waves; radio galaxies, quasars, and cosmology; methods of observation and data collection and analysis.

141.027 Hey, J. S. **The Radio Universe.** rev. 2nd ed. Oxford, UK; New York: Pergamon Press; 1976. 264p. (Pergamon Int. Popular Sci. Ser.). 10 chapters. no refs.

A guide for the intelligent layperson, covering radio waves; radio telescopes; radio emission from the Moon and planets; the radio Sun; radar astronomy; radio sources in the galaxy; quasars; cosmology; etc. A good overview and introduction, this well-written standard source was first published in 1971.

141.028 Jauncey, D. L. Radio surveys and source counts. **Annu. Rev. Astron. Astrophys.** 13: 23-44; 1975. 89 refs, 1955-75.

A review aimed at presenting a detailed examination of the published radio survey material, with the idea of constructing reliable counts from these surveys.

141.029 Jauncey, David L., ed. **Radio Astronomy and Cosmology.** Dordrecht, Holland;
C Boston: D. Reidel, 1977. 398p. (IAU Symp., 74). 40 papers.

Cambridge, UK: 16-20 Aug 1976. Sections included surveys of radio sources; source counts and anisotropies; spectral index distributions; angular diameter-redshift and flux density tests; optical spectra and identifications; interpretation of cosmological information on radio sources; and microwave background radiation.

141.030 Katgert, J. K. A catalogue of sources found at 610 MHz with the Westerbork
A Synthesis Radio Telescope: source parameters and identifications. **Astron. Astrophys. Suppl. Ser.** 31: 409-426; 1978. 11 refs, 1963-71.

Catalog and description of 347 sources detected in eight fields. Spectral index data are given for 182 sources.

141.031 Kesteven, M. J. L.; Bridle, A. H. Index of extragalactic radio-source catalogues. **J. R. Astron. Soc. Canada** 71: 21-39; 1977. 97 refs, 1960-76.

A guide to extragalactic radio-source nomenclature, this article lists over 50 catalogs of these objects along with 97 bibliographic references. Its purpose is to aid the reader in identifying any radio-source name likely to be encountered in the literature. Includes a description of the numbering system for catalogs adopted by IAU in 1973. A similar list of catalogs appears in E. R. Craine's *Handbook of Quasistellar and BL Lacertae Objects* (*141.011*).

141.032 Kinman, T. D. Variable quasi-stellar sources with particular emphasis on objects of the BL Lac type. **IAU Symp.** 67: 573-590; 1975. 80 refs, 1929-75.

A review concerned with optical variability and the possible explanations for such variability, along with theoretical implications thereof with regard to their relative distribution.

141.033 Kraus, J. D. The Ohio Sky Survey and other radio surveys. **Vistas Astron.** 20: 445-474; 1977. 93 refs, 1933-76.

Review of major radio astronomy surveys. Topics: importance of astronomical surveys; early radio surveys; description of the Ohio survey (>19,000 sources at 1415 MHz); relation of the Ohio survey to other efforts, including Cambridge and Parkes surveys; source nomenclature.

141.034 Kristian, J.; Minkowski, R. Optical identification of 3CR sources (October 1972). In: Sandage, Allan; Sandage, Mary; Kristian, Jerome, eds. **Galaxies and the Universe.** Chicago: University of Chicago Press; 1975: 199-210. 18 refs, 1963-74.

An appendix to "The identification of radio sources," by Minkowski (see *141.042*). Essentially a table of 328 3CR sources which have been optically identified, it includes flux density, galactic coordinates, identification (e.g., galaxy or quasar), magnitude, etc.

141.035 Krüger, A. **Introduction to Solar Radio Astronomy and Radio Physics.** Dordrecht, Holland; Boston: D. Reidel; 1979. 330p. (Geophys. Astrophys. Monogr., 16). 5 chapters. 1,694 refs, 1946-77.

An advanced monograph concerned with solar physics from the point of view of the results and activities of radio astronomy. The book provides an up-to-date and comprehensive look at both theory and observation in the field. Topics: instrumentation; types of solar radio emission (the Quiet Sun, the slowly varying component, continuum bursts, fast-drift and slow-drift bursts, noise storms, solar radio pulsations); theory of solar radio emission; integration of radio astronomy into solar and solar-terrestrial physics.

141.036 Large, M. I., et al. The Molonglo reference catalogue of radio sources. **Mon.**
A **Not. R. Astron. Soc.** 194: 693-704; 1981. 28 refs, 1963-80.

Catalog of 12,141 discrete sources of listed flux density ≥ 0.7 Jy. The survey covers 7.85 sr of the sky defined by $+18\overset{\circ}{.}5 \geq \delta (1950) \geq -85\overset{\circ}{.}0, |b| \geq 3°$.

141.037 Longair, M. S. The counts of radio sources. **Soviet Phys. Usp.** 12: 673-683; 1970. Transl. of *Usp. Fiz. Nauk.* 99: 229-248; 1969. 63 refs, 1957-70.

Concentration is on showing how much detail about cosmological evolution can be determined from radio observations.

141.038 Longair, M. S. Radio astronomy and cosmology. In: Maeder, Andre; Martinet, Louis; Tammann, Gustave, eds. **Observational Cosmology.** Sauverny, Switzerland: Geneva Observatory; 1978: 127-257. 94 refs, 1957-78.

A lecture/review paper describing the contributions of radio observations to cosmology in three areas: 1) the background radiation at meter and microwave wavelengths; 2) radio evidence on the properties of intergalactic gas; 3) the space distribution and cosmological evolution of extragalactic radio sources and quasars.

141.039 McEwan, N. J.; Browne, I. W. A.; Crowther, J. H. Accurate positions and
A identifications for a complete sample of 341 radio sources from the Parkes +4° survey. **Mem. R. Astron. Soc.** 80: 1-59; 1975. 85 refs, 1956-74.

For all sufficiently compact sources, radio positions have been measured with the RRE Malvern interferometer to an accuracy of about 1.5 arcsec rms, and frequently to about 0.6 arcsec rms. Optical positions have been measured for candidate identifications to an rms accuracy of 1.7 arcsec.

141.040 Machalski, J. The Green Bank Second (GB2) survey of extragalactic radio
A sources at 1400 MHz. The catalog of sources. **Acta Astron.** 28: 367-440; 1978. Also in: **NRAO Repr.** B495. 28 refs, 1957-78.

The catalog contains 2,022 sources down to a limiting flux density of 0.09 Jy.

141.041 Mayer, C. H. Thermal radio emission of the planets and Moon. In: Dollfus, A., ed. **Surfaces and Interiors of Planets and Satellites.** London: Academic; 1970: 169-224. 152 refs, 1924-68.

A review of observational data and its interpretation. All planets but Pluto, which had not been observed, are included, with an emphasis on Venus and Mars.

141.042 Minkowski, R. The identification of radio sources. In: Sandage, Allan; Sandage, Mary; Kristian, Jerome, eds. **Galaxies and the Universe.** Chicago: University of Chicago Press; 1975: 177-197. 150 refs, 1918-68.

An overview of efforts to locate the optical counterparts of various radio objects such as galaxies and quasars.

141.043 Moffet, A. T. Extragalactic radio astronomy. In: Chretien, M.; Deser, S.; Goldstein, J., eds. **Astrophysics and General Relativity** (v.1). New York: Gordon and Breach; 1969: 217-300. 144 refs, 1912-68.

A basic introductory lecture for the advanced student. Topics: historical review; basic concepts; observed characteristics of radio sources; identifications and intrinsic properties of radio sources; synchrotron radiation; etc.

141.044 Moffet, A. T. Strong nonthermal radio emission from galaxies. In: Sandage, Allan; Sandage, Mary; Kristian, Jerome, eds. **Galaxies and the Universe.** Chicago: University of Chicago Press; 1975: 211-281. 279 refs, 1912-72.

A description of the observed properties of radio galaxies and the inferred intrinsic properties and physical conditions. The evolution of radio galaxies and the theory of synchrotron radiation are also discussed.

141.045 Pacholczyk, A. G. **A Handbook of Radio Sources.** Tucson, AZ: Pachart
A Publishing; 1978. 234p. (Volume 1: Strong extragalactic sources, right ascension range 0 through 11 hours).

The first of a planned three-part work, this book includes most objects exceeding a flux density limit "index 0.7 set at the level of 15 flux units at a frequency of 408 MHz." The main catalogs used for selecting sources were *R73* at 408 MHz and *BDFL72, BF74,* and *E69* at 1400 MHz. Each entry, or "data sheet," includes a finding chart and radio map, integrated spectral data, radio structure data such as angular size of components, and a list of references used in compiling the information. The book also has a list of 374 radio catalogs and data compilations.

141.046 Pacholczyk, A. G. **Radio Astrophysics: Nonthermal Processes in Galactic and Extragalactic Sources.** San Francisco: W. H. Freeman; 1970. 269p. (Ser. Books Astron. Astrophys.). 8 chapters w/refs.

Intended for observational radio astronomers and graduate students working on the interpretation of radioastronomical data, this book emphasizes the physical processes believed to be responsible for the observed radio emissions from various types of celestial objects. Observational data *per se* are deemphasized. Topics: radio astronomical measurements; plasma in magnetic fields; synchrotron radiation; Compton scattering; interpretation of spectra from discrete radio sources; physical conditions in radio sources; spectral radio lines; etc.

141.047 Pacholczyk, A. G. **Radio Galaxies: Radiation Transfer, Dynamics, Stability and Evolution of a Synchrotron Plasmon.** Oxford, UK; New York: Pergamon Press; 1977. 293p. (Int. Ser. Nat. Philos., 89). 8 chapters w/refs.

Aimed primarily at radio astronomers and graduate students working on the interpretation of observational data, this book is concerned with "the physics of a region in space containing magnetic field and thermal and relativistic particles (a plasmon)." Topics: observational data and relationships; incoherent synchrotron spectra; hydromagnetics; confinement and structure of extended radio sources; radio source theories; etc.

141.048 Pacht, E. Radio two-color diagram for QSOs, spiral galaxies and BL Lac
A objects. **Astron. J.** 81: 574-581; 1976. 116 refs, 1964-76.

Contains redshift and radio spectral data on 311 QSOs with known redshifts and radio spectral data for 26 BL Lacs and 75 spiral galaxies. The advantages of plotting certain radio emitting objects on a radio two-color diagram are presented and discussed.

141.049 Perry, J. J.; Burbidge, E. M.; Burbidge, G. R. Absorption in the spectra of quasistellar objects and BL Lacertae objects. **Publ. Astron. Soc. Pacific** 90: 337-366; 1978. 148 refs, 1966-78.

An extensive review is given of the observations of absorption in the spectra of QSOs and BL Lacs. Table Ia summarizes all of the information available up to May 1, 1978 on objects which show absorption.

141.050 Petukhova, M. S. **Radiofizicheskie issledovaniia planet: bibliograficheskii ukazatel': 1960-1973.** [Radiophysical Investigations of the Planets: A Bibliography: 1960-1973.] Moscow: Nauka; 1976. 184p. 7 chapters. In Russian.

An extensive bibliography of both Russian and Western sources covering general topics, each of the planets, and the Moon. Each of the seven chapters is divided into two alphabetically arranged parts, first the Russian, then Western languages. Each entry includes author, title, bibliographic citation, and number of references. The non-Russian entries also include the title in Russian translation. Author indexes and lists of journal title abbreviations are also presented.

141.051 Radio astronomy. **IAU Trans. Rep. Astron.** 17A(3): 139-164; 1979. + refs, 1976-79.

An overview of research activity worldwide, with an emphasis on the previous three years, as reported by IAU Commission 40 (Radio Astronomy). Topics: solar radio emissions; continuum radiation from the galaxy; pulsars; extragalactic radio astronomy; instrumentation; line radiation from the galaxy; interstellar molecules; neutral hydrogen and radio recombination lines; etc. Additional reports: 14A: 455-479; 1970. 516 refs, 1964-70. 15A: 601-637; 1973. 1,199 refs, 1969-73. 16A(3): 159-194; 1976. 1,859 refs, 1971-75.

141.052 Rees, M. J. Cosmological evidence from QSOs and radio galaxies. **IAU Symp.** 44: 407-436; 1972. 101 refs, 1961-71.

Evidence provided by counts of radio and optical objects and evidence yielded by QSO redshifts are discussed. Consideration of the nature of QSO redshifts along with evolutional properties is also given.

141.053 Roberts, M. S. Radio observations of neutral hydrogen in galaxies. In: Sandage, Allan; Sandage, Mary; Kristian, Jerome, eds. **Galaxies and the Universe.** Chicago: University of Chicago Press; 1975: 309-357. 127 refs, 1952-71.

A review of 21-cm hydrogen line studies of galaxies and the resultant information gained about the distribution of hydrogen, total hydrogen content, total mass, etc. Also covered are properties of galaxies derived from 21-cm studies and the evolution of galaxies. Includes a table of 149 galaxies for which 21-cm studies have been made.

141.054 Robertson, J. G. The Molonglo deep sky survey of radio sources. I. Declination zone -20°. **Australian J. Phys.** 30: 209-230; 1977. 18 refs, 1965-77.
A

The catalog lists positions and flux densities for a total of 373 radio sources. Part II, declination zone -62°, lists positions and flux densities for 95 sources: *Australian J. Phys.* 30: 231-239; 1977. 4 refs, 1972-77.

141.055 Rowan-Robinson, M. Quasars and the cosmological distance scale. **Nature** 262: 97-101; 1976. 126 refs, 1921-76.

A review supporting the thesis that quasar redshifts are cosmological. Included is a table of the properties of BL Lac objects.

141.056 Schmidt, M. Quasars. In: Sandage, Allan; Sandage, Mary; Kristian, Jerome, eds. **Galaxies and the Universe.** Chicago: University of Chicago Press; 1975: 283-308. 122 refs, 1963-70.

A review covering history, identification, redshifts, emission lines, continuum energy distribution, absorption lines, models, and other topics.

141.057 Schmidt, M. Quasistellar objects. **Annu. Rev. Astron. Astrophys.** 7: 527-552; 1969. 296 refs, 1956-69.

A progress report, covering 1967-68, including identification, continuous energy distribution, spectra, redshifts, evolution, radio properties, and hypotheses on their exact nature.

141.058 Setti, G., ed. **The Physics of Non-Thermal Radio Sources.** Dordrecht, Holland;
C Boston: D. Reidel; 1976. 287p. (NATO Adv. Stud. Inst. Ser., Ser. C, Math. Phys. Sci., 28). 16 papers.

Third Course of the International School of Astrophysics. Urbino, Italy: 29 Jun—13 Jul 1975. A series of lectures aimed at providing an up-to-date account of the structure, origin, and evolution of non-thermal radio sources such as quasars, BL Lac objects, radio galaxies, pulsars, supernova remnants, and radio stars.

141.059 Shaver, Peter A., ed. **Radio Recombination Lines.** Dordrecht, Holland; Boston:
C D. Reidel; 1980. 284p. (Astrophys. Space Sci. Libr., 80). 22 papers.

Workshop. Ottawa, ON, Canada: 24-25 Aug 1979. (Proceedings). A comprehensive review of research in the field, of interest to graduate students and astrophysicists, covering the study of radio recombination lines of several elements observed in H II regions, planetary nebulae, molecular clouds, the diffuse interstellar medium, and other galaxies. Both single-dish and aperture synthesis techniques are covered. Topics: physics of radio recombination lines; radio recombination lines from H II regions and C II regions; large-scale properties of the galaxy; extragalactic radio recombination lines.

141.060 Slee, O. B. Culgoora-3 list of radio source measurements. **Australian J. Phys.,**
A **Astrophys. Suppl.** 43: 1-123; 1977. 51 refs, 1962-76.

Catalog of positions, flux densities, and beam broadening of 1,946 sources detected at 80 MHz and reexamined at 160 MHz, plus a further 99 sources observed at both frequencies.

141.061 Slee, O. B.; Higgins, C. S. Culgoora-2 list of radio source measurements at
A 80 MHz. **Australian J. Phys., Astrophys. Suppl.** 36: 1-60; 1975. 54 refs, 1958-75.

A survey of 1,748 previously cataloged sources in the declination range of -48° to +35°. Positions, flux densities, and beam broadening are given for 1,291 sources, while 457 undetected sources are listed separately.

141.062 Smith, H. E.; Spinrad, H.; Smith, E. O. The revised 3C catalog of radio sources:
A a review of optical identifications and spectroscopy. **Publ. Astron. Soc. Pacific** 88: 621-646; 1976. 144 refs, 1954-76.

Optical positions, redshifts, magnitudes, and identifications have been included, as well as radio flux densities and spectral indices for the sample of 297 extragalactic 3CR sources.

141.063 Smith, M. G. Quasars: observed properties of optically-selected objects at large redshifts. **Vistas Astron.** 22: 321-362; 1978. 150 refs, 1960-78.

Description of slitless spectroscopy use to discover high-redshift ($z > 2$) quasistellar objects by direct detection of Ly-α emission and a review of the observed properties of these objects.

141.064 Smith-Haenni, A. L. A comprehensive catalogue of quasi-stellar objects.
A **Astron. Astrophys. Suppl. Ser.** 27: 205-214; 1977. 173 refs, 1957-75.

The definition and designation of a QSO are discussed, and optical and radio data on 403 of them are cataloged.

141.065 Stannard, D. The radioastronomy of Jupiter. In: Formisano, V., ed. **The Magnetospheres of the Earth and Jupiter.** Dordrecht, Holland: D. Reidel; 1975: 221-236. 54 refs, 1955-74.

A review of our current understanding of Jovian radio emission, with references to recent observational data. The decrimetric synchrotron emission, decametric burst emission, and thermal emission are all briefly described, along with the radio rotation period, asymmetries in the emission, and other topics.

141.066 Stannard, D.; Neal, D. S. A comparative study of the properties of 3CR and 4C
A quasars. **Mon. Not. R. Astron. Soc.** 179: 719-740; 1977. 52 refs, 1962-77.

Measurements of the radio structure of 26 quasars are combined with previous structural and spectral data to investigate the radio properties of a sample of 112 quasars from the 3CR and 4C surveys.

141.067 Stein, W. A.; O'Dell, S. L.; Strittmatter, P. A. The BL Lacertae objects. **Annu. Rev. Astron. Astrophys.** 14: 173-195; 1976. 197 refs, 1962-76.

Review of recent study of BL Lacs leading to a better understanding of both BL Lacs and QSOs, including a summary of known sources.

141.068 Strittmatter, P. A.; Williams, R. E. The line spectra of quasi-stellar objects. **Annu. Rev. Astron. Astrophys.** 14: 307-338; 1976. 123 refs, 1960-76.

Review of observations and interpretation of line spectra of QSOs, with an emphasis on optical data. Topics include models of the emission line region, the interpretation of the absorption-line data, and attempts to relate the emission and absorption-line data in terms of mass-outflow models.

141.069 Thaddeus, P. The short-wavelength spectrum of the microwave background. **Annu. Rev. Astron. Astrophys.** 10: 305-334; 1972. 97 refs, 1937-72.

Review dealing with recent determinations of the intensity of the microwave background radiation at wavelengths too short for direct observation from the ground.

141.070 Ulfbeck, O., ed. Quasars and active nuclei of galaxies. **Phys. Scr.** 17: 130-385;
C 1978. 42 papers.

Copenhagen symposium on "Active Nuclei." 27 Jun—2 Jul 1977. Seen as continuing and updating the work of the 1971 Vatican conference on Nuclei of Galaxies.

141.071 Veron, M. P.; Veron, P. A catalogue of extragalactic radiosource identifica-
A tions. **Astron. Astrophys. Suppl. Ser.** 18: 309-403; 1974. 279 refs, 1888-1974.

A compilation of all published optical identifications of extragalactic radio sources is presented. The catalog contains 4,022 entries for 2,882 different sources published in 232 papers.

141.072 Veron, M. P.; Veron, P. Optical positions of radio sources. **Astron. Astrophys.**
A **Suppl. Ser.** 29: 149-159; 1977. 149 refs, 1926-77.

Report of measurements of 125 optical positions of radio sources on the prints of the Palomar Sky Survey, with an accuracy of 0".5 arcsec when the AGK3 is used as the reference catalog and of 0".8 if the SAO is used. Confirmation of 85 of the identifications is given.

141.073 Veron, M. P.; Veron, P.; Witzel, A. The spectra of 373 radio sources. **Astron.**
A **Astrophys. Suppl. Ser.** 13: 1-53; 1974. 128 refs, 1957-72.

This compilation of published flux densities between 10 and 10,000 MHz is used to plot radio spectra of 373 sources. Table 1 contains a list of catalogs of flux density used.

141.074 Veron, P. A study of the revised 3C catalogue. I. Confusion and resolution.
A **Astron. Astrophys. Suppl. Ser.** 30: 131-144; 1977. 55 refs, 1962-77.

A catalog is given of 205 radio sources with flux densities larger than 9.0 Jy at 178 MHz.

141.075 Verschuur, Gerrit L. **The Invisible Universe: The Story of Radio Astronomy.** London: English Universities Press; New York; Berlin: Springer-Verlag; 1974. 173p. (Heidelberg Sci. Libr., 20). 19 chapters. no refs.

Written for the educated layperson and student, this excellent book is one of the best available for this audience. Topics: instrumentation; radio sources; observational techniques; history; life in the universe. Radio sources covered include the Sun, the planets, the Milky Way, gas clouds, interstellar molecules, exploding stars and galaxies, quasars, radio stars, pulsars. Excellent illustrations.

141.076 Verschuur, Gerrit L.; Kellermann, Kenneth I., eds. **Galactic and Extragalactic Radio Astronomy.** New York; Berlin: Springer-Verlag; 1974. 402p. 13 papers.

A collection of papers dealing with radio astronomy beyond the solar system, with an emphasis on the various types of data and their interpretation. Theory and instrumentation are deemphasized. Aimed at graduate students and nonspecialists, the book provides a good review of what was known in the mid-1970s. Topics: galactic nonthermal continuum emission; interstellar neutral hydrogen; H II regions; distribution of neutral hydrogen in the galaxy; supernovae remnants; the galactic magnetic field; radio stars; pulsars; interstellar molecules; radio galaxies and quasars; cosmology; interferometry and aperture synthesis; mapping neutral hydrogen in galaxies.

141.077 Willis, A. G.; Oosterbaan, C. E.; de Ruiter, H. R. A Westerbork 1415 MHz survey of background radio sources. I. The catalogue. **Astron. Astrophys. Suppl. Ser.** 25: 453-505; 1976. 24 refs, 1959-76.
A

The catalog includes 1,075 radio sources in 96 fields.

141.078 Wills, B. J. Accurate spectra of 300 radio sources from the Parkes Catalogue: observational results. **Australian J. Phys. Astrophys. Suppl.** 38: 1-65; 1975. 59 refs, 1958-73.
A

A presentation of relative flux densities made between 468 and 5009 MHz for 300 radio sources. An analysis and discussion of accuracy is included.

141.079 Wolfe, A. M., ed. **Pittsburgh Conference on BL Lac Objects.** Pittsburgh, PA: Department of Physics and Astronomy, University of Pittsburgh; 1978. 428p. 33 papers.
C

Pittsburgh, PA, USA: 24-26 Apr 1978. Reports of recent research covering observations at radio, optical, UV, and x-ray wavelengths. Also included are papers on spectral line observations, distance determinations, emission and absorption line regions, distribution, and models for compact sources of radio and optical continuum radiation. A review paper with 59 references is included (W. A. Stein, pp. 1-20).

141.080 Zheleznyakov, V. V. **Radio Emission of the Sun and Planets.** Oxford, UK; New York: Pergamon Press; 1970. 697p. (Int. Ser. Monogr. Nat. Philos., 25). 10 chapters. 983 refs, 1904-1968.

An advanced text reviewing the various forms of solar system radio emission, their theory and study. Topics: physical conditions of the Sun, Moon, and planets; characteristics of radio emission and methods of study; solar radio emission results; radio observations of the Moon and planets; propagation, generation, and absorption of electromagnetic waves in the solar corona; solar thermal radio emission; origin of radio emission of the planets and Moon. Translation by H. S. H. Massey of *Radioizluchenie Solnt͡sa i planet* (Moscow: Nauka; 1964), edited by J. S. Hey.

Astrophysics; Part C: Radio Observations. 031.523.

External Galaxies and Quasi-stellar Objects. 158.012.

List of Radio and Radar Astronomy Observatories. R008.009.

Radio and radar astronomy. *033.007.*

Radio astronomy and cosmology. *162.028.*

Radio radiation from interstellar molecules. *131.084.*

Radio recombination lines. *132.001.*

Seyfert galaxies, quasars, and redshifts. *158.051.*

Statistical study of integral properties of galaxies measured in the 21-cm line. *158.003.*

141.5 Pulsars

141.501 Arons, J. Some problems of pulsar physics, or I'm madly in love with electricity. **Space Sci. Rev.** 24: 437-510; 1979. 170 refs, 1939-80.

Review of current topics in the theory of pulsar magnetospheres and their emission.

141.502 Bertotti, B. Electrodynamics of pulsars. **Nuovo Cimento, Riv.** 2: 102-118, 1970. 38 refs, 1950-70.

A technical review of the complex dynamics underlying the electromagnetic emissions of pulsars. This is a theoretical, physics-oriented paper, concerned with the concept of a pulsar as an oblique rotator.

141.503 Chiu, H.-Y. A review of theories of pulsars. **Publ. Astron. Soc. Pacific** 82: 487-533; 1970. 75 refs, 1928-70.

Discussion of the present theoretical status of pulsars based on observed data and current knowledge of physics.

141.504 Ginzburg, V. L. Pulsars (theoretical concepts). **Soviet Phys. Usp.** 14: 83-103; 1971. Transl. of *Usp. Fiz. Nauk.* 103: 393-429; 1971. Also in: **Highlights Astron.** 2: 20-62; 1971. 171 refs, 1918-71.

Review of the nature of pulsars, including their physical nature, mechanics, some theoretical models, and possible use in astronomy and physics.

141.505 Ginzburg, V. L.; Zheleznyakov, V. V. On the pulsar emission mechanisms. **Annu. Rev. Astron. Astrophys.** 13: 511-535; 1975. 76 refs, 1964-75.

Review of what is known about pulsar-emission mechanisms.

141.506 Groth, E. J. Observational properties of pulsars. In: Gursky, Herbert; Ruffini, Remo, eds. **Neutron Stars, Black Holes and Binary X-Ray Sources.** Dordrecht, Holland: D. Reidel; 1975: 119-173. 336 refs, 1934-74.

Among the topics covered by this review are basic parameters (e.g., coordinates, period, dispersion measure) for 105 known pulsars; pulse propagation; pulsar supernova remnant associations; distances and distribution; flux density spectra; the pulse phenomena; and pulse timing.

141.507 Hewish, A. Pulsars. **Annu. Rev. Astron. Astrophys.** 8: 265-296; 1970. 102 refs, 1965-70.

A review of observations and theory, with an emphasis on the former. Includes a pulsar catalog of 50 objects as well as charts and graphs illustrating galactic distribution and periods.

141.508 Lenchek, Allen M., ed. **The Physics of Pulsars.** New York: Gordon and
C Breach; 1972. 173p. (Top. Astrophys. Space Phys.). 15 papers.

Papers based on a series of lectures at the University of Maryland in 1969 include a variety of topics for the advanced reader: radio, optical, and x-ray observations; measurement of periods; searching for pulsars; rotating neutron stars; pulsars and the origin of cosmic rays; etc. Parameters of 55 pulsars are included in a table at the end.

141.509 Manchester, R. N.; Taylor, J. H. Parameters of 61 pulsars. **Astrophys. Lett.** 10:
A 67-70; 1972. 59 refs, 1968-71.

References to the literature are included with the following tabular data: right ascension, declination, galactic coordinates, period, period derivative, period epoch, dispersion, pulse widths, mean pulse energies.

141.510 Manchester, Richard N.; Taylor, Joseph H. **Pulsars.** San Francisco: W. H. Freeman; 1977. 281p. (Ser. Books Astron. Astrophys.). 10 chapters. 394 refs, 1934-77.

A comprehensive introduction for advanced undergraduates, graduate students, and nonspecialists. Topics: general properties; characteristics of integrated pulse profiles; characteristics of individual pulses; the Crab Nebula and its pulsars; x-ray pulsars and binary systems; pulse timing observations; pulsars as probes of the interstellar medium; pulsar statistics and galactic distribution; the rotating neutron star model; pulse emission mechanics.

141.511 Maran, S. P.; Cameron, A. G. W. Progress report on pulsars. **Earth Extraterr. Sci.** 1: 3-25; 1969. 213 refs, 1968-69.

Presents data on known pulsars including pulse frequency, radio spectra, coordinates and dispersion measures; plus a discussion of three possible pulsar theories (white dwarfs, neutron stars, and intense magnetic fields).

141.512 Matheny, William D. **Pulsars Bibliography.** Oak Ridge, TN: Atomic Energy Commission, Technical Information Division; 1971. 64p. (TID-3320-Suppl-1; N71-38631). 380 refs, 1970.

A listing of citations on pulsars, neutron stars, black holes, supernova remnants, and cosmic x-ray sources taken from *Nuclear Science Abstracts*. Includes author, subject, and report number indexes. Two additional editions were released: 1972: 97p. (TID-3320-Suppl-2; N73-12909). 629 refs, 1971. 1973: 129p. (TID-3320-Suppl-3; N74-14526). 770 refs, 1972.

141.513 Papers presented at the joint Australia-U.S.S.R.-U.S.A. symposium on pulsars
C and supernova remnants. **Australian J. Phys.** 32: 1-136; 1979. 15 papers.

Sydney, Australia: Apr 1978. Selected topics: galactic distribution of pulsars; pulsar microstructure; polarization of pulsar radio emission; new catalog of galactic supernova remnants; structure of the Crab Nebula; optical observations of supernova remnants in the Magellanic Clouds; etc.

141.514 Partridge, R. B. Pulse astronomy: short time scale phenomena in electromagnetic and gravitational wave astronomy. In: Gursky, Herbert; Ruffini, Remo, eds. **Neutron Stars, Black Holes and Binary X-Ray Sources.** Dordrecht, Holland: D. Reidel; 1975: 29-45. 54 refs, 1960-75.

An overview of one aspect of "astronomy on a time scale of a few seconds or less," pulsed electromagnetic and other emissions. Aspects covered include gravitational radiation, pulsars, pulsating x-ray sources, and gamma-ray bursts.

141.515 **Pulsating Stars: A Nature Reprint.** New York: Plenum Press; 1968. 92p. 51 papers.

A collection of papers from *Nature*, describing early pulsar studies in Great Britain and the United States. Two introductory pieces by F. G. Smith and A. Hewish describe the discovery of pulsars and their physical properties, providing excellent background. The reprinted papers are arranged by topic: discovery; signal characteristics; optical measurements; theories; applications.

141.516 **Pulsating Stars 2: A Nature Reprint.** London: Macmillan; 1969. 116p. 64 papers.
Additional papers from *Nature* (see previous entry) arranged into four groups: 1) PSR 0833-45 and NP 0532: the fast pulsars; 2) distributions and distances; 3) theories; 4) observations: radio, optical, gamma-ray, x-ray.

141.517 Radhakrishnan, V. Fifteen months of pulsar astronomy. **Proc. Astron. Soc. Australia** 1: 254-263; 1969. 102 refs, 1939-69.
A summary of research and results in the months immediately following the discovery of pulsars. Data on 36 known objects gathered by six observatories are presented.

141.518 Rees, M. J., ed. Pulsars, cosmic rays and background radiation. **Highlights**
C **Astron.** 2: 723-767; 1971. 6 papers.
Joint discussion at the Fourteenth General Assembly of the IAU. Topics: astrophysical aspects of pulsars; origin of cosmic rays; surface composition of neutron stars; cosmic background radiation; etc.

141.519 Ruderman, M. Pulsars: structure and dynamics. **Annu. Rev. Astron. Astrophys.** 10: 427-476; 1972. 170 refs, 1966-72.
Review of general theoretical notions of pulsar structure and their consequences. Topics include general properties, neutron stars, glitches, fluctuations in periods, and pulses.

141.520 Sieber, W. Pulsar spectra: a summary. **Astron. Astrophys.** 28: 237-252; 1973. 73 refs, 1961-73.
Spectra of 27 pulsars, derived using observed pulse energy values, are presented in tabular form, along with related data used in the derivations. Plots of energy vs. various radio frequencies are presented for 10 of the objects. Derivations and results are discussed at length. Four different models of the spectral behaviors are discussed.

141.521 Smith, F. G. **Pulsars.** Cambridge, UK: Cambridge University Press; 1977. 239p. (Cambridge Monogr. Phys.). 21 chapters w/refs.
A thorough, non-elementary review of the phenomena and the related research done, with an emphasis on descriptive information. Among the many topics covered are discovery; identification with rotating neutron stars; x-ray pulsars; internal structure; the pulse phenomena; emission mechanism; radiative processes; etc. Includes a table of 105 pulsars, their positions and periods, as well as many diagrams, charts, and references.

141.522 Taylor, J. H. Catalog of 37 pulsars. **Astrophys. Lett.** 3: 205-208; 1969. 31 refs,
A 1968-69.
An early catalog listing coordinates, galactic coordinates, period, dispersion measure, pulse width, and references.

141.523 Taylor, J. H.; Manchester, R. N. Observed properties of 147 pulsars. **Astron. J.**
A 80: 794-806; 1975. 99 refs, 1968-75.
A compilation of the principal known quantities for all known pulsars.

141.524 Taylor, J. H.; Manchester, R. N. Recent observations of pulsars. **Annu. Rev. Astron. Astrophys.** 15: 19-44; 1977. 119 refs, 1967-77.
Concentration is on existing observational material relevant to the pulse emission mechanism, and the problem of the origin and evolution of pulsars.

141.525 Tsakadze, D. S.; Tsakadze, S. D. Superfluidity in pulsars. **Soviet Phys. Usp.** 18: 242-250; 1975. Transl. of *Usp. Fiz. Nauk.* 115: 503-519; 1975. 48 refs, 1932-73.

Reviews the general properties of pulsars and describes experiments performed on liquid helium-filled vessels to determine the hydrodynamic features of pulsar rotation.

The Crab Nebula. 125.001; 125.010; 125.015.

Neutron Stars. 066.506.

142 UV Sources, X-ray Sources, X-ray Background

142.001 A Amnuel, P. R.; Guseinov, O. H.; Radhamimov, S. Y. A catalog of x-ray sources. **Astrophys. J. Suppl. Ser.** 41: 327-367; 1979. 462 refs, 1954-78.
The catalog contains 517 objects known as of February 1978. Optical and radio counterparts are suggested.

142.002 Apparao, K. M. V.; Chitre, S. M. Celestial binary x-ray sources. **Space Sci. Rev.** 19: 281-404; 1976. 351 refs, 1932-75.
An in-depth review of x-ray binary systems including sections on orbital parameters, observational data, and theoretical aspects such as accretion.

142.003 Avni, Y.; Bahcall, J. N. Short-time optical variability of x-ray sources. **Astrophys. J.** 191: 221-230; 1974. 62 refs, 1965-74.
A formalism for interpreting the short-time optical variability caused by x-ray heating in variable x-ray sources is presented and applied to seven known sources.

142.004 Bahcall, J. N. Masses of neutron stars and black holes in x-ray binaries. **Annu. Rev. Astron. Astrophys.** 16: 241-264; 1978. 110 refs, 1924-78.
Presents a coherent summary of the methods used to determine the masses of x-ray sources in binary systems, and describes the presently available results. Special attention is given to the importance of the assumed geometry and the nature of possible systematic distortions that may affect observational parameters.

142.005 Bahcall, J. N. Optical properties of binary x-ray sources. In: Giacconi, R.; Ruffini, R., eds. **Physics and Astrophysics of Neutron Stars and Black Holes.** Amsterdam: North-Holland; 1978: 63-110. 240 refs, 1951-75.
A detailed review of studies and results obtained from observations of the eight known optically identified sources. Included are identification of each source, spectroscopy, photometry, distance, mass, magnitudes, etc. The author also addresses the astronomical techniques used in optical studies. Includes a summary table of important data for each source.

142.006 Bahcall, J. N.; Bahcall, N. A. Optical properties of binary x-ray sources. In: **Astrophysics and Gravitation.** Bruxelles: Editions de l'Universite de Bruxelles; 1974: 73-96. 147 refs, 1955-74.
Primarily a discussion of data obtained from observations of individual systems, this paper also briefly addresses mass determination for these systems. Systems described include Her X-1, Cen X-3, 3U 1700-37, Vela XR-1, SMC X-1, Cyg X-1, and Cyg X-3.

142.007 Baity, W. A. An x-ray astronomy bibliographic system. **Bull. American Astron. Soc.** 8: 454; 1976. An abstract.
A brief description of a computer database of approximately 2,000 references pertaining to non-solar research. Entries include title, source, authors, subjects, and serial number.

142.008 Basinska-Grzesik, E. Pozagalaktyczne zrodla promieniowania. [Extragalactic x-ray sources.] **Postepy Astron.** 27: 25-46; 1979. 97 refs, 1958-78. In Polish.

Review of extragalactic x-ray sources. The distinct classes of the objects, Seyfert galaxies and clusters of galaxies, are described in detail.

142.009 Beuermann, K. P. Galaktische Röntgenastronomie. [Galactic X-Ray Astronomy.] In: Schmidt-Kaler, Th., ed. **Optische Beobachtungsprogramme zur Galaktischen Struktur und Dynamik.** [Conference on Optical Observing Programs on Galactic Structure and Dynamics.] Bochum, West Germany: Astron. Inst. Ruhr-Univ.; 1975: 191-229. 90 refs, 1956-75. In German.

Bochum, West Germany: 19-21 Feb 1975. A review of recent research and current understanding of galactic x-ray sources. The paper draws upon satellite data (especially Uhuru) in its coverage of x-ray surveys, x-source population, stellar x-ray sources, mean luminosities, supernova remnants, the interstellar medium, and the diffuse x-ray background emission. Future developments are also presented.

142.010 Blumenthal, G. R.; Tucker, W. H. Compact x-ray sources. **Annu. Rev. Astron. Astrophys.** 12: 23-46; 1974. Also in: **Lick Obs. Contr.** 399. 149 refs, 1941-74.

An outline of progress and direction of current research. Sections include the general properties of x-ray stars as revealed by observations; discussion of mass transfer binary models; and an interpretation of observations of certain well-studied x-ray stars.

142.011 Blumenthal, G. R.; Tucker, W. H. Mechanisms for the production of x-rays in a cosmic setting. In: Giacconi, Riccardo; Gursky, Herbert, eds. **X-ray Astronomy.** Dordrecht, Holland: D. Reidel; 1974: 99-153. 51 refs, 1912-73.

A consideration of the various physical processes which produce x-rays in space: synchrotron radiation, Thomson scattering, Coulomb bremsstrahlung, and line emission.

142.012 Cavaliere, A. Fisica delle sorgenti x extragalattiche. [Physics of extragalactic x-ray sources.] **Mem. Soc. Astron. Italiana** 44: 571-597; 1973. 71 refs, 1956-73. In Italian.

The nature and characteristics of extragalactic x-ray sources are discussed under three main classes: compact sources, normal galaxies, clusters of galaxies, along with a fourth class of "unidentified x-ray objects."

142.013 Cooke, B. A., et al. The Ariel V (SSI) catalogue of high galactic latitude
A ($|b| > 10°$) x-ray sources. **Mon. Not. R. Astron. Soc.** 182: 489-515; 1978. 67 refs, 1958-78.

The 2A catalog is the result of 10,000 orbits of observation by the Leicester University Sky Survey Instrument on the Ariel V satellite, and contains 105 x-ray sources.

142.014 COSPAR symposium on fast transients in x- and gamma-rays, held at Varna,
C Bulgaria, 29-31 May 1975. **Astrophys. Space Sci.** 42: 3-254; 1976. 32 papers.

Emphasis is on recent observational results.

142.015 Culhane, J. L. X-rays from supernovae remnants. In: Schramm, David N., ed. **Supernovae.** Dordrecht, Holland: D. Reidel; 1977: 29-51. 81 refs, 1950-77.

A discussion of "the evolution of supernova remnants with special reference to the production of x-radiation and to the effect of homogeneities in the interstellar medium." The Crab Nebula and other remnants are discussed in detail.

142.016 Felten, J. E. "Local" theories of the x-ray background. **IAU Symp.** 55: 258-275; 1973. 72 refs, 1959-72.

Recent theories of the origins of the diffuse background x-rays are reviewed, with an emphasis on theories of the soft flux in the galactic plane and at the poles.

142.017 Felten, J. E. Theories of discrete x-ray and γ-ray sources. **IAU Symp.** 37: 216-237; 1970. 103 refs, 1956-70.

Critical review of theories of known discrete x-ray sources, excluding the Crab.

142.018 Fichtel, Carl E.; Stecker, Floyd W., eds. **The Structure and Content of the Galaxy and Galactic Gamma Rays.** Washington, DC: NASA; 1977. 369p. (NASA CP-002). 24 papers.
C

Greenbelt, MD, USA: 2-4 Jun 1976. A summary of recent satellite observations of gamma rays, as well as an attempt to relate galactic gamma-ray astrophysics to other fields of galactic astronomy. There are papers covering observations in the radio, infrared, optical, and ultraviolet portions of the spectrum. Also included are γ-ray results from the second Small Astronomy Satellite (SAS-2) and the COS-B satellite which detects γ-rays with energies $>$30 MeV.

142.019 Forman, W., et al. The fourth Uhuru catalog of x-ray sources. **Astrophys. J. Suppl. Ser.** 38: 357-412; 1978. 144 refs, 1958-78.
A

Positions and intensities are presented for 339 x-ray sources observed by the Uhuru (SAS-A) x-ray satellite observatory.

142.020 Gershberg, R. E. **Flares of Red Dwarf Stars.** Armagh, Northern Ireland: Armagh Observatory; 1971[?]. 224p. 4 chapters. 203 refs, 1939-70.

A survey of observational results and theoretical work on the topic of red dwarfs of UV Ceti type, eruptive variables in the same class as novae, supernovae, symbiotic stars, etc. Aimed at advanced students and scientists, the book covers four major topics: observational results; stellar flare hypotheses; the nebular model of flares of UV Ceti stars; flares of UV Ceti stars and some general problems of stellar instability.

142.021 Giacconi, R. Binary x-ray sources. **IAU Symp.** 64: 147-180; 1974. 44 refs, 1963-73.

Observational data concerning binary x-ray sources is reviewed, with emphasis on Her X-1, Cen X-3, and Cyg X-1.

142.022 Giacconi, Riccardo; Gursky, Herbert. **X-ray Astronomy.** Dordrecht, Holland: D. Reidel; 1974. 450p. (Astrophys. Space Sci. Libr., 43). 9 papers.

A review of current theory and observation. Topics: observational techniques; production of x-ray emission; solar x-ray emission; compact x-ray sources; supernova remnants; interstellar medium; extragalactic sources; cosmic x-ray background; etc. Includes a catalog of 161 x-ray sources known as of 1974.

142.023 Gorenstein, P.; Tucker, W. H. Soft x-ray sources. **Annu. Rev. Astron. Astrophys.** 14: 373-416; 1976. 162 refs, 1959-76.

Review of 14 objects that have been detected as discrete soft x-ray sources, including supernova remnants, a recurrent nova, flare stars, the coronas of nearby stars, SS Cygni, a white dwarf, and a soft component of emission from binary x-ray stars.

142.024 Gratton, L., ed. **Non-solar X- and Gamma-ray Astronomy.** Dordrecht, Holland: D. Reidel; 1970. 425p. (IAU Symp., 37). 57 papers, 4 abstracts.
C

Rome, Italy: 8-10 May 1969. Review of recent results from the high-energy region of the electromagnetic spectrum, including soft and hard x-rays and gamma rays. Includes discussions of new observational techniques, discrete x-ray sources, the relation of x-ray stars to other galactic objects, the possible phenomena responsible for strong x-ray emission, and background radiation.

142.025 Greisen, Kenneth. **The Physics of Cosmic X-ray, γ-ray, and Particle Sources.** New York: Gordon and Breach; 1971. 115p. (Top. Astrophys. Space Phys., 8). 7 chapters. 209 refs, 1941-69. Also in: **Astrophysics and General Relativity** (New York: Gordon and Breach; 1971).

A brief introduction for nonspecialists to high-energy physics in astronomy, covering known data and recent research. Topics: sources of high-energy particles; Energetic Solar Particles (ESP); discrete sources remote from the Sun; interactions with the interstellar medium; origin of primary electrons; cosmic gamma rays; cosmic x-rays.

142.026 Gursky, H.; Schreier, E. The binary x-ray stars—the observational picture. **IAU Symp.** 67: 413-464; 1975. 137 refs, 1959-74.

Current observational evidence is presented, along with a discussion of distributional properties of the sources, where they appear in the galaxy, and certain average characteristics.

142.027 Gursky, H.; Schreier, E. The galactic x-ray sources. In: Gursky, Herbert; Ruffini, Remo, eds. **Neutron Stars, Black Holes and Binary X-Ray Sources.** Dordrecht, Holland: D. Reidel; 1975: 175-220. 122 refs, 1961-74.

A summary of the various objects within our galaxy and their physical characteristics, based on observational evidence, examining both astronomical and astrophysical aspects. The authors contend that binary star systems account for all galactic x-ray sources, and they present a model of such a system for analysis.

142.028 Gursky, H.; Schwartz, D. A. Extragalactic x-ray sources. **Annu. Rev. Astron. Astrophys.** 15: 541-568; 1977. 154 refs, 1954-77.

Discussion of extragalactic objects that are likely to be intrinsic x-ray sources, including active galaxies, unidentified high-latitude sources, x-ray emitting clusters of galaxies (including a table of x-ray source clusters), and the diffuse x-ray background.

142.029 Hiltner, W. A.; Mook, D. E. Optical observations of extrasolar x-ray sources. **Annu. Rev. Astron. Astrophys.** 8: 139-160; 1970. 80 refs, 1942-69.

Concerned with sources for which there are well-established optical counterparts. Methods of optical identification and descriptions of the individual objects are also included.

142.030 Horstman, H. M.; Cavallo, G.; Moretti-Horstman, E. The x and δ diffuse background. **Nuovo Cimento, Riv.** 5: 255-311; 1975. 187 refs, 1951-74.

Discussion of experimental results and theoretical interpretations of the electromagnetic diffuse background from 0.1 keV to 100 MeV, with an emphasis on the definition of the intensity and coarse spectra of the diffuse background.

142.031 Kellogg, E. M. Extragalactic x-ray sources. In: Giacconi, Riccardo; Gursky, Herbert, eds. **X-ray Astronomy.** Dordrecht, Holland: D. Reidel; 1974: 321-357. 55 refs, 1956-74.

An overview of the following x-ray emitting objects: normal galaxies; active galaxies; quasistellar sources; Seyfert galaxies; clusters of galaxies; x-ray galaxies.

142.032 C Lamb, F.; Pines, D., eds. **Compact Galactic X-ray Sources: Current Status and Future Prospects.** Urbana, IL: University of Illinois Physics Department; 1979. 285p. 21 papers.

Workshop. Washington, DC, USA: 20-21 Apr 1979. A discussion of current knowledge and presentation of recent observational data. Future areas of research are discussed in terms of what can be accomplished given certain observational methods and conditions.

142.033 Lewin, W. H. G.; Joss, P. C. X-ray burst sources. **Nature** 270: 211-216; 1977. 80 refs, 1971-77.

Review of the x-ray burst phenomena, with discussion of possible origins of such bursts, including the Earth's magnetosphere. Includes data on eight x-ray sources from which bursts have been observed.

142.034 Markert, T. H., et al. The MIT/OSO 7 catalog of x-ray sources: intensities,
A spectra, and long-term variability. **Astrophys. J. Suppl. Ser.** 39: 573-632; 1979. 36 refs, 1968-79.

Summary of the observations of the cosmic x-ray sky performed by the MIT 1-40 keV x-ray detectors on the OSO 7 between October 1971 and May 1973. Mean intensities or upper limits are given for 185 sources (all third Uhuru or OSO 7 cataloged sources) in the 3-10 keV range.

142.035 Miyamoto, S.; Matsuoka, M. Sco X-1. **Space Sci. Rev.** 20: 687-775; 1977. 195 refs, 1957-77.

Review of the physical properties of x-ray, optical, and radio emissions from Sco X-1.

142.036 Pounds, K. A. Cosmic x-ray spectra. **Space Sci. Rev.** 13: 871-889; 1972. (IAU Colloq., 14). 68 refs, 1961-72.

A review of the most probable emission processes for x-ray sources and the data associated with known celestial objects in this category. X-ray sources covered include Scorpius X-1, the Crab Nebula, extragalactic sources, and the diffuse background radiation.

142.037 Pounds, K. A. Some comments on the transient x-ray sources. **Comments Astrophys.** 6: 145-156; 1976. 55 refs, 1954-76.

Brief review of current knowledge of transient x-ray sources. Includes a table of the 12 then known sources with their main properties.

142.038 Reinhardt, M. X-ray sources in binary systems. **Naturwissenschaften** 60: 532-538; 1973. 155 refs, 1962-73.

Discussion of the observed properties obtained from the Uhuru satellite. Interpretation of the data in the framework of currently discussed models is given in the last section.

142.039 Schwartz, D. The cosmic x-ray background. In: Giacconi, Riccardo; Gursky, Herbert, eds. **X-ray Astronomy.** Dordrecht, Holland: D. Reidel; 1974: 359-388. 79 refs, 1935-73.

A summary, primarily theoretical, of diffuse background x-ray emission not attributable to a single source. Topics: observations of background radiation; the diffuse x-ray spectrum; x-ray emissivity function; theories of the x-ray background; isotropy of the x-ray background; relation to discrete sources.

142.040 Silk, J. Diffuse x and gamma radiation. **Annu. Rev. Astron. Astrophys.** 11: 269-308; 1973. 131 refs, 1959-73.

A synthesis and update of views of diffuse x and gamma radiation, particularly over the energy range spanning soft x-rays to hard gamma rays. The contribution of diffuse galactic emission to the overall background radiation is also considered.

142.041 Tanaka, Y.; Bleeker, J. A. M. The diffuse soft x-ray sky. Astrophysics related to cosmic soft x-rays in the energy range 0.1-2.0 keV. **Space Sci. Rev.** 20: 815-888; 1977. 142 refs, 1954-77.

Review of the current status of the investigation of the soft x-ray background in the energy range 0.1-2.0 keV.

142.042 van den Heuvel, E. P. J., ed. X-ray binaries and compact objects. **Highlights**
C **Astron.** 4(1): 71-171; 1977. 14 papers.

Joint discussion no. 2 at the Sixteenth General Assembly of the IAU. Selected topics: periods in x-ray sources; recent transient x-ray sources; x-ray bursts; x-ray source models; etc.

142.043 C Venkatesan, D., ed. **International Conference on X-rays in Space (Cosmic, Solar and Auroral X-rays).** Calgary, AB: Department of Physics, University of Calgary; 1975. 2v. 1,236p. 48 papers.

Calgary, AB, Canada: 14-21 Aug 1974. Reports on current research and some reviews on the three subtopics in the title. Selected subjects include black holes; galactic x-ray sources; gamma ray bursts; the solar x-ray continuum; solar flare x-rays and gamma rays; soft x-ray astronomy; satellite and rocket observations; high-energy plasmas. Panel discussions, topical summaries, and abstracts are also included.

142.044 Walker, A. B. C., Jr. Spectroscopic analysis of solar and cosmic x-ray spectra. I. Nature of cosmic x-ray spectra and proposed analytical techniques. **Space Sci. Instrum.** 2: 9-51; 1976. 171 refs, 1928-76.

A review of the techniques used and the results obtained in the spectroscopic study of the x-ray spectra of the interstellar medium and cosmic sources.

142.045 Watson, M. G. Galactic x-ray sources. **Proc. R. Soc. London, Ser. A.** 366: 329-344; 1979. 110 refs, 1971-78.

The current status of galactic x-ray sources is reviewed from both theoretical and observational points of view.

142.046 Willmore, A. P. The cosmic x-ray sources. **Rep. Prog. Phys.** 41: 511-585; 1978. 116 refs, 1921-78.

Review with a strong emphasis on observations in the photon energy range 0.5-10 keV. Excludes diffuse x-ray background.

142.047 Wilson, A. S. X-ray galaxies. **Proc. R. Soc. London, Ser. A.** 366: 461-489; 1979. 96 refs, 1967-79.

Review of currently identified type 1 Seyfert galaxies, galaxies with active nuclei and relatively sharp emission lines, BL Lac objects, and quasars. The known members of these categories are listed.

142.048 Wilson, Robert M. **A Cosmic X-ray Astronomy Bibliography: The Astrophysical Journal, 1962 to 1972.** Huntsville, AL: Marshall Space Flight Center, NASA; 1972. 116p. 395 refs.

An alphabetical listing by first author of articles from *Astrophysical Journal* with no notes or annotations. Includes author and subject indexes and a chronological listing of articles.

142.049 C **X-ray Astronomy and Related Topics.** Neuilly, France: ESRO; 1975. 214p. (ESA SP-110). 20 papers.

ESRO Colloquium. Noordwijk, The Netherlands: 25-26 Feb 1975. (Proceedings). Besides review papers on different types of x-ray sources, the primary topic of this volume is the EXOSAT satellite and the x-ray astronomy experiments to be carried out. Individual experiments and instrumentation are discussed.

142.050 **X-ray Binaries.** Washington, DC: NASA; 1976. 757p. (NASA SP-389). 87 papers.

Greenbelt, MD, USA: 20-22 Oct 1975. (Proceedings). Provides detailed descriptions of known and theorized x-ray double stars as well as the equipment and satellites used to study them. Data and its interpretation are discussed, too. Includes tabular and diagrammatical data, as well as hundreds of references to the literature.

Binary systems with an x-ray component. *114.024.*

Compact x-ray sources. *061.069.*

HEAO Science Symposium. 061.030.

Interstellar medium. *131.021.*

The interstellar medium in x- and gamma-ray astronomy. *131.017.*

Neutron Stars, Black Holes and Binary X-Ray Sources. 066.504.

Objects of High Redshift. 141.001.

Recognition of Compact Astrophysical Objects. 065.027.

Supernova remnants. *125.012.*

26.3 MHz radio source survey. *141.021.*

Ultraviolet astronomy—new results from recent space experiments. *061.032.*

X-ray Astronomy. 061.007.

X-ray astronomy—1968 vintage. *061.075.*

X-ray Imaging. 032.003.

142.5 Gamma-ray Sources, Gamma-ray Background

142.501 Chupp, E. L. **Gamma-Ray Astronomy: Nuclear Transition Region.** Dordrecht, Holland; Boston: D. Reidel; 1976. 317p. (Geophys. Astrophys. Monogr., 14). 6 chapters. 467 refs, 1895-1975.

A survey of the field "in the energy region 10 keV to ~100 MeV, where nuclear lines are expected." Reviewing the major theoretical and experimental efforts since 1963, the author covers mechanisms for gamma-ray line and continuum production; theoretical estimates of gamma-ray emission; interaction of gamma-rays with matter; gamma-ray flux observations; and experimental considerations for nuclear gamma-ray astronomy.

142.502 Pinkau, K. Observation of celestial gamma rays. In: Osborne, J. L.; Wolfendale, A. W., eds. **Origin of Cosmic Rays.** Dordrecht, Holland: D. Reidel; 1975: 335-370. 59 refs, 1955-74.

A review of equipment, techniques, and the inherent problems in making observations. A discussion of results and data, including line emission, the diffuse flux, galactic emission, and localized sources. Also includes a detailed study of the cosmic radiation emanating from the Vela supernova remnant.

142.503 Porter, N. A.; Weekes, T. C. Gamma ray astronomy from 10^{11} to 10^{14} eV using the atmospheric Cerenkov technique. **Smithsonian Astrophys. Obs., Spec. Rep.** 381; 1978. 99p. 140 refs, 1948-77.

Basic principles underlying the atmospheric Cerenkov technique, whereby gamma rays are detected by the Cerenkov light emission from the small air showers that they create in the atmosphere, are reviewed, with particular reference to the experiments performed to September 1977.

142.504 Prilutskiĭ, O. F.; Rozental', I. L.; Usov, V. V. Intense bursts of cosmic gamma radiation. **Soviet Phys. Usp.** 18: 548-558; 1976. Transl. of *Usp. Fiz. Nauk.* 116: 517-538; 1975. 86 refs, 1965-75.

This review discusses the history of the discovery of intense bursts of cosmic gamma rays in the Vela group and their theoretical interpretation as galactic or metagalactic sources.

142.505 Schönfelder, V. Die diffuse kosmische Gammastrahlung. [The diffuse cosmic gamma radiation.] **Forschr. Phys.** 23: 1-69; 1975. 95 refs, 1955-74. In German.

A critical survey of our knowledge on the more or less directionally isotropic gamma radiation, which has not been identified with a discrete source.

142.506 Stecker, F. W. Gamma ray astrophysics. In: Osborne, J. L.; Wolfendale, A. W., eds. **Origin of Cosmic Rays.** Dordrecht, Holland: D. Reidel; 1975: 267-334. 121 refs, 1948-74.

A review of the processes that produce and absorb cosmic radiation. The appropriate data is interpreted, and its applications for cosmology and cosmic ray origin are covered.

142.507 Stecker, Floyd W.; Trombka, Jacob I., eds. **Gamma-ray Astrophysics.** Washington, DC: NASA; 1973. 412p. (NASA SP-339). 29 papers.
C

Greenbelt, MD, USA: 30 Apr—2 May 1973. (Proceedings). An examination of the phenomenon and the possible mechanisms of its origin. Papers cover both observational data and theory, as well as the cosmological implications of gamma rays, and future directions in the field.

142.508 Stecker, Floyd William. **Cosmic Gamma Rays.** Baltimore, MD: Mono Book Corporation; 1971. 246p. 14 chapters. 203 refs, 1930-70. Also published by NASA (1971) as NASA SP-249.

The production and interaction of cosmic gamma rays and the physical processes involved are the subjects of this advanced volume for graduate students and nonspecialists. Topics: gamma-ray production (four methods); galactic gamma rays and their production spectra; extragalactic gamma rays and cosmology (including an introduction to relativistic cosmology). Includes an appendix on gamma-ray telescopes.

142.509 Taylor, B. G., ed. **The Context and Status of Gamma-ray Astronomy.** Noordwijk, The Netherlands: ESRO Scientific and Technical Information Branch; 1974. 371p. (ESRO SP-106). 48 papers.
C

Ninth ESLAB Symposium. Frascati, Italy: 10-12 Jun 1974. A review of data, theory, and observations. Topics: cosmic gamma-ray bursts; the diffuse gamma-ray background and low energy gamma-ray astronomy; galactic emission and discrete gamma-ray sources; the mission of ESRO's COS-B satellite for gamma-ray astronomy.

142.510 Tindo, I. P. Galactic and extragalactic x-ray astronomy. **Itogi Nauki Tekh. Ser. Astron.** 9: 167-272; 1974. 318 refs, 1953-73. In Russian.

A survey and analytic commentary under the main headings: apparatus and research methods, discrete galactic sources, extragalactic sources, and the x-ray background. Items 1-32 of the bibliography are in Russian, and the remainder are in English.

142.511 Wills, R. D.; Battrick, B., eds. **Recent Advances in Gamma-ray Astronomy.** Noordwijk, The Netherlands: ESA Scientific and Technical Publication Branch; 1977. 381p. (ESA SP-124). 53 papers.
C

Twelfth ESLAB Symposium. Frascati, Italy: 24-27 May 1977. (Proceedings). Reports on equipment, techniques, and results. Papers include discussions of localized sources of high-energy radiation; galactic gamma radiation; cosmic gamma rays in the galaxy; the diffuse gamma-ray background; astrophysical and cosmological processes; instrumentation; measurement and analysis techniques; gamma-ray burst measurements.

COSPAR symposium on fast transients in x- and gamma-rays. *142.014.*

Diffusc x and gamma radiation. *142.040.*

The interstellar medium in x- and gamma-ray astronomy. *131.017.*

Non-solar X- and Gamma-ray Astronomy. 142.024.

The Physics of Cosmic X-ray, γ-ray, and Particle Sources. 142.025.

Theories of discrete x-ray and γ-ray sources. *142.017.*

The x and γ diffuse background. *142.030.*

143 Cosmic Radiation

143.001 Apparao, Krishna M. V. **Composition of Cosmic Radiation.** London; New York: Gordon and Breach; 1975. 86p. (Top. Astrophys. Space Phys., 11). 7 chapters. 281 refs, 1900-1971.

A review of the nature of the cosmic ray phenomena, its source and composition, based on data measured near the Earth. Topics: composition of cosmic rays near the Earth; cosmic radiation in the solar system; cosmic radiation in interstellar space; composition of cosmic rays at their source; cosmic rays and nucleosynthesis; origin of cosmic rays.

143.002 Basov, N. G., ed. **Cosmic Rays in the Stratosphere and in Near Space.** New York; London: Consultants Bureau; 1978. 179p. (Proc. [Tr.] P. N. Lebedev Phys. Inst., 88). 8 papers.

Reports of recent research in the Soviet Union. Topics: photon component of cosmic rays in the atmosphere; electron component of high-energy primary cosmic rays; 27-day cosmic-ray variations; solar-flare x-rays; emission of nuclei on the Sun; etc. Translation by James S. Wood of *Kosmicheskie luchi v stratosfere i v okolozemnom kosmicheskom prostranstve.*

143.003 Charakhch'yan, A. N.; Charakhch'yan, T. N. Investigation of cosmic rays in the stratosphere. In: Skobel'tsyn, D. V., ed. **Primary Cosmic Radiation.** New York; London: Consultants Bureau; 1975: 1-50. (Proc. [Tr.] P. N. Lebedev Phys. Inst., 64). 91 refs, 1939-70.

A review of the interaction between cosmic-ray particles and atomic nuclei in the atmosphere. Covered are energy spectra of cosmic-ray electrons and gamma rays in the stratosphere; energy spectrum of galactic cosmic rays; the origin of cosmic rays; and the 11-year modulation in the cosmic-ray intensity. Experiments and data are presented.

143.004 C Daniel, R. R.; Lavakare, P. J.; Ramadurai, S., eds. **Cosmic Ray Studies in Relation to Recent Developments in Astronomy and Astrophysics.** Bombay: Tata Institute of Fundamental Research; 1969. 374p. 18 papers.

Colloquium. Bombay, India: 11-16 Nov 1968. (Proceedings). A series of review papers and panel discussions exploring the physics of the cosmic-ray phenomenon and its relation to various astronomical phenomena such as x-rays, gamma rays, interstellar matter, quasars, solar neutrinos, etc.

143.005 Dorman, L. I. **Cosmic Rays: Variations and Space Explorations.** Amsterdam: North Holland; distr., New York: American Elsevier; 1974. 675p. 7 chapters. 1,061 refs, 1926-68.

An advanced text reviewing past research, current theory, and techniques of study. Emphasis is on cosmic-ray variation, i.e., changes in intensity. Topics: variations in cosmic rays as a means for investigating the cosmos; experimental methods for investigating cosmic-ray variations; meteorological cosmic-ray effects; the method of coupling coefficients; geomagnetic separation of cosmic rays; cosmic-ray variations of geomagnetic origin; problem of determining extraterrestrial variations.

143.006 Fabbri, R.; Melchiorri, F. The cosmological background radiation. **Mem. Soc. Astron. Italiana** 49: 153-195; 1978. 186 refs, 1826-1977.

Review of the properties of the background radiation in relation to recent developments in experimental and theoretical cosmology. Topics include Friedmann cosmology, background spectrum, the large-scale anisotropy, the fine-scale anisotropy, and linear polarization of the cosmological background.

143.007 Gaisser, T. K., ed. **Cosmic Rays and Particle Physics—1978.** New York: American Institute of Physics; 1978. 513p. (AIP Conf. Proc., 49). (Part. Fields Subser., 16). (DOE CONF-7810116). 42 papers.
C

Newark, DE, USA: 16-21 Oct 1976. (Proceedings). Scientific papers reporting on recent research, emphasizing the interaction between the two fields, especially particle interactions above 10 TeV. The question of cosmic-ray composition above 10^{13} eV is also covered, along with plans for new cosmic-ray and accelerator experimental facilities.

143.008 Harrison, E. R. Radiation in homogeneous and isotropic models of the universe. **Vistas Astron.** 20: 341-409; 1977. Also in: **Five College Obs. Contr.** 181. 220 refs, 1720-1974.

Covers the historical development of the theory of radiation in homogeneous and isotropic models of the universe. General properties of wave propagation in an expanding universe, containing a dispersive medium, are considered.

143.009 Hayakawa, Satio. **Cosmic Ray Physics: Nuclear and Astrophysical Aspects.** New York; London: John Wiley & Sons; 1969. 774p. (Intersci. Monogr. Texts Phys. Astron., 22). 6 chapters w/refs.

An advanced text for graduate students and scientists containing basic source data and hundreds of references. Topics: historical survey; interactions of high-energy particles with matter; very high energy interactions; behavior of cosmic rays in the atmosphere and underground; extensive air showers; origin of cosmic rays (astrophysical aspects).

143.010 Hillas, A. M. **Cosmic Rays.** Oxford, UK; New York: Pergamon Press; 1972. 297p. (Commonwealth Int. Libr. Sel. Readings Phys.). 9 chapters. 143 refs, 1901-1969.

Aimed at university students and nonspecialists, this book presents basic background information followed by 16 selected key papers from the literature. The first part of the volume covers discovery; the nature of the radiation; particles produced by cosmic rays; primary cosmic radiation; galactic radio waves; air showers; origin of cosmic rays; etc. The key papers cover observations, techniques of study, physics, and important discoveries.

143.011 **International Conference on Cosmic Rays.** 10th. Parts A & B published as follows: **Part A:** Prescott, J. R., et al., eds. *Invited and Rapporteur Papers.* Calgary, AB, Canada: University of Calgary; 1967. 572p. 18 papers. **Part B:** Wilson, M. D., ed. *Canadian J. Phys.* 46(10): parts 2, 3, 4; 1968. 1,171p. 286 papers. (part 1 is a regular part of journal).
C

Calgary, AB, Canada: 19-30 Jun 1967. Very brief papers and abstracts on the following topics: origin and galactic phenomena; modulation and geophysical effects; energetic solar particles; photons; muons and neutrinos; ultra-high-energy interactions; extensive air showers; techniques and instrumentation; gamma-ray astronomy; galactic x-ray sources; etc. Succeeding conferences cover the same or similar topics.

143.012 **International Conference on Cosmic Rays.** 12th. Hobart, Tasmania[?]: University of Tasmania[?]; 1971. 7v. 3,036p. 464 papers.
C

Hobart, Tasmania, Australia: 16-25 Aug 1971. See *143.011* for topics. Also a volume entitled *Invited Papers and Rapporteur Talks* (1971), 482p.

143.013 **International Conference on Cosmic Rays.** 15th. Budapest: Central Research Institute for Physics; 1977.
C

Plovdiv, Bulgaria: Aug 1977. No information available. See *143.011* for probable topics.

143.014 **International Cosmic Ray Conference.** 13th. Denver, CO: [?]; 1973. 5v. 3,750p. 648 papers.

Denver, CO, USA: 17-30 Aug 1973. See *143.011* for topics. Volume 5 includes miscellany, invited lectures, memorial sessions, rapporteur papers. Note variance in title from previous conference.

143.015 **International Cosmic Ray Conference.** 14th. Munich: Max-Planck-Institut für Extraterrestrische Physik; 1975. 12v. 4,461p. 806 papers.
C

Munich, West Germany: 15-29 Aug 1975. For topics, see *143.011.*

143.016 Meyer, P. Cosmic rays in the galaxy. **Annu. Rev. Astron. Astrophys.** 7: 1-38; 1969. 159 refs, 1948-68.

Selective review with an emphasis on general problems of galactic physics. Discusses the extrapolation of the cosmic radiation to interstellar space, followed by a summary of experimental facts and consideration of the implications.

143.017 Moraal, H. Observations of the eleven-year cosmic-ray modulation cycle. **Space Sci. Rev.** 19: 845-920; 1976. 331 refs, 1912-76.

Review of knowledge of the long-term modulation of the intensity of galactic cosmic rays.

143.018 Osborne, J. L.; Wolfendale, A. W., eds. **Origin of Cosmic Rays.** Dordrecht, Holland; Boston: D. Reidel; 1975. 466p. 17 papers.
C

NATO Advanced Study Institute. Durham, UK: 26 Aug–6 Sep 1974. A review of possible sources for cosmic rays, including supernovae, collapsed stars, and extragalactic sources.

143.019 Peterson, L. E. Hard cosmic x-ray sources. **IAU Symp.** 55: 51-73; 1973. 58 refs, 1967-72.

Review of the observational status of x-ray sources detected in the 2 $\simeq$ 500 keV range.

143.020 Rochester, G. D.; Wolfendale, A. W., eds. A discussion on the origin of the cosmic radiation. **Philos. Trans. R. Soc. London, Ser. A.** 277: 317-501; 1974. 12 papers.
C

Held 20-21 Feb 1974. Review of the present knowledge in this field and an attempt at synthesis thereof. Particular attention is on discoveries from balloon and space explorations as well as radio astronomy.

143.021 Rosen, Stephen, ed. **Selected Papers on Cosmic Ray Origin Theories.** New York: Dover Publications; 1969. 439p. 76 papers.

A collection of reprinted key articles from 1932 to 1966, tracing the development and documentation of the wide variety of origin theories. Included is the "Resource Letter CR-1 on Cosmic Rays" from the *American Journal of Physics* (35: 1-11; 1967), which includes 106 annotated references on cosmic rays useful for college-level work.

143.022 Shapiro, M. M. Composition and galactic confinement of cosmic rays. **Highlights Astron.** 2: 740-756; 1971. 88 refs, 1952-71.

Brief review of the distribution in abundance of the cosmic-ray elements arriving at the Earth, touching mainly upon work done at the Naval Research Laboratory.

143.023 Shapiro, M. M.; Silberberg, R. Cosmic-ray nuclei up to 10^{10} eV/u in the galaxy. **Philos. Trans. R. Soc. London, Ser. A.** 277: 319-347; 1974. 91 refs, 1959-74.

Recent progress in probing the composition and spectrum of ultra-heavy nuclei is outlined.

143.024 Shapiro, M. M.; Silberberg, R. Heavy cosmic ray nuclei. **Annu. Rev. Nucl. Sci.** 20: 323-392; 1970. 312 refs, 1934-70.

Critical summary of what is known of the chemical composition of the heavy primaries. Also discusses origin, propagation, and primordial composition of the radiation.

143.025 Sitte, K. The cosmic radiation and its significance. **Vistas Astron.** 19: 235-276; 1975. 127 refs, 1948-75.

Review of the Big Bang theory through a discussion of cosmic radiation, and a general review of the propagation of cosmic-ray particles.

143.026 Skobel'tsyn, D. V., ed. **Cosmic Rays and Nuclear Interactions at High Energies.** New York: Consultants Bureau; 1971. 228p. (Proc. [Tr.] P. N. Lebedev Phys. Inst., 46). 9 papers.

Reports on research in the Soviet Union. Topics: apparatus and methods of measuring high-energy particles; extensive air showers; investigation of the nuclear component of cosmic rays in a period of minimum solar activity; etc. Translation by Frank L. Sinclair of *Kosmicheskie luchi i i͡adernye vzaimodeĭstvii͡a vycokoĭ ėnergii.*

143.027 Somogyi, A., ed. **International Conference on Cosmic Rays.** 11th. Budapest:
C Akadémiai Kiado; 1970. 4v. 2,708p. 429 papers. (Supplement to *Acta Physica Academiae Scientiarum Hungaricae*, v.29).

Budapest, Hungary: 25 Aug—4 Sep 1969. See *143.011* for topics. An additional volume was also published: Bozóki, G., et al., eds. *Invited Papers and Rapporteur Talks.* Budapest: Central Research Institute for Physics; 1970. 612p.

143.028 Völk, H. J. Cosmic ray propagation in interplanetary space. **Rev. Geophys. Space Phys.** 13: 547-566; 1975. 125 refs, 1935-75.

Review of propagation characteristics as connected with structure of the solar wind plasma and electromagnetic fields.

143.029 Watson, A. A. Energy spectrum and mass composition of cosmic ray nuclei from 10^{12} to 10^{20} eV. In: Osborne, J. L.; Wolfendale, A. W., eds. **Origin of Cosmic Rays.** Dordrecht, Holland: D. Reidel; 1975: 61-95. 92 refs, 1953-74.

A revised view of the changes in primary energy intensity for the cosmic-ray energy spectrum above 10^{12} eV. The paper discusses in detail the composition and spectrum measurements of cosmic rays in three areas: 1) 10^{12} to 10^{14} eV; 2) 10^{14} to 10^{17} eV; and 3) above 10^{17} eV.

143.030 Wentzel, D. G. Cosmic-ray propagation in the galaxy: collective effects. **Annu. Rev. Astron. Astrophys.** 12: 71-96; 1974. 80 refs, 1960-74.

Review concentrating on the collective effects of cosmic rays on both cosmic-ray propagation and interstellar gas dynamics.

143.031 Wilson, J. G. **Cosmic Rays.** London: Wykeham Publications; New York: Springer-Verlag; 1976. 137p. (Wykeham Sci. Ser., 40). 7 chapters. no refs.

An introduction for students and laypersons emphasizing the physical processes involved in the production of cosmic particle radiation. Topics: discovery and identification; formation of secondary cosmic rays; solar modulation; charge and energy spectra of primary cosmic rays; cosmology of cosmic rays; galactic cosmic rays; origin of cosmic rays.

143.032 Worrall, D. M.; Wolfendale, A. W. Some aspects of the origin of cosmic rays. **Vistas Astron.** 19: 277-297; 1975. 85 refs, 1912-75.

Brief description of the main components of cosmic radiation, their energy spectra and energy densities, plus a discussion of the origin problem.

Gamma ray astrophysics. *142.506.*

Proceedings of the symposium on the techniques of solar and cosmic x-ray spectroscopy. *076.001.*

Pulsars, cosmic rays and background radiation. *141.518.*

STELLAR SYSTEMS, GALAXY, EXTRAGALACTIC OBJECTS, COSMOLOGY

151 Stellar Systems (Kinematics, Dynamics, Evolution)

151.001 Aarseth, S. J. Computer simulations of star cluster dynamics. **Vistas Astron.** 15: 13-37; 1973. 53 refs, 1938-71.

Intended as an introduction to numerical studies of small, self-gravitating stellar systems.

151.002 Aarseth, S. J.; Lecar, M. Computer simulations of stellar systems. **Annu. Rev. Astron. Astrophys.** 13: 1-21; 1975. 161 refs, 1942-74.

Concentrates mainly on collisional systems. Topics include N-body problems, galactic and globular clusters, clusters of galaxies, and interacting galaxies.

151.003 Astronomie extragalactique rencontre Sol-Espace. [Extragalactic Astronomy—Meeting "Sol-Espace."] **Ann. Physique** 4: 109-236; 1979. 18 papers. In French.

Paris, France: 11-13 Dec 1978. Reports of recent research on nearby extragalactic objects and active galaxies and quasars. Topics: theory of spiral structure; optical observations of bright nearby galaxies; the Hipparcos astrometric satellite; active nuclei of galaxies; cosmology; etc.

151.004 Audouze, J.; Tinsley, B. M. Chemical evolution of galaxies. **Annu. Rev. Astron. Astrophys.** 14: 43-79; 1976. 244 refs, 1955-76.

Summarizes observational and theoretical material that make up the present patchwork of galactic evolutionary understanding, leading to a discussion of models, constraints, and unsolved problems for chemical evolution in the solar neighborhood and in galaxies.

151.005 Balazs, B. A., ed. **The Role of Star Clusters in Cosmogony and in the Study of Galactic Structure.** Budapest: Roland Eötvös University; 1978. 158p. Russian/English.
C

Stellar Physics and Evolution. No. 6 Subcommission Symposium. Budapest, Hungary: 12-14 Sep 1977. A review of observational results and their meaning.

151.006 Barbanis, B.; Hadjidemetriou, J. D., eds. **Galaxies and Relativistic Astrophysics.** Berlin: Springer-Verlag; 1974. 247p. 31 papers.
C

First European Astronomical Meeting. Athens, Greece: 4-9 Sep 1972. (Proceedings, v.3). Reports of research by European astronomers on radio studies; neutral hydrogen; galactic nuclei; galactic dynamics; black holes; pulsars; and supernovae.

151.007 Berkhuijsen, Elly M.; Wielebinski, Richard, eds. **Structure and Properties of Nearby Galaxies.** Dordrecht, Holland; Boston: D. Reidel; 1978. 307p. (IAU Symp., 77). 28 papers, 46 abstracts.
C

Bad Münstereifel, Federal Republic of Germany: 22-26 Aug 1977. The aim of the symposium was to compare recent high-resolution radio continuum and H I-line results in nearby galaxies with both new and established optical material and with recent theoretical work. Sections included the smooth background, spiral structure and star formation, nearby galaxies of large angular size, nearby active galaxies and their nuclei, and the outskirts of galaxies.

151.008 Brosche, P.; Einasto, J.; Rümmel, U. **Bibliography on the Structure of Galaxies.** Heidelberg: Veröff. Astron. Rech.-Inst.; 1974. 104p. (Rep., 26). 13+ refs, 1909-1973.

"A basic list of references for studies of the inner structure of galaxies." Arranged in ascending NGC number order, galaxies are included if there has been observation of inner motion, if data exists allowing analysis of the structure, and if the distance does not exceed 20 Mpc. Each galaxy includes a number of references to papers (from as few as 2 to as many as 221 for M31), each coded as to the type of data or paper (e.g., review). The usual bibliographic data is included: author, source, beginning page, and year.

151.009 C Contopoulos, G.; Henon, M.; Lynden-Bell, D. **Dynamical Structure and Evolution of Stellar Systems.** Sauverny, Switzerland: Geneva Observatory; 1973. 260p. 4 papers.

Third Advanced Course of the Swiss Society of Astronomy and Astrophysics. Saas-Fee, Switzerland: 2-7 Apr 1973. Four highly mathematical lectures on recent developments in the field of stellar dynamics: 1) The density wave theory of spiral structure (Contopoulos, pp. 1-51); 2) Topological methods in stellar dynamics (Contopoulos, pp. 52-90); 3) Topics in the dynamics of stellar systems (Lynden-Bell, pp. 91-182); 4) Collisional dynamics of spherical stellar systems (Henon, pp. 183-260).

151.010 de Vaucouleurs, G. Structure, dynamics and statistical properties of galaxies. **IAU Symp.** 58: 1-53; 1974. 223 refs, 1847-1974.

Review of empirical and physical parameters describing individual and statistical properties of galaxies.

151.011 Einasto, J. Galactic models and stellar orbits. In: Mavridis, L. N., ed. **Stars and the Milky Way System.** Berlin: Springer-Verlag; 1974: 291-325. 127 refs, 1915-73.

A description of the construction of composite galactic models and their use, and a consideration of stellar orbits and kinematical properties of galactic populations. The paper also reviews new observational data and evolution of galaxies. Various physical parameters of galactic models are applied to six known star systems, including the Milky Way.

151.012 Faber, S. M.; Gallagher, J. S. Masses and mass-to-light ratios of galaxies. **Annu. Rev. Astron. Astrophys.** 17: 135-187; 1979. 264 refs, 1914-79.

A review of the evidence and theory supporting the contention that galaxies are more massive than optical data might indicate. Relying on mass-to-light ratios, the authors examine the "missing mass" problem for the Milky Way, spiral galaxies, E and S0 galaxies, binary galaxies, and groups and clusters of galaxies. A table of physical data on 51 galaxies with extended rotation curves is included.

151.013 Field, G. B. The formation and early dynamical history of galaxies. In: Sandage, Allan; Sandage, Mary; Kristian, Jerome, eds. **Galaxies and the Universe.** Chicago: University of Chicago Press; 1975: 359-407. 171 refs, 1928-69.

A review of the study of galaxy formation with an emphasis on work based on the Jeans hypothesis, first mentioned in 1928, which conjectured that galaxies condense from a gaseous background which is uniform and at rest.

151.014 Freeman, K. C. Stellar dynamics and the structure of galaxies. In: Sandage, Allan; Sandage, Mary; Kristian, Jerome, eds. **Galaxies and the Universe.** Chicago: University of Chicago Press; 1975: 409-507. 176 refs, 1907-1972.

An examination of dynamical properties of star systems is used to draw conclusions about the formation and early history of galaxies.

151.015 Galaxies. **IAU Trans. Rep. Astron.** 17A(3): 1-36; 1979. + refs, 1976-79.
An overview of research activity worldwide, with an emphasis on the previous three years, as reported by IAU Commission 28 (Galaxies). Additional reports: 14A: 301-318; 1970. + refs, 1967-70. 15A: 357-386; 1973. + refs, 1970-73. 16A(3): 1-35; 1976. + refs, 1973-76.

151.016 Gott, J. R., III. Recent theories of galaxy formation. **Annu. Rev. Astron. Astrophys.** 15: 235-266; 1977. 120 refs, 1926-77.
Review of observational data, description of how galaxies may originate in a standard Big Bang model, and consideration of several theories of formation.

151.017 Hayli, A., ed. **Dynamics of Stellar Systems.** Dordrecht, Holland; Boston:
C D. Reidel; 1975. 461p. (IAU Symp., 69). 38 papers, 4 abstracts.
Besancon, France: 9-13 Sep 1974. Topics include spherical systems, flattened systems, nuclei, and relativistic stellar dynamics.

151.018 Heggie, D. C. Binary evolution in stellar dynamics. **Mon. Not. R. Astron. Soc.** 173: 729-787; 1975. 90 refs, 1918-75.
A comprehensive theoretical picture of the behavior of binaries in n-body systems. Mainly concerned with the dynamics of encounters between binaries and other members of the system.

151.019 Heidmann, J.; Heidmann, N.; de Vaucouleurs, G. Inclination and absorption effects on the apparent diameters, optical luminosities and neutral hydrogen radiation of galaxies—I. Optical and 21-cm line data. II. Empirical properties. III. Theory and applications. **Mem. R. Astron. Soc.** 75: 85-141; 1971. 102 refs, 1940-71.
Study of the systematic effects, depending on inclination and on galactic and internal absorption, on observed optical luminosities and apparent diameters of galaxies and their 21-cm H I line emission.

151.020 Hunter, C. Self-gravitating gaseous disks. **Annu. Rev. Fluid Mech.** 4: 219-242; 1972. 65 refs, 1686-1970.
Review of the theoretical dynamical investigations inspired by the study of galaxies, especially spirals, and how they are bound gravitationally.

151.021 Kaplan, S. A.; Pikel'ner, S. B. Large-scale dynamics of the interstellar medium. **Annu. Rev. Astron. Astrophys.** 12: 113-133; 1974. 101 refs, 1950-73.
Review of the propagation of density waves in galaxies in relation to spiral structure, and large-scale interstellar shocks and their role in star formation.

151.022 Kaplan, S. A.; Tsytovich, V. N. Plasma processes in the universe. **Phys. Rep. Phys. Lett. C.** 7C: 1-33; 1973. 63 refs, 1940-72.
Discussion of the role of collective plasma processes in generating electromagnetic radiation and in accelerating relativistic particles in astrophysical objects such as galactic nuclei, quasars, or pulsars.

151.023 Lightman, A. P.; Shapiro, S. L. The dynamical evolution of globular clusters. **Rev. Mod. Phys.** 50: 437-481; 1978. 165 refs, 1902-1978.
Fundamental physical ideas underlying the dynamical behavior of globular star clusters are reviewed. The paper is intended as an introduction to recent developments in stellar dynamical theory.

151.024 Lin, C. C. Theory of spiral structure. **Highlights Astron.** 2: 88-121; 1971. 125 refs, 1932-70.

Discusses the nature of the problem of spiral structure, general spiral features observed, the theory of density waves, and application of the theory to the Milky Way system.

151.025 Lin, C. C. Theory of spiral structure. In: Koiter, W. T., ed. **Theoretical and**
C **Applied Mechanics.** Amsterdam: North-Holland; 1977: 57-69. 61 refs, 1955-76.

Fourteenth IUTAM Congress. Delft, The Netherlands: 30 Aug–4 Sep 1976. (Proceedings). A review of current developments in the density wave theory of galactic spiral structure. Observational data is compared to theory, and certain dynamical problems of galaxies are reviewed.

151.026 Lin, C. C.; Lau, Y. Y. Density wave theory of spiral structure of galaxies. **Stud. Appl. Math.** 60: 97-163; 1979. 110 refs, 1907-1979.

The density wave theory is studied as a dynamical problem; i.e., the gravitational instability of a galactic disk with respect to spiral modes. The asymptotic theory for tightly wound spirals is presented in detail.

151.027 Martinet, L.; Mayor, M., eds. **Galaxies.** Sauverny, Switzerland: Geneva Obser-
C vatory; 1976. 252p. 3 papers.

Sixth Advanced Course of the Swiss Society of Astronomy and Astrophysics. Saas-Fee, Switzerland: 29 Mar–3 Apr 1976. Contains three lectures of interest to advanced students: 1) Observational determination of the overall features (K. C. Freeman, pp. 1-65); 2) The formation of galaxies (R. B. Larson, pp. 67-154); 3) The evolution of chemical abundances and stellar populations (B. M. Tinsley, pp. 155-252).

151.028 Rohlfs, K. Die Dichtewellentheorie der Spiralstruktur. Erfolge und offene Probleme. [The density wave theory of spiral structure: results and unsolved problems.] **Mitt. Astron. Ges.** 43: 48-69; 1977. 239 refs, 1960-78. In German.

Concise review of the density wave theory, the successes of the theory in explaining a large class of observations, and also the controversies that still exist concerning the foundations.

151.029 Sandage, Allan; Sandage, Mary; Kristian, Jerome, eds. **Galaxies and the Universe.** Chicago: University of Chicago Press; 1975. 818p. (Stars Stellar Syst., 9). 19 papers. 2,388 refs, 1620-1975.

A collection of review articles covering characteristics and types of galaxies, distribution of galaxies in the universe, cosmology, and more.

151.030 Saslaw, W. C. The dynamics of dense stellar systems. **Publ. Astron. Soc. Pacific** 85: 5-23; 1973. 104 refs, 1809-1973.

Review of the main physical processes in dense stellar systems. Emphasis is on physical theory rather than observations.

151.031 Saslaw, W. C. Galactic nuclei and compact supermassive objects. **IAU Symp.** 69: 379-393; 1975. 56 refs, 1910-75.

Review of the observational evidence concerning compact supermassive objects, their formation and evolution, and their dynamical interaction with dense stellar systems in galactic nuclei.

151.032 Saslaw, W. C. Theory of galactic nuclei. **IAU Symp.** 58: 305-334; 1974. 124 refs, 1873-1974.

Review of current ideas about the possible constituents of galactic nuclei and the mechanisms for ejecting gas and massive objects.

151.033 C Setti, G., ed. **Structure and Evolution of Galaxies.** Dordrecht, Holland; Boston: D. Reidel; 1975. 334p. (NATO Adv. Stud. Inst. Ser. C., Math. Phys. Sci., 21). 15 papers.

NATO Advanced Study Institute. Erice, Sicily: 22 Jun–9 Jul 1974. A series of advanced lectures on the various aspects of galactic composition and evolution. Some of the subjects covered are stellar populations, stellar dynamics, models of galaxies, interstellar matter, radio continuum emission, galactic nuclei.

151.034 C Shakeshaft, J. R., ed. **The Formation and Dynamics of Galaxies.** Dordrecht, Holland: D. Reidel; 1974. 441p. (IAU Symp., 58). 32 papers, 2 abstracts.

Canberra, Australia: 12-15 Aug 1973. Topics include structure, dynamics and statistical properties of galaxies, intergalactic matter and radiation, galactic history, galactic nuclei, and spiral structure.

151.035 Shapley, Harlow. **Galaxies.** 3rd ed. Cambridge, MA: Harvard University Press; 1972. 232p. (Harvard Books Astron.). 9 chapters. no refs.

Aimed at the student and layperson, this standard text provides a basic overview and introduction in a nontechnical, highly readable manner. Topics: the Magellanic Clouds; the period-luminosity relation and light curves of Cepheids; the Milky Way, its features and constituents; nearby galaxies (including Andromeda); evolution of galaxies; types of star systems; clusters of galaxies; cosmology; etc. This edition was revised by Paul W. Hodge.

151.036 Tayler, R. J. **Galaxies: Structure and Evolution.** London: Wykeham Publications; distr., New York: Crane, Russak & Co.; 1978. 203p. (Wykeham Sci. Ser., 49). 9 chapters. 21 refs, no dates.

An intermediate text for university students and educated laypersons. Topics: observations of the galaxy; properties of external galaxies; stellar dynamics; masses of galaxies; the interstellar medium; chemical evolution of galaxies; galaxies and the universe.

151.037 Tinsley, B. M. Galactic evolution: program and initial results. **Astron. Astrophys.** 20: 383-396; 1972. 102 refs, 1952-72.

Description of a program to study the evolution of galaxies by tracing the evolution of stars in a model galaxy, starting with a stellar birthrate, which is a chosen function of stellar and galactic properties and time.

151.038 C Tinsley, B. M.; Larson, R. B., eds. **The Evolution of Galaxies and Stellar Populations.** New Haven, CT: Yale University Observatory; 1977. 449p. 14 papers.

New Haven, CT, USA: 19-21 May 1977. Scientific papers addressing the following subjects: systematic properties of galaxies; star formation in galaxies; chemical and stellar evolution in galaxies; origin and dynamical evolution of galaxies. An overview of current knowledge and data.

151.039 Toomre, A. Theories of spiral structure. **Annu. Rev. Astron. Astrophys.** 15: 437-478; 1977. 145 refs, 1850-1977.

Review of the various theories put forward to explain spiral arms of galaxies and spiral dynamics.

151.040 Unsöld, A. The chemical evolution of the galaxies. In: Barbanis, B.; Hadjidemetriou, J. D., eds. **Galaxies and Relativistic Astrophysics.** Berlin: Springer-Verlag; 1974: 84-103. 65 refs, 1957-73.

A brief review of the following topics: quantitative analysis of stellar spectra; abundances in "normal" stars (including a table of elemental abundances); anomalous abundance distributions produced by stellar evolution; origin of the elements; etc.

151.041 van der Kruit, P. C.; Allen, R. J. The kinematics of spiral and irregular galaxies. **Annu. Rev. Astron. Astrophys.** 16: 103-139; 1978. 270 refs, 1899-1978.

Review of the present status of kinematical studies of intermediate- and late-type galaxies, with an emphasis on observations that result in extensive maps of velocity fields.

151.042 C Weliachew, L., ed. **La dynamique des galaxies spirales.** [The Dynamics of Spiral Galaxies.] Paris: Editions du Centre National de la Recherche Scientifique; 1975. 534p. (Colloq. Int. Cent. Natl. Rech. Sci., 241). 43 papers. In English.

Bures-sur-Yvette, France: 16-20 Sep 1974. Reports of research as well as invited review papers on the following topics: 1) general concept of spiral structure; 2) spiral structure of gas; 3) spiral structure and the formation of stars; 4) spiral distribution of stars in galaxies; 5) barred spirals; 6) origin of spiral structure.

151.043 Whitford, A. E. Integrated energy distribution of galaxies. In: Sandage, Allan; Sandage, Mary; Kristian, Jerome, eds. **Galaxies and the Universe.** Chicago: University of Chicago Press; 1975: 159-176. 73 refs, 1934-73.

A review of the observational measurements of the spectral energy distribution of the integrated light from normal galaxies. The types of stars that make up a galaxy are presented in light of these studies.

151.044 Wielen, R. Density-wave theory of the spiral structure of galaxies. **Publ. Astron. Soc. Pacific** 86: 341-362; 1974. 166 refs, 1952-74.

Review of the gravitational interpretation of the spiral structure of galaxies in terms of density waves.

Lectures on Density Wave Theory. 022.071.

The Milky Way in comparison to other galaxies. *155.024.*

Stars and Galaxies from Observational Points of View. 061.044.

Stars and Star Systems. 061.078.

Stars and the Milky Way System. 061.054.

152 Stellar Associations

[No entries; cross-references only]

A photometric study of the Orion OB 1 association. *113.023.*

Star Clusters. 153.003.

Star clusters and associations. 153.004.

153 Open Clusters

153.001 A Fenkart, R. P.; Binggeli, B. A catalogue of galactic clusters observed in three colours. **Astron. Astrophys. Suppl. Ser.** 35: 271-275; 1979. 77 refs, 1958-78.

New photometric data for 190 galactic clusters all observed in UBV or RGU; a further extension of the Basle catalogs.

153.002 Hagen, G. L. An atlas of open cluster colour-magnitude diagrams. **Publ. David**
A **Dunlap Obs.** 4; 1970. 62p. 189 plates. 322 refs, 1921-69.

The purpose of the atlas is to draw together observations of galactic open clusters that have been studied on the UBV system, by presenting cluster color-magnitude diagrams in a uniform format. Data are included on 189 clusters.

153.003 Hesser, J. E.; ed. **Star Clusters.** Dordrecht, Holland; Boston: D. Reidel; 1980.
C 516p. (IAU Symp., 85). 23 papers, 64 abstracts.

Victoria, BC, Canada: 27-30 Aug 1979. Topics include associations, galactic structure, open clusters, and globular clusters.

153.004 Star clusters and associations. **IAU Trans. Rep. Astron.** 17A(3): 115-138; 1979. 116 refs, 1977-78.

An overview of research activity worldwide, with an emphasis on the previous three years, as reported by IAU Commission 37 (Star Clusters and Associations). Topics: galactic globular clusters; open clusters; associations; etc. Additional reports: 14A: 437-450; 1970. - refs, 1967-69. 15A: 571-597; 1973. 74 refs, 1968-73. 16A(3): 117-136; 1976. + refs, 1973-76.

153.005 Trimble, V. L. Binary stars in globular and open clusters. **IAU Symp.** 85: 259-279; 1980. 83 refs, 1937-80.

An overview of theory, with observations of binaries in globular clusters and open clusters, and recent conclusions that have been drawn from such work.

The ESO/Uppsala survey of the ESO(B) atlas of the southern sky. *158.020.*

Formation of open clusters. *065.004.*

154 Globular Clusters

154.001 Alcaino, G. Basic data for galactic globular clusters. **Publ. Astron. Soc. Pacific**
A 89: 491-503; 1977. 44 refs, 1927-77.

Forty-seven basic parameters are tabulated for 132 globular clusters in the galaxy. Partially replaces the Arp globular cluster catalog.

154.002 Harris, W. E. Spatial structure of the globular cluster system and the distance to the Galactic Center. **Astron. J.** 81: 1095-1116; 1976. 146 refs, 1918-76.

A list of distance moduli for 111 globular clusters is included.

154.003 Harris, W. E.; Racine, R. Globular clusters in galaxies. **Annu. Rev. Astron. Astrophys.** 17: 241-274; 1979. 202 refs, 1918-79.

An assessment of current knowledge about globular cluster systems as viewed as subsystems of their parent galaxies. Topics include galaxy halos, luminosity and color distribution of clusters, kinematics of cluster systems, and cluster populations. Included is a table of fundamental data for the 131 known galactic globular clusters.

154.004 Kraft, R. P. On the nonhomogeneity of metal abundances in stars of globular clusters and satellite subsystems of the galaxy. **Annu. Rev. Astron. Astrophys.** 17: 309-343; 1979. 155 refs, 1939-79.

Covers techniques for estimating cluster metal abundances; helium-abundance determinations; and the chemical composition of cluster stars. Includes a table of 38 galactic clusters and their metal abundances.

154.005 Philip, A. G. D. UBV color-magnitude diagrams of galactic clusters. **Vistas Astron.** 21: 407-445; 1977. 28+ refs, 1955-76.

Describes a catalog of over 37,000 UBV observations available in 165 color-magnitude and color-color plates. A bibliography of references is given, as well as a table of data for 115 clusters summarizing much of the current information concerning globular clusters.

154.006 Woltjer, L. The galactic halo: globular clusters. **Astron. Astrophys.** 42: 109-118; 1975. 54 refs, 1958-74.

New evaluation of data on globular clusters to study the suggestion that galactic halos may contain a large fraction of the total masses of galaxies.

Atlas of Galactic Globular Clusters with Colour Magnitude Diagrams. 115.001.

Binary stars in globular and open clusters. *153.005.*

The ESO/Uppsala survey of the ESO(B) atlas of the southern sky. *158.020.*

Star Clusters. 153.003.

Star clusters and associations. *153.004.*

UBV color-magnitude diagrams of galactic globular clusters. *113.016.*

Variable Stars in Globular Clusters and in Related Systems. 122.008.

155 Galaxy (Structure, Evolution)

155.001 Atanasijević, I. **Selected Exercises in Galactic Astronomy.** Dordrecht, Holland: D. Reidel; 1971. 144p. (Astrophys. Space Sci. Libr., 26).

A set of eight lengthy exercises for advanced students concerned with the structure and dynamics of the Milky Way. Problems include determining the position of the galactic equator; the distribution of globular clusters; non-cluster star fields; galactic rotation; etc. All problems but one can be solved using a hand calculator; the other requires a computer.

155.002 Basinska-Grzesik, E.; Mayor, M., eds. **Chemical and Dynamical Evolution of**
C **Our Galaxy.** Sauverny, Switzerland: Geneva Observatory; 1977. 319p. (IAU Colloq., 45). 43 papers.

Torun, Poland: 7-9 Sep 1977. Review papers and reports of recent research on both theoretical and observational aspects of the topic are presented. Topics: formation and evolution of disk galaxies; evolution of the central parts of the galaxy; chemical properties of the disk and halo; kinematics and dynamical evolution of the galaxy.

155.003 Becker, W.; Contopoulos, G., eds. **The Spiral Structure of Our Galaxy.**
C Dordrecht, Holland: D. Reidel; New York: Springer-Verlag; 1970. 478p. (IAU Symp., 38). 55 papers, 31 abstracts.

Basel, Switzerland: 29 Aug—4 Sep 1969. Sections included spiral structure in galaxies, observations of spiral structure in our galaxy, theory of spiral structure, and comparison of theory and observations.

155.004 Bok, B. J. Observational evidence for galactic spiral structure. **Highlights Astron.** 2: 63-87; 1971. 63 refs, 1951-70.

Review of observations that point to a spiral structure for the Milky Way.

155.005 Bok, Bart J.; Bok, Priscilla F. **The Milky Way.** 4th ed., rev. & enl. Cambridge, MA: Harvard University Press; 1974. 273p. (Harvard Books Astron.). 11 chapters. no refs.

This standard and well-known volume for laypersons and students gives a highly readable and informative picture of our galaxy. Including both optical and radio data, the book covers the galaxy as a whole, and looks at its individual constituents: stars, clusters (open and globular), nebulae, dust clouds, variable stars, gas, etc. Galactic evolution, dynamics, and features are presented, too, along with telescopes and the types of data they collect concerning the Milky Way.

155.006 Burton, W. B., ed. **The Large-Scale Characteristics of the Galaxy.** Dordrecht,
C Holland; Boston: D. Reidel; 1979. 611p. (IAU Symp., 84). 72 papers, 28 abstracts.

College Park, MD, USA: 12-17 Jun 1978. Sections include the disk component, spiral structure, galactic kinematics and distances, physical properties of the interstellar medium, the galactic nucleus, comparisons of our galaxy with other galaxies, the spheroidal component, galactic warp, and high-velocity clouds and the Magellanic Stream.

155.007 Burton, W. B. The morphology of hydrogen and of other tracers in the galaxy. **Annu. Rev. Astron. Astrophys.** 14: 275-306; 1976. 155 refs, 1955-76.

The focus of the review is on the relative overall morphological characteristics of constituents of the interstellar medium which can be observed along transgalactic paths: neutral atomic hydrogen, CO, H_2, ionized hydrogen, gamma radiation, synchrotron radiation, supernova remnants, and pulsars.

155.008 Caloi, V., ed. Chemical inhomogeneities in the galaxy. **Mem. Soc. Astron.
C Italiana** 50: 1-185; 1979. 18 papers.

Frascati, Italy: 29 May–2 Jun 1978. A workshop covering the chemical evolution and make-up of our star system and its constituents.

155.009 Chiu, H.-Y.; Muriel, A., eds. **Galactic Astronomy.** New York; London: Gordon
C and Breach; 1970. 2v. v.1: 334p. 6 papers; v.2: 300p. 8 papers.

Second Summer Institute for Astronomy and Astrophysics. Stony Brook, NY, USA: Summer 1968. (Proceedings). A series of basic lectures, with few or no references, covering a variety of topics. Volume 1 includes an introduction, followed by observational data resulting from both optical and radio studies. Volume 2 is concerned primarily with theory and the comparison of theory with observation.

155.010 Dickens, R. J.; Perry, Joan E., eds. **The Galaxy and the Local Group.**
C Herstmonceux, UK: Royal Greenwich Observatory; 1976. 293p. (R. Greenwich Obs. Bull., 182). 29 papers.

Tercentary Symposim. Herstmonceux Castle, UK: 22-25 Jul 1975. A wide-ranging compendium of papers on the structure, composition, and evolution of the Milky Way and the local group of galaxies near the Milky Way. Both theory and observational results for radio and optical studies are presented.

155.011 Gliese, W.; Jaschek, C.; McCarthy, M. Recent bibliography on the galactic polar caps. **Bull. Inf. Cent. Données Stellaires** 12: 11-25; 1977. 173 refs, 1964-76.

Entries include authors, title, year, source, volume, page, and *Astronomy and Astrophysics Abstracts* number or *Astronomischer Jahresbericht* number. Includes an author index. Errata appeared in 13: 27; 1977.

155.012 Green, A. J. The structure of the Milky Way. I. A radio continuum survey of
A the Galactic Plane at 408 MHz. **Astron. Astrophys. Suppl. Ser.** 18: 267-307; 1974. 89 refs, 1958-74.

Complete survey of the Galactic Plane as seen from the southern sky. Includes a table of 117 supernova remnants.

155.013 Heiles, C.; Jenkins, E. B. An almost complete survey of 21-cm line radiation for $|b| \geq 10°$ V. Photographic presentation and qualitative comparison with other data. **Astron. Astrophys.** 46: 333-360; 1976. 80 refs, 1959-75.

The main thrust of this study is to synthesize pictures and compare detailed surveys of data on the interstellar medium over broad areas of the sky. Photographs are presented that show the distribution in galactic coordinates of 21-cm line intensities in three velocity ranges.

155.014 Kerr, F. J. The large-scale distribution of hydrogen in the galaxy. **Annu. Rev. Astron. Astrophys.** 7: 39-66; 1969. 87 refs, 1957-69.

A survey of the known facts about the hydrogen distribution, stressing the many difficulties in the way of a complete understanding of the distribution and velocity field.

155.015 Lynden-Bell, D.; Yallop, B. D., eds. The galaxy and the distance scale. **Q. J. R.**
C **Astron. Soc.** 13: 130-302; 1972. 22 papers.

Herstmonceux Castle, UK: 17-20 Aug 1971. A conference in honor of Richard Woolley.

155.016 McCarthy, M. F.; Philip, A. G. D. Galactic structure in the direction of the
C polar caps. **Highlights Astron.** 4(2): 3-98; 1977. 30 papers.

Joint discussion no. 1 of the Sixteenth General Assembly of the IAU (Commissions 25, 33, and 45). General problems under discussion were the "missing mass" in the solar neighborhood and the variation in the luminosity function, and the different stellar populations and chemical abundances encountered at higher z distances from the plane of the galaxy.

155.017 Mavridis, L. N., ed. **Structure and Evolution of the Galaxy.** Dordrecht, Hol-
C land: D. Reidel; 1971. 312p. (Astrophys. Space Sci. Libr., 22). 16 papers.

NATO Advanced Study Institute. Athens, Greece: 8-19 Sep 1969. (Proceedings). Basic lectures giving a general outline of the field. Selected specific topics: systems and catalogs of proper motions; photometry; interstellar dust; distribution and motions of stars; radio emission; stellar evolution; evolution of the galaxy; x-ray sources in the Milky Way.

155.018 Mihalas, Dimitri; Routly, Paul M. **Galactic Astronomy.** San Francisco; London: W. H. Freeman; 1968. 257p. (Ser. Books Astron. Astrophys.). 14 chapters. 16 refs, 1914-65.

A basic text for advanced undergraduates emphasizing the structure and dynamics of the Milky Way. Chapters include astronomical terminology; physical properties of stars; stellar distribution; solar motion; galactic rotation; galactic dynamics; star clusters; spiral structure; etc. A technical, non-descriptive book.

155.019 Oort, J. H. The galactic center. **Annu. Rev. Astron. Astrophys.** 15: 295-362; 1977. 115 refs, 1956-77.

Coverage includes the gravitational field and mass distribution, expulsive phenomena, and the infrared core.

155.020 Peimbert, M. The early history of our galaxy: chemical evolution. **IAU Symp.** 58: 141-156; 1974. 112 refs, 1957-74.

General review of chemical abundance determinations, with particular emphasis on abundances of galactic and extragalactic metal-poor objects.

155.021 Perek, L., ed. Kinematics and ages of stars near the sun. **Highlights Astron.** 3:
C 365-465; 1974. 14 papers.

Joint discussion at the Fifteenth General Assembly of the IAU. Topics: age distributions; kinematical properties; solar neighborhood make-up; etc. Some of the types of stars considered are red variables, low luminosity stars, white dwarfs, M-dwarfs, etc.

155.022 Schmidt-Kaler, T. The spiral structure of our galaxy—a review of current studies. **Vistas Astron.** 19: 69-89; 1975. 107 refs, 1955-75.

The principal methods of determining spiral structure in the galaxy are described, and the value of different spiral tracers is assessed. The second part of the review deals with the internal structure of spiral features.

155.023 Structure and dynamics of the galactic system. **IAU Trans. Rep. Astron.** 17A(3): 37-72; 1979. + refs, 1976-79.

An overview of research activity worldwide, with an emphasis on the previous three years, as reported by IAU Commission 33 (Structure and Dynamics of the Galactic System). Topics: basic data and calibration problems; local galactic structure; overall structure; kinematics; etc. Additional reports: 14A: 357-385; 1970. + refs, 1967-69. 15A: 431-466; 1973. 142 refs, 1964-73. 16A(3): 37-72; 1976. + refs, 1973-76.

155.024 Tammann, G. A. Die Milchstrasse im Vergleich zu andern Galaxien. [The Milky Way in comparison to other galaxies.] In: Schmidt-Kaler, Th., ed. **Optische Beobachtungsprogramme zur Galaktischen Struktur und Dynamik.** [Conference on Optical Observing Programs on Galactic Structure and Dynamics.] Bochum, West Germany: Astron. Inst. Ruhr-Univ.; 1975: 1-58. 254 refs, 1942-75. In German.

A thorough review of the physical properties of our galaxy. Topics: galaxy type; expansion; luminosity; rotation and mass; gas content; structure; stellar populations; chemical composition; and place in the universe. The bibliography is quite complete.

155.025 Weaver, H. F. Some characteristics of interstellar gas in the galaxy. **IAU Symp.** 39: 22-50; 1970. 58 refs, 1952-70.

Discussion of structural features in the galaxy as they may involve or relate to gas dynamics.

Galaxies. 151.035.

The Role of Star Clusters in Cosmogony and in the Study of Galactic Structure. 151.005.

Stellar populations in galaxies. *158.047.*

A survey of OH near the galactic plane. *131.068.*

Theory of spiral structure. *151.024.*

156 Galaxy (Magnetic Field, Radio and Infrared Radiation)

156.001 Hyland, A. R. Galactic infrared astronomy. **Proc. Astron. Soc. Australia** 2: 14-20; 1971. 87 refs, 1920-71.

Brief review of some aspects of infrared techniques and astronomical observations of current importance.

156.002 Kerr, F. J.; Simonson, S. C., III, eds. **Galactic Radio Astronomy.** Dordrecht,
C Holland; Boston: D. Reidel; 1974. 654p. (IAU Symp., 60). 55 papers, 18 abstracts.

Queensland, Australia: 3-7 Sep 1973. Topics include the interstellar medium, galactic H II regions, supernova remnants, stellar and circumstellar sources, the galactic center, and large-scale galactic structure.

Cosmic Electrodynamics. 106.009.

157 Galaxy (UV, X, Gamma Radiation)

157.001 Daniel, R. R.; Stephens, S. A. Propagation of cosmic rays in the galaxy. **Space Sci. Rev.** 17: 45-158; 1975. 430 refs, 1904-1974.

Review of models in which cosmic rays are generated, propagated, and stored in the galaxy.

157.002 Gavazzi, G.; Sironi, G. Cosmic-ray electrons and galactic radio noise: some problems. **Nuovo Cimento, Riv.** 5: 155-186; 1975. 159 refs, 1950-74.

Interpretation of the electronic component of cosmic rays and the galactic component of the celestial radio background, both separately and together.

157.003 Ginzburg, V. L. The origin of cosmic rays (past, present, future). **Soviet Phys. Usp.** 21: 155-170; 1978. Transl. of *Usp. Fiz. Nauk.* 124: 307-331; 1978. 65 refs, 1934-78.

Discussion of historical aspects, followed by a number of questions concerning the present state of the problem of the origin of cosmic rays. Appendices present some information on the work of the Fifteenth International Conference on Cosmic Rays (Plovdiv, Bulgaria, August 1977).

157.004 Ginzburg, V. L.; Ptuskin, V. S. On the origin of cosmic rays: some problems in high-energy astrophysics. **Rev. Mod. Phys.** 48: 161-189; 1976. Also in: **Usp. Fiz. Nauk.** 117: 585-636; 1975. Transl. in *Soviet Phys. Usp.* 18: 931-959; 1975. 135 refs, 1949-76.

A review of the origin of cosmic rays. Primary attention is paid to galactic diffusion models with a halo and questions of cosmic-ray chemical composition, electron component, and synchrotron galactic radio emission.

157.005 Parker, E. N. Galactic effects of the cosmic-ray gas. **Space Sci. Rev.** 9: 651-712; 1969. 140 refs, 1935-69.

An exposition of some of the basic theoretical dynamical properties of the cosmic-ray gas, and of some of the observable physical effects that follow.

157.006 Willmore, A. P. The galactic x-ray sources. **Q. J. R. Astron. Soc.** 17: 400-421; 1976. 57 refs, 1962-76.

Review of the nature and discovery of galactic x-ray sources, excluding supernova remnants, based on satellite observations.

Extragalactic high energy astrophysics. *066.047.*

Non-solar X- and Gamma-ray Astronomy. 142.024.

Ultraviolet astronomy—new results from recent space experiments. *061.032.*

158 Single and Multiple Galaxies, Peculiar Objects

158.001 Adams, T. F. A survey of the Seyfert galaxies based on large-scale image-tube plates. **Astrophys. J. Suppl. Ser.** 33: 19-34; 1977. 109 refs, 1956-76.

Survey of the forms of the main bodies of Seyfert galaxies. Data on 80 Seyfert galaxies are summarized in Table 2.

158.002 Ambartsumian, V. A. Galaxies and their nuclei. **Highlights Astron.** 3: 51-66; 1974. 12 refs, 1957-73.

A brief overview of current knowledge of some properties of galaxies which are immediately connected with the activity of nuclei, and ultimately with the nature of nuclei themselves.

158.003 Balkowski, C. Statistical study of integral properties of galaxies measured in the 21-cm line. **Astron. Astrophys.** 29: 43-55; 1973. 53 refs, 1958-73.

Determination of the hydrogen masses, indicative total masses, and luminosities and ratios between these quantities have been put together to study statistically the integral properties of 149 galaxies measured in the 21-cm line.

158.004 Burbidge, E. M.; Burbidge, G. R. The masses of galaxies. In: Sandage, Allan; Sandage, Mary; Kristian, Jerome, eds. **Galaxies and the Universe.** Chicago: University of Chicago Press; 1975: 81-121. 154 refs, 1927-68.

A review including theories of mass determination, methods of observation, selected results, and discussions of mass-to-light ratios, masses of galaxies in clusters, and angular momentum. Includes four tables of galaxies for which complete or partial mass determinations have been made.

158.005 Burbidge, G. R. The nuclei of galaxies. **Annu. Rev. Astron. Astrophys.** 8: 369-460; 1970. 442 refs, 1908-1970.

A thorough review of observational evidence and theories of origin and evolution, followed by consideration of the effect of galactic nuclei on general galactic evolution and on cosmological problems.

158.006 Burton, W. B., ed. Nuclei of normal galaxies. **Highlights Astron.** 5: 129-225;
C 1980. 15 papers.

Topics include structure, atomic and molecular gas, infrared observations, radio observations, kinematics, optical spectra, stellar content, and star formation.

158.007 Buta, R. J. Revised classifications for 412 NGC and IC galaxies in the declina-
A tion zone -33° to -45°. **Univ. Texas Publ. Astron.** 12; 1978. 55p. 15 refs, 1888-1975.

Revised Hubble types are presented from studies of the Whiteoak extension.

158.008 Comte, G. Observations optiques des galaxies brillantes proches. [Optical observations of nearby bright galaxies.] **Ann. Physique** 4: 125-137; 1979. 75 refs, 1930-79.

Review of recent observational results on nearby bright spiral and elliptical galaxies.

158.009 Conference on problems of extragalactic research. **Astron. Nachr.** 297: 263-
C 322; 1976. 14 papers.

Potsdam, Germany: 29-31 Oct 1975. Organized by the Zentralinstitut für Astrophysik der Akademie der Wissenschaften der DDR. Participants were from Bulgaria, Hungary, and the USSR.

158.010 de Vaucouleurs, G. Contributions to galaxy photometry. I. Standard total
A magnitudes, luminosity curves, and photometric parameters of 115 bright galaxies in the B system from detailed surface photometry. **Astrophys. J. Suppl. Ser.** 33: 211-218; 1977. 54 refs, 1930-77.

Brief review of galaxy photometry over the past 30 years and its present status. Total magnitudes and related photometric parameters are defined and derived for 115 bright galaxies.

158.011 de Vaucouleurs, G.; de Vaucouleurs, A.; Corwin, H. G., Jr. **Second Reference**
A **Catalogue of Bright Galaxies.** Austin, TX; London: University of Texas Press; 1976. 396p. 595 refs, 1932-76.

This work contains information on 4,364 galaxies with references to papers published between 1964 and 1975. The present catalog should be considered as a companion volume to the original, not as a complete substitute. Includes all coordinate, classification, diameter, magnitude, color, radial velocity, and radio data in tabular form. Objects are listed in order of increasing 1950 right ascension.

158.012 Evans, David S., ed. **External Galaxies and Quasi-stellar Objects.** Dordrecht,
C Holland: D. Reidel; 1972. 549p. (IAU Symp., 44). 54 papers, 28 abstracts.

Uppsala, Sweden: 10-14 Aug 1970. Topics include galactic structure and composition, classification of compact objects, optical spectra, radio emission, theoretical considerations, intergalactic matter and evolution, and cosmological theories.

158.013 Fernie, J. D. The historical quest for the nature of the spiral nebulae. **Publ. Astron. Soc. Pacific** 82: 1189-1230; 1970. 193 refs, 1836-1970.

An interesting historical treatment of spiral nebulae, beginning with 1612 and going through Hubble's discovery in 1925 that they were undoubtedly external galaxies.

158.014 Fricke, K. J.; Reinhardt, M. Aktive Galaxien. [Active Galaxies.] **Naturwissenschaften** 62: 309-320; 1975. 357 refs, 1908-1974. In German.

Different forms of active galaxies are reviewed, including M82-type galaxies, Seyferts, the NGC phenomenon, N-galaxies, active elliptical galaxies, and tadpole galaxies.

158.015 Gisler, Galen R.; Friel, Eileen D. **Index of Galaxy Spectra.** Tucson, AZ: Pachart
A Publishing; 1979. 190p. (Ref. Works Astron.). (Astron. Astrophys. Ser., 10). 359 refs, 1956-78.

A list of 2,004 galaxies and their redshifts. Arranged by UGC number, this index lists the following data for each galaxy: NGC number; 1950 coordinates; photographic magnitude; Hubble type; distinguishing features; cluster and/or group (if applicable); recession velocity; spectrum; dispersion; references to the literature; notes. Appendices give data for 432 clusters of galaxies (position, characteristics, population, radius).

158.016 Gordon, K. J. History of our understanding of a spiral galaxy: Messier 33. **Q. J. R. Astron. Soc.** 10: 293-307; 1969. 130 refs, 1734-1968.

Historical study of M33.

158.017 Hazard, C.; Mitton, S., eds. **Active Galactic Nuclei.** Cambridge, UK; New
C York: Cambridge University Press; 1979. 317p. 15 papers.

NATO Advanced Study Institute on Active Galactic Nuclei. Cambridge, UK: Aug 1977. Papers based on lectures dealing with the astrophysics of nuclear regions of active galaxies, with an emphasis on theoretical considerations. Topics: quasars; Seyfert galaxies; black holes; optical spectra; etc.

158.018 Hodge, P. W. Dwarf galaxies. **Annu. Rev. Astron. Astrophys.** 9: 35-66; 1971. 89 refs, 1925-71.

Discussion of galaxies of small intrinsic size, small absolute luminosity, and low surface brightness. Dwarf elliptical and irregular galaxies are described, along with a review of the distribution of these objects.

158.019 Holmberg, E. Magnitudes, colors, surface brightness, intensity distributions, absolute luminosities, and diameters of galaxies. In: Sandage, Allan; Sandage, Mary; Kristian, Jerome, eds. **Galaxies and the Universe.** Chicago: University of Chicago Press; 1975: 123-157. 88 refs, 1926-69.

An overview of certain characteristics of different classes of galaxies based upon photometric observations. The problems encountered in the photometry of galaxies are briefly discussed.

158.020 Holmberg, E. B., et al. The ESO/Uppsala survey of the ESO(B) atlas of the
A southern sky. I. **Astron. Astrophys. Suppl. Ser.** 18: 463-489; 1974. II. **Astron. Astrophys. Suppl. Ser.** 18: 491-509; 1974. III. **Astron. Astrophys. Suppl. Ser.** 22: 327-402; 1975. IV. **Astron. Astrophys. Suppl. Ser.** 27: 295-342; 1977. V. **Astron. Astrophys. Suppl. Ser.** 31: 15-54; 1978. VI. **Astron. Astrophys. Suppl. Ser.** 34: 285-340; 1978. VII. **Astron. Astrophys. Suppl. Ser.** 39: 173-195; 1980. VIII. **ESO Sci. Prepr.** 105.

The aim of this program is to perform a systematic, thorough, and homogeneous search of the 606 plates contained in the ESO(B) atlas. The objects being searched are NGC + IC galaxies, all galaxies with a diameter larger than about $1'.0$, all disturbed galaxies, all star clusters in the Budapest Catalogue, and all listed planetary nebulae.

158.021 Huchra, J. P. The nature of Markarian galaxies. **Astrophys. J. Suppl. Ser.** 35:
A 171-195; 1977. 78 refs, 1931-77.

Integrated UBV photometry is given for 196 non-Seyfert Markarian galaxies, and compared with existing photometry of field galaxies.

158.022 Humphreys, R. M. Studies of luminous stars in nearby galaxies. I. Supergiants
A and O stars in the Milky Way. **Astrophys. J. Suppl. Ser.** 38: 309-350; 1978. 75 refs, 1953-78.

Includes a catalog of over 1,000 supergiants and O stars in associations and clusters.

158.023 Karachentseva, V. E. The catalogue of isolated galaxies. **Soobshch. Spets.**
A **Astrofiz. Obs.** 8: 1-72; 1973. 53 refs, 1950-72. In Russian.

This catalog, based on the Palomar Sky Survey, contains 1,051 isolated galaxies with apparent magnitudes $m \leq 15.7$ and with declination $\delta \geq -3°$.

158.024 Karachentseva, V. E.; Karachentseva, I. D.; Shcherbanovskij, A. L. Isolated
A triplets of galaxies. I. A list. **Astrofiz. Issled. Izv. Spets. Astrofiz. Obs.** 11: 3-39; 1979. 35 refs, 1937-77. In Russian.

A list of 84 isolated triple galaxies of the northern hemisphere with apparent magnitudes $m_p \leq 15.7$, prepared as a result of a total scanning of the Palomar Sky Survey.

158.025 Kellermann, K. I. Radio emission from compact objects. **IAU Symp.** 44: 190-213; 1972. 84 refs, 1961-71.

Brief discussion of observed radio emission from all of the various types of compact extragalactic objects. Topics covered include radio spectra, radio intensity variations, structure of compact radio sources, and energetics.

158.026 Khachikian, E. Y.; Weedman, D. W. An atlas of Seyfert galaxies. **Astrophys.**
A **J.** 192: 581-589; 1974. 60 refs, 1961-74.

Data for 71 Seyfert galaxies, along with representative spectra, are summarized and discussed.

158.027 King, I. R. Stellar populations in galaxies. **Publ. Astron. Soc. Pacific** 83: 377-400; 1971. 157 refs, 1922-71.

A historical outline of the concept of populations is followed by a summary of the present picture of stellar populations and the relation of populations to the forms of galaxies.

158.028 Kraan-Korteweg, R. C.; Tammann, G. A. A catalogue of galaxies within 10
A MPC. **Astron. Nachr.** 300: 181-194; 1979. Also in: **ESO Sci. Prepr.** no. 42, 1979. 78 refs, 1950-78.

The catalogue contains 179 known galaxies within 10 Mpc believed to be nearly complete for galaxies brighter than approximately $-18^{m}.5$. Represents the first attempt at compiling a distance-limited sample of galaxies.

158.029 Lewis, B. M. Systematic errors in velocities of galaxies. **Mem. R. Astron. Soc.** 78: 75-100; 1975. 260 refs, 1929-74.

The 21-cm estimate V_{21} of the systematic velocities of 202 galaxies is compared with all available optical estimates for the same objects.

158.030 Marochnik, L. S.; Suchkov, A. A. The structure of spiral galaxies. **Soviet Phys. Usp.** 17: 85-102; 1974. Transl. of *Usp. Fiz. Nauk.* 112: 275-308; 1974. 150 refs, 1928-74.

Review of observational data on the structure of spiral galaxies and the wave theory of the spiral structure.

158.031 O'Connell, D. J. K., ed. **Study Week on Nuclei of Galaxies.** Amsterdam; Lon-
C don: North-Holland; New York: American Elsevier; 1971. 795p. (Pontificiae Acad. Sci. Scr. Varia, 35). 28 papers.

Vatican City: 13-18 Apr 1970. Reports of research and reviews covering four major areas: observations and their interpretations; theory; observational cosmology; and galaxy evolution. Included are papers describing studies of various types of emission from galactic centers: radio, infrared, optical, etc. Several other papers center on quasars, objects which also emit great amounts of energy.

158.032 Osterbrock, D. E. Physical state of the emission-line region. **Phys. Scr.** 17: 285-292; 1978. Also in: **Lick Obs. Bull.** 775. 50 refs, 1958-77.

Short review of the physical properties of the ionized gas in the active nuclei of Seyfert and radio galaxies derived from their emission-line spectra.

158.033 Ozernoy, L. M. Galactic nuclei. In: Barbanis, B.; Hadjidemetriou, J. D., eds. **Galaxies and Relativistic Astrophysics.** Berlin: Springer-Verlag; 1974: 65-83. 111 refs, 1942-73.

A critical review of "the present state of the problem concerning the nature of ultimate energy source for active nuclei and quasars." Possible energy sources are compact star systems, an accreting black hole, and a supermassive body. Observational data is presented in support of theory.

158.034 Page, T. Binary galaxies. In: Sandage, Allan; Sandage, Mary; Kristian, Jerome, eds. **Galaxies and the Universe.** Chicago: University of Chicago Press; 1975: 541-556. 71 refs, 1919-68.

An overview of physical and dynamical characteristics, observational data, and theories of evolution and formation.

158.035 Richter, L.; Richter, N. Astrophysikalische statistik kompakter galaxien in der
A nähe des galaktischen nordpoles. Teil I. Katalog und auffindungskarten für 745 kompakte galaxien. [Astrophysical statistics of compact galaxies near the galactic north pole. Part I. Catalogue and finding charts of 745 compact

galaxies.] **Mitt. Karl-Schwarzschild-Obs. Tautenburg** 80; 1977. 16p. 16 refs, 1956-77. In German.

The catalog contains coordinates, integral magnitudes in V, U-B, B-V, and B-r, equidensitometric diameters, in the case of elliptical objects the major and minor axes, ellipticities, mean surface brightness in V, concentrating classes, separation in cluster and field galaxies, and finding charts.

158.036 Sandage, A. Classification and stellar content of galaxies obtained from direct photography. In: Sandage, Allan; Sandage, Mary; Kristian, Jerome, eds. **Galaxies and the Universe.** Chicago: University of Chicago Press; 1975: 1-35. 116 refs, 1620-1974.

A review of several classification systems and their bases, including selected photographs of different galaxy types for illustrative purposes.

158.037 Sandage, A. Optical redshifts for 719 bright galaxies. **Astron. J.** 83: 904-937;
A 1978. 106 refs, 1920-77.

New redshifts measured from approximately 1,000 plates taken at Stromlo and Palomar since 1968. These, together with new positions and galaxy types, are listed as a step toward completion of the redshift coverage of the Shapley-Ames catalog.

158.038 Sandage, A. The redshift. In: Sandage, Allan; Sandage, Mary; Kristian, Jerome, eds. **Galaxies and the Universe.** Chicago: University of Chicago Press; 1975: 761-785. 91 refs, 1899-1975.

A review "concerned in part with a few particular aspects of the measurement of redshifts, and in part with a simplified Friedmann theory of the redshift-apparent-magnitude and diameter relations in a nonstatic universe."

158.039 Spinrad, H.; Peimbert, M. The stellar and gaseous content of normal galaxies as derived from their integrated spectra. In: Sandage, Allan; Sandage, Mary; Kristian, Jerome, eds. **Galaxies and the Universe.** Chicago: University of Chicago Press; 1975: 37-80. 158 refs, 1935-74.

A summary of research on stellar populations and gaseous regions in galaxies, based on the interpretation of composite spectra.

158.040 Sramek, R. A.; Tovmassian, H. M. A radio survey of Markarian galaxies at 6 centimeters. **Astrophys. J.** 196: 339-345; 1975. 55 refs, 1963-75.

Discussion of a survey of all 506 galaxies in Markarian's first five lists at 6 cm wavelength with a 3 σ detection limit of about 30 mJy. Twenty-eight galaxies were detected.

158.041 Strom, S. E.; Strom, K. M. Disk and spheroidal components of external galaxies: an overview. **IAU Symp.** 84: 9-26; 1979. 63 refs, 1959-78.

Brief review of theory on formation, and a summary of recent optical observations of external galaxies. The primary emphasis is a discussion of the large-scale distribution and chemical composition of the stellar and gaseous constituents of relatively isolated galaxies.

158.042 Sulentic, J. W.; Tifft, W. G. **The Revised New General Catalogue of Nonstellar**
A **Astronomical Objects.** Tucson: University of Arizona Press; 1973. 384p. 127 refs, 1864-1971.

A working list of fundamental data on 8,167 bright nonstellar objects of all types. Includes for each object: identification, type, position, brightness, and description. Keyed to the Palomar Sky Survey.

158.043 Tammann, G. A. Hubble-konstante und Hubble-fluss. [Hubble constants and Hubble flow.] **Mitt. Astron. Ges.** 42: 42-58; 1977. 117 refs, 1950-77. In German.

The distance indicators within the Local Group are discussed, as well as the calibration and application of other distance indicators with wider range.

158.044 van den Bergh, S. The classification of active galaxies. **J. R. Astron. Soc. Canada** 69: 105-125; 1975. 97 refs, 1908-1975.

Discusses the classification of and the interrelations between Haro galaxies, Markarian galaxies, compact galaxies, N galaxies, Seyfert galaxies, and quasars. Based on a talk given at the Kingston (ON, Canada) Workshop on Theoretical Astronomy, January 16, 1975.

158.045 van den Bergh, S. The classification of normal galaxies. **J. R. Astron. Soc. Canada** 69: 57-76; 1975. 74 refs, 1918-75.

Review of galaxy classification systems, with special emphasis on the relations between galaxy classification, galactic evolution, and stellar content of galaxies.

158.046 van den Bergh, S. The extragalactic distance scale. In: Sandage, Allan; Sandage, Mary; Kristian, Jerome, eds. **Galaxies and the Universe.** Chicago: University of Chicago Press; 1975: 509-539. 117 refs, 1908-1969.

A review of methods for determining distances to galaxies, using comparisons of like objects in the Milky Way and in other star systems. Cited are the classical Cepheids, novae, RR Lyrae variables, globular clusters, and more.

158.047 van den Bergh, S. Stellar populations in galaxies. **Annu. Rev. Astron. Astrophys.** 13: 217-255; 1975. 179 refs, 1936-75.

A review of observational work leading to a model that the dominant stellar population in the nuclear bulges of the galaxy and M31 consists of old metal-rich stars, and that these regions contain only a sprinkling of true globular cluster stars.

158.048 van der Kruit, P. C.; Allen, R. J. The radio continuum morphology of spiral galaxies. **Annu. Rev. Astron. Astrophys.** 14: 417-445; 1976. 102 refs, 1933-76.

Review of radio continuum observations of galaxies that show structural details in the brightness distribution. Limited to normal galaxies.

158.049 Vorontsov-Velyaminov, B. A. Atlas of interacting galaxies, Part II, and the con-
A cept of fragmentation of galaxies. **Astron. Astrophys. Suppl. Ser.** 28: 1-117; 1977. 162 refs, 1958-76.

Includes 700 photographs of 600 objects found by the author on the Palomar Sky Survey and published in the Morphological Catalogue of Galaxies.

158.050 Weedman, D. W. Seyfert galaxies. **Annu. Rev. Astron. Astrophys.** 15: 69-95;
A 1977. 243 refs, 1943-77.

Review emphasizing the description of the Seyfert galaxy phenomenon as synthesized from recent observational studies; mostly on observations since 1970. Includes a listing of 88 Seyfert galaxies.

158.051 Weedman, D. W. Seyfert galaxies, quasars, and redshifts. **Q. J. R. Astron. Soc.** 17: 227-262; 1976. 51 refs, 1908-1976.

Review of the sequence of discovery for quasars and Seyfert galaxies together with recently obtained astrophysical data.

A catalogue of absorption lines in QSO spectra. *141.019.*

Masses and mass-to-light ratios of galaxies. *151.012.*

A Master List of Nonstellar Optical Astronomical Objects. R012.006.

Physics of Stars and Stellar Systems. R011.023.

Quasars and active nuclei of galaxies. *141.070.*

159 Magellanic Clouds

159.001 Feitzinger, J. V.; Weiss, G. A catalogue of radial velocities in the Large Magel-
A lanic Cloud. **Astron. Astrophys. Suppl. Ser.** 37: 575-589; 1979. 61 refs, 1953-77.

Presently known radial velocities of planetary nebulae, star clusters, H I and H II regions, and stars belonging in the LMC have been collected into a catalog containing 1,323 radial velocities.

159.002 Hodge, P. W.; Wright, F. W. **The Small Magellanic Cloud.** Seattle: University
A of Washington Press; 1977. 69p. 202 photographic charts. 35 refs, 1908-1974.

Prepared as a source of information regarding the nature and location of various kinds of objects so far discovered in this galaxy. Charts include variable stars, star clusters, bright H II regions, emission-line stars, and bright supergiants. Thirteen tables contain data necessary for working with the atlas, the most extensive being the one for variable stars.

159.003 Kilmartin, P. M., comp. **The Magellanic Clouds: A Bibliography 1951-1972.** Wellington, New Zealand: Carter Observatory; 1973. 68p. (Carter Obs. Astron. Bull., 79). 438 refs.

Primarily a compilation of journal article citations, this pamphlet also contains 5 entries under "charts" and 18 more under the heading "Review Papers and Symposia." Entries include author, title, source, volume, date, pagination, series, number of references, and an abstract. Arrangement is chronological, and by author for each year. A subject index, including named stars, is presented.

159.004 Muller, A. B., ed. **The Magellanic Clouds.** Dordrecht, Holland: D. Reidel;
C 1971. 189p. (Astrophys. Space Sci. Libr., 23). 24 papers.

Santiago de Chile: Mar 1969. A combination of short review papers, brief reports of observational results, and theoretical considerations, this volume provides a good overview of research in the field. The review papers concentrate on the physical properties of the clouds, including color-magnitude diagrams of different stellar populations.

159.005 Westerlund, B. E. The Magellanic Clouds. In: Barbanis, B.; Hadjidemetriou, J. D., eds. **Galaxies and Relativistic Astrophysics.** Berlin: Springer-Verlag; 1974: 39-61. 93 refs, 1959-72.

A review of recent observational data, including information on the brightest stars, x-rays, variable stars, dust content, and chemical composition. There are also discussions of distance moduli, cloud structure, and rotation of the clouds.

UBV photometry for supergiants of the Large Magellanic Cloud. *113.001.*

160 Groups of Galaxies, Clusters of Galaxies, Superclusters

160.001 Abell, G. O. Clusters of galaxies. In: Sandage, Allan; Sandage, Mary; Kristian, Jerome, eds. **Galaxies and the Universe.** Chicago: University of Chicago Press; 1975: 601-645. 171 refs, 1902-1974.

A review of observed properties, dynamics, and distribution. Includes references to catalogs of galaxies and clusters.

160.002 Bahcall, N. A. Clusters of galaxies. **Annu. Rev. Astron. Astrophys.** 15: 505-540; 1977. 231 refs, 1933-77.

Review of the observational properties of clusters of galaxies and some of the theoretical interpretations. Topics covered include catalogs, static properties, dynamics, x-ray emission, and radio emission.

160.003 Baier, F. W. Compilation of clusters of galaxies with known number-density
A distribution. **Astron. Nachr.** 299: 311-323; 1978. 84 refs, 1931-79.

Compilation of data on 145 clusters as well as the corresponding references. So far as known, data includes distance and richness, redshift, cluster type, color systems and limiting magnitudes, core radii, total radii, and total populations.

160.004 Braid, M. K.; MacGillivray, H. T. A finding list of southern clusters of galax-
A ies—I. **Mon. Not. R. Astron. Soc.** 182: 241-248; 1978. 10 refs, 1956-77.

A finding list for 474 rich clusters of galaxies in 99 fields of the southern sky.

160.005 Culhane, J. L. X-rays from clusters of galaxies. **Highlights Astron.** 4(1): 293-309; 1977. 39 refs, 1968-76.

Brief review of the available information on cluster spectra and structure, and a list of proposed identifications of x-ray sources with clusters.

160.006 de Vaucouleurs, G. The large-scale distribution of galaxies and clusters of galaxies. **Publ. Astron. Soc. Pacific** 83: 113-143; 1971. 152 refs, 1864-1971.

Mathematical tests and criteria used to distinguish between an observed distribution and a random, statistically uniform distribution are outlined. Observational evidence for and against hierarchical clustering is reviewed.

160.007 de Vaucouleurs, G. Nearby groups of galaxies. In: Sandage, Allan; Sandage, Mary; Kristian, Jerome, eds. **Galaxies and the Universe.** Chicago: University of Chicago Press; 1975: 557-600. 81 refs, 1932-74.

An examination of gravitationally bound groups of more than two galaxies within 20 Mpc. Among the groups cited are the Local Group and 14 other associations within 10 Mpc. Also contains several tables of data covering dozens of nearby groups, including information on populations, magnitudes, distances, etc.

160.008 Duus, A.; Newell, B. A catalog of southern groups and clusters of galaxies.
A **Astrophys. J. Suppl. Ser.** 35: 209-219; 1977. 11 refs, 1956-76.

The catalog contains all previously identified groups and clusters that have $\delta \lesssim -27°$, plus 710 newly located clusters out of a total of 770 identified. The survey was done on 97 ESO B film copies.

160.009 Jones, C.; Forman, W. X-ray clusters of galaxies and the luminosity-richness
A relation. **Astrophys. J.** 224: 1-13; 1978. 62 refs, 1958-78.

The second Ariel and fourth Uhuru catalogs of x-ray sources are combined to generate a list of x-ray sources associated with clusters of galaxies.

160.010 Layzer, D. Galaxy clustering: its description and its interpretation. In: Sandage, Allan; Sandage, Mary; Kristian, Jerome, eds. **Galaxies and the Universe.** Chicago: University of Chicago Press; 1975: 665-723. 51 refs, 1692-1975.

An overview of research concerned with an analysis of galaxy counts, dynamics of clustering, and the origin of clustering.

160.011 Longair, M. S.; Einasto, J., eds. **The Large Scale Structure of the Universe.**
C Dordrecht, Holland; Boston: D. Reidel; 1978. 464p. (IAU Symp., 79). 32 papers, 46 abstracts.

Tallinn, Estonia, USSR: 12-16 Sep 1977. Topics include galaxies in small groups, clusters of galaxies, large-scale systems, observational evidence for cosmological evolution, and the formation of structure in the universe.

160.012 Longair, M. S.; Riley, J. M., eds. Clusters of galaxies, cosmology and
C intergalactic matter. **Highlights Astron.** 4(1): 243-342; 1977. 13 papers.

Joint discussion no. 4 of the Sixteenth General Assembly of the IAU. Selected topics: density profiles of clusters of galaxies; the galaxy content of clusters; dynamical evolution of clusters of galaxies; x-ray and radio sources in clusters of galaxies; microwave background radiation in the direction of clusters; etc.

160.013 Ozernoy, L. M.; Reinhardt, M. Some selection effects in catalogues of groups of galaxies. **Astron. Astrophys.** 52: 31-42; 1976. 33 refs, 1933-76.

A critical examination of the catalogs of groups of galaxies by de Vaucouleurs (1965, 1973) and Karachentsev (1970). Special attention is paid to selection effects.

160.014 Rose, J. A. A catalogue of southern clusters of galaxies. **Astron. Astrophys.**
A **Suppl. Ser.** 23: 109-114; 1976. 9 refs, 1958-75.

A catalog of 124 probable clusters of galaxies in selected areas around the south galactic pole.

160.015 Shane, C. D. Distribution of galaxies. In: Sandage, Allan; Sandage, Mary; Kristian, Jerome, eds. **Galaxies and the Universe.** Chicago: University of Chicago Press; 1975: 647-663. 44 refs, 1784-1970.

A review including discussions of observations, galactic extinction, clouds of galaxies, and distribution in depth.

160.016 Tarter, J.; Silk, J. Current constraints on hidden mass in the Coma Cluster. **Q. J. R. Astron. Soc.** 15: 122-140; 1974. 56 refs, 1942-74.

Various possible means of accounting for the dynamical mass of the Coma Cluster are reviewed, including a summary of what current observational limits and theoretical concepts of cluster dynamics allow in the way of missing mass.

160.017 Turner, E. L.; Gott, J. R., III. Groups of galaxies. I. A catalog. **Astrophys. J.**
A **Suppl. Ser.** 32: 409-427; 1976. 19 refs, 1937-76.

A catalog of small groups of galaxies, generated by identifying regions of the north galactic cap in which the surface density of galaxies is enhanced. A total of 737 galaxies are assigned to 103 separate groups, while 350 galaxies are assigned to the field.

160.018 van den Bergh, S. Clusters of galaxies. **Vistas Astron.** 21: 71-92; 1977. 150 refs, 1932-77.

Available data on clusters of galaxies are reviewed. Properties of cD galaxies are described, and the origin of the angular momentum of galaxies is discussed.

Extragalactic x-ray sources. 142.008.

The Galaxy and the Local Group. 155.010.

H II regions in galaxies of the Local Group. *132.004.*

Objects of High Redshift. 141.001.

Studies of luminous stars in nearby galaxies. I. Supergiants and O stars in the Milky Way. *158.022.*

161 Intergalactic Matter

161.001 Burbidge, G. R. Intergalactic matter and radiation. **IAU Symp.** 44: 492-517; 1972. 115 refs, 1936-71.

A critical discussion of present knowledge of intergalactic matter and radiation as it relates to cosmology.

161.002 Field, G. B. Intergalactic matter. **Annu. Rev. Astron. Astrophys.** 10: 227-260; 1972. 153 refs, 1948-72.

Discussion of continuing experimental studies of the intergalactic matter and assessment of the significance of recent data.

Clusters of galaxies, cosmology and intergalactic matter. *160.012.*

162 Universe (Structure, Evolution)

162.001 Alpher, R. A.; Herman, R. Big Bang cosmology and the cosmic black-body radiation. **Proc. American Philos. Soc.** 119: 325-348; 1975. 218 refs, 1916-75.

Review of the physics of the Big Bang model of the universe, particularly describing the underlying cosmological models and the role played by non-equilibriuim theories of nucleosynthesis.

162.002 Balkowski, C.; Westerlund, B. E., eds. **Decalages vers le rouge et expansion de**
C **l'univers—l'evolution des galaxies et ses implications cosmologiques.** Paris: Centre National de la Recherche Scientifique; 1977. 619p. (IAU Colloq., 37). (Colloq. Int. Cent. Natl. Rech. Sci., 263). 50 papers.

Paris, France: 6-7 Sep 1976 (IAU Colloquium); 8-9 Sep 1976 (Colloquium International du CNRS). Review papers and reports of research on the expansion of the universe, the evolution of galaxies, and cosmology. Topics: redshifts; quasars; clusters of galaxies; the Hubble constant; radio sources; etc.

162.003 Barrow, J. D. Modern cosmological models. **Sci. Prog.** 65: 129-160; 1978. 105 refs, 1902-1977.

An outline of the observational and theoretical approaches to the cosmological problem is given. The Big Bang theory is described, with special reference to the microwave background radiation field and light element abundances. The second half of the article discusses various problems associated with this superficially successful approach to the cosmological problem.

162.004 Berry, Michael. **Principles of Cosmology and Gravitation.** Cambridge, UK; New York: Cambridge University Press; 1976. 179p. 8 chapters. 7 refs, 1959-74.

An intermediate undergraduate text aimed at describing the universe through observation, and presenting a theoretical framework for further, more advanced study. Topics: cosmography; physical basis for general relativity; curved spacetime and the physical mathematics of general relativity; general relativity near massive objects; cosmic kinematics; cosmic dynamics; in the beginning.

162.005 Bulletin GRG, no. 35. Cosmology: list of publications. **Gen. Relativ. Gravitation** 8: 1003-1037; 1977. 558 refs, 1965-76.

Consisting primarily of journal articles and conference papers, this worldwide bibliography covers all aspects of the topic. Its usefulness is greatly limited, however, because the arrangement (by authors' last names, with no subject arrangement or index) makes it difficult to locate information on a particular aspect of cosmology. This list covers authors with last names A through K. Topics: Hubble diagram; the microwave background; black holes; etc.

162.006 Bulletin GRG, no. 36. Cosmology: list of publications. **Gen. Relativ. Gravitation** 9: 721-756; 1978. 589 refs, 1965-76.
See previous entry. Authors K through Z.

162.007 Burbidge, G. Was there really a Big Bang? **Nature** 233: 36-40; 1971. 79 refs, 1922-71.
Review of the major scientific events which make up the case for a Big Bang.

162.008 Canuto, V. M. New trends in cosmology. **Nuovo Cimento, Riv.** 1: 1-42; 1978. 49 refs, 1946-77.
A brief review of historical cosmology, including the author's justifications for studying a scale-covariant theory of gravitation.

162.009 Cosmology. **IAU Trans. Rep. Astron.** 17A(3): 203-228; 1979. 138 refs, 1968-79.
An overview of research activity worldwide, with an emphasis on the previous three years, as reported by IAU Commission 47 (Cosmology). Topics: Hubble constant; radio astronomy and cosmology; space astronomy and cosmology; formation of galaxies; etc. Additional reports: 15A: 729-735; 1973. + refs, 1970-73. 16A(3): 137-157; 1976. + refs, 1973-76.

162.010 **Cosmology + 1.** San Francisco: W. H. Freeman; 1977. 113p. 11 chapters.
A collection of articles from *Scientific American* (1956-77) exploring some of the various aspects of our universe as a whole. Topics: stellar redshifts; curvature of space; cosmology and quasars; black holes; the expanding universe; SETI; etc. Contains biographical notes on the authors and selected references to the literature.

162.011 Criss, R. B., et al. Modern theoretical and observational cosmology. In: Shaviv, G.; Rosen, J., eds. **General Relativity and Gravitation.** New York: John Wiley & Sons; 1975: 33-107. 231 refs, 1897-1975.
"A review is given of recent observational and theoretical results in cosmology in light of general relativistic models of the Universe." Topics: isotropic cosmology; anisotropic homogeneous cosmology; inhomogeneities and galaxy formation; singularities and the origin of matter. The emphasis throughout is on solutions obtained from Einstein's theory of gravitation.

162.012 Demiánski, M., ed. **Physics of the Expanding Universe.** Berlin; New York:
C Springer-Verlag; 1979. 210p. (Lecture Notes Phys., 109). 8 papers.
Cracow School on Cosmology. Jodłowy Dwór, Poland: Sep 1978. Lectures discussing the "physical processes playing an important role in development and revealing the structure of the universe." Aimed at advanced students and scientists, this volume presents papers on mathematics of anisotropic spatially homogeneous cosmologies; creation of particles by gravitational fields; homogeneous cosmological models; formation of galaxies; cosmological microwave background blackbody radiation; clustering of galaxies; etc.

162.013 Ellis, G. F. R. Relativistic cosmology. In: Sachs, R. K., ed. **General Relativity and Cosmology.** New York: Academic; 1971: 104-182. 139 refs, 1917-70.
A rigorous paper in which the author constructs a cosmological model which predicts "the results of astronomical observations, and whose behaviour is determined by those physical laws which describe the behaviour of matter on scales up to those of clusters of galaxies." In obtaining the dynamical relations valid in any cosmological model, the author compares fluid dynamics in Newtonian theory with general relativity. Emphasis here is on theory rather than observational evidence.

162.014 Fall, S. M. Galaxy correlations and cosmology. **Rev. Mod. Phys.** 51: 21-42; 1979. 169 refs, 1896-1979.

A review of the efforts to estimate and interpret galaxy correlations, meant to provide both an introductory overview and a critical assessment of some recent developments.

162.015 Harrison, E. R. Standard model of the early universe. **Annu. Rev. Astron. Astrophys.** 11: 155-186; 1973. 186 refs, 1898-1973.

Discussion of the early universe during the period when the temperature was $10^{12} \geq T \geq 10^{3}$°K.

162.016 Hawking, S. W.; Ellis, G. F. R. **The Large Scale Structure of Space-Time.** Cambridge, UK: Cambridge University Press; 1973. 391p. 10 chapters. 176 refs, 1918-72.

An advanced, rigorous survey of the universe concerned primarily with the following: 1) "that the final fate of massive stars is to collapse behind an event horizon to form a 'black hole' which will contain a singularity" and 2) "that there is a singularity in our past which constitutes, in some sense, a beginning to the universe." Topics: gravity; differential geometry; general relativity; curvature; spacetime singularities; gravitational collapse and black holes; etc.

162.017 Jones, B. J. T. The origin of galaxies: a review of recent theoretical developments and their confrontation with observation. **Rev. Mod. Phys.** 48: 107-149; 1976. 387 refs, 340 B.C.-1976.

Lengthy discussion of the gravitational instability picture and the cosmic turbulence theory, with particular attention to the question of the origin of cosmic angular momentum and the nature of the initial conditions.

162.018 Landsberg, Peter T.; Evans, David A. **Mathematical Cosmology: An Introduction.** Oxford, UK: Clarendon Press; Oxford University Press; 1977. 309p. 11 chapters. 70 refs, 1744-1976.

An intermediate text with problem sets for university science students. Emphasis is on theory rather than observational evidences. Topics: fundamentals; Newtonian gravitation; the cosmological differential equation (the particle model and the continuum model); Friedmann models; optical observations and cosmological models; etc.

162.019 Maeder, Andre; Martinet, Louis; Tammann, Gustave, eds. **Observational Cos-**
C **mology.** Sauverny, Switzerland: Geneva Observatory; 1978. 321p. 3 papers.

Eighth Advanced Course of the Swiss Society of Astronomy and Astrophysics. Saas-Fee, Switzerland: 10-15 Apr 1978. The text of three lectures for graduate students on the observational evidence or basis for cosmological theory: 1) The Friedmann models and optical observations in cosmology (J. E. Gunn, pp. 1-124); 2) Radio astronomy and cosmology (M. S. Longair, pp. 125-257); 3) Growth and fate of inhomogeneities in a "Big Bang" cosmology (M. J. Rees, pp. 259-321).

162.020 Mirzoyan, L. V., ed. **Problems of Physics and Evolution of the Universe.** Yerevan, Armenia: Publishing House of the Armenian Academy of Sciences; 1978. 444p. 20 papers. English/Russian/French.

A collection of papers on a variety of subjects honoring V. A. Ambartsumian on his seventieth birthday. Topics: cosmology; galaxy formation; Herbig-Haro objects; radio emission from flare stars; novae physics; Markarian galaxies; etc. Translation of *Voprosy fiziki i evoli͡ut͡sii kosmosa.*

162.021 Novikov, I. D.; Zel'dovich, Y. B. Physical processes near cosmological singularities. **Annu. Rev. Astron. Astrophys.** 11: 387-412; 1973. 97 refs, 1925-73.

Review of the problem of the cosmological singularity, followed by a discussion of the implications of the theory for the creation of baryons, the equation of state of superdense matter, and the possibility of "bounce" at the singularity.

162.022 Partridge, R. B. Observational cosmology. In: Ruffini, Remo, ed. **Marcel Grossman Meeting on General Relativity.** 1st. Amsterdam; New York: North-Holland; 1977: 617-648. 83 refs, 1948-75.
C

Trieste, Italy: 7-12 Jul 1975. (Proceedings). A discussion of the Big Bang model, with a review of observational support for its occurrence. Covered are evidence for the expanding universe, the primeval fireball, and an isotropic and homogeneous universe. Standard and non-standard models are addressed.

162.023 Peebles, P. J. E. **Physical Cosmology.** Princeton, NJ: Princeton University Press; 1971. 282p. (Princeton Ser. Phys.). 8 chapters w/refs.

Aimed at the scientist, this advanced text is a guide to the popular questions and answers on the topic, addressing the following problems: classical cosmology; homogeneity and isotropy of the universe; Hubble's constant and the cosmic time scale; the mean mass density of the universe; the microwave background and the primeval fireball hypothesis; possible histories of the universe; the theoretical basis of the general cosmological models; helium production in the Big Bang.

162.024 Raychaudhuri, A. K. **Theoretical Cosmology.** Oxford, UK: Clarendon Press, Oxford University Press; 1979. 198p. (Oxford Stud. Phys.). 15 chapters. 833 refs, 1896-1978.

An advanced text aimed at graduate students, specialists, and nonspecialists, which reviews the field in some detail, examining the major and lesser-known theories. Topics: general relativity and relativistic cosmology; analysis of observational data; relativistic models; the microwave background radiation; the gravitational constant; cosmological models; formation of galaxies; etc.

162.025 Reines, Frederick, ed. **Cosmology, Fusion and Other Matters: George Gamow Memorial Volume.** Boulder, CO: Colorado Associated Universities Press; 1972. 320p. 19 papers.

A collection of semi-technical papers honoring the many varied interests of George Gamow, from atomic physics to cosmology. Written by colleagues, the articles cover such topics as the Big Bang; microwave astronomy and cosmology; the expanding universe; nuclear particles and reactions; nuclear fusion energy; etc.

162.026 Ryan, Michael. **Hamiltonian Cosmology.** Berlin; New York: Springer-Verlag; 1972. 169p. (Lecture Notes Phys., 13). 7 chapters. 63 refs, 1897-1970.

An advanced work considering "the study of cosmological models by means of equations of motion in Hamiltonian form." A review of past, present, and future work, the book covers ADM formalism applied to homogeneous cosmologies; the Hamiltonian formalism; applications to Bianchi-type universes; superspace; quantization.

162.027 Ryan, M. P., Jr.; Shepley, L. C. Resource letter RC-1: cosmology. **American J. Phys.** 44: 223-230; 1976. 106 refs, 1934-75.

A bibliography, with briefly annotated entries, divided into books, conference reports, current research topics, journals, and suggestions for courses of study. The useful introduction will lead the reader to the appropriate sources in the literature.

162.028 Scheuer, P. A. G. Radio astronomy and cosmology. In: Sandage, Allan; Sandage, Mary; Kristian, Jerome, eds. **Galaxies and the Universe.** Chicago: University of Chicago Press; 1975: 725-760. 130 refs, 1948-72.

A description of three ways "in which observations can be used in attempts to test cosmological models: by determining limits for the mean density of the Universe; by measurement of any residual radiation which may be left over from an early highly condensed state in an exploding cosmology; by studying the development of the radio source population."

162.029 Sciama, D. W. Astrophysical cosmology. In: Sachs, R. K., ed. **General Relativity and Cosmology.** New York: Academic; 1971: 183-236. 145 refs, 1917-70.

A review of the observational evidence for cosmological theory from the early twentieth century to the present. Topics: Robertson-Walker models of the universe; the intergalactic medium; the microwave background radiation; the molecular measurements; the thermal history of the universe; isotropy of the background; relation to Mach's principle; the helium problem; and high-energy interactions with black-body radiation.

162.030 Sciama, D. W. Early stages of the universe. **Highlights Astron.** 3: 21-35; 1974. 64 refs, 1938-74.

Review of theoretical cosmology in light of current knowledge from radio astronomy, general relativity, quantum field theory, and elementary particle physics.

162.031 Sciama, D. W. **Modern Cosmology.** Cambridge, UK: Cambridge University Press; 1971. 212p. 16 chapters. no refs.

An introduction for students and laypersons with backgrounds in mathematics and physics. The book first describes the components of the universe: stars, galaxies, quasars, etc., then moves to red shifts and finally on to cosmological theory and models of the universe. The Big Bang theory and cosmic microwave radiation are also discussed.

162.032 Segal, Irving Ezra. **Mathematical Cosmology and Extragalactic Astronomy.** New York; London: Academic; 1976. 204p. (Pure Appl. Math. Ser. Monogr. Textbooks). 5 chapters. 127 refs, 1920-75.

An advanced monograph which does not attempt to survey the field, but rather expounds the author's own theory, supported by observational evidence that is different from the expanding universe concept. "It is in large part an attempt to lay rational foundations for cosmology on the basis of the most elementary types of casuality and related symmetry considerations."

162.033 Steigman, G. Observational tests of antimatter cosmologies. **Annu. Rev. Astron. Astrophys.** 14: 339-372; 1976. 118 refs, 1928-76.

Discussion of the question of the amount and astrophysical role of antimatter in the universe, with an emphasis on the question of the symmetry of the universe.

162.034 Wesson, Paul S. **Cosmology and Geophysics.** Bristol, UK: Adam Hilger; 1978. 240p. (Monogr. Astron. Subj., 3). 8 chapters. 643 refs, 1922-78.

A review of "those cosmologies that have salient consequences for geophysics, and those theories of gravity which can draw observational support or censure from certain non-Newtonian processes observable in the Earth." Topics: variable-G cosmologies; gravitational theory; the Eddington numbers; the rotation of the Earth; the expanding Earth; expansion in gravity.

162.035 Wilson, R. W. The cosmic microwave background radiation. **Rev. Mod. Phys.** 51: 433-445; 1979. 54 refs, 1937-79.

Nobel lecture, 1978. Discussion of the discovery of the cosmic microwave background radiation.

Black Holes, Gravitational Waves and Cosmology: An Introduction to Current Research. 066.035.

Clusters of galaxies, cosmology and intergalactic matter. *160.012.*

Confrontation of Cosmological Theories with Observational Data. 066.026.

The counts of radio sources. *141.037.*

Extragalactic high energy astrophysics. *066.047.*

The formation of galaxies in Friedmannian universes. *066.011.*

Galaxies and the Universe. 151.029.

General Relativity and Cosmology. 066.039.

Gravitation and Cosmology: Principles and Applications of the General Theory of Relativity. 066.049.

The Large Scale Structure of the Universe. 160.011.

Lectures on General Relativity and Cosmology. 066.029.

Quantum Gravity. 066.021.

Radiation in homogeneous and isotropic models of the universe. *143.008.*

Radio astronomy and cosmology. *141.038.*

The redshift. *158.038.*

Relativity, Astrophysics and Cosmology. 066.022.

Unified Field Theory; General Relativity—Gravitation Theory—Cosmology; Volume 1. 066.046.

The Universe. 022.054.

APPENDIX I
INTERNATIONAL ASTRONOMICAL UNION COLLOQUIA

1. Caceres, O. Problem of variation of the geographical coordinates in the southern hemisphere. La Plata, Argentina, November 1968. La Plata Observatory.

2. Groth, H. G.; Wellman, P. Spectrum formation in stars with steady state extended atmospheres. Munich, April 1969. National Bureau of Standards SP 332; 1970.

3. Interstellar dust. Jena, August 1969. *Astron. Nachr.* 293: 1-77; 1971.

4. Slettebak, A. Stellar rotation. Columbus, OH, September 1969. D. Reidel; Gordon and Breach, 1970.

5. Heintz, W. D. La cohesion entre les procedes d'observation des etoiles doubles visuelles. Nice, September 1969. *Astrophys. Space Sci.* 11: 1-188; 1971.

6. Gyldenkerne, K.; West, R. M. Mass loss and evolution in close binaries. Elsinore, Denmark, September 1969. Copenhagen University Publications Fund, 1970.

7. Luyten, W. J. Proper motions. Minneapolis, April 1970. University of Minnesota, 1970.

8. Shore, B. W. Experimental techniques for the determination of fundamental spectroscopic data. London, September 1970. *Q. J. R. Astron. Soc.* 12: 48-60; 1971 (summary only).

9. Lederle, T.; Emerson, B. IAU system of astronomical constants. Heidelberg, August 1970. *Celest. Mech.* 4: 128-280; 1971.

10. Lecar, M. Gravitational n-body problem. Cambridge, August 1970. D. Reidel, 1972. Also in: *Astrophys. Space Sci.* 14: 3-178; 1971 and *Earth Extraterr. Sci.* 1: 185-191; 1971.

11. Seddon, H.; Smyth, M. J. Automation in astrophysics. Edinburgh, August 1970. *Publ. R. Obs. Edinburgh* 8: 1-211; 1971.

12. Gehrels, T. Physical studies of minor planets. Tucson, March 1971. National Aeronautics and Space Administration SP 267, 1971.

13. Hemenway, C. L.; Millman, P. M.; Cook, A. F. Evolutionary and physical properties of meteoroids. Albany, June 1971. National Aeronautics and Space Administration SP 319, 1973.

14. Gabriel, A. H. Ultraviolet and x-ray spectroscopy of astrophysical and laboratory plasmas. Utrecht, August 1971. *Space Sci. Rev.* 13: 488-889; 1972.

15. Strohmeier, W. New directions and new frontiers in variable star research. Bamberg, September 1971. Kleine Veröff. Remeis-Sternw. Bamberg, nr. 100; 1971.

16. Analytical procedures for elipsing binary light curves. Philadelphia, September 1971. *Astrophys. Space Sci.* 21: 7-12; 1973 (one article) and Villanova Univ. Obs. Contrib. no. 1, 1975 (one article).

17. Cayrel de Strobel, G.; Delplace, A. M. Age des etoiles. Paris, September 1972. Observatoire de Paris-Meudon, 1972.

18. Heintz, W. D. Orbital and physical parameters of double stars. Swarthmore, PA, April 1972. *J. R. Astron. Soc. Canada* 67: 49-87; 1971.

19. Jordan, S. D.; Avrett, E. H. Stellar chromospheres. Greenbelt, MD, February 1972. National Aeronautics and Space Administration SP 317, 1973.

20. Meridian astronomy. Copenhagen, September 1972. *IAU Symp.* 61: 7-21; 1974 (report only).

21. Fernie, J. D. Variable stars in globular clusters and in related systems. Richmond Hill, ON, August 1972. D. Reidel, 1973.

22. Cristescu, C.; Klepczynski, W. J.; Milet, B. Asteroids, comets, meteoric matter. Nice, April 1972. Academiei Republicii Socialiste Romania, 1974.

23. Gehrels, T. Planets, stars, and nebulae: studied with photopolarimetry. Tucson, November 1972. University of Arizona Press, 1974.

24. Moutsoulas, M. Lunar dynamics and observational coordinate systems. Houston, January 1973. *Moon* 8: 433-556; 1973. Revised abstracts published as Lunar Sci. Inst. Contrib. 135.

25. Donn, B. D., et al. Study of comets. Greenbelt, MD, November 1974. National Aeronautics and Space Administration SP 393, 1976.

26. Kolaczek, B.; Weiffenbach, G. On reference coordinate systems for earth dynamics. Torun, Poland, August 1974. Wykonano w Zakladzie Graficznym Politechniki Warszawskiej, 1975. Report in *Bull. Geod.* 112: 219-223; 1974.

27. Garton, W. R. S. UV and x-ray spectroscopy of astrophysical and laboratory plasmas. Cambridge, MA, September 1974. *Astrophys. Space Sci.* 38: 167-190, 313-380; 1975.

28. Burns, J. A. Planetary satellites. Ithaca, NY, August 1974. *Icarus* 24: 395-524; 25: 371-491, 569-594; 1975. *Celest. Mech.* 12: 1-110; 1975. University of Arizona Press, 1977.

29. Fitch, W. S. Multiple periodic variable stars. Budapest, September 1975. D. Reidel, 1976. Contributed papers published by Akadémiai Kiadó, 1976.

30. Gehrels, T. Jupiter. Tucson, May 1975. *Icarus* 27: 171-179, 335-459; 29: 165-328; 1976. *J. Geophys. Res.* 81: 3373-3422; 1976. Review papers, Pioneer papers, and a few contributed papers in book titled: *Jupiter: Studies of the Interior, Atmosphere, Magnetosphere and Satellites*, published by University of Arizona Press, 1976.

31. Elsaesser, H.; Fechtig, H. Interplanetary dust and zodiacal light. Heidelberg, June 1975. Springer-Verlag, 1976.

32. Weiss, W. W.; Jenkner, H.; Wood, H. J. Physics of Ap stars. Vienna, September 1975. Universitätssternwarte Wien, 1976.

33. Franz, O. G.; Pismis, P. Observational parameters and dynamical evolution of multiple stars. Oaxtepec, Mexico, October 1975. *Rev. Mexicana Astron. Astrofis.* 3: numero especial, 1977.

34. Morrison, D. Mercury. Pasadena, June 1975. *Icarus* 28: 429-609; 1976.

35. Jaschek, C.; Wilkins, G. A. Compilation, critical evaluation and distribution of stellar data. Strasbourg, August 1976. D. Reidel, 1977.

36. Bonnet, R. M.; Delache, P. Energy balance and hydrodynamics of the solar chromosphere and corona. Nice, September 1976. Clermont-Ferrand, 1977.

37. Balkowski, C.; Westerlund, B. E. Decalages vers le rouge et expansion de l'univers. Paris, September 1976. Centre National de la Recherche Scientifique, 1977.

38. Spiegel, E. A.; Zahn, J. P. Problems of stellar convection. Nice, August 1976. Springer-Verlag, 1978.

39. Delsemme, A. H. Relationship between comets, minor planets, and meteorites. Lyon, August 1976. University of Toledo Press, 1977, titled *Comets, Asteroids, Meteorites—Interrelations, Evolution and Origins.* Meeting review in *Icarus* 32: 251-254; 1977 and *Earth Extraterr. Sci.* 3: no. 3, 4; 1977.

40. Duchesne, M.; Lelievre, G. Astronomical applications of image detectors with linear response. Paris, September 1976. Observatoire de Paris-Meudon, 1977.

41. Szebehely, V. Dynamics of planets and satellites and theories of their motion. Cambridge, August 1976. D. Reidel, 1978.

42. Kippenhahn, R. The interaction of variable stars with their environment. Bamberg, September 1977. Veröff. Remeis-Sternw. Bamberg, nr. 121, 1977.

43. Fifth conference on ultraviolet and x-ray spectroscopy of astrophysical and laboratory plasmas. London, July 1977. Abstracts prepared by Physics Dept., Imperial College, London.

44. Jensen, E.; Maltby, P.; Orrall, F. Q. Physics of solar prominences. Oslo, August 1978. Blindern: Institute of Theoretical Astrophysics, 1979.

45. Basinska-Grzesik, E.; Mayor, M. Chemical and dynamical evolution of our galaxy. Torun, Poland, September 1977. Geneva Observatory, 1977.

46. Bateson, F. M.; Smak, J.; Urch, I. H. Changing trends in variable star research. Hamilton, NZ, November 1978. University of Waikato, 1979.

47. McCarthy, M. F.; Philip, A. G. D.; Coyne, G. V. Spectral classification of the future. Vatican City, July 1978. Vatican Observatory, 1979.

48. Prochazha, F. V.; Tucker, R. H. Modern astrometry. Vienna, September 1978. Vienna University Observatory, 1978.

49. Van Schooneveld, C. Formation of images from spatial coherence functions in astronomy. Groningen, August 1978. D. Reidel, 1979.

50. Davis, J.; Tango, W. J. High angular resolution stellar interferometry. College Park, MD, August 1978. University of Sydney, 1979.

51. Gray, D. F.; Linsky, J. L. Stellar turbulence. London, ON, August 1978. Springer-Verlag, 1980.

52. Gehrels, T. Protostars and planets. Tucson, January 1978. *Moon Planets* 19: 106-315; 1978 and 20: 2-101; 1979. Review papers in book from University of Arizona Press, 1978.

53. Van Horn, H. M.; Weidemann, V. White dwarfs and variable degenerate stars. Rochester, July 1979. University of Rochester, 1979.

54. Longair, M. S.; Warner, J. W. Scientific research with the space telescope. Princeton, August 1979. National Aeronautics and Space Administration CP 2111, 1980.

55. Sixth international colloquium on ultraviolet and x-ray spectroscopy of astrophysical and laboratory plasmas. Toronto, July 1980. No published proceedings.

APPENDIX II
INTERNATIONAL ASTRONOMICAL UNION SYMPOSIA
1969-1979

All listed are published by the D. Reidel Publishing Company. Parenthetical numbers following entries are to the section of this book wherein the full description appears.

37 Gratton, L. Non-solar X- and Gamma Ray Astronomy. (142.0)

38 Becker, W.; Contopoulos, G. The Spiral Structure of Our Galaxy. (155)

39 Habing, H. J. Interstellar Gas Dynamics. (062)

40 Sagan, C.; Owen, T. C.; Smith, H. J. Planetary Atmospheres. (093)

41 Labuhn, F.; Lüst, R. New Techniques in Space Astronomy. (061)

42 Luyten, W. J. White Dwarfs. (065)

43 Howard, R. Solar Magnetic Fields. (075)

44 Evans, D. S. External Galaxies and Quasi-stellar Objects. (150)

45 Chebotarev, G. A.; Kazimirchak-Polonskaya, E. I.; Marsden, B. G. The Motion, Evolution of Orbits, and Origin of Comets. (102)

46 Davies, R. D.; Smith, F. G. The Crab Nebula. (125)

47 Runcorn, S. K.; Urey, H. C. The Moon. (094)

48 Melchior, P.; Yumi, S. Rotation of the Earth. (044)

49 Bappu, M. K. V.; Sahade, J. Wolf-Rayet and High-Temperature Stars. (114)

50 Fehrenbach, Ch.; Westerlund, B. E. Spectral Classification and Multicolour Photometry. (114)

51 Batten, A. H. Extended Atmospheres and Circumstellar Matter in Spectroscopic Binary Systems. (120)

52 Greenberg, J. M.; van de Hulst, H. C. Interstellar Dust and Related Topics. (112)

53 Hansen, C. J. Physics of Dense Matter. (022)

54 Hauck, B.; Westerlund, B. E. Problems of Calibration of Absolute Magnitudes and Temperature of Stars. (113)

55 Bradt, H.; Giacconi, R. X- and Gamma-Ray Astronomy. (061)

56 Athay, R. G. Chromospheric Fine Structure. (073)

57 Newkirk, G., Jr. Coronal Disturbances. (062)

58 Shakeshaft, J. R. The Formation and Dynamics of Galaxies. (155)

59 Ledoux, P.; Noels, A.; Rodgers, A. W. Stellar Instability and Evolution. (065)

60 Kerr, F. J.; Simonson, S. C., III. Galactic Radio Astronomy. (112)

61 Gleise, W.; Murray, C. A.; Tucker, R. H. New Problems in Astrometry. (041)

62 Kozai, Y. The Stability of the Solar System and of Small Stellar Systems. (042)

63 Longair, M. S. Confrontation of Cosmological Theories with Observational Data. (066)

64 DeWitt-Morette, C. Gravitational Radiation and Gravitational Collapse. (066)

65 Woszcyk, A.; Iwaniszewska, C. Exploration of the Planetary System. (091)

66 Tayler, R. J. Late Stages of Stellar Evolution. (065)

67 Sherwood, V. E.; Plaut, L. Variable Stars and Stellar Evolution. (065)

68 Kane, S. R. Solar Gamma-, X- and EUV Radiation. (076)

69 Hayli, A. Dynamics of Stellar Systems. (151)

70 Slettebak, A. Be and Shell Stars. (064)

71 Bumba, V.; Kleczek, J. Basic Mechanisms of Solar Activity. (072)

72 Hauck, B.; Keenan, P. C. Abundance Effects in Classification. (114)

73 Eggleton, P.; Mitton, S.; Whelan, J. Structure and Evolution of Close Binary Systems. (117)

74 Jauncey, D. L. Radio Astronomy and Cosmology. (141)

75 De Jong, R.; Maeder, A. Star Formation. (131)

76 Terzian, Y. Planetary Nebulae. Observations and Theory. (135)

77 Berkhuijsen, E. M.; Wielebinski, R. Structure and Properties of Nearby Galaxies. (151)

78 Fedorov, E. P.; Smith, M. L.; Bender, P. L. Nutation and the Earth's Rotation. (043)

79 Longair, M. S.; Einasto, J. The Large Scale Structure of the Universe. (160)

80 Philip, A. G. D.; Hayes, D. S. The HR Diagram. The 100th Anniversary of Henry Norris Russell. (115)

81 Duncombe, R. L. Dynamics of the Solar System. (091)

82 McCarthy, D. D.; Pilkington, J. D. H. Time and the Earth's Rotation. (044)

83 Conti, P. S.; de Loore, C. W. H. Mass Loss and Evolution of O-type Stars. (064)

84 Burton, W. B. Thc Large-Scale Characteristics of the Galaxy. (155)

85 Hesser, J. E. Star Clusters. (152)

86 Kundu, M. R.; Gergely, T. E. Radio Physics of the Sun. (077)

87 Andrew, B. H. Interstellar Molecules. (131)

88 Plavec, M. J.; Popper, D. M.; Ulrich, R. K. Close Binary Stars: Observations and Interpretation. (117)

89 Tengström, E.; Teleki, G. Refractional Influences in Astrometry and Geodesy. (041)

90 Halliday, I.; McIntosh, B. A. Solid Particles in the Solar System. (104)

91 Dryer, M.; Tandberg-Hanssen, E. Solar and Interplanetary Dynamics. (074)

92 Abell, G. O.; Peebles, P. J. E. Objects of High Redshift. (141)

APPENDIX III
ABBREVIATIONS

AAAS	American Association for the Advancement of Science
AAS	American Astronautical Society; American Astronomical Society
AC	Astrographic Catalogue
AG	Catalog der Astronomischen Gesellschaft
AIP	American Institute of Physics
AJ	Astronomical Journal
ATM	Apollo Telescope Mount
AU	Astronomical Unit
AV	Audiovisual
BD	Bonner Durchmusterung
BIIEA	Bureau International d'Information sur les Ephemerides Astronomiques
BSI	Bibliographical Star Index
CERN	Conseil Europeen pour la Recherche Nucleaire
CETI	Communication with Extraterrestrial Intelligence
CIRP	Conference on Infrared Physics
CNO	Carbon, Nitrogen, Oxygen
CNRS	Centre National de la Recherche Scientifique
COS B	European x-ray and gamma-ray satellite
COSPAR	Committee on Space Research (ICSU)
DDO	David Dunlap Observatory
Dec	Declination
DM	Durchmusterung
ELF	Extremely Low Frequency

ESA	European Space Agency
ESLAB	European Space Laboratory
ESO	European Southern Observatory
ESP	Energetic Solar Particles
ESRIN	European Space Research Institute (ESRO)
ESRO	European Space Research Organization
ESTEC	European Space Technology Institute
EUV	Extreme Ultraviolet
eV	Electron Volt
FK3	Dritter Fundamentalkatalog
FK4	Fourth Fundamental Catalogue
FTS	Fourier Transform Spectroscopy
GC	General Catalogue
GHz	Gigahertz
GSFC	Goddard Space Flight Center
HD	Henry Draper Catalogue
HEAO	High Energy Astronomy Observatory
IAA	International Aerospace Abstracts
IAF	International Astronautical Federation
IAGA	International Association of Geomagnetism and Aeronomy
IAU	International Astronomical Union
ICO	International Commission for Optics
ICSU	International Council of Scientific Unions
IEE	Institution of Electrical Engineers
IEEE	Institute of Electrical and Electronic Engineers
IGY	International Geophysical Year
IMF	Interstellar Magnetic Field
IMS	International Magnetospheric Study

IQSY	International Years of the Quiet Sun
IR	Infrared
IRE	Institute of Radio Engineers
ISAS	Institute of Space and Aeronautical Science
ISM	Interstellar Medium
IUGG	International Union of Geodesy and Geophysics
IUPAP	International Union of Pure and Applied Physics
IUTAM	International Union of Theoretical and Applied Mechanics
JOSO	Joint Organization for Solar Observation
JPL	Jet Propulsion Laboratory
Jy	Jansky (unit of flux density)
keV	10^3 Electron Volts
LHS	Luyten Half-Second
LMC	Large Magellanic Cloud
LPL	Lunar and Planetary Laboratory
LST	Large Space Telescope
LTE	Local Thermodynamic Equilibrium
MeV	10^6 Electron Volts
MHD	Magnetohydrodynamics
MIT	Massachusetts Institute of Technology
MK/MKK	Morgan, Keenan, Kellman Spectral Classification System
Mpc	Megaparsec
NAS	National Academy of Sciences
NASA	National Aeronautics and Space Administration
NATO	North Atlantic Treaty Organization
NBS	National Bureau of Standards
NGC	New General Catalogue
NOAA	National Oceanic and Atmospheric Administration

NRC	National Research Council (Canada)
NRL	Naval Research Laboratory
NSRDS	National Standard Reference Data Series
NSSDC	National Space Science Data Center (GSFC)
NTIS	National Technical Information Service
OAO	Orbiting Astronomical Observatory
OH	Hydroxyl Radical (Interstellar Molecule)
OSO	Orbiting Solar Observatory
pc	Parsec
PCA	Polar Cap Absorption
PSS	Palomar Sky Survey
QSO	Quasistellar Object
RA	Right Ascension
RAE	Radio Astronomy Explorer satellite
RGU	Rot, Gelb, Ultraviolet
RPN	Reverse Polish Notation
SAO	Smithsonian Astrophysical Observatory
SAS	Small Astronomy Satellite
SCI	Science Citation Index
SETI	Search for Extraterrestrial Intelligence
SID	Sudden Ionospheric Disturbance
SNR	Supernova Remnant
SPARMO	Solar Particle and Radiation Monitoring Organization
SPIE	Society of Photo-optical Instrumentation Engineers
SPIN	Searchable Physics Information Notices
SRC	Space Research Council
SRS	Southern Reference Stars
SSI	Sky Survey Instrument

STAR Scientific and Technical Aerospace Reports

TeV 10^{12} Electron Volts

UB Ultraviolet, Blue photometric system

UBV Ultraviolet, Blue, Visual photometric system

UCLA University of California at Los Angeles

UFO Unidentified Flying Object

UK United Kingdom

URSI Union Radio Şcientifique Internationale

USI Ultraviolet Spectroheliographic Instrument

UV Ultraviolet

VINITI Vsesoiuznoe Institut Nauchnoĭ i Tekhnicheskoe Informatsiĭ

VLBI Very Long Baseline Interferometry

VLF Very Low Frequency

WDC World Data Centres

AUTHOR INDEX

Numbers refer to subject classification and entry number; e.g., 162.014 refers to the fourteenth entry in section 162. For multiple entries in the same classification for the same author, the classification number is not repeated; e.g., 151.001, .002 refers to entries one and two in section 151.

MONOGRAPHIC TITLE/CONFERENCE PROCEEDINGS INDEX

Entries are to the primary entry for a monograph or a specific conference. For monographs with a personal author or authors, the first is listed parenthetically following the title. Therefore, if no such author is mentioned, one may assume that the title listed is either a conference proceeding, a collection of review papers, or a compilation rather than an original work. References are to subject classification and entry number. Acronyms and initialisms are filed at the beginning of the letter section (e.g., ESLAB precedes Early).

SUBJECT INDEX

The subject index includes references to individual topics and their subdivisions and to entire sections. As with the other two indexes, numbers refer to the section and item number within the section. If reference is made to an entire section, the entry will take the form *Sec. n*, where *n* is the three-digit section number found in the table of contents. If there is more than one item for a particular subject within a section, the entries will be abbreviated as in the author index (e.g., 131.001 .002 .006).